Neurological Disorders

As the burden of neurological conditions continues to grow globally, bridging the gap between neuroscience and medicine becomes imperative. This book provides a roadmap for researchers, neuroscientists, physicians, and healthcare professionals seeking to comprehend the underlying biological mechanisms of these disorders. It serves as a valuable resource for students, educators, and anyone intrigued by the fascinating intersection of neuroscience and disease. (1) Targets a wide audience, including professionals and researchers in neuroscience, medicine, and allied disciplines. (2) Provides a comprehensive understanding of the complex molecular and cellular mechanisms that underlie these illnesses. (3) Will facilitate more knowledgeable diagnosis and treatment options. (4) Is an indispensable resource for medical students, researchers, and educators. (5) Provides a thorough examination of the molecular underpinnings of different neurological diseases.

Jitendra Kumar Sinha, Ph.D., is the Chief Scientific Officer of GloNeuro Academy and a visiting faculty member of the School of Studies in Neurosciences at Jiwaji University. Sinha received the ALBA-CHS Travel Award from Chica & Heinz Schaller Foundation (CHS) at 2024 FENS Forum, Vienna, Austria, and the IBRO-APRC Associate School on Advances in Neurodegenerative Disorders and Therapeutics at Amity Institute of Neuropsychology and Neurosciences, Amity University UP, India. He is a member of the International Brain Research Organisation (IBRO), the International Society for Developmental Neuroscience (ISDN), and the International Behavioural Neuroscience Society (IBNS).

Shampa Ghosh, M.Phil., Ph.D., is a Neuroscientist at GloNeuro Academy, India, and was previously a Senior Research Fellow at the National Institute of Nutrition, ICMR, Hyderabad. Ghosh received his Ph.D. in Biochemistry at Osmania University in Hyderabad, India, and has also received the Gold Medal Topper in B.Sc. (Environmental Sciences) at Mount Carmel College, Bangalore. He is a member of the International Brain Research Organization (IBRO), the Indian Academy of Neurosciences (IAN), and the Nutrition Society of India (NSI).

Neuroscience of Neurodegenerative Disorders
Series Editor Wael Mohamed

Wael Mohamed Yousef Mohamed
BBCH (Egypt), MMSc (Egypt), MD/PhD (USA)

Dr. Mohamed earned his PhD from Penn State University (USA) and currently is an Associate Professor at IIUM Medical School in Malaysia. He serves as an editor for *Frontiers in Neurology* and PLOS-ONE Journals. His research is well supported by extramural funding. He is the founder and chair of the AfrAbia-PD-Genomic Consortium.

About This Series
This book series aims to publish volumes focusing on both basic and clinical research in the field of Neurodegenerative Disorders with a focus on delineating the recent advances in various disorders including Parkinson's Disease, Alzheimer's Disease, ALS, Huntington's Disease, among other common Neurodegenerative disorders. The books will examine different disorders aspects including pathophysiology, diagnostic, treatment, and precision medicine. Volumes in the series will also highlight the recent advances in the management of Neurodegenerative disorders especially in OMICS and Genomic era and precision medicine. Volumes will be useful for neuroscientists, neurologists, psychiatrists, psychologists, neurosurgeons, and those interested in preventive medicine.

Integrating Neuroimaging, Computational Neuroscience, and Artificial Intelligence
Edited by Indranath Chatterjee and Nasrallah Moradikor

Non-Pharmacological Interventions for Dementia: A Comprehensive Approach
Edited by Abraham Olufemi Asuku, Ebele Joan Ajagun, and Gbonjubola Oyinlola Ogungbangbe

Burn-Induced Neurodegeneration: Subtitle Therapies for Neuromotor Complications
Edited by Amany Atiaa and Shehab E-Kader

Neurological Disorders: Insights into Disease Biology
Edited by Jitendra Kumar Sinha and Shampa Ghosh

The Gut-Brain Axis in Neurological Disorders: Clinical Implications and Holistic Management
Edited by Vishal Ahuja, Preeti Solanki, and Harmail Singh

Neurological Disorders
Insights into Disease Biology

Edited by
Jitendra Kumar Sinha and Shampa Ghosh

CRC Press
Taylor & Francis Group
Boca Raton London New York

CRC Press is an imprint of the
Taylor & Francis Group, an **informa** business

Designed cover image: Jitendra Kumar Sinha

First edition published 2027
by CRC Press
2385 NW Executive Center Drive, Suite 320, Boca Raton FL 33431

and by CRC Press
4 Park Square, Milton Park, Abingdon, Oxon, OX14 4RN

CRC Press is an imprint of Taylor & Francis Group, LLC

ISBN: 978-1-032-81343-1
ISBN: 978-1-032-81469-8
ISBN: 978-1-003-50002-5

DOI: 10.1201/9781003500025

Typeset in Times
by KnowledgeWorks Global Ltd.

Dedication

This volume is dedicated, first and foremost, to our parents, whose steadfast commitment and sacrifices shaped our education and values. In particular, we dedicate this book to Mr. Sahdeo Prasad Sinha and Mrs. Prem Lata Sinha (parents of JKS), and to Mr. Ganesh Chandra Ghosh and the Late Mrs. Jaya Ghosh (parents of SG). Through their hard work, encouragement, and unwavering belief in us, they inspired us to pursue scientific inquiry and equipped us to contribute to research that can ultimately serve those with the greatest health needs.

We also dedicate this book to the patients and families living with neurological disorders, whose resilience and endurance give meaning to scientific work and deepen our responsibility to translate knowledge into better care. In the same spirit, we acknowledge the clinicians, caregivers, and researchers who labor, often quietly and persistently, to reduce suffering, preserve dignity, and improve outcomes across diverse neurological conditions.

Finally, this volume is dedicated to students of neuroscience and medicine. Their curiosity, rigor, and determination will shape the next generation of discoveries and, with them, the future of neurological health.

Contents

Foreword

Neurological disorders are generally permanent and incurable. Recent medical advances allow us to treat the symptoms and slow down the progress. Neurodegeneration also affects the very faculties like memory, movement, language, affect, and autonomy. At the same time, they represent some of the most intellectually challenging problems in science, because the nervous system is not a single organ in isolation but is comprised of an integrated network embedded within vascular, immune, endocrine, and metabolic systems. A meaningful understanding of neurological disease therefore requires an extensive battery of diagnoses. It also demands a coherent biological framework that can connect molecules to cells, circuits to behavior, and systemic physiology to brain vulnerability. This volume, *Neurological Disorders: Insights into Disease Biology,* is timely, precisely because it attempts integration with rigor and breadth.

The book's central strength lies in its mechanism-driven approach. Contemporary neuroscience has made it clear that most neurodegenerative disorders share common biological threads: disrupted proteostasis, mitochondrial dysfunction and oxidative stress, neuroinflammation, synaptic failure, altered excitability, and failures of repair. These processes determine clinical progression and are also the rational targets for therapy. By organizing neurological diseases around a biological narrative, this book is expected to help readers develop a clear understanding of the various disorders, their genesis, progression, care, and possible cure. Students of biosciences, medicine, and neurosciences in particular would make the best out of this.

The other important contribution of this volume is its explicit attention to the interface between neurology, metabolism, and nutrition. It is more relevant at a time when we are living through a period in which obesity, metabolic syndrome, and related cardiometabolic disorders are rising globally, alongside undernutrition and micronutrient deficiency. The neurological consequences of this are increasingly evident. Metabolic dysregulations are known to influence vascular integrity, blood–brain barrier function, glial activation, synaptic plasticity, and even protein-handling pathways that intersect with neurodegenerative mechanisms. Nutritional status, undernutrition, malnutrition and overeating, modifies brain development, resilience, and aging trajectories through effects on mitochondrial function, antioxidant defenses, and neurotransmitter synthesis. The inclusion of chapters on obesity, metabolic syndrome, micronutrients, and malnutrition makes this compendium special. The content of the book also substantiates that brain health is inseparable from systemic biology, and prevention-oriented neurology, must grapple seriously with these modifiable determinants.

The coverage on the major disease domains is appropriate and comprehensive. The chapters on Alzheimer's disease and Parkinson's disease summarize hallmark pathologies simultaneously engaging with evolving concepts of disease initiation, selective vulnerability, and emerging therapeutic interventions. Epilepsy is addressed both at the level of understanding seizure circuits and through a discussion of contemporary and emerging therapeutic approaches, reflecting the field's progress toward individualized, mechanism-informed patient care. The paired treatment of ischemic and hemorrhagic stroke is valuable not only for clinical completeness, as it enables comparison of shared and distinct injury cascades that influence the outcome and recovery. The spinal cord injury chapters further reinforce essential translational aspects. Understanding pathology is incomplete without understanding repair, rehabilitation, and neuroplasticity domains, where biological insight must meet functional goals. The inclusion of ALS and glioblastoma, including motor neuron degeneration and neuro-oncology, emphasizes the complexity of central nervous system disease in the context of genetic factors, cellular heterogeneity, and therapeutic resistance.

The book's attention to aging and geriatric neurology is commendable. Aging is the dominant risk factor for many neurological disorders, but it is too often treated as a background rather than its biological influences. It is now known that aging shapes neuroinflammatory tone, vascular flexibility, metabolic efficiency, and the capacity for synaptic and structural plasticity. It also accelerates clinical complexities: comorbidity, polypharmacy, frailty, and heterogeneity of treatment response. With the chapters on cognitive decline and on geriatric neurology as a discipline with distinct challenges and opportunities, the book aligns itself with the demographic realities that will define neurology in the coming decades.

From an academic and training perspective, this volume is well-positioned to serve multiple audiences. For graduate students and early-career researchers, it provides a structured entry into major disorder classes with an emphasis on biological mechanisms and indications for future research directions. For clinicians and allied health professionals, it offers a mechanistic context that can sharpen the interpretation of symptoms, prognosis, and therapeutic logic. For educators, the chapter structure allows modular adoption across courses in neuroscience, neurobiology of disease, translational medicine, rehabilitation science, and neuro-metabolic health. The integrative scope is especially suited to interdisciplinary programs, where students must learn to navigate boundaries between basic and clinical science.

A foreword is, by nature, an endorsement, but it is also an invitation. With the Editors, I invite you to think across traditional partitions: to recognize shared mechanisms across disorders; to visualize metabolism and nutrition as integral to neurological health; to regard aging as an active biological context; and to pursue translational rigor that links mechanistic claims to outcomes that matter for patients and caregivers. I expect this book will help readers adopt an integrative mindset and equip them with a clear, evidence-oriented understanding of neurological disease biology.

I commend the editors, Jitendra and Shampa, and the contributing authors for assembling a volume of 18 chapters on this critical and exciting subject: Neurodegeneration. I am confident it will be a useful resource for students, researchers, and clinicians, and would also contribute meaningfully to education and scholarship in the neurosciences.

Prof. Ishan Patro
FIAN, FCINP, FZS, FAMS, FNASc.
ICMR-Emeritus Scientist
School of Studies in Zoology, Jiwaji University, Gwalior 474011
(Founder Coordinator-Head *of School of Studies in Neuroscience, Jiwaji University, Gwalior;* Immediate-Past President, *Indian Academy of Neuroscience; and* Former Vice Chancellor, *Ravenshaw University, Cuttack, Odisha)*

Preface

Neurological disorders constitute one of the most formidable challenges to human health in the twenty-first century. Their impact is measured not only in mortality, but also more importantly in the years lived with disability, caregiver burden, and the progressive erosion of independence and quality of life. Across the life course, the nervous system is shaped by tightly regulated genetic programs, metabolic demands, immune surveillance, vascular integrity, and environmental exposures. When these layers of regulation acutely fail, as in stroke or spinal cord injury, or chronically, as in neurodegenerative diseases, the consequences can be profound and long-lasting. A central motivation for this volume is the conviction that meaningful progress in prevention and therapy requires a rigorous understanding of disease biology: the molecular and cellular events that initiate pathology, the systems-level mechanisms that sustain it, and the compensatory processes that can be leveraged for recovery.

Neurological Disorders: Insights into Disease Biology has been developed to provide an integrated, mechanistic account of diverse neurological conditions, spanning neurodegeneration, epilepsy, cerebrovascular disease, metabolic and nutritional influences on brain health, traumatic injury, motor neuron disease, and primary brain tumors. The breadth of topics is intentional. Contemporary neurology increasingly demands a systems view that does not isolate the brain from the body, or development from aging, or genes from environment. Accordingly, the book positions neurological illness within a broader biological landscape that includes obesity, metabolic syndrome, and nutritional deficiency. These factors are now recognized as powerful modifiers of neural resilience and vulnerability.

The scientific and clinical rationale for such an integrative approach is straightforward. Alzheimer's disease, Parkinson's disease, Huntington's disease, and amyotrophic lateral sclerosis are often presented as distinct entities, each with characteristic proteinopathies, patterns of circuit dysfunction, and genetic risk architectures. Yet across these conditions, shared mechanistic themes recur: impaired proteostasis, mitochondrial dysfunction and oxidative stress, dysregulated neuroinflammation, synaptic failure, altered axonal transport, and selective cellular vulnerability. Similarly, in epilepsy, stroke, and spinal cord injury, one encounters convergent biology: excitotoxicity, ionic imbalance, blood-brain barrier disruption, glial reactivity, and maladaptive plasticity. All these are unfolding on different time scales and in different anatomical contexts. Brain cancers, including glioblastoma, introduce an additional dimension: the interaction of aberrant cellular proliferation with neural microenvironments, immune evasion, metabolic rewiring, and therapy resistance. By assembling these topics in one volume, we aim to support comparison across disorders and encourage readers to identify transferable concepts, methods, and therapeutic strategies.

A distinctive emphasis of this book is the interface between neurology and systemic physiology. Obesity and metabolic syndrome, once treated primarily as cardiovascular and endocrine problems, are now understood to influence cognition, mood, cerebrovascular risk, and neurodegenerative trajectories through multiple biological channels. These include chronic low-grade inflammation, insulin resistance and altered insulin signaling in the brain, dyslipidemia, vascular dysfunction, gut-brain axis signaling, and changes in neuronal energy metabolism and synaptic plasticity. The nervous system's energetic requirements are exceptionally high, and even modest perturbations in glucose handling, mitochondrial efficiency, or cerebral perfusion can have outsized effects on network stability and cognitive function. The inclusion of dedicated chapters on obesity and metabolic syndrome therefore reflects more than epidemiological relevance; it reflects a mechanistic reality: metabolic state is an active determinant of neuronal health and a plausible target for prevention and adjunctive therapy.

Aging provides the second major integrative axis of the volume. As life expectancy increases, the demographic shift toward older populations has made age-associated neurological disorders a global priority. Yet aging is not merely the passage of time; it is a set of biological processes including the genomic instability, epigenetic drift, impaired autophagy, senescent cell accumulation, altered immune tone, and reduced vascular and metabolic flexibility. All these can amplify disease mechanisms or erode compensatory capacity. The chapters addressing cognitive decline and geriatric neurology are positioned to help readers think about neurological disorders through a life-course lens: how early-life vulnerabilities and midlife exposures shape late-life outcomes; why the same pathological burden may yield different clinical phenotypes across individuals; and how geriatric care introduces distinct challenges, including multimorbidity, polypharmacy, frailty, and heterogeneity of treatment response.

The organization of the book is designed to be both comprehensive and navigable. Following an opening section that addresses the scope and importance of studying neurological disorders and situates current work within historical perspectives on neurological research, the text moves through major disease domains in a progression that mirrors both clinical prevalence and mechanistic continuity. Neurodegenerative disorders are addressed early, with chapters dedicated to Alzheimer's disease, Parkinson's disease, and Huntington's disease, each emphasizing mechanisms, pathogenesis, genetic basis, and emerging therapies. The epilepsy section then bridges molecular excitability and circuit dynamics, pairing foundational neurobiology of seizures with therapeutic advances and novel management strategies. Stroke and cerebrovascular disorders follow, with complementary treatments of ischemic and hemorrhagic mechanisms, an arrangement intended to sharpen understanding of shared cascades (e.g., inflammation and edema) alongside crucial differences in tissue injury and clinical management.

Subsequent sections expand the integrative perspective. The metabolic syndrome and neurological consequences chapters highlight how systemic physiology intersects with neural function, while the spinal cord injury chapters focus on molecular and cellular responses to injury alongside rehabilitation strategies and neuroplasticity. Nutritional deficiency and malnutrition are treated not as ancillary concerns but as biologically meaningful drivers of neurological dysfunction, particularly in contexts where micronutrient availability, developmental timing, and socioeconomic determinants shape brain health trajectories. The dedicated ALS chapter addresses motor neuron degeneration through the lens of genetic factors, molecular mechanisms, and therapeutic advances, while the brain cancer chapter turns to glioblastoma and emerging therapies for brain tumors. The volume concludes with chapters explicitly centered on neurological disorders in aging, reinforcing the concept that age modifies both risk architecture and mechanistic expression.

This structure is intended to support multiple modes of use. Readers seeking a disease-specific entry point can engage a relevant chapter directly. Readers interested in cross-cutting mechanisms like neuroinflammation, mitochondrial dysfunction, synaptic pathology, vascular compromise, or maladaptive plasticity can read across sections to identify convergent biology. Educators may find the organization conducive to modular teaching, where foundational principles in one disorder illuminate others. Clinicians and translational researchers may use the text to connect bedside phenotypes to biological pathways and to survey emerging therapeutic directions across disease classes.

The intended audience for this volume is necessarily broad, reflecting the interdisciplinary nature of modern neuroscience and neurological medicine. It is written for neuroscientists, physicians, healthcare practitioners, researchers, and trainees who seek a biologically grounded understanding of neurological disorders and their systemic determinants. We have aimed to balance depth with clarity: to present mechanistic detail without losing sight of integrative explanations; to highlight genetic and molecular insights while emphasizing the importance of circuits, glia, vasculature, immunity, and metabolism; and to connect basic biology to therapeutic logic, including both established treatments and emerging strategies.

As editors, we also view this volume as an invitation to methodologically pluralistic neuroscience. Many of the key advances in disease biology now emerge at the interfaces: single-cell and spatial omics that map vulnerable cell states; human genetics that identifies causal pathways; in vivo imaging that links pathology to circuit dysfunction; induced pluripotent stem cell models that capture patient-specific biology; and computational approaches that integrate multiscale data into mechanistic hypotheses. Progress will depend on robust experimental design, careful interpretation, and the willingness to test mechanistic claims across model systems and human data. The chapters in this book are therefore oriented toward conceptual frameworks that can be interrogated experimentally and translated clinically.

Finally, we acknowledge that a volume of this scope is only possible through collaborative scholarship. We are grateful to the contributing authors for synthesizing complex literature and for articulating where evidence is strong, where controversies remain, and where the most promising opportunities lie. We also acknowledge the broader scientific community whose work underpins every chapter and whose ongoing efforts continue to transform neurological science. It is our hope that *Neurological Disorders: Insights into Disease Biology* will serve as a durable reference, one that supports learning, stimulates cross-disciplinary thinking, and accelerates biologically informed approaches to preventing and treating neurological disease.

—Dr. Jitendra Kumar Sinha, Ph.D.
Dr. Shampa Ghosh, Ph.D.

Acknowledgments

The editors gratefully acknowledge all chapter authors for their scholarly contributions and for accommodating revisions aimed at ensuring conceptual consistency and a uniform academic style across the volume. We also thank the many clinicians and scientists whose foundational work made it possible to synthesize current knowledge on neurological disorders within an integrated framework.

We are grateful to the peer colleagues who supported the writing and editing process. We also acknowledge the technical and administrative assistance that facilitated coordination across multiple chapters and contributors. We appreciate the efforts of the editorial and production teams for their guidance and help in managing the complex workflow involved in assembling a multi-author academic book.

About the Editors

Dr. Jitendra Kumar Sinha, Ph.D., is a neuroscientist and Associate Professor in the Department of Behavioural Health Sciences at the Symbiosis Institute of Health Sciences (SIHS), Symbiosis International (Deemed University), Pune, India. He obtained his PhD from the ICMR–National Institute of Nutrition (ICMR-NIN). After receiving prestigious international and national grants, he carried out collaborative research studies at the University of Verona, Italy, the University of New South Wales, Australia, and CSIR–Centre for Cellular and Molecular Biology (CSIR-CCMB), Hyderabad, India. He previously worked as an Assistant Professor at Amity University, Noida, served as Chief Scientific Officer at GloNeuro Academy, Noida, and was also associated as Visiting Faculty in Neuroscience at Jiwaji University, Gwalior, India. Dr. Sinha's research focuses on the neurobiology of ageing and obesity, nutritional neuroscience, neuropsychiatric disorders, neuropharmacology, and healthy ageing research. With more than 20 years of experience in research and teaching, he has contributed to both scientific investigation and academic programme development. His work has particularly advanced the use of animal models in ageing and epilepsy research, with emphasis on oxidative protein damage, DNA damage, and biomarkers of ageing. He has also contributed to rodent model development for studying the molecular basis of accelerated ageing. He is deeply committed to mentoring students, especially in the paramedical sciences, and to fostering research training among young scholars. Dr. Sinha has published more than 150 articles and book chapters in the areas of ageing, epilepsy, obesity, sleep, and spinal cord injury. He has also served the academic community as a guest editor, reviewer, and editorial board member for numerous journals. His contributions have been recognised through several national and international awards, including those from the Government of India, IBRO, ISN, SfN, and UNESCO.

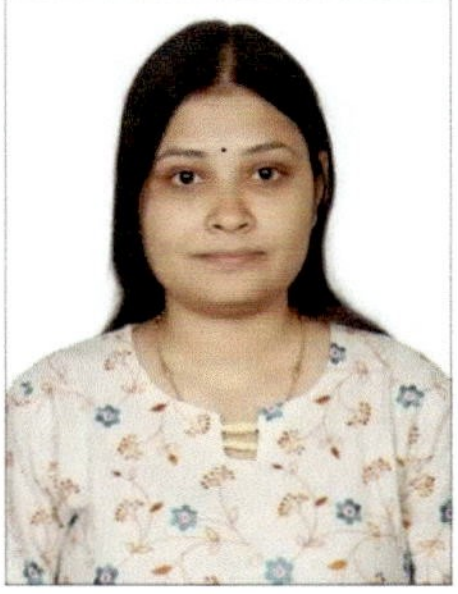

Dr. Shampa Ghosh, M.Phil., Ph.D., is a neuroscientist and biochemist currently serving as Director and Senior Scientist at GloNeuro, Noida, India. She has over eight years of leadership experience in translational neuroscience research, with a strong focus on neurodegeneration, neurodevelopment, and metabolic influences on brain health. Dr. Ghosh obtained her PhD in Biochemistry from Osmania University, Hyderabad, and her M.Phil. in Neurophysiology from the National Institute of Mental Health and Neurosciences (NIMHANS), Bangalore, where she was awarded the Gold Medal for securing first rank. She also completed her MSc in Biochemistry from Bangalore University. During her early research career, she was a Senior Research Fellow at the ICMR–National Institute of Nutrition (NIN), Hyderabad, where she contributed to studies on nutritional neuroscience and metabolic regulation of brain function. Her research interests span neurodegeneration, fetal programming of adult disorders, epigenetic regulation of neurodevelopmental, and neuropsychiatric conditions. She has made significant contributions to understanding the molecular links between metabolic dysfunction, oxidative stress, and brain ageing. Dr. Ghosh has an extensive publication record, with numerous research articles, reviews, and book chapters in high-impact peer-reviewed journals, including The Lancet, Ageing Research Reviews, Drug Discovery Today, and Journal of Alzheimer's Disease. Her work addresses critical areas such as Alzheimer's disease, epilepsy, neuroinflammation, metabolic disorders, and advanced drug delivery systems targeting the central nervous system. In addition to her research contributions, she is actively involved in academic editing and scholarly publishing. She has served as Guest Associate Editor for Frontiers in Aging Neuroscience and as Special Issue Editor for journals such as Brain Sciences and Current Topics in Medicinal Chemistry. She is also co-editing international volumes with leading publishers including CRC Press (Taylor & Francis) and Academic Press (Elsevier). Dr. Ghosh has received several prestigious national and international fellowships and travel awards from organisations such as the International Brain Research Organisation (IBRO), International Society for Neurochemistry (ISN), and the Indian Council of Medical Research (ICMR). She is an active member of multiple scientific societies, including IBRO, ISDN, IBNS, and the Indian Academy of Neurosciences. Through her interdisciplinary work integrating neuroscience, metabolism, and therapeutic innovation, Dr. Ghosh continues to contribute to advancing the understanding of brain disorders and the development of translational strategies for improved neurological health.

List of Contributors

Prachi Agarwal
Manipal Centre for Biotherapeutic Research
Manipal Academy of Higher Education
Manipal, Karnataka, India

Alina Arulsamy
Neuropharmacology Research Laboratory, Jeffrey Cheah School of Medicine and Health Sciences
Monash University Malaysia
Subang Jaya, Selangor, Malaysia

Rakesh Bhaskar
School of Chemical Engineering
Yeungnam University
Gyeongsan, South Korea

Pallab Bhattacharya
Department of Pharmacology and Toxicology
National Institute of Pharmaceutical Education and Research (NIPER)
Gandhinagar, Gujarat, India

Sanjay Das
Molecular and Behavioral Neuroscience Laboratory, Institute of Health Sciences
Presidency University
Kolkata, West Bengal, India

Aishika Datta
Department of Pharmacology and Toxicology
National Institute of Pharmaceutical Education and Research (NIPER)
Gandhinagar, Gujarat, India

Vinay Shankar Dubey
Department of Anatomy
All India Institute of Medical Sciences
New Delhi, India

Akash Gautam
Centre for Neural and Cognitive Sciences
University of Hyderabad
Hyderabad, Telangana, India

Bijoyani Ghosh
Department of Pharmacology and Toxicology
National Institute of Pharmaceutical Education and Research (NIPER)
Gandhinagar, Gujarat, India

Shampa Ghosh
GloNeuro
Noida, Uttar Pradesh, India

Sung Soo Han
School of Chemical Engineering
Yeungnam University
Gyeongsan, South Korea

Harsha
Department of Psychiatry
All India Institute of Medical Sciences—Central Armed Police Forces Institute of Medical Sciences (AIIMS-CAPFIMS)
New Delhi, India

Srishti Jain
Department of Anatomy
All India Institute of Medical Sciences
New Delhi, India

Charu Jaiswal
Manipal Centre for Biotherapeutics Research
Manipal Academy of Higher Education
Manipal, Karnataka, India

Kulsum Jan
Department of Food Technology
Jamia Hamdard University
Jamia Hamdard, New Delhi, India

Iman Imtiyaz Ahmed Juvale
Neuropharmacology Research Laboratory, Jeffrey Cheah School of Medicine and Health Science
Monash University Malaysia
Subang Jaya, Selangor, Malaysia

Tejasvani Khanna
Maulana Azad Medical College
Delhi, India

Pragya Komal
Department of Biological Sciences
Birla Institute of Technology and Science-Pilani (BITS-Pilani) Hyderabad campus
Hyderabad, Telangana, India.

Arpita Konar
Molecular and Behavioral Neuroscience Laboratory, Institute of Health Sciences
Presidency University
Kolkata, West Bengal, India

Ashish Kumar
HumaneRx
Kolkata, West Bengal, India

Nikhil Maheshwari
Department of Anatomy
All India Institute of Medical Sciences
New Delhi, India

Nikita Malik
Department of Pharmacology and Toxicology
National Institute of Pharmaceutical Education and Research (NIPER)
Gandhinagar, Gujarat, India

SK Venkata Manjari
Department of Biological Sciences
Birla Institute of Technology and Science-Pilani (BITS-Pilani) Hyderabad campus
Hyderabad, Telangana, India

Kamakshi Mehta
University of Pittsburgh, School of Medicine
UPMC Vision Institute
Pittsburgh, Pennsylvania

Ushmita Mukherjee
Department of Pharmacology and Toxicology
National Institute of Pharmaceutical Education and Research (NIPER)
Gandhinagar, Gujarat, India

Kannan Badri Narayanan
School of Chemical Engineering
Yeungnam University
Gyeongsan, South Korea

Renu Negi
Department of Pharmacology and Experimental Therapeutics
Boston University School of Medicine
Boston, Massachusetts

Irma Wati Ngadimon
Department of Pharmaceutical Life Sciences, Faculty of Pharmacy
Universiti Malaya
Kuala Lumpur, Malaysia

Varsha Pai
Manipal Centre for Biotherapeutics Research
Manipal Academy of Higher Education
Manipal, Karnataka, India

Amit Kumar Pandey
Amity University Uttar Pradesh
Noida, Uttar Pradesh, India

Kamlesh Kumar Pandey
OM Diagnostic Lab, Inc.
Norwood, Massachusetts

Surya Prakash Pandey
Department of Zoology, School of Science
IFTM University
Moradabad, Uttar Pradesh, India

Soumya S Pattnaik
Department of Pharmacology
All India Institute of Medical Sciences—Central Armed Police Forces Institute of Medical Sciences (AIIMS-CAPFIMS)
New Delhi, India

Anand Prakash
Department of Biotechnology
Mahatma Gandhi Central University
Motihari, Bihar, India

Hannah Priscilla
Department of Anatomy
ESIC Medical College and Hospital
Hyderabad, Telangana, India

Deepak Rathi
Department of Anatomy
All India Institute of Medical Sciences
New Delhi, India

Syed Shadab Raza
Laboratory for Stem Cell & Restorative Neurology, Department of Anatomy
Era University
Lucknow, Uttar Pradesh, India

Mohd. Farooq Shaikh
School of Dentistry and Medical Sciences
Charles Sturt University
Orange, New South Wales, Australia

Bhaskar Sharma
Department of Anatomy
All India Institute of Medical Sciences
New Delhi, India

Yogesh Sharma
Department of Anatomy
All India Institute of Medical Sciences
New Delhi, India

Abhishek Kumar Singh
Manipal Centre for Biotherapeutics Research
Manipal Academy of Higher Education
Manipal, Karnataka, India

Ishika Singh
Manipal Centre for Biotherapeutics Research
Manipal Academy of Higher Education
Manipal, Karnataka, India

Krishna Kumar Singh
Symbiosis International (Deemed University)
Pune, Maharashtra, India

Maharudra Pratap Singh
Department of Biotechnology
Mahatma Gandhi Central University
Motihari, Bihar, India

Mukund P Singh
Graphic Era (Deemed to be University)
Dehradun, Uttarakhand, India

Nibha Singh
Department of Biochemistry
Jamia Hamdard University
New Delhi, India

Prabhakar Singh
Department of Anatomy
All India Institute of Medical Sciences
New Delhi, India

Sakshi Singh
Department of Food Technology
Jamia Hamdard University
New Delhi, India

Anjana Sinha
GloNeuro
Noida, Uttar Pradesh, India

Jitendra Kumar Sinha
Symbiosis Institute of Health Sciences (SIHS)
Symbiosis International (Deemed University)
Pune, Maharashtra, India

Shreya Sridhar
Sapthagiri Institute of Medical Sciences and Research Centre
Bangalore, Karnataka, India

Kapil Suchal
Department of Pharmacology
All India Institute of Medical Sciences—Central Armed Police Forces Institute of Medical Sciences (AIIMS-CAPFIMS)
New Delhi, India

Rajesh Tamatta
Manipal Centre for Biotherapeutics Research
Manipal Academy of Higher Education
Manipal, Karnataka, India

Kirthanashri S Vasanthan
Manipal Centre for Biotherapeutic Research
Manipal Academy of Higher Education
Manipal, Karnataka, India

Arshi Waseem
Laboratory for Stem Cell & Restorative Neurology, Department of Anatomy
Era University
Lucknow, Uttar Pradesh, India

1 Introduction to Neurological Disorders

Jitendra Kumar Sinha and Shampa Ghosh

1.1 INTRODUCTION

Neurological disorders represent a wide range of diseases or disorders involving the nervous system. They may arise due to different reasons and with different origins. The neurological disorders can be due to genetic defects, neurodegeneration, trauma, infection, vascular injury, autoimmune reactions, or metabolic imbalance.[1] Their presentation varies from mild cognitive dysfunction to clear motor, sensory, and/or behavioral dysfunctions. These are frequently resulting in chronic disability and decreased quality of life. The World Health Organization (WHO) considers neurological disorders as "diseases of the central and peripheral nervous system."[2] These cover epilepsy, Alzheimer's disease and other dementias, cerebrovascular diseases such as stroke, migraine, and other headache disorders, multiple sclerosis, Parkinson's disease, neuroinfections, brain tumors, traumatic disorders of the nervous system, and neurological disorders related to malnutrition.[2] In one of the ways, various neurological disorders can be broadly categorized (as shown in Figure 1.1).

Neurological diseases are a major cause of global morbidity and mortality. In the Global Burden of Disease (GBD) 2019 study, neurological diseases accounted for 9 million deaths and 349 million disability-adjusted life years (DALYs), ranking them as the second major cause of death and the major cause of disability globally.[3,4] Stroke, Alzheimer's disease and other dementias, and migraine

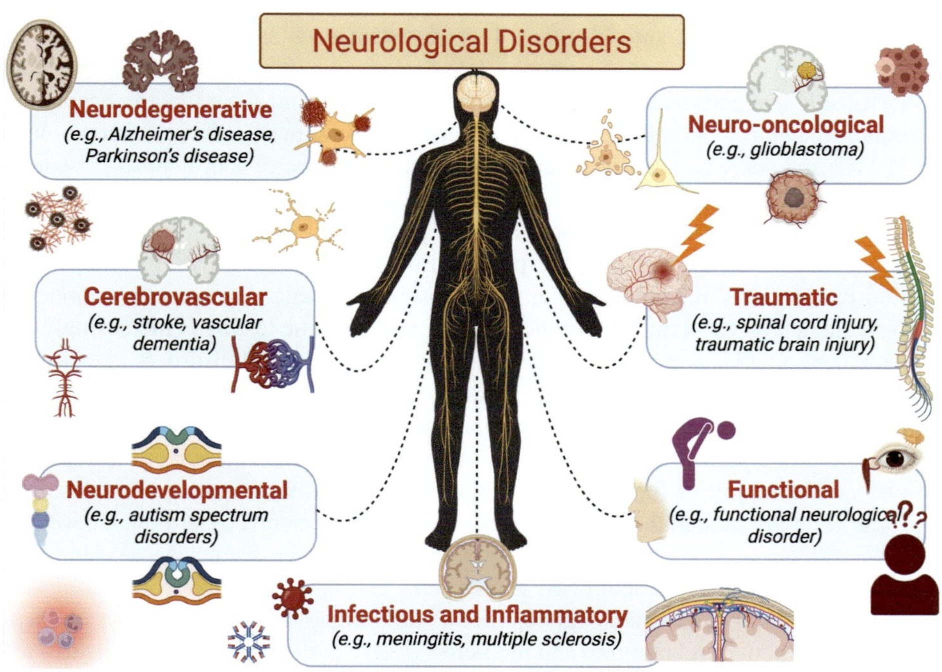

FIGURE 1.1 Various categories of neurological disorders.

DOI: 10.1201/9781003500025-1

were the three most burdensome disorders worldwide.[4] One of the reasons for increased occurrence of neurological illness is that the world is aging.[5] Additionally, more people are surviving most of the diseases, and diagnostic skills are now more advanced. For instance, dementia affects more than 55 million people in the world, with almost 10 million new cases annually.[6] Epilepsy has more than 50 million sufferers, whereas stroke has more than 100 million living patients, most with some deficits.[7] Neurodegenerative diseases' prevalence is increasing as the world's population is aging, and this is happening particularly in high- and middle-income nations. This increasing burden results in a plethora of challenges to our healthcare systems and caregivers. While neurological diseases exhibit diverse causes and impacted systems, they share several clinical characteristics. The examples include motor signs like paralysis, tremors, spasticity, ataxia, weakness, or sensory abnormalities like paresthesia, pain, or numbness.[8] On the other hand, it can be cognitive impairments including executive dysfunction, memory loss, language impairment, and/or seizures, altered consciousness.[9] In some cases, psychiatric syndromes like depression, psychosis, and anxiety can be there or autonomic dysfunction including bowel/bladder dysfunction, instability of blood pressure. Most disorders have a subtle presentation, making them difficult to diagnose and early intervention. On the other hand, few like seizures or stroke have an acute presentation that needs to be treated on an urgent basis.[8,9]

Diagnosing a neurological problem is demanding and multidimensional strategic process. It can encompass both careful clinical assessment as well as high-end investigational techniques. As a first step, neurological examination is done wherein doctors evaluate cranial nerve function, motor strength, reflexes, coordination, and sensory responses. Also, cognitive screening for memory, attention, and executive function is done. Bedside examinations offer critical clues and direct subsequent investigations. Neuroimaging has become an essential part in contemporary neurology due to its non-invasive nature. Traditional techniques like magnetic resonance imaging (MRI) and computed tomography (CT) remain the cornerstone for detecting structural abnormalities, vascular lesions, tumors, and degenerative changes.[10] Recent developments in functional imaging modalities like functional MRI (fMRI), diffusion tensor imaging (DTI), and positron emission tomography (PET) enable clinicians not only to see anatomical structures. This also helps in correlating the connectivity patterns, metabolic functions, and molecular pathology.[11] They are especially crucial in dementia, epilepsy, and neuro-oncology, among others. The reason includes early and differential diagnosis are urgently required in these cases. Electrophysiological examinations still form the cornerstone of diagnosing disorders of neuronal activity and neuromuscular transmission. Electroencephalography (EEG) remains the gold standard for assessing epilepsy and seizure disorders,[12] whereas electromyography (EMG) and nerve conduction studies shed light on peripheral neuropathies, diseases of the neuromuscular junction, and myopathies.[13] These methods supplement the imaging test by recording dynamic electrical activity that can be missed by a structural scan.

Laboratory studies further enhance the precision of diagnosis. The cerebrospinal fluid (CSF) examination can uncover infections, inflammatory signs, or neurodegenerative biomarkers like amyloid beta and tau proteins in AD.[14] In the last few years, genetic testing is increasingly incorporated into medical practice and most notably for inherited neurological disorders like Huntington's disease, spinal muscular atrophy (SMA), or familial amyotrophic lateral sclerosis (ALS).[15] On the other hand, blood-based biomarkers are also emerging as inexpensive methods of screening and tracking disease progression.[16] Nevertheless, neuropsychological examination offers objective measurement of various cognitive functions like language, memory, visuospatial, and executive functions.[17] These tests are essential in the diagnosis of various dementia syndromes, developmental disorders, and brain injury outcomes. They help in informing better diagnosis and planning appropriate management. The diagnostic process is also being revolutionized by incorporating AI/ML.[18] These technologies are being increasingly applied to neuroimaging and electrophysiological data. And that facilitates pattern recognition and predictive analytics that can be more sensitive and specific than human interpretation. Importantly, such advancements promise early diagnosis, personalized prognosis, and more directed interventions in neurological management.[18,19]

Neurological disorders are not only biological processes but also economic and social problems. These disorders are mostly progressive, chronic, and disabling. And therefore they create long-term burdens for patients, families, and health systems at large. In the low- and middle-income countries (LMICs), there is an added disadvantage of delayed diagnosis, restricted access to care, and stigma. Global health agencies have been moved into action by the expanding burden. The WHO's adopted Intersectoral Global Action Plan on Epilepsy and Other Neurological Disorders 2022–2031 will decrease the burden by enhancing access, awareness, surveillance, and financing for research.[20]

1.2 AGING AND THE BRAIN

Aging is a universal biological phenomenon characterized by progressive physiological and structural changes in the body.[5,21] The most intricate of all changes that happen is taking place in the human brain. Although previously believed the brain to be relatively immune from aging, but now it is known that brain undergoes a host of neurochemical, anatomical, and vascular changes.[22] These changes add substantially to the cognitive and functional diminution that emerges in later life. They not only constitute normal senescence but also enhance susceptibility to a variety of neurological diseases, particularly neurodegenerative conditions and cerebrovascular disease. The brain also develops various macroscopic as well as microscopic changes during aging.[23] At structural level, there is a progressive decline in brain volume mainly in the prefrontal cortex and hippocampal region. These are two of the areas important for executive function and memory, respectively.[24] Such atrophy is also associated with loss of synapses, dendritic retraction, and decreased neurogenesis in areas like the dentate gyrus. Neurochemically, aging is linked to decreased levels of dominant neurotransmitters (specifically dopamine, serotonin, and acetylcholine).[25,26] These alterations result in the slowing down of cognitive processing, decreased attentional ability, and compromised short-term memory (commonly found in older individuals).[25]

The pace of cognitive aging is not the same in all individuals. Although some may maintain high-functioning cognitive abilities even in their 80s or 90s but others show significant deterioration in the absence of clinical dementia.[27] Such heterogeneity is there may be due to variations in genetic influences, cognitive reserve, educational backgrounds, and lifestyle factors. Cognitive areas most vulnerable to age-related impairment are working memory, speed of processing, and episodic memory.[28] On the other hand, semantic memory and vocabulary are protected. Physical alterations, such as decreases in motor coordination, gait speed, and balance, also appear with aging and can be signs of early neurodegenerative processes. The processes responsible for age-related changes in the brain are multicausal. One core hypothesis is that oxidative stress drives neuronal injury, as a result of the rise in reactive oxygen species (ROS) beyond the antioxidant capacity of the brain.[29] Mitochondrial impairment further adds to oxidative injury resulting in energy deficits.[30] Neuroinflammation is defined by chronic activation of microglia and increased pro-inflammatory cytokines, becomes a major contributor to the aging brain.[31] Such processes do not just impair the integrity of neurons but also enhance pre-existing pathology in diseases like AD and PD.[31]

One of the most important causes of cognitive impairment in aging is different vascular factors. The cerebral high metabolic rate renders the brain extraordinarily vulnerable to even slight impairments in cerebral blood flow.[32] Arteriosclerosis associated with aging, hypertension, and microvascular disease can lead to chronic cerebral hypoperfusion, endothelial disturbance, and breakdown of the blood–brain barrier.[33] These vascular insults form the core of the pathophysiology of vascular cognitive impairment and dementia (VCID). In many of the cases, patients present a mixed pathology having both vascular and neurodegenerative lesions, which makes diagnosis and management more challenging.

Recent studies also indicate that aging causes changes in glymphatic function (waste clearance mechanism of the brain) which might reduce the clearance of toxic proteins like beta-amyloid and tau, both features of AD.[34] Similarly, the depreciation in trophic factors like brain-derived neurotrophic factor (BDNF) lowers neuronal plasticity and resilience.[35] And this makes the brain more susceptible to damage. Apart from biological alterations, the psychosocial processes like social isolation, depression, and diminished physical activity can also speedup cognitive and functional deterioration. Older adult frailty has been linked with a vicious cycle of sarcopenia, falls, cognitive impairment, and hospitalization.[36] Importantly, these processes are not fixed or unalterable. Evidence from epidemiological research highlights the modifiable nature of many aging risk factors. Physical activity on a regular basis, mental stimulation, good nutrition, and control of vascular risk factors have all been associated with healthier brain aging.[37] The aging brain is therefore both a physiological process and an insult state for a broad spectrum of neurological diseases. It is critical not just for the early diagnosis and prevention of illnesses such as dementia and stroke but also for the development of interventions to support healthy cognitive aging throughout life to comprehend the interaction between healthy aging and disease.

1.3 KEY NEUROLOGICAL DISORDERS

Neurological disorders constitute the most disabling and challenging conditions in the field of medicine. They range from acute conditions like stroke to chronic, progressive diseases such as dementia and PD. Although these disorders are heterogeneous but all share profound effects on cognition, mobility, behavior as well as general quality of life. They are also the significant causes of GBD, responsible for millions of deaths and DALYs annually.[4] It becomes crucial to discuss the most clinically and socially important neurological disorders. And this is starting with dementia and cognitive impairment, followed by stroke, PD, motor neuron disorders, epilepsy, and brain tumors.

1.3.1 Dementia and Cognitive Impairment

Dementia is a clinical syndrome of a progressive loss of cognitive functions that can severely interfere with a person's daily life.[38] It is considered as one of the most urgent public health issues of the present times. There are over 57 million affected individuals across the globe, having almost 10 million new cases annually.[39] The syndrome consists of a collection of disorders with various etiologies, of which AD is most prevalent, occurring in 60%–70% of cases.[39] It is followed by vascular dementia, dementia with Lewy bodies and then frontotemporal dementia.[39] Mild cognitive impairment (MCI) has been identified as a prodromal condition, where quantifiable deficits are present without overt loss of functional independence.[40] Nevertheless, a significant number of patients with MCI develop dementia eventually. Its etiologies are multidetermined, including genetic, vascular, metabolic, and lifestyle determinants. In AD, pathological accumulation of beta-amyloid plaques and hyperphosphorylated tau tangles impair neuronal communication and induce neurodegeneration.[41] Vascular dementia results from ischemic or hemorrhagic brain damage, most frequently secondary to small vessel disease, large vessel stroke, or cerebral amyloid angiopathy.[42] Dementia with Lewy bodies has similarities with PD, with the characteristic pathological hallmark being aggregation of alpha-synuclein, resulting in cognitive impairment accompanied by attentional fluctuations, parkinsonism, and visual hallucinations.[43] Frontotemporal dementia, on the other hand, is linked to frontal and temporal lobe degeneration and primarily features with personality changes, language impairment, and executive dysfunction, usually at an earlier age than AD.[44] Therefore clinically speaking, dementia is expressed through impairment in memory, language, problem-solving ability, visuospatial function as well as judgment. Psychological and behavioral symptoms of apathy, depression, agitation, delusions, and sleep disturbances are also frequent that further complicate management and care. The pace of progression varies according to etiology. But in the majority of forms the decline is unyielding, ending eventually in severe disability, loss of independence, and rising mortality.

Diagnosis of dementia requires a thorough and multidimensional strategy. The neuropsychological examination is necessary to record cognitive deficits.[45] Whereas, neuroimaging methods like MRI and CT facilitate the detection of patterns of atrophy and exclusion of secondary factors.[46]

More modern equipment such as the PET scans can identify amyloid and tau deposition in AD.[47] Then with dopaminergic imaging the differentiation of dementia with Lewy bodies from AD or PD dementia can be facilitated. Laboratory tests such as thyroid function, vitamin B12, and syphilis serology continue to be relevant to exclude reversible causes. Nevertheless, new biomarkers of CSF tau and amyloid levels and plasma neurofilament light chain are revolutionizing diagnostic capacity and timing.

Although there is progress but still treatment of dementia continues to be mainly symptomatic. Cholinesterase inhibitors (donepezil, rivastigmine, galantamine) and NMDA receptor antagonists (memantine) mildly improve symptoms and delay AD progression.[48] Non-pharmacologic interventions like cognitive stimulation therapy, regular exercise, and caregiver training are just as important. New disease-modifying therapies in the form of monoclonal antibodies against amyloid-beta like aducanumab and lecanemab provide hope but are controversial owing to variable efficacy and safety.[49] In vascular dementia, management of risk factors like hypertension, diabetes, and dyslipidemia is essential both for prevention and worsening.[50] Palliative strategies that are focused on quality of life, behavioral management, and caregiver support are the foundation of late-dementia care.

We need to understand the public health scenario of dementia, as it is still poor after decades of research and development. In addition to the individual cost for patients, dementia places enormous emotional, financial, and physical stress on caregivers and health systems.[50] The total cost of dementia worldwide annually is estimated to be more than US$1.3 trillion with costs likely to escalate rapidly as populations age.[39] Dementia needs to be addressed not only for the need for progress in diagnosis and treatment but also for societal interventions aimed at caregiving support, reduction of stigma, and preparedness of health systems.[39] An understanding of the interaction between biological and social determinants is crucial for the prevention of this ruinous syndrome's future burden.

1.3.2 Stroke

Stroke is one of the most important causes of chronic disability and death. It is a significant public health problem because of its frequent occurrence, recurrence, and long-term complications. Stroke is a sudden neurological impairment caused by the interruption of cerebral blood supply, either due to vascular blockage (ischemic stroke) or rupture (hemorrhagic stroke).[51] Another related condition is the transient ischemic attack (TIA), which is defined by the presence of temporary focal neurological deficits that have returned to baseline in 24 hours without permanent tissue damage but still is a potent predictor of future stroke. Talking about the numbers, stroke worldwide affects over 12 million individuals yearly and is responsible for about 6.5 million deaths, with most of the burden occurring in low- and middle-income nations.[4] Data show that ischemic stroke is responsible for approximately 87% of all strokes.[51] It most frequently results from thrombotic or embolic cerebral artery obstruction. Large artery atherosclerosis, small vessel disease, and cardioembolism (especially due to atrial fibrillation) are leading mechanisms. Hemorrhagic stroke, on the other hand, includes intracerebral and subarachnoid hemorrhage, is due to rupture of diseased vessels. This usually happens against the background of chronic hypertension, cerebral amyloid angiopathy, or aneurysmal disease. Although less frequent, hemorrhagic strokes have a greater mortality and morbidity because of mass effect, increased intracranial pressure, and direct parenchymal damage.[52]

The clinical picture of stroke is determined by the location and size of the lesion but most commonly presents as sudden onset hemiparesis, facial paresis, dysarthria, sensory change, visual field defect, or loss of coordination. In the case of more severe strokes, it may result in coma/death. Fast recognition of these symptoms is the basis of the "time is brain" principle, as millions of neurons are destroyed with each passing minute in the absence of reperfusion.[53] It has been public awareness campaigns like the "FAST" (Face, Arms, Speech, Time) acronym that have helped enhance early detection and on-time intervention.

Diagnosis is based on the emerging neuroimaging, with non-contrast CT for exclusion of hemorrhage and examination of early ischemic changes.[54] MRI, especially diffusion-weighted imaging is better for acute ischemia and posterior circulation strokes. Vascular imaging with CT or MR angiogram and carotid Doppler examination aids in the identification of stenosis or occlusion. Laboratory evaluation is also necessary to assess coagulation status, metabolic factors, and secondary risk factors.[55]

Stroke management has undergone significant changes in the last three decades. For acute ischemic stroke, intravenous thrombolysis with tissue plasminogen activator (tPA) is still the gold standard if given within 4.5 hours of symptom onset.[56] Recently, mechanical thrombectomy has made a dramatic impact on the management of large vessel occlusion.[57] This is significant because it helps with an extended window of up to 24 hours in highly selected cases depending on advanced imaging. Hemorrhagic stroke treatment involves blood pressure management, reversal of anticoagulation, neurosurgical evacuation in suitable patients, and treatment of complications like hydrocephalus.[58] In addition to acute management, secondary prevention is essential and involves antiplatelet or anticoagulant therapy, statin, blood pressure control, smoking cessation, and lifestyle change. Rehabilitation is crucial for enhancing better outcomes. This is because after the incident of stroke, a majority of survivors have residual motor, cognitive, and speech impairment.[59] Multidisciplinary care comprising physiotherapy and speech-language therapy need to be done to optimize recovery and independence. Long-term results like depression, post-stroke dementia, and future vascular events need continued care and management. Notably, stroke can be considered as a preventable condition because almost 90% of the disease burden is related to modifiable risk factors like hypertension, diabetes, obesity, smoking, physical inactivity, and unhealthy diet.[60] Whatsoever about the progress, inequities remain in access to rehabilitation and acute care, especially in limited-resource environments. Improving stroke systems of care, broadening access to imaging and reperfusion therapies, and optimizing public health approaches to risk factor control are critical steps toward reducing the worldwide burden.

1.3.3 Parkinson's Disease and Parkinsonism

Parkinson's disease is the second most common neurodegenerative condition after AD. More than 11.77 million people are affected by PD worldwide and contribute heavily to disability and reduced quality of life.[61] Clinically, it has a combination of motor and non-motor symptoms, and pathologically as the progressive degeneration of dopaminergic neurons in the substantia nigra pars compacta. The more generic term parkinsonism refers to a clinical syndrome of bradykinesia, rigidity, resting tremor, and postural instability due to PD, atypical parkinsonian syndromes, or secondary causes such as drugs or cerebrovascular disease.

The motor features of PD begin insidiously and evolve gradually. Slowness of movement is the most incapacitating defining feature, often associated with accompanying resting tremor, cogwheel rigidity, and gait shuffling. As disease evolves, postural instability results in frequent falls, a principal cause of morbidity. Equally significant are the non-motor symptoms, which can occur years before motor features. The non-motor symptoms consist of hyposmia, constipation, depression, sleep disturbances such as REM sleep behavior disorder and autonomic dysfunction.[62] Additionally, many PD patients show cognitive dysfunction. The 80% of all patients eventually come up with PD dementia in later years.[63]

The pathophysiology of PD involves selective neuronal degeneration in the nigrostriatal dopaminergic pathway and extensive Lewy body deposition, intracellular inclusions composed predominantly of aggregated alpha-synuclein.[64] The pathologic process reaches outside the substantia nigra to affect cortical, limbic, and autonomic systems and is responsible for the severity of non-motor manifestations. Genetic causes account for a percentage of cases, with mutations in the SNCA, LRRK2, PARK2, and PINK1 genes associated with familial or early-onset disease.[65] Environmental exposures with pesticides, heavy metals, and rural residence have also been identified as modifiers of PD risk.[66] Diagnosis of PD is still largely clinical, from the identification of cardinal motor signs and exclusion of secondary causes. Responsiveness to dopaminergic agents confirms the diagnosis. Ancillary methods, for example, dopamine transporter (DAT) single-photon emission computed tomography (SPECT), may distinguish PD from mimetic disorders of parkinsonism, e.g., drug-induced parkinsonism or essential tremor.[67] MRI is usually not helpful in early PD but is useful in excluding structural lesions and for the diagnosis of atypical parkinsonian syndromes like multiple system atrophy or progressive supranuclear palsy.

The treatment of PD is symptomatic and directed at replacing dopaminergic function. Levodopa, usually with carbidopa or benserazide supplementation to increase central bioavailability, is the established treatment and is extremely effective for suppressing motor symptoms.[68] For the treatment of motor fluctuations, dopamine agonists like monoamine oxidase-B inhibitors and catechol-O-methyltransferase inhibitors are used as therapies. There are few complications that may include wearing-off of benefits and sometimes even dyskinesias develop with progressive disease. This necessitates a cautious prescription and progression of therapy paradigm. Subthalamic nucleus or globus pallidus interna deep brain stimulation (DBS) is a proven surgical treatment in well-selected patients with chronic motor complications, with long-term improvement in motor control.[69] In this case, non-motor symptoms management is also required because they most significantly contribute to disability and caregiver burden. Pharmacologic therapy of depression, anxiety, sleep disorder as well as autonomic dysfunction is individually customized. On the other hand, cognitive impairment necessitates multidisciplinary therapies. Also, rehabilitative modalities such as physical, speech, and occupational therapy preserve function and overall quality of life. Investigational studies on disease-modifying therapy are ongoing but are mostly in experimental phase. This is including alpha-synuclein aggregation, mitochondrial impairment, and neuroinflammation treatment. From a public health point of view, PD's disease burden is increasing very rapidly and largely on account of population aging worldwide.[70] The prevalence of PD increased more than twofold between 1990 and 2019, and this accelerated the rapid development of scalable prevention, diagnostic, and treatment options in high demand.[70] Although there is currently no cure available but advances in early diagnosis, personalized therapeutics, and neuroprotective treatments promise hope to reduce the future burden of this debilitating disease.

1.3.4 Motor Neuron Disorders (ALS, MND, SMA)

Motor neuron diseases (or MNDs) are a set of degenerative neurological disorders involving the loss of upper and/or lower motor neurons.[71] They result in weakness, atrophy, and ultimately paralysis. Although not very common but when compared to other neurological disorders, they have serious clinical and social implications.[71] The reason is their relentless course and absence of any cure. The MND spectrum encompasses ALS, which is the most frequent adult-onset form showing progressive muscular atrophy and primary lateral sclerosis.[72] Then there is SMA, which is a genetic condition typically manifesting in childhood.[73] As a group, these diseases illustrate the crippling effect of motor system degeneration on life expectancy and quality.

Amyotrophic lateral sclerosis is the classical motor neuron disease involving both the upper and lower motor neurons.[72] ALS manifests between 50 and 70 years of age with a predominance in males. Clinical features include progressive weakness, wasting of the muscles, fasciculations, spasticity, and hyperreflexia. Bulbar presentation results in dysarthria, dysphagia as well as compromised respiratory function. Median survival is 3–5 years after onset of symptoms, primarily secondary to respiratory failure. The ALS pathology involves abnormal cytoplasmic protein aggregation by proteins like TDP-43, FUS, and SOD1.[74] This causes disruption of RNA processing and cellular homeostasis. Although the majority of cases are sporadic but around 10% are familial that are typically associated with mutations within genes like SOD1, C9orf72, and TARDBP.[75] On the other hand, SMA is a childhood genetic disorder that results from mutations in the SMN1 gene to cause deficiency of survival motor neuron (SMN) protein.[73] SMA has an unpredictable presentation of severity like Type I (Werdnig–Hoffmann disease) presents in early infancy with extreme weakness and hypotonia with a high likelihood of death before 2 years of age if left untreated.[76] Whereas, the Type IV presents[77] in adulthood with slower progression.[77] This condition mainly involves lower motor neurons in the anterior horn of the spinal cord that leads to weakness of muscles, atrophy, and also contractures.[77]

Diagnosis of MND is a combination of careful clinical assessment as well as supportive investigations. In ALS, diagnosis is set according to revised El Escorial criteria with evidence of both upper and lower motor neuron dysfunction in two or more regions.[78] And there should be a progression over a period of time and exclusion of mimics. Corroborating the diagnosis, EMG confirms diffuse denervation and reinnervation. In SMA, primarily the molecular genetic analysis for SMN1 deletion is considered diagnostic.[79] Neuroimaging is normal in these conditions but usually is done to rule out other structural etiology. Treatment is currently limited, although progress has been made. In ALS, riluzole (glutamate release blocker) modestly prolongs survival, whereas edaravone (an antioxidant) may delay functional decline in few of the patients.[80] Supportive therapy using non-invasive ventilation, gastrostomy feeds, and multidisciplinary rehabilitation has remained the cornerstone of treatment. And this is seen to significantly improve the survival and quality of life. Additionally, the genetic therapies have transformed the disease process in SMA. Nusinersen (an antisense oligonucleotide to correct SMN2 splicing) improves survival and motor function.[81] Onasemnogene abeparvovec is a pediatric gene replacement therapy that delivers functional SMN1 using adeno-associated virus vector AAV9.[82] Additionally, small molecules like risdiplam (an oral splicing corrector) offer further therapeutic options.[83] The treatments have changed SMA from a condition always resulting in death in infancy to a treatable/manageable disease with significantly improved outcomes.

Even with many developments, there are serious challenges that persist. The extremely high cost of novel therapies limits access, especially in LMICs. In ALS, modifying treatments remain inaccessible, and research continues to focus on clarifying pathogenic mechanisms such as RNA dysfunction, protein misfolding, excitotoxicity, and neuroinflammation.[84] Simultaneously, patient-led multidisciplinary management, early palliative involvement, and support for caregivers remain paramount to improving quality of life. MNDs, while less common than dementia or stroke, represent the worst of neurodegeneration's consequences. They underscore the imperative to develop both fundamental studies of mechanisms of neuronal susceptibility and universal access to new genetic therapies. As therapeutic technology continues to accelerate, the next decades promise to be transformative for these once-incurable disorders.

1.3.5 Epilepsy

Epilepsy is a prevalent chronic neurological disease, occurring in more than 50 million individuals across the globe, and is defined by a chronic propensity to produce seizures in addition to the cognitive, psychological, neurobiological, and social effects of the disorder.[7,85] Epilepsy is different from single seizures. The main difference is its repetition and spontaneity. It indicates persistent brain network abnormalities that reduce the threshold for synchronous discharges of neuronal cells.[85] Unfortunately, epilepsy is still a stigmatized and health disparity-laden condition despite dramatic progress in diagnosis and management.[86] Particularly this is true as most patients reside in LMICs. The pathophysiology of epilepsy is heterogeneous and includes genetic, structural, metabolic, infectious, and autoimmune processes.[84] In a large percentage of cases, an underlying cause is not identifiable and the disorder is labeled as idiopathic. Seizures are generally classified as focal, from a single cortical area, or generalized, affecting both hemispheres simultaneously. Focal seizures can present as sensory phenomena, automatisms, or motor signs, often secondarily generalizing to bilateral tonic–clonic seizures. Generalized seizures include absence seizures, myoclonic jerks, and generalized tonic–clonic seizures. The spectrum is broad in the clinical setting, and accurate classification using the International League Against Epilepsy (ILAE) system is important to inform treatment and prognosis.[24,87]

It is imperative that epilepsy diagnosis takes a careful combination of clinical history, eyewitness testimony, and confirmatory tests.[88] In this pursuit, EEG continues to be the cornerstone. EEG is rightly enabling the detection of interictal epileptiform discharges and facilitating seizure classification. Neuroimaging like high-resolution MRI is crucial to demonstrate structural lesions like cortical dysplasia, hippocampal sclerosis, tumors, or vascular malformations.[88] Functional imaging with PET or SPECT is very helpful and offers ancillary localization for surgical purposes. Laboratory studies and genetic testing might elucidate etiology in children and familial cases. Notably, epilepsy should be separated from disorders like syncope, migraines, and psychogenic non-epileptic seizures, whose manifestations are similar to those of a seizure.[88,89]

Epilepsy treatment is mainly with antiseizure medications (ASMs), with seizure control in about two-thirds of patients.[90] Some of the widely used ASMs include sodium valproate, levetiracetam, lamotrigine, and carbamazepine.[90] The choice of drug is determined by the seizure type, patient, and side effects.[7,88,91] For drug-resistant epilepsy, as specified by failure of satisfactory trials of two suitable ASMs, other treatments are considered.[7,91] Surgery for epilepsy, as with temporal lobectomy for mesial temporal lobe epilepsy, can offer long-term remission in well-selected candidates. Neurostimulation treatments like vagus nerve stimulation and DBS, are additional choices in the refractory cases.[92] Importantly, diet-based treatments (like ketogenic diet) have been seen to be effective in drug-resistant epilepsy in children.[93] Along with seizure control, there is a need for complete care so that the psychosocial and cognitive aspects of epilepsy is also taken care of. The associated neuropsychological issues including depression, anxiety, memory impairment, and even stigma severely compromise quality of life.[94] Sudden unexpected death in epilepsy (SUDEP) continues to be a tragic complication, emphasizing strict follow-up and risk avoidance measures. A treatment gap is added to the global burden with over 70% of those with epilepsy in resource-constrained environments deprived of access to beneficial therapies.[95] Public health efforts to enhance awareness, stigma reduction, and access to ASMs are important to control the situation. Epilepsy shows the dilemma of a complex neurological condition that is both curable and socially challenging. With the latest advances of neuroimaging, genetics, and neurostimulation, the promise of the future lies in more targeted and individualized strategies.

1.3.6 Brain Tumors and Central Nervous System Cancers

Brain tumors are a heterogenous group of neoplasms of the brain, spinal cord, and their supportive structures.[96,97] These tumors can be either benign or malignant. They can have widely variable behavior ranging from slow progressing to very aggressive.[97] Even though compared to overall cancer, they are relatively less frequent, brain tumors are out of proportion in morbidity and mortality due to their location within the CNS where even small lesions can interfere with important neurological functions. The global incidence of primary brain tumors has been estimated to be 3–6 cases/100,000/year and are most commonly malignant gliomas.[98] Even the five-year survival is approximately 67%.[98] Advances in molecular genetics have allowed for better classification and treatment planning, and neuro-oncology is among the most rapidly advancing areas of neurology. Integrated diagnosis with molecular markers and histology is the focus of the 2021 revised WHO classification of CNS tumors.[99] Historically, tumors were grouped into gliomas (astrocytomas, oligodendrogliomas, glioblastomas), meningiomas, embryonal tumors (medulloblastomas), nerve sheath tumors (acoustic neuromas), pituitary adenomas, and others. The incorporation of molecular parameters, e.g., IDH mutation status, 1p/19q codeletion, and TERT promoter mutations, has enhanced prognostication and treatment decision-making.[100] For instance, IDH-mutant vs IDH-wildtype gliomas is important for both prognosis and treatment with respect to survival.

Glioblastoma (GBM) remains the most common and most lethal primary malignant brain tumor in adults, occurring in nearly half of all malignant CNS neoplasms. GBM usually presents with headaches, seizures, focal neurological impairments, or intellectual decline depending on the tumor location.[101,102] Despite vigorous management with maximal safe surgical resection, radiotherapy, and chemotherapy with temozolomide, median survival is approximately 12–15 months, with a two-year survival of less than 30%.[102] Novel modalities, including tumor-treating fields, immunotherapy, and molecular-targeted inhibitors, are under investigation but are still yet to significantly alter long-term outcomes.

Meningiomas, the most frequent primary CNS tumors in general, are typically benign but can be highly debilitating by causing compression of nearby structures.[103] They represent more than one-third of primary CNS tumors and more than half of all benign CNS tumors.[104] It also has an increasing prevalence by age and higher incidence in women. While the majority of meningiomas are incidentally discovered on physical examination, symptomatic ones may manifest with seizures, focal deficits, or signs of raised intracranial pressure. Surgical excision is often curative, although recurrence is a risk, particularly with atypical or anaplastic forms.[105]

Medulloblastomas, or cerebellar embryonal tumors, are the most common malignant childhood brain tumors.[106] Molecular subgrouping, including WNT-activated and SHH-activated medulloblastomas, has greatly improved prognosis and planning of treatment.[107] More than 70% survival occurs in some of these subgroups, with multimodal therapy with surgery, craniospinal irradiation, and chemotherapy commonly employed and having serious long-term cognitive and endocrine sequelae.[107]

Other important neoplasms include acoustic neuromas (vestibular schwannomas), benign nerve sheath tumors presenting with progressively diminishing hearing and balance; pituitary adenomas, which cause endocrine secretions or mass effects; and less common tumors such as pineal gland neoplasias or choroid plexus neoplasias. Each of these requires individually tailored diagnostic and management strategies. They typically require multimodal interdisciplinary consultation from neurologists, neurosurgeons, oncologists as well as endocrinologists. Brain tumor diagnosis relies on MRI that provides high-resolution contrast-enhanced anatomical images to delineate tumor borders. MR spectroscopy, perfusion imaging, and PET scans are the most advanced imaging techniques that can provide helpful information regarding metabolic activity as well as the tumor grade.[108] The diagnosis can only be established with histopathological and increasingly molecular evaluation acquired by biopsy or surgical resection. The current treatment approaches try to maximize tumor control with preservation of neurological functions. Surgery remains the basis of treatment for most of the resectable tumors. Adjuvant therapy with radiotherapy and chemotherapy is based on the tumor type, grade, and molecular features. Targeted agents and immunotherapy have given new expectations in the shape of precision medicine strategies in recent years. Nevertheless, challenges such as tumor heterogeneity, the blood–brain barrier, and resistance to treatment remain the major challenges. From a public health perspective, CNS tumors contribute disproportionately to DALYs in proportion to their lower incidence compared with systemic cancer. The devastation is most acutely appreciated in children, where survival comes at the cost of permanent neurocognitive impairment.[109] Continuing investment in uncovering molecular pathways, earlier diagnosis, and emerging treatment modalities is essential for improving outcome and mitigating the devastating impact of brain tumors.

1.4 SYSTEMIC CONDITIONS AND NEUROLOGICAL CONSEQUENCES

Systemic health profoundly affects the neurological function. Increasingly, it has become clear that chronic systemic disorders like obesity, metabolic syndrome, or nutritional deficiency exert strong effects on our brain.[29,110] This causes both cognitive dysfunction and heightened vulnerability to major neurological problems. These conditions were previously regarded primarily as cardiometabolic illnesses. But today they are recognized to be major risk factors for dementia, stroke, and other neurological disorders. The relationship between systemic health and neurological status, beginning with obesity and its impact on neurocognition is crucial in understanding holistic neurological health.

1.4.1 Obesity and Neurocognitive Function

Obesity is defined as being overweight with a body mass index (BMI) of 30 or higher.[111] It has become a pandemic around the globe affecting all age groups. Obesity is not only recognized for its cardiovascular and metabolic effects, but also it is increasingly regarded as a significant brain health determinant.[111] Increasing evidence points to obesity as having adverse effects on cognition and brain structure. It is rendering individuals at risk for dementia and other neurocognitive diseases.[112] The link is mediated through a complex interrelationship of vascular, metabolic, inflammatory, and hormonal processes. Neuroimaging studies have indicated that adults with obesity have reduced brain volume, particularly in the hippocampus and frontal lobes, which are memory and executive functioning areas.[113] These structural changes correlate with diminished cognitive function and earlier cognitive aging. An fMRI study in obese individuals shows distinct patterns of connectivity in various cognitive networks.[113] This clearly suggests that even in young adults, obesity can reduce overall neural efficiency. Crucially, obesity is identified as one of the most powerful modifiable risk factors for dementia in midlife, particularly AD.[114]

The cognitive impairment pathophysiology due to obesity comprises multifaceted interactions. And central to this is chronic systemic inflammation where pro-inflammatory cytokines like TNF-α and IL-6 that are secreted by adipose tissue.[115] These can get into the brain and trigger neuroinflammation. One of the characteristics of obesity can be insulin resistance (IR) that can also disturb cerebral glucose metabolism. IR can deteriorate insulin signaling pathways in neurons that further contributes to synaptic dysfunction and cognitive impairments.[115] Besides, obesity is causally linked with atherogenic vascular risk factors such as hypertension, dyslipidemia, and diabetes. All of these are also implicated in cerebrovascular disease and vascular dementia. Disordered breathing during sleep and specifically obstructive sleep apnea are also more common in obese individuals.[116] This further worsens cognitive impairment through intermittent hypoxia and oxidative stress.

Obesity's effects are not just limited to the risk of dementia but also to other less apparent neurocognitive effects. Obese children and adolescents have also been demonstrated to score lower in attention, memory, and executive function tasks with long-term educational and social implications.[117] In adults, obesity has been associated with not just cognitive impairment but also affective disorders like depression and anxiety. This is adding further neurologic insult to its effects. From a clinical perspective, weight loss and lifestyle change have been correlated with improved cognitive performance.[115] Caloric restriction, exercise, and bariatric surgery not only enhance metabolic well-being but evidence of reversing certain obesity-associated changes in the brain. These results highlight the acute significance of public health programs addressing obesity prevention as well as management. This is not merely to lower cardiometabolic disease burden but to maintain cognitive and neurological health throughout life. Obesity thus represents a classic example of how systemic conditions can exert widespread influence on brain structure and function. Recognizing and addressing obesity as a neurological as well as a metabolic disorder has profound implications for both clinical practice and preventive strategies.

1.4.2 Metabolic Syndrome and the Brain

Metabolic syndrome can be referred to as a cluster of interrelated problems or risk factors such as abdominal obesity, hypertension, IR, hyperglycemia, and dyslipidemia.[118] Nevertheless, it has become a potent predictor of neurological well-being. While it is largely regarded as a precursor to cardiovascular disease and type 2 diabetes, evidence suggests it has broad significance for the CNS. Metabolic syndrome patients are at a much increased risk of cognitive impairment, stroke, and dementia. This demonstrates the close interrelationships between systemic metabolic disease and brain disease.[119] Metabolic syndrome causes brain dysfunction through multifactorial mechanisms.[119] IR that is chronic alters neuronal glucose metabolism, damages synaptic plasticity, and causes neurodegeneration. This has prompted some researchers to refer to AD as a type of "type 3 diabetes," highlighting the significance of insulin signaling in the preservation of cognitive integrity.[120] In addition, metabolic syndrome is pro-atherogenic and pro-small-vessel disease, leading to cerebral hypoperfusion, microinfarcts, and white matter lesions.[121] All these are important causes of vascular cognitive impairment. Systemic inflammation is one of the characteristics of metabolic syndrome that exacerbates neuroinflammation and increases amyloid and tau pathology in AD models.[122]

Epidemiological evidence repeatedly demonstrates that all the components of metabolic syndrome as well as metabolic syndrome itself, are related to adverse cognitive outcomes.[123] Midlife hypertension and diabetes are two of the most potent predictors of late-onset dementia.[124] Dyslipidemia and obesity contribute synergistically to both AD and vascular dementia risks.[125] Importantly, the hazard appears to be cumulative, i.e., the more elements of metabolic syndrome, the higher the risk of dementia. This almost dose-response relationship clearly emphasizes the importance of early and compulsory control of risk factors.[125] Clinically speaking, individuals with metabolic syndrome appear subtly to have impairments in executive function, attention, and processing speed. This may antecede the frank dementia. These changes in cognition are frequently subtle but may impair occupational performance and quality of life. Neuroimaging research shows that metabolic syndrome is linked to smaller hippocampal volume, cortical thinning, and white matter hyperintensities.[126] These structural findings are consistent with reported impairments in cognition also.

From the therapeutic point of view, interventions against metabolic syndrome also benefit the brain. Lifestyle modifications like exercising regularly, healthy diet, and weight management improve metabolic profiles and reduce dementia risk.[127] Pharmacologic control of hypertension, diabetes, and dyslipidemia has been found to protect against cognitive deterioration.[128] Although the level of benefit depends on timing and disease severity. Increasingly, integrated strategies addressing the syndrome as a whole, rather than isolated risk factors, are advocated to maximize neurological protection. The identification of metabolic syndrome as both a neurological and cardiometabolic disorder has significant implications for public health. Prevention of the neurological effects of metabolic syndrome necessitates collaborative action among neurology, endocrinology, and preventive medicine. Preserving brain health therefore not just depends on the management of established neurological disease but on actively intervening in systemic risk factors (well before clinical presentation).

1.4.3 Nutritional Deficiencies and Neurological Disorders

Nutrition has a significant role in brain health. On the other hand, deficiencies of important micronutrients have dramatic neurological effects throughout the lifetime.[129] The nervous system is particularly susceptible to poor nutrition due to its high metabolic rate, dependence on certain cofactors to synthesize neurotransmitters, and dependence on vitamins and minerals for myelination and neuronal repair.[130] Nutrient deficiencies are still an important etiology of avoidable neurological illness, especially in LMIC regions. But are also increasingly identified in developed nations in the setting of aging, restrictive eating, or chronic disease.

Vitamin B12 deficiency is perhaps the most classic example and is associated with a wide range of neurological findings from peripheral neuropathy and myelopathy to cognitive impairment and dementia.[131] This vitamin is needed for myelin production and methylation processes. Its deficiency results in demyelination of spinal cord and cerebral white matter. Paresthesias, gait abnormalities, memory impairment, or psychiatric disturbances may be the presenting features at the bedside.[131] The early detection of vitamin B12 deficiency is important since neurological impairment can become permanent if parenteral cobalamin therapy is not given in time.[132] Thiamine (vitamin B1) deficiency is another important cause of neurological illness, most commonly encountered in chronic alcoholism but also in malnutrition and hyperemesis gravidarum. It presents as Wernicke's encephalopathy (a triad of confusion, ataxia, and ophthalmoplegia) which, if left untreated, can evolve into Korsakoff's syndrome with severe amnesia and confabulation.[133] Thiamine is an essential cofactor for glucose metabolism and the deficiency results in cerebral energy impairment with consequent death of vulnerable neurons in the mammillary bodies. Early recognition and supplementation are life-saving and may forestall long-term cognitive disability.

Many other micronutrients also play essential roles in neurology. Deficiency of vitamin D has been linked to an increased risk of multiple sclerosis and impaired cognitive function.[134] This is perhaps due to its role in immunomodulation and neuroprotection. Deficiency of folate causes megaloblastic anemia as well as neuropsychiatric symptoms. Maternal low folate status before pregnancy is an important risk factor for neural tube defects.[135] Iron deficiency, the most prevalent dietary disorder in the world, has been associated with impaired neurodevelopment in children and resultant disabilities in attention, learning, and behavior.[136] Similarly, deficiency of iodine remains the leading cause of preventable intellectual disability worldwide, highlighting the vast neurological importance of adequacy of micronutrients.[137] In countries like India, the common salt is sold in iodized form. And that has helped to sort out deficiency issues to a huge level.

Diagnosis of nutritional deficiency relies on a combination of clinical suspicion, dietary history, and laboratory determination of serum vitamin or mineral level. Neuroimaging may be able to show pathognomonic findings in some, such as mammillary body swelling in Wernicke's encephalopathy or spinal cord signal abnormality in B12 deficiency.[138] Importantly, the majority of such entities can be reversed if treated early. Therefore, clinicians must beware during the evaluation of patients with unusual neurological complaints. Preventive measures like diversification of diet, food fortification, and specific supplementation have been highly effective in cutting down the disease burden of deficiency-induced neurological disorders. Public health measures like folic acid enrichment of grains and universal iodization of salt are among the most effective neurological prevention strategies of the last century. However, high-risk groups like the elderly or those with gastrointestinal illness, vegetarians, and patients with alcohol use disorder continue to be at risk and need active screening and management. Nutrient deficiencies are of the highest relevance since they are avoidable and reversible. Therapy of these deficiencies not only improves neurological result but also serves further to promote healthy development and successful aging. In global neurology, they emphasize the critical interface of diet, healthcare systems, and neurological health.

1.5 CONCLUSION

Neurological disorders are very heterogeneous and multifaceted array of conditions that significantly impact not only individuals and families but also the whole societies globally. Ranging from dementia and PD to stroke, epilepsy, MND, and brain tumors, these disorders play a significant part in the global burden of disability and death. Importantly, recent technological advances in neuroimaging, electrophysiology, biomarker identification, and genetics have considerably improved diagnosis. It has also improved the classification of these diseases with accuracy, while new technologies like AI keep redefining practice. Nevertheless, with all advancements, most neurological diseases are still beyond cure. This clearly underpins the necessity for continuous research into pathophysiology and the development of new therapeutics. There is a close relationship between systemic health and neurological function. Metabolic syndrome, obesity, and malnutrition demonstrate that the brain cannot be separated from metabolic, vascular, and dietary processes within the body. Confronting these entities with preventive interventions, lifestyle change, and policy is a potent way of preventing the future burden of neurological disease. The way forward must be multidisciplinary and multi-level. This also calls clinicians, researchers, and policymakers to collaborate and advance diagnostic capabilities, and develop disease-modifying therapies. Early diagnosis, tailored treatment, caregiver support, and community-based interventions are integral components of complete neurological well-being. At the same time, investment in preventive interventions targeting systemic risk factors such as obesity, hypertension, diabetes, and malnutrition shall reap high rewards for brain health across the complete life course. Therefore, neurological disease occupies the intersection of biology, behavior, and public health. By integrating breakthroughs in neuroscience with prevention medicine and global health methods, we can curb their devastating burden and optimize outcomes for millions worldwide.

REFERENCES

1. Steinmetz JD, Seeher KM, Schiess N, et al. Global, regional, and national burden of disorders affecting the nervous system, 1990–2021: a systematic analysis for the Global Burden of Disease Study 2021. *The Lancet Neurology*. 2024;23(4):344–381. doi:10.1016/s1474-4422(24)00038-3
2. World Health Organization. *Neurological Disorders: Public Health Challenges*. World Health Organization; 2006.
3. Ding C, Wu Y, Chen X, et al. Global, regional, and national burden and attributable risk factors of neurological disorders: the Global Burden of Disease study 1990–2019. *Frontiers in Public Health*. 2022;10. doi:10.3389/fpubh.2022.952161
4. Feigin VL, Vos T, Nichols E, et al. The global burden of neurological disorders: translating evidence into policy. *The Lancet Neurology*. 2020;19(3):255–265. doi:10.1016/s1474-4422(19)30411-9
5. Nagayach A, Bhaskar R, Ghosh S, et al. Interplay between circadian rhythm, ageing and neurodegenerative disorder. *Open Biology*. 2025;15(7). doi:10.1098/rsob.240161
6. World Health Organization. Dementia: Fact Sheet. 2025/03/31; 2025.
7. World Health Organization. Epilepsy: Fact Sheet. 2024/02/07; 2024.
8. Mukherjee S, Ali S, Hashmi S, Jahan S. History, Origin and Types of Neurological Disorders. *Applications of Stem Cells and derived Exosomes in Neurodegenerative Disorders*. 2023:1–32:chap Chapter 1. Springer, Singapore. https://doi.org/10.1007/978-981-99-3848-3_1
9. Dev P, Pathak A. The Current Status of Cognitive Disorders and Their Diagnosis. *Cognitive Predictive Maintenance Tools for Brain Diseases*. 2023:10–35. Taylor & Francis. https://doi.org/10.1201/9781003245346-2 Group
10. Ruiz Santiago F, Láinez Ramos-Bossini AJ, Wáng YXJ, Martínez Barbero JP, Espinosa JG, Martínez Martínez A. The value of magnetic resonance imaging and computed tomography in the study of spinal disorders. *Quantitative Imaging in Medicine and Surgery*. 2022;12(7):3947–3986. doi:10.21037/qims-2022-04
11. Chandrasekar SK, Arthanari J, Chandrasekar KK, et al. Advanced imaging techniques (PET, fMRI, DTI) in early detection of neurodegenerative diseases: a systematic review. *Health Science Reports*. 2025;8(7). doi:10.1002/hsr2.70855
12. Cohen MX. Where does EEG come from and what does it mean? *Trends in Neurosciences*. 2017;40(4):208–218. doi:10.1016/j.tins.2017.02.004
13. Tankisi H, Burke D, Cui L, et al. Standards of instrumentation of EMG. *Clinical Neurophysiology*. 2020;131(1):243–258. doi:10.1016/j.clinph.2019.07.025
14. Papaliagkas V, Kalinderi K, Vareltzis P, Moraitou D, Papamitsou T, Chatzidimitriou M. CSF biomarkers in the early diagnosis of mild cognitive impairment and Alzheimer's disease. *International Journal of Molecular Sciences*. 2023;24(10). doi:10.3390/ijms24108976
15. Dratch L, Azage M, Baldwin A, et al. Genetic testing in adults with neurologic disorders: indications, approach, and clinical impacts. *Journal of Neurology*. 2023;271(2):733–747. doi:10.1007/s00415-023-12058-6
16. Varesi A, Carrara A, Pires VG, et al. Blood-based biomarkers for Alzheimer's disease diagnosis and progression: an overview. *Cells*. 2022;11(8). doi:10.3390/cells11081367
17. Kaur J, Ghosh S, Singh P, Dwivedi AK, Sahani AK, Sinha JK. Cervical spinal lesion, completeness of injury, stress, and depression reduce the efficiency of mental imagery in people with spinal cord injury. *American Journal of Physical Medicine & Rehabilitation*. 2022;101(6):513–519. doi:10.1097/phm.0000000000001955
18. Khalighi S, Reddy K, Midya A, Pandav KB, Madabhushi A, Abedalthagafi M. Artificial intelligence in neuro-oncology: advances and challenges in brain tumor diagnosis, prognosis, and precision treatment. *npj Precision Oncology*. 2024;8(1). doi:10.1038/s41698-024-00575-0
19. Kumar R, Waisberg E, Ong J, et al. Artificial intelligence-based methodologies for early diagnostic precision and personalized therapeutic strategies in neuro-ophthalmic and neurodegenerative pathologies. *Brain Sciences*. 2024;14(12). doi:10.3390/brainsci14121266

20. World Health Organization. *Intersectoral Global Action Plan on Epilepsy and Other Neurological Disorders 2022–2031.* World Health Organization; 2023.
21. Kopp W. Aging and "age-related" diseases – what is the relation? *Aging and Disease.* 2024. doi:10.14336/ad.2024.0570
22. Fülöp T, Witkowski JM. Immunity, Immunosenescence, Brain and Cognitive/Mental Health. *Brain and Mental Health in Ageing.* 2024:95–117:chap Chapter 6. *Healthy Ageing and Longevity.* Springer, Singapore. https://doi.org/10.1007/978-3-031-68513-2_6
23. Hille M, Kühn S, Kempermann G, Bonhoeffer T, Lindenberger U. From animal models to human individuality: integrative approaches to the study of brain plasticity. *Neuron.* 2024;112(21):3522–3541. doi:10.1016/j.neuron.2024.10.006
24. Zhou H, Hu Y, Li G, et al. Obesity is associated with progressive brain structural changes. *Obesity.* 2025;33(4):709–719. doi:10.1002/oby.24251
25. Gasiorowska A, Wydrych M, Drapich P, et al. The biology and pathobiology of glutamatergic, cholinergic, and dopaminergic signaling in the aging brain. *Frontiers in Aging Neuroscience.* 2021;13. doi:10.3389/fnagi.2021.654931
26. Wang S, Cai Y, Zhao Y, Yu Y, Chen X, Wang L. Altered resting-state cerebral blood flow and its relationship with molecular architecture in tremor dominant Parkinson's disease. *Brain Research Bulletin.* 2025;222. doi:10.1016/j.brainresbull.2025.111237
27. Sinha JK, Ghosh S. Scoring more than ten plus century – antiquity in gerontology? *Indian Journal of Medical Research.* 2010;131:586–587.
28. Sümer E, Kaynak H. Age-related decline in source and associative memory. *Cognitive Processing.* 2024;26(1):1–13. doi:10.1007/s10339-024-01230-z
29. Sinha JK, Ghosh S, Swain U, Giridharan NV, Raghunath M. Increased macromolecular damage due to oxidative stress in the neocortex and hippocampus of WNIN/Ob, a novel rat model of premature aging. *Neuroscience.* 2014;269:256–264. doi:10.1016/j.neuroscience.2014.03.040
30. Sinha JK, Jorwal K, Singh KK, Han SS, Bhaskar R, Ghosh S. The potential of mitochondrial therapeutics in the treatment of oxidative stress and inflammation in aging. *Molecular Neurobiology.* 2024;62(6):6748–6763. doi:10.1007/s12035-024-04474-0
31. Ghimire A, Rehman SA, Subhani A, et al. Mechanism of microglia-mediated neuroinflammation, associated cognitive dysfunction, and therapeutic updates in Alzheimer's disease. *hLife.* 2025;3(2):64–81. doi:10.1016/j.hlife.2024.11.006
32. Lu H, Xu F, Rodrigue KM, et al. Alterations in cerebral metabolic rate and blood supply across the adult lifespan. *Cerebral Cortex.* 2011;21(6):1426–1434. doi:10.1093/cercor/bhq224
33. Rajeev V, Fann DY, Dinh QN, et al. Pathophysiology of blood brain barrier dysfunction during chronic cerebral hypoperfusion in vascular cognitive impairment. *Theranostics.* 2022;12(4):1639–1658. doi:10.7150/thno.68304
34. Kopeć K, Koziorowski D, Szlufik S. The therapeutic potential of glymphatic system activity to reduce the pathogenic accumulation of cytotoxic proteins in Alzheimer's disease. *International Journal of Molecular Sciences.* 2025;26(15). doi:10.3390/ijms26157552
35. Mansoor AK, Thomas S, Sinha JK, Alladi PA, Ravi V, Raju TR. Olfactory tract transection reveals robust tissue-level plasticity by cellular numbers and neurotrophic factor expression in olfactory bulb. *Indian Journal of Experimental Biology.* 2012;50(11):765–770.
36. Kozak-Szkopek E, Wdowiak K, Imiela AM, et al. Frailty and cognitive decline in hospitalized elderly patients: associations with clinical and laboratory parameters. *Medical Science Monitor.* 2025;31. doi:10.12659/msm.947354
37. Tari AR, Walker TL, Huuha AM, Sando SB, Wisloff U. Neuroprotective mechanisms of exercise and the importance of fitness for healthy brain ageing. *The Lancet.* 2025;405(10484):1093–1118. doi:10.1016/s0140-6736(25)00184-9
38. Ghosh S, Durgvanshi S, Agarwal S, Raghunath M, Sinha JK. Current status of drug targets and emerging therapeutic strategies in the management of Alzheimer's disease. *Current Neuropharmacology.* 2020;18(9):883–903. doi:10.2174/1570159x18666200429011823
39. World Health Organization. Dementia. 2025/09/26; 2025.
40. DeCarli C. Mild cognitive impairment: prevalence, prognosis, aetiology, and treatment. *The Lancet Neurology.* 2003;2(1):15–21. doi:10.1016/s1474-4422(03)00262-x
41. Singh A, Sethi P, Ghosh S, et al. Current status of prevention strategies adapted for Alzheimer's disease. *A New Era in Alzheimer' s Research.* 2025:91–108. Academic Press. https://doi.org/10.1016/B978-0-443-15540-6.00004-5

42. Inoue Y, Shue F, Bu G, Kanekiyo T. Pathophysiology and probable etiology of cerebral small vessel disease in vascular dementia and Alzheimer's disease. *Molecular Neurodegeneration*. 2023;18(1). doi:10.1186/s13024-023-00640-5
43. Borghammer P, Okkels N, Weintraub D. Parkinson's disease and dementia with Lewy bodies: one and the same. *Journal of Parkinson's Disease*. 2024;14(3):383–397. doi:10.3233/jpd-240002
44. Warren JD, Rohrer JD, Rossor MN. Frontotemporal dementia. *BMJ*. 2013;347(August):f4827. doi:10.1136/bmj.f4827
45. Weintraub S. Neuropsychological assessment in dementia diagnosis. *Continuum*. 2022;28(3):781–799. doi:10.1212/con.0000000000001135
46. Chouliaras L, O'Brien JT. The use of neuroimaging techniques in the early and differential diagnosis of dementia. *Molecular Psychiatry*. 2023;28(10):4084–4097. doi:10.1038/s41380-023-02215-8
47. Rabinovici GD, Knopman DS, Arbizu J, et al. Updated appropriate use criteria for amyloid and tau PET: a report from the Alzheimer's association and society for nuclear medicine and molecular imaging workgroup. *Journal of Nuclear Medicine*. 2025;66(Supplement 2):S5–S31. doi:10.2967/jnumed.124.268756
48. Miculas DC, Negru PA, Bungau SG, Behl T, Hassan SSu, Tit DM. Pharmacotherapy evolution in Alzheimer's disease: current framework and relevant directions. *Cells*. 2022;12(1). doi:10.3390/cells12010131
49. Cummings J, Osse AML, Cammann D, Powell J, Chen J. Anti-amyloid monoclonal antibodies for the treatment of Alzheimer's disease. *BioDrugs*. 2023;38(1):5–22. doi:10.1007/s40259-023-00633-2
50. Arshad F, Mm S, Paplikar A, Rajendran S, Kalkonde Y, Alladi S. Vascular cognitive impairment in India: challenges and opportunities for prevention and treatment. *Cerebral Circulation – Cognition and Behavior*. 2022;3. doi:10.1016/j.cccb.2021.100034
51. Murphy SJX, Werring DJ. Stroke: causes and clinical features. *Medicine*. 2023;51(9):602–607. doi:10.1016/j.mpmed.2023.06.003
52. Gross BA, Jankowitz BT, Friedlander RM. Cerebral intraparenchymal hemorrhage. *JAMA*. 2019;321(13). doi:10.1001/jama.2019.2413
53. Saver JL. Time is brain—quantified. *Stroke*. 2006;37(1):263–266. doi:10.1161/01.STR.0000196957.55928.ab
54. Nukovic JJ, Opancina V, Ciceri E, et al. Neuroimaging modalities used for ischemic stroke diagnosis and monitoring. *Medicina*. 2023;59(11). doi:10.3390/medicina59111908
55. Feinbloom D, Bauer KA. Assessment of hemostatic risk factors in predicting arterial thrombotic events. *Arteriosclerosis, Thrombosis, and Vascular Biology*. 2005;25(10):2043–2053. doi:10.1161/01.ATV.0000181762.31694.da
56. Dewar B, Shamy M. tPA for acute ischemic stroke and its controversies: a review. *The Neurohospitalist*. 2019;10(1):5–10. doi:10.1177/1941874419838961
57. Bhogal P, AlMatter M, Hellstern V, et al. Mechanical thrombectomy for recurrent large vessel occlusion. *Journal of Clinical Neuroscience*. 2019;66:107–112. doi:10.1016/j.jocn.2019.05.010
58. Boccardi E, Cenzato M, Curto F, Motto C. Diagnosis and therapy in the acute phase of hemorrhagic stroke: latest developments. *Hemorrhagic Stroke*. 2017:1–97:chap Chapter 1. *Emergency Management in Neurology*.
59. Rost NS, Brodtmann A, Pase MP, et al. Post-stroke cognitive impairment and dementia. *Circulation Research*. 2022;130(8):1252–1271. doi:10.1161/circresaha.122.319951
60. Ozturk S. Stroke and stroke risk factors as disease burden. *Health Risk Analysis*. 2021;(4):146–150. doi:10.21668/health.risk/2021.4.16.eng
61. Luo Y, Qiao L, Li M, Wen X, Zhang W, Li X. Global, regional, national epidemiology and trends of Parkinson's disease from 1990 to 2021: findings from the Global Burden of Disease Study 2021. *Frontiers in Aging Neuroscience*. 2025;16. doi:10.3389/fnagi.2024.1498756
62. Lee HM, Koh S-B. Many faces of Parkinson's disease: non-motor symptoms of Parkinson's disease. *Journal of Movement Disorders*. 2015;8(2):92–97. doi:10.14802/jmd.15003
63. Savica R, Grossardt BR, Rocca WA, Bower JH. Parkinson disease with and without dementia: a prevalence study and future projections. *Movement Disorders*. 2018;33(4):537–543. doi:10.1002/mds.27277
64. Calabresi P, Mechelli A, Natale G, Volpicelli-Daley L, Di Lazzaro G, Ghiglieri V. Alpha-synuclein in Parkinson's disease and other synucleinopathies: from overt neurodegeneration back to early synaptic dysfunction. *Cell Death & Disease*. 2023;14(3). doi:10.1038/s41419-023-05672-9

65. Nuytemans K, Theuns J, Cruts M, Van Broeckhoven C. Genetic etiology of Parkinson disease associated with mutations in the SNCA, PARK2, PINK1, PARK7, and LRRK2 genes: a mutation update. *Human Mutation*. 2010;31(7):763–780. doi:10.1002/humu.21277
66. Shan L, Heusinkveld HJ, Paul KC, et al. Towards improved screening of toxins for Parkinson's risk. *npj Parkinson's Disease*. 2023;9(1). doi:10.1038/s41531-023-00615-9
67. Tsang K, Walker R. Dopamine transporter single photon emission computed tomography (DaT-SPECT) use in the diagnosis and clinical management of parkinsonism: an 8-year retrospective study. *Journal of Neurology*. 2023;270(5):2550–2558. doi:10.1007/s00415-023-11563-y
68. van Vliet EF, Knol MJ, Schiffelers RM, Caiazzo M, Fens MHAM. Levodopa-loaded nanoparticles for the treatment of Parkinson's disease. *Journal of Controlled Release*. 2023;360:212–224. doi:10.1016/j.jconrel.2023.06.026
69. Zhang C, Pan Y, Wang L, et al. Globus pallidus internus deep brain stimulation improves axial symptoms of Parkinson patients after long-term subthalamic nucleus stimulation: a case series study. *Interdisciplinary Neurosurgery*. 2019;18. doi:10.1016/j.inat.2019.100516
70. Li M, Ye X, Huang Z, Ye L, Chen C. Global burden of Parkinson's disease from 1990 to 2021: a population-based study. *BMJ Open*. 2025;15(4). doi:10.1136/bmjopen-2024-095610
71. Mora G, Chiò A. Disorders of upper and lower motor neurons. *Prognosis of Neurological Diseases*. 2015:261–272:chap Chapter 21.
72. Hardiman O, Al-Chalabi A, Chio A, et al. Amyotrophic lateral sclerosis. *Nature Reviews Disease Primers*. 2017;3(1). doi:10.1038/nrdp.2017.71
73. Mercuri E, Bertini E, Iannaccone ST. Childhood spinal muscular atrophy: controversies and challenges. *The Lancet Neurology*. 2012;11(5):443–452. doi:10.1016/s1474-4422(12)70061-3
74. Ratti A, Buratti E. Physiological functions and pathobiology of TDP-43 and FUS/TLS proteins. *Journal of Neurochemistry*. 2016;138(S1):95–111. doi:10.1111/jnc.13625
75. Vildan C, Sule D, Turker B, Hilmi U, Sibel KB, Spurkland A. Genetic alterations of C9orf72, SOD1, TARDBP, FUS, and UBQLN2 genes in patients with Amyotrophic Lateral Sclerosis. *Cogent Medicine*. 2019;6(1). doi:10.1080/2331205x.2019.1582400
76. Audic F, Barnerias C. Spinal muscular atrophy (SMA) type I (Werdnig–Hoffmann disease). *Archives de Pédiatrie*. 2020;27(7):7S15–7S17. doi:10.1016/s0929-693x(20)30271-2
77. Mercuri E, Sumner CJ, Muntoni F, Darras BT, Finkel RS. Spinal muscular atrophy. *Nature Reviews Disease Primers*. 2022;8(1). doi:10.1038/s41572-022-00380-8
78. Shefner JM, Al-Chalabi A, Baker MR, et al. A proposal for new diagnostic criteria for ALS. *Clinical Neurophysiology*. 2020;131(8):1975–1978. doi:10.1016/j.clinph.2020.04.005
79. Rouzier C, Chaussenot A, Paquis-Flucklinger V. Molecular diagnosis and genetic counseling for spinal muscular atrophy (SMA). *Archives de Pédiatrie*. 2020;27(7):7S9–7S14. doi:10.1016/s0929-693x(20)30270-0
80. Jaiswal MK. Riluzole and edaravone: a tale of two amyotrophic lateral sclerosis drugs. *Medicinal Research Reviews*. 2018;39(2):733–748. doi:10.1002/med.21528
81. Neil EE, Bisaccia EK. Nusinersen: a novel antisense oligonucleotide for the treatment of spinal muscular atrophy. *The Journal of Pediatric Pharmacology and Therapeutics*. 2019;24(3):194–203. doi:10.5863/1551-6776-24.3.194
82. Day JW, Finkel RS, Chiriboga CA, et al. Onasemnogene abeparvovec gene therapy for symptomatic infantile-onset spinal muscular atrophy in patients with two copies of SMN2 (STR1VE): an open-label, single-arm, multicentre, phase 3 trial. *The Lancet Neurology*. 2021;20(4):284–293. doi:10.1016/s1474-4422(21)00001-6
83. Ratni H, Scalco RS, Stephan AH. Risdiplam, the first approved small molecule splicing modifier drug as a blueprint for future transformative medicines. *ACS Medicinal Chemistry Letters*. 2021;12(6):874–877. doi:10.1021/acsmedchemlett.0c00659
84. Toader C, Tataru CP, Munteanu O, et al. Decoding neurodegeneration: a review of molecular mechanisms and therapeutic advances in Alzheimer's, Parkinson's, and ALS. *International Journal of Molecular Sciences*. 2024;25(23). doi:10.3390/ijms252312613
85. Mishra P, Sinha JK, Rajput SK. Efficacy of Cicuta virosa medicinal preparations against pentylenetetrazole-induced seizures. *Epilepsy & Behavior*. 2021;115. doi:10.1016/j.yebeh.2020.107653
86. Silva MHLe, Lima JVS, Lopes IPdS, et al. Stigmas in epilepsy: systematic review and meta-analysis. *Journal of Epilepsy Research*. 2025;15(1):23–32. doi:10.14581/jer.25003
87. Specchio N, Wirrell EC, Scheffer IE, et al. International League Against Epilepsy classification and definition of epilepsy syndromes with onset in childhood: position paper by the ILAE Task Force on Nosology and Definitions. *Epilepsia*. 2022;63(6):1398–1442. doi:10.1111/epi.17241

88. National Institute for Health and Care Excellence. *Epilepsies: Diagnosis and Management.* NICE; 2021.
89. Toffa DH, Poirier L, Nguyen DK. The first-line management of psychogenic non-epileptic seizures (PNES) in adults in the emergency: a practical approach. *Acta Epileptologica.* 2020;2(1). doi:10.1186/s42494-020-00016-y
90. Endayen D, Ayalew M, Getahun M. Multiple anti-seizure medications use and pattern of seizure control in children with epilepsy at neurology follow up clinic, Tikur Anbessa specialized hospital, Addis Ababa, Ethiopia. *African Health Sciences.* 2023;23(2):732–42. doi:10.4314/ahs.v23i2.84
91. Löscher W, Klein P. The pharmacology and clinical efficacy of antiseizure medications: from bromide salts to cenobamate and beyond. *CNS Drugs.* 2021;35(9):935–963. doi:10.1007/s40263-021-00827-8
92. Uysal U, Shamim D, Nwabueze O. Beyond resection: neuromodulation and minimally invasive epilepsy surgery. *Archives of Neuropsychiatry.* 2022.
93. Mishra P, Singh SC, Ramadass B. Drug resistant epilepsy and ketogenic diet: a narrative review of mechanisms of action. *World Neurosurgery: X.* 2024;22. doi:10.1016/j.wnsx.2024.100328
94. Novak A, Vizjak K, Rakusa M. Cognitive impairment in people with epilepsy. *Journal of Clinical Medicine.* 2022;11(1). doi:10.3390/jcm11010267
95. Kim JB. Sudden unexpected death in epilepsy: from short-term risk prediction to mechanistic insights and preventive strategies. *Journal of Clinical Neurology.* 2025;21(5). doi:10.3988/jcn.2025.0397
96. Singh SR, Bhaskar R, Ghosh S, et al. Exploring the genetic orchestra of cancer: the interplay between oncogenes and tumor-suppressor genes. *Cancers.* 2025;17(7). doi:10.3390/cancers17071082
97. National Institute of Neurological Disorders and Stroke. Brain and spinal cord tumors. 2025/09/26; 2025.
98. Schaff LR, Mellinghoff IK. Glioblastoma and other primary brain malignancies in adults. *JAMA.* 2023;329(7). doi:10.1001/jama.2023.0023
99. Torp SH, Solheim O, Skjulsvik AJ. The WHO 2021 Classification of Central Nervous System tumours: a practical update on what neurosurgeons need to know—a minireview. *Acta Neurochirurgica.* 2022;164(9):2453–2464. doi:10.1007/s00701-022-05301-y
100. Arita H, Matsushita Y, Machida R, et al. TERT promoter mutation confers favorable prognosis regardless of 1p/19q status in adult diffuse gliomas with IDH1/2 mutations. *Acta Neuropathologica Communications.* 2020;8(1). doi:10.1186/s40478-020-01078-2
101. Stojchevski R, Sutanto EA, Sutanto R, et al. Translational advances in oncogene and tumor-suppressor gene research. *Cancers.* 2025;17(6). doi:10.3390/cancers17061008
102. Sipos D, Raposa BL, Freihat O, et al. Glioblastoma: clinical presentation, multidisciplinary management, and long-term outcomes. *Cancers.* 2025;17(1). doi:10.3390/cancers17010146
103. Haider S, Taphoorn MJB, Drummond KJ, Walbert T. Health-related quality of life in meningioma. *Neuro-Oncology Advances.* 2021;3(1). doi:10.1093/noajnl/vdab089
104. Ogasawara C, Philbrick BD, Adamson DC. Meningioma: a review of epidemiology, pathology, diagnosis, treatment, and future directions. *Biomedicines.* 2021;9(3). doi:10.3390/biomedicines9030319
105. Yarabarla V, Mylarapu A, Han TJ, McGovern SL, Raza SM, Beckham TH. Intracranial meningiomas: an update of the 2021 World Health Organization classifications and review of management with a focus on radiation therapy. *Frontiers in Oncology.* 2023;13. doi:10.3389/fonc.2023.1137849
106. Cooney T, Lindsay H, Leary S, Wechsler-Reya R. Current studies and future directions for medulloblastoma: a review from the pacific pediatric neuro-oncology consortium (PNOC) disease working group. *Neoplasia.* 2023;35. doi:10.1016/j.neo.2022.100861
107. Choi JY. Medulloblastoma: current perspectives and recent advances. *Brain Tumor Research and Treatment.* 2023;11(1). doi:10.14791/btrt.2022.0046
108. Overcast WB, Davis KM, Ho CY, et al. Advanced imaging techniques for neuro-oncologic tumor diagnosis, with an emphasis on PET-MRI imaging of malignant brain tumors. *Current Oncology Reports.* 2021;23(3). doi:10.1007/s11912-021-01020-2
109. Marusak HA, Iadipaolo AS, Harper FW, et al. Neurodevelopmental consequences of pediatric cancer and its treatment: applying an early adversity framework to understanding cognitive, behavioral, and emotional outcomes. *Neuropsychology Review.* 2017;28(2):123–175. doi:10.1007/s11065-017-9365-1
110. Maiuolo J, Gliozzi M, Musolino V, et al. From metabolic syndrome to neurological diseases: role of autophagy. *Frontiers in Cell and Developmental Biology.* 2021;9. doi:10.3389/fcell.2021.651021
111. World Health Organization. Obesity and overweight: Fact Sheet. 2024/03/26; 2024.

112. Stillman CM, Weinstein AM, Marsland AL, Gianaros PJ, Erickson KI. Body–brain connections: the effects of obesity and behavioral interventions on neurocognitive aging. *Frontiers in Aging Neuroscience*. 2017;9. doi:10.3389/fnagi.2017.00115
113. Li G, Hu Y, Zhang W, et al. Brain functional and structural magnetic resonance imaging of obesity and weight loss interventions. *Molecular Psychiatry*. 2023;28(4):1466–1479. doi:10.1038/s41380-023-02025-y
114. Flores-Cordero JA, Pérez-Pérez A, Jiménez-Cortegana C, Alba G, Flores-Barragán A, Sánchez-Margalet V. Obesity as a risk factor for dementia and Alzheimer's disease: the role of leptin. *International Journal of Molecular Sciences*. 2022;23(9). doi:10.3390/ijms23095202
115. Neto A, Fernandes A, Barateiro A. The complex relationship between obesity and neurodegenerative diseases: an updated review. *Frontiers in Cellular Neuroscience*. 2023;17. doi:10.3389/fncel.2023.1294420
116. Schwartz AR, Patil SP, Laffan AM, Polotsky V, Schneider H, Smith PL. Obesity and obstructive sleep apnea: pathogenic mechanisms and therapeutic approaches. *Proceedings of the American Thoracic Society*. 2008;5(2):185–192. doi:10.1513/pats.200708-137MG
117. Reinert KRS, Po'e EK, Barkin SL. The relationship between executive function and obesity in children and adolescents: a systematic literature review. *Journal of Obesity*. 2013;2013:1–10. doi:10.1155/2013/820956
118. Islam MS, Wei P, Suzauddula M, et al. The interplay of factors in metabolic syndrome: understanding its roots and complexity. *Molecular Medicine*. 2024;30(1). doi:10.1186/s10020-024-01019-y
119. Zouridis S, Nasir AB, Aspichueta P, Syn W-K. The link between metabolic syndrome and the brain. *Digestion*. 2024;106(3):203–211. doi:10.1159/000541696
120. Vashisth K, Sinha JK, Jha NK, et al. Neuropharmacological insights into type 3 diabetes: molecular mechanisms, therapeutic advances, and future directions. *Journal of Biological Regulators and Homeostatic Agents*. 2024;38(5):3645–3655.
121. Alfaro FJ, Gavrieli A, Saade-Lemus P, Lioutas V-A, Upadhyay J, Novak V. White matter microstructure and cognitive decline in metabolic syndrome: a review of diffusion tensor imaging. *Metabolism*. 2018;78:52–68. doi:10.1016/j.metabol.2017.08.009
122. Tyagi A, Mirita C, Taher N, et al. Metabolic syndrome exacerbates amyloid pathology in a comorbid Alzheimer's mouse model. *Biochimica et Biophysica Acta (BBA) – Molecular Basis of Disease*. 2020;1866(10). doi:10.1016/j.bbadis.2020.165849
123. Yang Y, Li Q, Long Y, Yuan J, Zha Y. Associations of metabolic syndrome, its severity with cognitive impairment among hemodialysis patients. *Diabetology & Metabolic Syndrome*. 2023;15(1). doi:10.1186/s13098-023-01080-3
124. Nagar SD, Pemu P, Qian J, et al. Investigation of hypertension and type 2 diabetes as risk factors for dementia in the all of us cohort. *Scientific Reports*. 2022;12(1). doi:10.1038/s41598-022-23353-z
125. Yue C, Fu Y, Zhao Y, Ou Y, Sun Y, Tan L. Association between Alzheimer's disease and metabolic syndrome: unveiling the role of dyslipidemia mechanisms. *Brain Network Disorders*. 2025;1(1):21–27. doi:10.1016/j.bnd.2024.10.006
126. Legdeur N, Visser PJ, Woodworth DC, et al. White matter hyperintensities and hippocampal atrophy in relation to cognition: the 90+ study. *Journal of the American Geriatrics Society*. 2019;67(9):1827–1834. doi:10.1111/jgs.15990
127. Dhana K, Evans DA, Rajan KB, Bennett DA, Morris MC. Healthy lifestyle and the risk of Alzheimer dementia. *Neurology*. 2020;95(4). doi:10.1212/wnl.0000000000009816
128. Kuang Z-M. Effect of combined antihypertensive and lipid-lowering therapies on cognitive function: a new treatment strategy? *Cardiology Research and Practice*. 2020;2020:1–10. doi:10.1155/2020/1484357
129. Ghosh S, Sinha JK, Khandelwal N, Chakravarty S, Kumar A, Raghunath M. Increased stress and altered expression of histone modifying enzymes in brain are associated with aberrant behaviour in vitamin B12 deficient female mice. *Nutritional Neuroscience*. 2018;23(9):714–723. doi:10.1080/1028415x.2018.1548676
130. Orywal K, Socha K, Iwaniuk P, et al. Vitamins in the prevention and support therapy of neurodegenerative diseases. *International Journal of Molecular Sciences*. 2025;26(3). doi:10.3390/ijms26031333
131. Moore E, Mander A, Ames D, Carne R, Sanders K, Watters D. Cognitive impairment and vitamin B12: a review. *International Psychogeriatrics*. 2012;24(4):541–556. doi:10.1017/s1041610211002511
132. Wolffenbuttel BHR, Wouters HJCM, Heiner-Fokkema MR, van der Klauw MM. The many faces of cobalamin (vitamin B12) deficiency. *Mayo Clinic Proceedings: Innovations, Quality & Outcomes*. 2019;3(2):200–214. doi:10.1016/j.mayocpiqo.2019.03.002

133. Ota Y, Capizzano AA, Moritani T, Naganawa S, Kurokawa R, Srinivasan A. Comprehensive review of Wernicke encephalopathy: pathophysiology, clinical symptoms and imaging findings. *Japanese Journal of Radiology*. 2020;38(9):809–820. doi:10.1007/s11604-020-00989-3
134. Spiezia AL, Falco F, Manganelli A, et al. Low serum 25-hydroxy-vitamin D levels are associated with cognitive impairment in multiple sclerosis. *Multiple Sclerosis and Related Disorders*. 2023;79. doi:10.1016/j.msard.2023.105044
135. Schlotz W, Jones A, Phillips DIW, Gale CR, Robinson SM, Godfrey KM. Lower maternal folate status in early pregnancy is associated with childhood hyperactivity and peer problems in offspring. *Journal of Child Psychology and Psychiatry*. 2010;51(5):594–602. doi:10.1111/j.1469-7610.2009.02182.x
136. Lozoff B, Beard J, Connor J, Felt B, Georgieff M, Schallert T. Long-lasting neural and behavioral effects of iron deficiency in infancy. *Nutrition Reviews*. 2006;64(5):34–43. doi:10.1301/nr.2006.may.S34-S43
137. Delange F, de Benoist B, Pretell E, Dunn JT. Iodine deficiency in the world: where do we stand at the turn of the century? *Thyroid*. 2001;11(5):437–447. doi:10.1089/105072501300176390
138. Zuccoli G, Pipitone N. Neuroimaging findings in acute Wernicke's encephalopathy: review of the literature. *American Journal of Roentgenology*. 2009;192(2):501–508. doi:10.2214/ajr.07.3959

2 Alzheimer's Disease
Unraveling the Mechanism, Pathogenesis, Genetic Basis, and Emerging Therapies

Maharudra Pratap Singh, Arshi Waseem, Anand Prakash, and Syed Shadab Raza

2.1 INTRODUCTION

Dementia has become one of the most pressing global health concerns, with Alzheimer's disease (AD) being its most common and devastating form. As of 2022, approximately 55 million people worldwide were living with dementia—a number projected to soar to 78 million by 2030 and 139 million by 2050. The aging population, especially in high-income countries, is driving this rapid rise, which is further intensified by the massive economic burden. Global costs related to dementia—including medical care, long-term support, and unpaid caregiving—are estimated to exceed \$2.8 trillion annually. AD is responsible for 60–80% of all dementia cases and is characterized by gradual, irreversible neurodegeneration (Zhang et al. 2024). Clinically, the disease follows a continuum that begins silently with no visible symptoms, progresses to mild cognitive impairment (MCI), and eventually advances to full-blown dementia. Early manifestations may involve memory lapses, mood changes such as depression or anxiety, and disruptions in sleep-wake cycles. In the later stages, patients often experience severe behavioral disturbances, personality changes, and extensive cognitive dysfunction. Despite ongoing research efforts, early detection remains rare, and no curative treatment currently exists (Alzheimer's Association 2023). At the biological level, AD is typified by the accumulation of amyloid-β (Aβ) plaques outside neurons and neurofibrillary tangles (NFTs) made of hyperphosphorylated tau protein within neurons (Zhang et al. 2021). However, emerging evidence points to a multifactorial nature, involving more than just these hallmark features. Factors like oxidative stress, mitochondrial damage, synaptic and glial dysfunction, energy metabolism failure, insulin resistance, disrupted autophagy, gut-brain axis imbalances, and abnormal metal ion regulation are increasingly recognized as contributors to disease progression. These complex interactions open avenues for biomarker discovery and multi-targeted therapeutic interventions (Kamatham et al. 2024). Notably, modifiable risk factors such as vascular health, lifestyle habits, physical activity, diet, education, and metabolic disorders play a significant role in disease development. Research suggests that nearly one-third of Alzheimer's cases could be prevented by addressing these risk factors—prompting the development of comprehensive prevention strategies that include cognitive training, lifestyle changes, and cardiovascular risk control (Dhana et al. 2020). Although current drug treatments—such as cholinesterase inhibitors (e.g., donepezil, rivastigmine) and NMDA receptor blockers (e.g., memantine)—only alleviate symptoms, recent years have seen the rise of monoclonal antibody therapies like aducanumab, lecanemab, and donanemab, which target amyloid pathology. While these have demonstrated modest success in slowing early-stage decline, concerns over safety, cost, and limited effectiveness windows have restricted their widespread use (Singh et al. 2024). To address these issues, researchers are looking at non-amyloid therapeutic targets, investigating

DOI: 10.1201/9781003500025-2

substances such as fatty acid synthase inhibitors, microglial modulators, and TREM2-activating antibodies. To get around current therapeutic restrictions and improve clinical results, advanced drug design methods like dual-target inhibitors, allosteric modulators, PROTACs, and protein–protein interaction inhibitors are also being developed (Cummings et al. 2023). This chapter offers an in-depth exploration of AD, examining its molecular and genetic underpinnings, key pathological mechanisms, and the emerging landscape of therapeutic innovation.

2.2 GENETIC INSIGHTS INTO ALZHEIMER'S DISEASE: FROM RISK FACTORS TO PROTECTIVE MECHANISMS

Recent progress in genetics, particularly through large-scale genome-wide association studies (GWAS), has identified over a 100 single nucleotide polymorphisms (SNPs) across multiple genes linked to sporadic late-onset Alzheimer's disease (LOAD). These discoveries have broadened our understanding of the disease's molecular roots, especially concerning amyloid plaque formation and other neurodegenerative processes (Misra et al. 2018). In early-onset familial Alzheimer's disease (EOFAD), mutations in Presenilin-1 (PSEN1), Presenilin-2 (PSEN2), and the Amyloid Precursor Protein (APP) genes are well-established drivers. PSEN1, a crucial component of the γ-secretase enzyme complex, is the most common culprit, often causing symptoms to appear in patients as early as their 30s. These mutations disrupt the normal cleavage of APP, leading to excessive formation of the Aβ42 peptide, a major component of amyloid plaques. PSEN2 mutations, albeit less common, also result in amyloid accumulation but typically induce milder symptoms than APP mutations near the γ-secretase site, which also enhance the generation of Aβ42 (Giau et al. 2019). The function of lipid metabolism in AD is clarified by additional genetic risk factors in addition to amyloid-centric genes. The Apolipoprotein E (APOE) gene stands out as the most consistently implicated factor in both familial and sporadic cases. Of its three common alleles—ε2, ε3, and ε4—the ε4 variant greatly increases the likelihood and lowers the age of disease onset. Individuals carrying ε4 show greater accumulation of amyloid plaques and tau tangles, along with a higher sensitivity to metabolic and vascular stressors. On the other hand, ε2 appears protective, while ε3 is considered neutral. ApoE is vital for lipid and cholesterol transport in the brain, supporting neural function and synaptic integrity. In ε4 carriers, disrupted lipid balance accelerates amyloid buildup and neurodegeneration (Yang et al. 2023). Several other genes linked to lipid transport and immune regulation have also been tied to Alzheimer's. In microglia, SORL1, which is involved in endosomal transport, influences protein clearance and lipid management. Amyloid buildup is facilitated by the loss of SORL1 function, which also makes it more difficult to eliminate harmful proteins. Likewise, phagocytosis and lipid export depend on ABCA7, another lipid transporter that is present in brain immune cells. Variants that reduce ABCA7 function are associated with higher Alzheimer's risk due to poor amyloid clearance (Mishra et al. 2022). Clusterin (CLU), also known as ApoJ, plays a multifaceted role in lipid trafficking, immune response, and cellular stress regulation. Genetic changes in CLU are linked to altered brain lipid dynamics and inflammatory responses, placing it among the top genetic risk factors for LOAD, second only to APOE ε4 (Foster et al. 2019). Microglial function in Alzheimer's is further shaped by TREM2, a receptor that senses lipids and regulates debris clearance in the brain. Loss-of-function mutations in TREM2 sharply elevate Alzheimer's risk, similar to that seen with APOE ε4 heterozygosity. These variants impair microglial response to injury and disrupt lipid metabolism, fueling disease progression (Qu and Li 2023). It's interesting to note that not every genetic discovery indicates risk. Resilience to Alzheimer's is provided by certain uncommon variations. One such example is RELN-COLBOS, a mutation in the RELN gene that is associated with delayed onset of cognitive symptoms in people with high-risk PSEN1 mutations. RELN produces the protein Reelin, which promotes synaptic communication and may lessen tau disease in areas of the brain that are susceptible (Mouofo and Spires-Jones 2023). Another remarkable example is a woman with a highly pathogenic PSEN1-E280A mutation who remained symptom-free for nearly three decades longer than expected. This delay was traced to her carrying two copies of a rare APOE3 Christchurch (APOECh) variant, which likely alters lipid transport and reduces neurodegeneration despite a heavy amyloid load (Arboleda-Velasquez et al. 2019). In conclusion, the genetic landscape of AD reveals a tightly woven network of pathways involving amyloid processing, lipid metabolism, and immune function. Key genes—PSEN1, APP, APOE, SORL1, TREM2, and RELN—are central to this web and offer promising targets for future research and therapeutic innovation.

2.3 MECHANISMS BEHIND ALZHEIMER'S DISEASE

AD represents a multifactorial puzzle involving numerous overlapping pathways. Far from being caused by a single malfunction, AD arises from a complex and evolving interplay of molecular and cellular disturbances. Although tau tangles and amyloid plaques are well-known hallmarks, a wider network of dysfunction (Figure 2.1), including neuroinflammation, oxidative stress, mitochondrial damage, neurotransmitter imbalances, metal ion toxicity, excitotoxicity, autophagy impairment, and

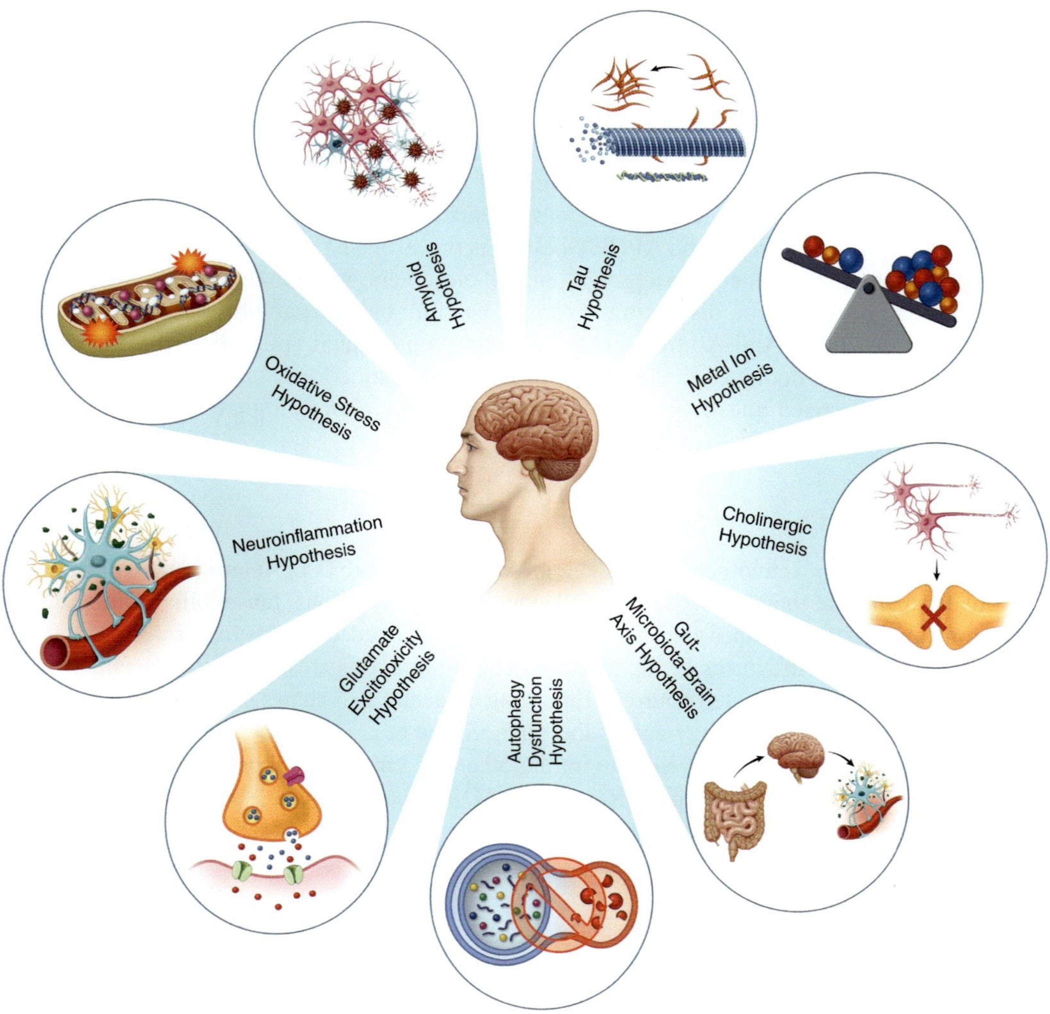

FIGURE 2.1 Alzheimer's disease is a multifactorial neurodegenerative disorder.

even disruptions of the gut–brain axis, drives the disease's progression. This intricate web of pathology highlights the need for multi-targeted, personalized, and systems-level approaches to treatment and diagnosis (Jadhav et al. 2024).

2.3.1 Cholinergic Hypothesis

One of the earliest proposed mechanisms of AD, the cholinergic hypothesis, focuses on the decline of acetylcholine—a neurotransmitter critical for memory and cognition. In AD, cholinergic neurons in the basal forebrain, especially in the nucleus basalis of Meynert, undergo significant degeneration. This leads to reduced activity of choline acetyltransferase (ChAT), the enzyme that synthesizes acetylcholine, resulting in impaired communication between neurons in the hippocampus and cortex (Machado et al. 2020). The disruption of signaling exacerbates these cholinergic impairments, which are intimately linked to cognitive decline. The necessity to consider more than one mechanism in therapeutic development is highlighted by the fact that while medications such as acetylcholinesterase inhibitors (e.g., donepezil and rivastigmine) temporarily alleviate symptoms, they do not stop the progression of the disease (Penke et al. 2020).

2.3.2 Amyloid Hypothesis

The amyloid cascade hypothesis remains one of the most studied frameworks in AD research. It centers around the buildup of Aβ peptides, which are derived from the cleavage of APP by β- and γ-secretase enzymes. The longer Aβ42 variant, prone to aggregation, accumulates into extracellular plaques believed to disrupt neuronal communication and trigger inflammation. Genetic mutations in PSEN1 and PSEN2 (components of the γ-secretase complex) are linked to early-onset familial AD, while the APOE ε4 allele is the strongest genetic risk factor in sporadic AD, influencing Aβ clearance and aggregation. Aβ oligomers, more than plaques, are now seen as the most neurotoxic form, disrupting synapses, calcium balance, and mitochondrial function (Kurkinen et al. 2023). Anti-Aβ treatments, including monoclonal antibodies like aducanumab and lecanemab, have demonstrated promise in lowering plaque load, but their modest clinical improvement has raised questions about whether Aβ is the underlying cause of the illness or merely one piece of the puzzle (Wu et al. 2022).

2.3.3 Tau Hypothesis

Tau protein pathology is another defining feature of AD, with NFTs correlating more closely with cognitive impairment than amyloid plaques. Under normal conditions, tau stabilizes microtubules in neurons. In AD, tau becomes abnormally hyperphosphorylated, detaches from microtubules, and aggregates into toxic structures inside neurons. These tau aggregates disrupt intracellular transport and eventually kill affected neurons. Emerging evidence also shows that tau can spread from cell to cell, propagating pathology across the brain (Simic et al. 2016). New therapies aim to halt tau pathology through approaches like tau immunotherapy, kinase inhibitors, antisense oligonucleotides, and proteolysis-targeting chimeras (PROTACs), though clinical success remains limited thus far (Congdon and Sigurdsson 2018).

2.3.4 Neuroinflammation Hypothesis

Chronic inflammation in the brain—primarily driven by overactive microglia and astrocytes—is now recognized as a key contributor to AD. Microglia, the brain's immune cells, become persistently activated in response to Aβ and tau, releasing inflammatory mediators that can damage neurons and exacerbate disease progression. Inflammasome activation, especially of the NLRP3 complex, promotes the release of ASC specks that enhance Aβ aggregation and spread. Memory loss may also result from hyperactive microglia accidentally pruning healthy synapses. Microglial activity is influenced by genetic variables such as TREM2 and APOE4, which further connect inflammation to the risk of disease. According to recent research, peripheral infections and gut microorganisms may also be involved in this neuroinflammatory cascade, creating novel treatment options (Wong-Guerra et al. 2023).

2.3.5 Oxidative Stress Hypothesis

Oxidative stress plays a pivotal role in the development of AD by damaging neuronal lipids, proteins, and DNA. It results from an imbalance between free radicals—such as reactive oxygen species (ROS)—and the brain's antioxidant defenses. Mitochondrial dysfunction is a major source of ROS, especially when disrupted by Aβ accumulation. Oxidative stress promotes the misprocessing of APP, enhances tau phosphorylation, activates inflammatory pathways, and creates a vicious cycle of cellular injury. While antioxidant therapies have shown protective effects in preclinical models, clinical trials have produced mixed results. Early intervention and precision-based antioxidant strategies may hold greater potential (Buccellato et al. 2021).

2.3.6 Metal Ion Hypothesis

Metals like iron, copper, and zinc are essential for brain function, but their dysregulation in AD contributes to protein aggregation, oxidative damage, and neuronal toxicity. These metals are often found within amyloid plaques and tau tangles. For example, excess iron can catalyze ROS production via the Fenton reaction, leading to lipid peroxidation and ferroptosis. Copper imbalance may result in a distinct form of cell death called cuproptosis. Zinc deficiency can influence synaptic transmission and increase excitotoxicity. Although chelators provide a potential therapeutic approach, there are still issues with properly transporting them across the blood–brain barrier (BBB) (Babic Leko et al. 2023).

2.3.7 Glutamate Excitotoxicity Hypothesis

Glutamate, the brain's main excitatory neurotransmitter, becomes neurotoxic when overactivated—particularly through NMDA receptors. In AD, excessive glutamate signaling leads to calcium overload in neurons, triggering oxidative stress, mitochondrial failure, and apoptosis. This process, known as excitotoxicity, may occur even before amyloid and tau pathology becomes apparent. NMDA receptor antagonists like memantine provide modest cognitive benefits in AD, but more targeted strategies are needed to minimize excitotoxic damage without impairing normal neurotransmission (Wang and Reddy 2017).

2.3.8 Microbiota–Gut–Brain Axis Hypothesis

A growing body of research is exploring how gut microbiota may influence brain health through the microbiota–gut–brain axis. Immune, hormonal, and neurological signals are all involved in this two-way exchange. Microbial imbalance, or dysbiosis, can weaken the integrity of the intestinal barrier, letting dangerous substances enter the bloodstream and ultimately the brain, where they may cause neuroinflammation. This theory proposes that gut health could be a modifiable risk factor for AD, influencing everything from microglial activation to neurotransmitter regulation. Though still in early stages, microbiome-targeted therapies like probiotics or fecal microbiota transplantation are being explored as potential AD treatments (Loh et al. 2024).

2.3.9 Autophagy Dysfunction Hypothesis

Autophagy is the cell's waste management system, responsible for degrading and recycling misfolded proteins and damaged organelles. In AD, this system becomes overwhelmed or dysfunctional, leading to a buildup of toxic proteins like Aβ and tau, along with defective mitochondria. Deficits in autophagy can affect all stages—from vesicle formation to fusion with lysosomes—causing cellular stress and death. Genetic mutations, lysosomal impairments, and reduced expression of autophagy-regulating proteins contribute to this dysfunction. Boosting autophagy with small molecules or gene therapies has shown promise in preclinical models, but translating these findings to clinical success remains a challenge (Angst et al. 2025).

2.4 SIGNALING PATHWAYS LINKED TO AD PATHOGENESIS

2.4.1 Neuroinflammatory Signaling and Microglial Activation in Alzheimer's Disease

In AD, chronic inflammation within the brain—known as neuroinflammation—is a major contributor to disease progression. Microglia, the brain's immune cells, become persistently activated in response to harmful stimuli such as Aβ deposits, oxidative stress, and inflammatory cytokines. Once activated, they initiate signaling cascades—primarily through MAPK, NF-κB, and PI3K/Akt pathways—that lead to the production of inflammatory molecules like TNF-α, IL-1β, and nitric oxide, all of which contribute to neuronal dysfunction (Leng and Edison 2021). The most important of them is the NF-κB pathway, which controls the expression of numerous genes linked to inflammation. It also encourages the activation of the NLRP3 inflammasome, a protein complex that triggers pyroptosis, an inflammatory cell death mechanism, and causes the production of inflammatory cytokines. Furthermore, pro-inflammatory microglia can be shifted by microRNAs (miRNAs) like miR-25802, which exacerbates cognitive deterioration (Zhao et al. 2024). Tau protein abnormalities, another core feature of AD, further activate microglia via toll-like receptors, creating a feedback loop that sustains inflammation. The cGAS–STING pathway, triggered by cytosolic DNA, also adds to the inflammatory burden by driving interferon production (Wang et al. 2022). Crosstalk between microglia, neurons, and astrocytes, along with infiltration of T and B cells, shapes an evolving neuroimmune landscape. The chemokine axis CXCL10/CXCR3, for instance, supports the recruitment of CD8+ T cells, which may disrupt neuronal gene expression (Jorfi et al. 2023). One promising target is the microglial receptor TREM2. It plays a protective role by modulating inflammation and supporting Aβ clearance through the PI3K–Akt–FoxO3a pathway, while also interacting with complement components like C1q to prevent excessive synaptic pruning (Zhong et al. 2023).

2.4.2 Lysosomal Dysfunction in Alzheimer's Disease

Lysosomes act as the cellular recycling centers, breaking down waste and damaged components. In the context of AD, these organelles lose their ability to function properly, leading to the accumulation of toxic proteins like Aβ and tau. A key factor in this dysfunction is impaired acidification of the lysosomal interior, which is necessary for activating the enzymes that perform degradation (Cao et al. 2021). The V-ATPase pump, which keeps this acidic environment going, is frequently interfered with in AD because of abnormalities in the PS1 gene, which is connected to familial AD. Mutations in PS1 prevent the pump's subunits from properly assembling and trafficking, which results in less acidification and ineffective Aβ breakdown (Lo and Zeng 2023). Additionally, lysosomal calcium channels such as TRPML1 and TPCN2 also become dysregulated, contributing to calcium imbalances that weaken lysosomal stability. Attempts to restore calcium homeostasis alone are often insufficient; indicating that lysosomal dysfunction in AD is driven by a complex interplay of factors (Lee et al. 2015). Another player is the β-C-terminal fragment of APP, which can bind to V-ATPase subunits and further impair their function. Altogether, these disturbances reinforce a cycle where inefficient degradation accelerates the build-up of Aβ and tau, contributing to cellular stress and neurodegeneration (Im et al. 2023).

2.4.3 Cholesterol Metabolism Disruption in Alzheimer's Disease

Cholesterol is vital for brain health, supporting neuronal membranes, synaptic activity, and signaling. However, when its regulation is disrupted—as seen in AD—it contributes to amyloid and tau pathology as well as neuroinflammation (Ahmed et al. 2024). Neuronal membranes rich in cholesterol form lipid rafts, which serve as sites for the amyloidogenic processing of APP, increasing Aβ production. APOE, particularly the APOE4 variant, plays a key role in cholesterol transport and is strongly linked to AD. APOE4 impairs the clearance of Aβ, promotes its deposition, and disrupts cholesterol distribution within brain cells. It also increases tau phosphorylation and drives microglial activation, promoting inflammatory responses and neurodegeneration (Feringa and van der Kant 2021). Additionally, cholesterol buildup can activate the NLRP3 inflammasome, linking metabolic stress to chronic inflammation. Although initial inflammasome activation might be protective, sustained activation leads to oxidative stress, cell death, and impaired immune clearance, worsening AD pathology. Therapeutic efforts targeting APOE and lipid raft dynamics offer potential for interrupting this cycle and restoring healthy cholesterol homeostasis in the brain (de Dios et al. 2023).

2.4.4 Mitochondrial Dysfunction in Alzheimer's Disease

Mitochondria power neurons by producing energy, regulating oxidative balance, and maintaining calcium levels. In AD, these organelles become dysfunctional, particularly in regions of the brain vulnerable to damage. The energy crisis in AD brains is reflected by impaired glucose metabolism and disrupted activity of mitochondrial enzymes (Wang et al. 2020). Amyloid-beta impairs several mitochondrial enzymes—including cytochrome c oxidase and pyruvate dehydrogenase—and binds to mitochondrial proteins like ABAD, increasing ROS production and oxidative stress (Misrani et al. 2021). By compromising mitochondrial transport along neurons, tau disease exacerbates the damage by resulting in energy deficits in vital regions. Aβ's effects on calcium channels and endoplasmic reticulum (ER) calcium release also contribute to mitochondrial calcium overload. Cell death results from this overload's activation of cyclophilin D and opening of the mitochondrial permeability transition pore (mPTP) (Calvo-Rodriguez and Bacskai 2020). The balance between mitochondrial fission and fusion is also disrupted in AD, with proteins like DRP1 and MFN2 showing abnormal activity. This results in fragmented, damaged mitochondria that are poorly cleared due to defective mitophagy. The failure of the PINK1/parkin pathway—responsible for removing dysfunctional mitochondria—leads to their accumulation, further impairing neuronal function and accelerating disease progression (Chen et al. 2023).

2.4.5 Calcium Signaling Imbalance in Alzheimer's Disease

Calcium ions are essential messengers in neurons, regulating communication, memory formation, and cell survival. In AD, calcium homeostasis becomes disrupted, contributing to excitotoxicity, cell death, and synaptic dysfunction. Amyloid-beta increases calcium entry by overactivating L-type calcium channels and interacting with NMDA and α7 nicotinic receptors. This leads to calcium overload, which damages synapses and promotes Aβ internalization (McDaid et al. 2020). Cellular stress and apoptosis are simultaneously brought on by the ER's excessive calcium release, which is caused by malfunctioning pumps and overactive IP3 and ryanodine receptors. Excessive mitochondrial calcium uptake, decreased ATP synthesis, and increased ROS generation result from an overactive connection between the ER and mitochondria at contact areas known as mitochondria-associated membranes (MAMs) (Yu et al. 2021). Lysosomal function is also impaired by calcium overload, which interferes with acidification and enzyme activity, thereby enhancing tau accumulation. This calcium dysregulation amplifies mitochondrial damage, hampers protein clearance, and deepens neurodegeneration in AD (Mustaly-Kalimi et al. 2022).

2.4.6 Impaired Insulin Signaling in Alzheimer's Disease

Insulin is not just important for blood sugar control—it also supports brain health by promoting synaptic plasticity and memory. In AD, insulin signaling is significantly disrupted, leading many to refer to the condition as "type 3 diabetes." The PI3K–Akt pathway, which mediates insulin's effects in the brain, becomes less active in AD. This allows GSK-3 to become overactive, driving tau hyperphosphorylation and increasing Aβ production (Nowell et al. 2023). When dysregulated, the downstream regulator mTORC1 inhibits autophagy and adds to the accumulation of misfolded proteins. By altering IRS-1, a crucial adaptor protein, inflammatory cytokines and stress kinases like JNK and IKK also disrupt insulin signaling, resulting in decreased insulin sensitivity in neurons. This cascade contributes to synaptic deficits and cognitive impairment. Correcting these insulin-related abnormalities is a promising therapeutic strategy, potentially capable of addressing both metabolic and neurodegenerative aspects of AD (Querfurth and Lee 2021).

2.4.7 Neurotrophic Factor Signaling Deficits in Alzheimer's Disease

Neurotrophic factors like NGF and BDNF are crucial for neuron survival, synaptic maintenance, and cognitive performance. In AD, this support system deteriorates. NGF availability decreases due to impaired conversion from its precursor form and accelerated breakdown, reducing its protective signaling through TrkA receptors (Xhima et al. 2022). Rather, proNGF binds more and more to p75NTR, a receptor that causes cell death, particularly in cholinergic neurons in the basal forebrain. In AD, BDNF levels also decrease, which impairs TrkB receptor signaling and increases the activation of pathways linked to stress. The δ-secretase enzyme, which cleaves tau and APP into neurotoxic pieces, is activated by this imbalance (Xiang et al. 2019). Moreover, Aβ directly interferes with BDNF-related pathways, impairing memory and accelerating neurodegeneration. Restoring neurotrophic signaling—particularly by enhancing BDNF-TrkB activity—holds potential for reversing synaptic deficits and improving cognition in AD (Wang et al. 2019).

2.4.8 Blood–Brain Barrier (BBB) Breakdown in Alzheimer's Disease

The BBB protects the brain from harmful substances and maintains its internal environment. In AD, the BBB becomes increasingly compromised, allowing toxic agents from the bloodstream to enter the brain. Structural disruptions—including loss of tight junctions, pericyte damage, and increased activity of enzymes like matrix metalloproteinases—make the BBB more permeable (Alkhalifa et al. 2023). This breach fosters oxidative stress and inflammation, accelerating neuronal damage. Transport mechanisms that usually clear amyloid-beta from the brain also become impaired. Key efflux proteins like LRP1 and P-glycoprotein are downregulated, while receptors such as RAGE that bring Aβ into the brain are upregulated, exacerbating plaque buildup (Nehra et al. 2022). Additionally, reduced glucose transport across the BBB due to diminished GLUT1 expression leads to energy deficits, further weakening BBB integrity. Together, these changes not only worsen Aβ and tau pathology but also disrupt metabolic balance and vascular health, highlighting BBB dysfunction as a central player in AD progression (Winkler et al. 2015).

2.5 TARGETING AMYLOIDOGENIC PATHWAYS WITH SMALL MOLECULE THERAPEUTICS IN ALZHEIMER'S DISEASE

Recent progress in AD research has sparked growing interest in small molecule therapies (Table 2.1) aimed at altering disease progression rather than just alleviating symptoms. A major therapeutic focus has been the modulation of APP processing, specifically targeting the enzymatic actions of α-, β-, and γ-secretases involved in Aβ generation. These strategies are designed either to favor the production of non-toxic fragments or to inhibit the formation and accumulation of harmful Aβ peptides (Singh et al. 2024).

TABLE 2.1
Major Small Molecule Therapeutics in AD—Mechanisms, Status, Efficacy, and Challenges

Drug Name	Target/Mechanism	Phase/ Approval	Therapeutic Outcome	Limitations/ Challenges	Reference(s)
Donepezil	Acetylcholinesterase inhibitor	FDA Approved	Modest improvement in cognition/global function	Symptomatic only; gastrointestinal side effects	Moreta et al. (2021)
Memantine	NMDA receptor antagonist	FDA Approved	Slows functional decline in moderate-to-severe AD	Limited benefit in early stages	Zhang et al. (2024)
Rivastigmine	Acetylcholinesterase and butyrylcholinesterase inhibitor	FDA Approved	Enhances cognition and daily living activities	Nausea, vomiting; requires dose titration	Fortea and Ferrer-Picón (2025)
Blarcamesine (ANAVEX2-73)	Sigma-1 receptor modulator; improves mitochondrial function/ neuroinflammation	Phase II/III	Promising cognitive and functional benefits	Awaiting larger confirmatory trials	Macfarlane et al. (2025)
Tramiprosate	Prevents soluble Aβ oligomerization	Phase III (discontinued)	Reduced Aβ aggregation; no cognitive impact	Failed to meet cognitive endpoints	Manzano et al. (2020)
Sodium Oligomannate (GV-971)	Modulates gut microbiota and reduces neuroinflammation	Approved in China	Statistically significant cognitive improvements	Mechanism unclear; not FDA approved	Xiao et al. (2021)
Hydromethylthionine Mesylate (HMTM)	Tau aggregation inhibitor, oral	Phase III (ongoing review)	Slows cognitive decline and reduces brain atrophy	Regulatory review ongoing	Wischik et al. (2022)
Valiltramiprosate (ALZ-801)	Aβ oligomerization inhibitor, oral	Phase III (ongoing)	Promising disease-modifying potential	Awaiting confirmatory data	Abushakra et al. (2024)
IGC-AD1	Cannabinoid-based small molecule for neuropsychiatric symptoms	Phase II	Potential disease-modifying for agitation symptoms	Early stage; trial ongoing	Rodriguez-Soacha et al. (2025)

(*Continued*)

TABLE 2.1 (*Continued*)
Major Small Molecule Therapeutics in AD—Mechanisms, Status, Efficacy, and Challenges

Drug Name	Target/Mechanism	Phase/ Approval	Therapeutic Outcome	Limitations/ Challenges	Reference(s)
Semaglutide	GLP-1 receptor agonist (diabetes drug repurposed)	Phase III	Cognitive benefits under investigation	Trial results pending	Al Qassab et al. (2025)
Metformin	AMPK activator, insulin sensitizer (antidiabetic)	Phase III	Being tested for neuroprotective effects	Results awaited	Isop et al. (2023)
Escitalopram	Selective serotonin reuptake inhibitor	Phase II	Investigated for neuropsychiatric symptoms	Trial ongoing	Ehrhardt et al. (2019)
Caffeine	CNS stimulant, adenosine receptor antagonist	Phase II	Studied for cognitive enhancement	Mixed results; trial ongoing	Merighi et al. (2023)

2.5.1 Boosting α-Secretase Activity: Promoting a Protective Pathway

Activation of α-secretases—especially ADAM10—redirects APP processing away from the amyloidogenic pathway, generating a neuroprotective fragment called sAPPα. Experimental drugs like *Etazolate* and *Acitretin* have demonstrated the ability to enhance α-secretase activity and increase sAPPα levels in early studies. *Bryostatin-1*, known to activate protein kinase C (PKC), also encourages this non-toxic processing and supports synaptic health. Even *Rivastigmine*, a common symptomatic treatment for AD, shows potential in stimulating α-secretase activity. Yet, despite these early findings, no therapy in this category has achieved consistent success in large-scale clinical trials (Tang 2020).

2.5.2 γ-Secretase Modulation: Shifting from Broad Inhibition to Precision Targeting

γ-Secretase contributes to Aβ formation by cleaving APP into peptide fragments. Initial attempts to block this enzyme with γ-secretase inhibitors (GSIs) like *Semagacestat* were met with significant side effects, especially due to disruption of Notch signaling. In response, researchers developed γ-secretase modulators (GSMs) such as *PF-06648671* and *NGP 555*, which selectively reduce production of toxic Aβ42 while preserving other essential cellular functions. Among these, *NGP 555* showed encouraging results in early trials, reducing plaque-forming peptides without compromising the BBB (Luo and Li 2022).

2.5.3 Inhibiting β-Secretase (BACE1): Stopping Aβ Formation at Its Source

β-Secretase initiates the amyloidogenic cascade. While early BACE1 inhibitors like *LY2811376* were abandoned due to safety concerns, newer candidates such as *LY2886721* and *LY3202626* achieved better Aβ suppression, but failed to produce measurable clinical benefits. A novel compound, *ALZ-801*—a prodrug of *Tramiprosate*—is currently in Phase III trials and shows potential in stabilizing Aβ monomers. Additional agents like *Blarcamesine* and *Buntanetap* are being investigated for their dual action against both Aβ and tau, reflecting a shift toward broader neuroprotective effects (Ulku et al. 2024).

2.5.4 Multi-Target Directed Ligands (MTDLs): Tackling AD's Complexity

Given AD's multifactorial nature involving oxidative damage, inflammation, and protein misfolding, there is increasing emphasis on compounds that hit multiple targets. *Ladostigil*, which combines cholinesterase and monoamine oxidase inhibition, is currently in late-stage trials for MCI. Despite ongoing efforts, the development of safe and effective multi-target agents remains a considerable challenge, with no MTDL yet approved for clinical use (Kumar et al. 2022).

2.6 ANTIBODY-BASED THERAPIES FOR ALZHEIMER'S DISEASE

One of the most effective methods for removing amyloid deposits and delaying the course of the disease is immunotherapy, especially monoclonal antibody techniques (Table 2.2). A handful of antibody candidates have demonstrated promise, reigniting interest in the amyloid cascade theory, despite the fact that many have not been successfully translated into clinical practice (Cummings et al. 2024). Given Aβ's limitations as a target, therapies aimed at tau—a protein closely associated with disease progression—have gained attention. Despite the growing roster of tau-directed antibodies, none have yet achieved regulatory approval, underscoring the complexity of tau as a therapeutic target (Congdon et al. 2023).

2.7 COMBINATION STRATEGIES IN ALZHEIMER'S THERAPY

A single pathological process does not drive AD, making combination therapies an attractive and potentially more effective approach (Table 2.3) (Kabir et al. 2020).

2.7.1 Antibody Plus Small Molecule Combinations

Research shows that a toxic variant of Aβ, known as pyroglutamate-Aβ (pGlu-Aβ), contributes significantly to neurodegeneration. *Varoglutamstat (PQ912)*, a glutaminyl cyclase (QC) inhibitor, prevents pGlu-Aβ formation. When paired with *PBD-C06*, a monoclonal antibody targeting this modified Aβ, studies showed a synergistic effect, achieving greater plaque reduction than either agent alone (Hoffmann et al. 2021).

2.7.2 Antibody Plus Scanning Ultrasound (SUS)

Low-intensity focused ultrasound can temporarily open the BBB, enhancing drug delivery. In AD mouse models, combining SUS with *Aducanumab* significantly reduced amyloid plaques and improved cognitive performance, suggesting a non-invasive enhancement to antibody efficacy (Leinenga et al. 2021).

2.7.3 Bispecific Antibodies and Brain Delivery Innovation

One major limitation of antibody therapy is poor penetration across the BBB. New bispecific antibodies—targeting both Aβ and the transferrin receptor (TfR)—have dramatically improved brain uptake. Denali Therapeutics and others are advancing such constructs, including multivalent antibodies like *HexaRmAb158-scFv8D3*, which exhibit strong brain entry and Aβ binding in preclinical studies. These engineered antibodies hold promise for more effective and targeted treatment delivery (Rofo et al. 2022).

TABLE 2.2
Major Monoclonal Antibodies in AD—Mechanisms, Status, Efficacy, and Challenges

Antibody	Target/Species	Mechanism/Description	Clinical Phase/ Status	Efficacy/Findings	Challenges/Safety	Reference(s)
Aducanumab (BIIB037)	Aggregated Aβ	Monoclonal antibody targeting aggregated Aβ; reduces amyloid plaques, modest cognitive benefit at high dose	FDA Approved (2021, accelerated)	Plaque reduction; modest benefit in high dose	ARIA risk; efficacy remains controversial	Budd Haeberlein et al. (2022)
Donanemab (LY3002813)	Pyroglutamate-modified Aβ	Targets pyro-Glu Aβ, a pathogenic N-terminally modified form; significant plaque clearance	Phase III (positive)	Significant amyloid clearance; mild clinical improvement	Best effect in early AD; ARIA; limited broader efficacy	Mintun et al. (2021)
Lecanemab (BAN2401)	Soluble Aβ protofibrils	Preferential binding to large, soluble toxic Aβ protofibrils; reduces plaques	FDA Approved (2023)	Slowed cognitive decline in early AD	ARIA risk; monitoring, cost	Vitek et al. (2023)
Solanezumab	Soluble Aβ monomers (mid-region)	Targets mid-Aβ region to promote peripheral sink clearance of monomers	Phase III (failed)	No significant clinical benefit	Development was discontinued due to inefficacy	Honig et al. (2018)
Gantenerumab	Fibrillar and plaque Aβ	Human IgG1 promotes Fc-mediated phagocytosis of amyloid plaques	Phase III (terminated)	No significant impact on cognition	High ARIA incidence; halted development	Tonegawa-Kuji et al. (2025)
Crenezumab	Oligomeric and fibrillar Aβ	Humanized IgG4 mAb supposed to minimize inflammation; binds multiple forms of Aβ	Phase III (terminated)	No significant benefit in clinical outcomes	Failed phase III, discontinued	Tonegawa-Kuji et al. (2025)
Bapineuzumab	Fibrillar Aβ	Passive immunotherapy binds to the N-terminus of Aβ	Phase III (terminated)	No cognitive benefit; plaque reduction	Vasogenic edema (ARIA-E); halted in 2012	Tonegawa-Kuji et al. (2025)
Semorinemab	Tau (N-terminal, all isoforms)	Humanized mAb aiming to block extracellular tau spread	Phase II (negative)	Minimal clinical efficacy observed	No cognitive benefit, well tolerated	Imbimbo et al. (2023)

Gosuranemab	Phosphorylated tau (N-terminus)	Targets extra-neuronal N-terminal tau to halt propagation	Phase II (negative)	Failed to slow cognitive decline	Safe, but ineffective	Imbimbo et al. (2023)
Bepranemab	Mid-region tau	Under evaluation for slowing/ spreading of tau pathology	Phase II (ongoing)	Early trial data pending	Evolving safety/efficacy profile	Chen & Yu (2023)
Zagotenemab	Conformational tau	Seeks to block tau seeding and propagation	Trials discontinued	No clinical benefit seen so far	Terminated trials due to inefficacy	Chen & Yu (2023)
JNJ-63733657	C-terminal tau	Designed to hinder tau aggregation and cell-to-cell transfer	Phase II (ongoing)	Data pending	Early clinical development	Chen & Yu (2023)
E2814	Microtubule-binding region tau	Binds to mid-domain aggregation region to prevent tau spread	Phase Ib/II (ongoing)	Early trial data pending	Safety and long-term benefit unclear	Chen & Yu (2023)
PNT001	cis-phosphorylated tau	Targets pathological tau conformations tied to neurodegeneration	Phase I/II (ongoing)	Early clinical testing	Proof-of-concept stage	Song et al. (2022)
Lu AF87908	Pathological phospho-tau	Targets disease-associated phospho-tau for immunotherapy	Phase I/II (ongoing)	Early clinical trial	Early-stage; safety, efficacy data awaited	Song et al. (2022)

TABLE 2.3
Major Combination Strategies in AD—Strategy, Components, Mechanism, Status, and Drawbacks

Strategy	Components	Mechanism of Synergy	Status	Benefits	Drawbacks	Reference
Donepezil + Memantine	Cholinesterase inhibitor + NMDA antagonist	Enhances neurotransmission while preventing excitotoxicity	FDA Approved	Superior cognitive and functional benefit vs monotherapy	Mild to moderate AD only	Litus et al. (2022)
Aβ mAb + NSAID	Aducanumab + Ibuprofen (theoretical)	Plaque clearance with anti-inflammatory effect	Preclinical	May reduce neuroinflammation synergistically	Risk of bleeding, toxicity, untested in trials	Tampi et al. (2021)
Aβ + Tau Antibodies	Lecanemab + Anti-tau mAb	Targets upstream (Aβ) and downstream (tau) pathologies	Investigational	Holistic targeting of AD core pathology	Complex immune response; high cost	van Dyck et al. (2023)
Curcumin + Donepezil	Phytochemical + ChEI	Anti-inflammatory, antioxidant, and cholinergic support	Preclinical/ clinical trials	Potential synergistic neuroprotection	Poor bioavailability; inconsistent trial outcomes	Ogunsuyi et al. (2023)
Vitamin E + Cognitive Drug	Antioxidant + Cholinesterase inhibitor	Combats oxidative stress and neurotransmitter loss	Limited clinical trials	Some benefit in slowing decline	High doses may be harmful	Dysken et al. (2014)
Triple Drug Cocktails	Anti-Aβ + Anti-tau + NLRP3 inhibitor	Multi-targeted intervention (amyloid, tau, neuroinflammation)	Conceptual/ Early trials	May address complex pathology holistically	Regulatory challenges, side effect profiles	Ising et al. (2019)

2.8 FUTURE PERSPECTIVES AND EMERGING THERAPIES IN ALZHEIMER'S DISEASE

2.8.1 Stem Cell Therapy: Regenerating Hope

Stem cell therapy is gaining momentum as a novel and promising option for treating AD, offering potential to restore lost neuronal function and slow cognitive decline. Studies in animal models and early-stage clinical research indicate that stem cell-based interventions can improve memory and cognition, partly by replenishing acetylcholine—a key neurotransmitter reduced in AD (Ahmad and Sachdeva 2022). Exogenous transplantation (introducing external stem cells) and endogenous stimulation (triggering the body's innate repair mechanisms) are the two main categories of therapeutic methods. Various types of stem cells have been investigated, including neural stem cells (NSCs), mesenchymal stem cells (MSCs), induced pluripotent stem cells (iPSCs), and embryonic stem cells (ESCs). Transplantation of these cells into AD models has been shown to reduce Aβ plaques, decrease neuroinflammation, and enhance neuronal survival (Kwak et al. 2018). For example, in transgenic mice models, stem cells derived from human amniotic membranes and umbilical cord tissue have shown significant increases in cognitive function. Furthermore, it has been demonstrated that substances such as stromal cell-derived factor 1 (SDF-1) attract immune cells that develop into cells that resemble Aβ-clearing microglia (Vasic et al. 2019). Another promising finding involves human amniotic epithelial cells (HAECs), which not only improve memory but also boost cholinergic signaling and hippocampal health. Additionally, transplantation of cholinergic neurons derived from mouse and human ESCs into AD models has led to meaningful recovery in learning and memory, underscoring the regenerative potential of this approach (Fleifel et al. 2018).

2.8.2 MicroRNA Therapeutics: Targeted Gene Modulation

miRNAs, which regulate gene expression by controlling mRNA translation, have emerged as both diagnostic markers and potential therapeutic agents in AD. Many miRNAs are crucial for maintaining neuronal health, and altered expression patterns have been linked to key AD processes, including amyloid formation, tau pathology, mitochondrial dysfunction, and neuroinflammation. For example, miRNAs like miR-132 and miR-212, found to be reduced in the blood exosomes of AD patients, are being explored as non-invasive biomarkers (Cha et al. 2019). Others, like miR-455-3p, can limit Aβ aggregation by stabilizing mitochondrial dynamics (Kumar et al. 2019). Importantly, miRNAs like miR-298 and miR-15b control enzymes like BACE1, which are implicated in the synthesis of Aβ, making these small RNAs interesting therapeutic targets (Chopra et al. 2021; Gong et al. 2017). New drug discovery platforms, like Inforna, are now being used to design small molecules that modulate miRNA activity. These MTDLs offer an innovative therapeutic strategy, especially for complex diseases like AD, where single-target drugs have limited success (Sun et al. 2023).

2.8.3 PROTACs: A New Paradigm in Protein Degradation

PROTACs are changing how we approach disease treatment by enabling selective degradation of disease-causing proteins. By joining the target protein to an E3 ubiquitin ligase, these modified molecules function as a bridge, causing the ubiquitin–proteasome system to tag and eliminate the protein. PROTACs completely remove the protein, in contrast to conventional inhibitors that only reduce protein activity (Gu et al. 2018). In AD, where accumulation of misfolded proteins like tau and α-synuclein plays a central role, PROTACs have shown encouraging results. One peptide-based PROTAC, TH006, successfully reduced tau levels in key brain regions in AD models (Wang et al. 2021). Other PROTACs, like C004019, used a VHL-based degradation system to clear pathological tau and improve cognition in AD mice without noticeable side effects (Liang et al. 2022). Similarly, bifunctional PROTACs targeting α-synuclein have demonstrated success in reducing toxic aggregates in cellular models, offering broader application for related neurodegenerative disorders. While still in experimental stages, PROTACs represent a cutting-edge strategy that could potentially overcome the limitations of current drug therapies (Kargbo 2020).

2.8.4 Antibody–Drug Conjugates: Precision Delivery to the Brain

Although traditionally used in cancer treatment, antibody–drug conjugates (ADCs) are gaining interest in AD therapy due to their ability to deliver treatments directly to diseased brain tissue. ADCs are composed of three parts: a monoclonal antibody that recognizes AD-specific proteins (like Aβ or tau), a therapeutic payload, and a chemical linker that binds them together (Punyakoti et al. 2023). With the help of this precision system, medications can target illness pathways and cross the BBB with few systemic adverse effects. Although ADCs for AD are still in the early stages of development, they present an intriguing opportunity for further study due to their capacity to deliver neuroprotective or anti-aggregating substances straight to neurons (Catania et al. 2022).

2.8.5 Peptide-Based Therapies: Small Molecules with Big Impact

Synthetic peptides offer another innovative path in AD treatment, particularly for disrupting the formation and toxicity of amyloid plaques. These short chains can mimic natural sequences or block harmful interactions between proteins. One notable study explored a peptide derived from a protective amyloid-beta variant, AβA2V. A synthetic version of this peptide, administered intranasally to mice, successfully blocked amyloid aggregation and preserved synaptic function. The potential for creating disease-modifying therapies is increased by this non-invasive delivery system and the peptide's capacity to function early in the illness cascade (Catania et al. 2022).

2.8.6 Vaccines: Activating the Immune System Against AD

To stop or delay the beginning of AD, vaccines that target the tau and Aβ proteins are being investigated. The goal is to encourage the immune system to identify and eliminate harmful proteins before they have a chance to do serious harm. Initial vaccine efforts, such as AN1792, showed promise in reducing plaque buildup in mouse models but encountered safety issues in human trials due to unwanted T-cell responses. As a result, newer generations of vaccines, like ACC-001 and CAD106, were designed to minimize T-cell activation while still promoting a strong antibody response (Wiessner et al. 2011). Multi-target vaccinations that target both tau and Aβ diseases, such as AV-1980R and AV-1959R, are examples of recent strategies. In preclinical studies, these have demonstrated synergistic effects in lowering brain pathology. However, it's important to note that most vaccines aim to prevent rather than reverse disease, making them more suitable for early-stage intervention or at-risk populations (Davtyan et al. 2019).

2.9 CONCLUSION

AD represents one of the most intricate and devastating neurodegenerative disorders, marked by progressive cognitive deterioration, memory loss, and functional impairment. Despite decades of intensive research, a definitive cure remains elusive. The pathophysiology of AD is extremely complex, combining cellular dysfunctions, molecular dysregulations, and hereditary predispositions. In order to comprehend the disease's course, this intricacy throws into question the traditional single-pathway theories and necessitates a more comprehensive, systems-level viewpoint. Core pathological hallmarks—including the accumulation of extracellular Aβ plaques, intracellular tau NFTs, synaptic and mitochondrial dysfunction, chronic neuroinflammation, and BBB compromise—collectively drive neuronal death and cognitive decline. Symptom-based therapies are insufficient to adequately address the cascade of damage caused by these overlapping pathways.

Even though current therapies like cholinesterase inhibitors and NMDA receptor antagonists provide some clinical relief, they are unable to stop the disease's progression. However, the field is undergoing a paradigm shift. Emerging research in genomics, transcriptomics, proteomics, and advanced neurobiology has significantly expanded our understanding of AD's molecular underpinnings and revealed new therapeutic avenues. Innovative strategies, including stem cell-based interventions, miRNA-regulated gene modulation, PROTACs, and personalized immunotherapies, are being explored for their potential to modify disease progression at a fundamental level. These next-generation approaches emphasize early intervention, precision targeting, and patient-specific treatment modalities. Effective management of AD in the future will necessitate a comprehensive, multi-targeted approach that integrates early and precise diagnosis, biomarker-guided monitoring, and individualized therapeutic interventions. Advancements in our understanding of the complex molecular and cellular mechanisms underlying AD pathogenesis—coupled with the development of innovative, targeted treatment modalities—are essential to improving clinical outcomes. Sustained efforts in translational research, drug development, and precision medicine hold promise for altering the trajectory of this debilitating neurodegenerative disorder.

ACKNOWLEDGMENT

The author is thankful to https://www.biorender.com/ for generating the figure.

ABBREVIATIONS

PKC	Activate protein kinase C
AD	Alzheimer's disease
APP	Amyloid Precursor Protein
Aβ	Amyloid-β
APOECh	APOE3 Christchurch
APOE	Apolipoprotein E
BBB	Blood–Brain Barrier
ChAT	Choline acetyltransferase
CLU	Clusterin
EOFAD	Early-onset familial Alzheimer's disease
ESCs	Embryonic stem cells
GSIs	γ-Secretase inhibitors
GSMs	γ-Secretase modulators
LOAD	Late-onset Alzheimer's disease
MSCs	Mesenchymal stem cells
miRNAs	MicroRNAs
MCI	Mild cognitive impairment
MAMs	Mitochondria-associated membranes
mPTP	Mitochondrial Permeability Transition Pore
MTDLs	Multi-Target Directed Ligands
NGF	Nerve growth factor
NSCs	Neural stem cells
NFTs	Neurofibrillary tangles
PSEN	Presenilin
PROTACs	Proteolysis-targeting chimeras
ROS	Reactive Oxygen Species
SNPs	Single nucleotide polymorphisms
SDF-1	Stromal cell-derived factor 1
TfR	Transferrin receptor

REFERENCES

Abushakra S, Porsteinsson AP, Sabbagh M, et al. APOLLOE4 Phase 3 study of oral ALZ-801/valiltramiprosate in *APOE* ε4/ε4 homozygotes with early Alzheimer's disease: Trial design and baseline characteristics. *Alzheimers Dement (N Y)*. 2024;10(3):e12498.

Ahmad F, Sachdeva P. A consolidated review on stem cell therapy for treatment and management of Alzheimer's disease. *Aging Med (Milton)*. 2022;5(3):182–190.

Ahmed H, Wang Y, Griffiths WJ, Levey AI, Pikuleva I, Liang SH, Haider A. Brain cholesterol and Alzheimer's disease: Challenges and opportunities in probe and drug development. *Brain*. 2024;147(5):1622–1635.

Al Qassab M, Mneimneh M, Jradi A, Derbas B, Dabboussi D, Khoury Baini J, Katrib N, Chaarani N, Attieh P, Kanaan A, et al. The expanding role of GLP-1 receptor agonists: Advancing clinical outcomes in metabolic and mental health. *Curr Iss Mol Biol*. 2025; 47(4):285.

Alkhalifa AE, Al-Ghraiybah NF, Odum J, Shunnarah JG, Austin N, Kaddoumi A. Blood-brain barrier breakdown in Alzheimer's disease: Mechanisms and targeted strategies. *Int J Mol Sci*. 2023;24(22):16288.

Alzheimer's Association. Alzheimer's disease facts and figures. *Alzheimers Dement*. 2023;19:1598–1695.

Angst G, Jia N, Esqueda LET, Fan Y, Cai Q, Wang C. Autophagy in Alzheimer disease pathogenesis and its therapeutic values. *Autophagy Rep*. 2025;4(1):2471677.

Arboleda-Velasquez JF, Lopera F, O'Hare M, et al. Resistance to autosomal dominant Alzheimer's disease in an APOE3 Christchurch homozygote: A case report. *Nat Med*. 2019;25(11):1680–1683.

Babic Leko M, Langer Horvat L, Spanic Popovacki E, Zubcic K, Hof PR, Simic G, et al. Metals in Alzheimer's disease. *Biomedicines*. 2023;11:1161.

Buccellato FR, D'Anca M, Fenoglio C, Scarpini E, Galimberti D. Role of oxidative damage in Alzheimer's disease and neurodegeneration: From pathogenic mechanisms to biomarker discovery. *Antioxidants (Basel)*. 2021;10:1353.

Budd Haeberlein S, Aisen PS, Barkhof F, et al. Two randomized phase 3 studies of aducanumab in early Alzheimer's disease. *J Prev Alzheimers Dis*. 2022;9(2):197–210.

Calvo-Rodriguez M, Bacskai BJ. High mitochondrial calcium levels precede neuronal death *in vivo* in Alzheimer's disease. *Cell Stress*. 2020;4(7):187–190.

Cao M, Luo X, Wu K, He X. Targeting lysosomes in human disease: From basic research to clinical applications. *Signal Transduct Target Ther*. 2021;6(1):379.

Catania M, Colombo L, Sorrentino S, et al. A novel bio-inspired strategy to prevent amyloidogenesis and synaptic damage in Alzheimer's disease. *Mol Psychiatry*. 2022;27(12):5227–5234.

Cha DJ, Mengel D, Mustapic M, et al. miR-212 and miR-132 are downregulated in neurally derived plasma exosomes of Alzheimer's patients. *Front Neurosci*. 2019;13:1208.

Chen W, Zhao H, Li Y. Mitochondrial dynamics in health and disease: Mechanisms and potential targets. *Signal Transduct Target Ther*. 2023;8(1):333.

Chen Y, Yu Y. Tau and neuroinflammation in Alzheimer's disease: Interplay mechanisms and clinical translation. *J Neuroinflammation*. 2023;20(1):165.

Chopra N, Wang R, Maloney B, et al. MicroRNA-298 reduces levels of human amyloid-β precursor protein (APP), β-site APP-converting enzyme 1 (BACE1) and specific tau protein moieties. *Mol Psychiatry*. 2021;26(10):5636–5657.

Congdon EE, Ji C, Tetlow AM, Jiang Y, Sigurdsson EM. Tau-targeting therapies for Alzheimer disease: Current status and future directions. *Nat Rev Neurol*. 2023;19(12):715–736.

Congdon EE, Sigurdsson EM. Tau-targeting therapies for Alzheimer disease. *Nat Rev Neurol*. 2018;14(7):399–415.

Cummings J, Osse AML, Cammann D, Powell J, Chen J. Anti-amyloid monoclonal antibodies for the treatment of Alzheimer's disease. *BioDrugs*. 2024;38(1):5–22.

Cummings JL, Osse AML, Kinney JW. Alzheimer's disease: Novel targets and investigational drugs for disease modification. *Drugs*. 2023;83(15):1387–1408.

Davtyan H, Hovakimyan A, Kiani Shabestari S, et al. Testing a MultiTEP-based combination vaccine to reduce Aβ and tau pathology in Tau22/5xFAD bigenic mice. *Alzheimers Res Ther*. 2019;11(1):107.

de Dios C, Abadin X, Roca-Agujetas V, Jimenez-Martinez M, Morales A, Trullas R, Mari M, Colell A. Inflammasome activation under high cholesterol load triggers a protective microglial phenotype while promoting neuronal pyroptosis. *Transl Neurodegener*. 2023;12(1):10.

Dhana K, Evans DA, Rajan KB, Bennett DA, Morris MC. Healthy lifestyle and the risk of Alzheimer dementia: Findings from 2 longitudinal studies. *Neurology*. 2020;95(4):e374–e383.

Dysken MW, Sano M, Asthana S, et al. Effect of vitamin E and memantine on functional decline in Alzheimer disease: The TEAM-AD VA cooperative randomized trial. *JAMA*. 2014;311(1):33–44.

Ehrhardt S, Porsteinsson AP, Munro CA, et al. Escitalopram for agitation in Alzheimer's disease (S-CitAD): Methods and design of an investigator-initiated, randomized, controlled, multicenter clinical trial. *Alzheimers Dement.* 2019;15(11):1427–1436.

Feringa FM, van der Kant R. Cholesterol and Alzheimer's disease; from risk genes to pathological effects. *Front Aging Neurosci.* 2021;13:690372.

Fleifel D, Rahmoon MA, AlOkda A, Nasr M, Elserafy M, El-Khamisy SF. Recent advances in stem cells therapy: A focus on cancer, Parkinson's and Alzheimer's. *J Genet Eng Biotechnol.* 2018;16(2):427–432.

Fortea J, Ferrer-Picón E. An overview of the rivastigmine 13.3 mg/24h transdermal patch as a treatment option for Alzheimer's disease. *Expert Rev Neurother.* Published online July 3, 2025, 25(9):1119-1129. https://doi.org/10.1080/14737175.2025.2527211.

Foster EM, Dangla-Valls A, Lovestone S, Ribe EM, Buckley NJ. Clusterin in Alzheimer's disease: Mechanisms, genetics, and lessons from other pathologies. *Front Neurosci.* 2019;13:164.

Giau VV, Bagyinszky E, Youn YC, An SSA, Kim S. *APP, PSEN1,* and *PSEN2* mutations in Asian patients with early-onset Alzheimer disease. *Int J Mol Sci.* 2019; 20(19):4757.

Gong G, An F, Wang Y, Bian M, Yu LJ, Wei C. miR-15b represses BACE1 expression in sporadic Alzheimer's disease. *Oncotarget.* 2017;8(53):91551–91557.

Gu S, Cui D, Chen X, Xiong X, Zhao Y. PROTACs: An emerging targeting technique for protein degradation in drug discovery. *Bioessays.* 2018;40(4):e1700247.

Hoffmann T, Rahfeld JU, Schenk M, et al. Combination of the glutaminyl cyclase inhibitor PQ912 (Varoglutamstat) and the murine monoclonal antibody PBD-C06 (m6) shows additive effects on brain Aβ pathology in transgenic mice. *Int J Mol Sci.* 2021;22(21):11791.

Honig LS, Vellas B, Woodward M, et al. Trial of Solanezumab for mild dementia due to Alzheimer's disease. *N Engl J Med.* 2018;378(4):321–330.

Im E, Jiang Y, Stavrides PH, Darji S, Erdjument-Bromage H, Neubert TA, Choi JY, Wegiel J, Lee JH, Nixon RA. Lysosomal dysfunction in Down syndrome and Alzheimer mouse models is caused by v-ATPase inhibition by Tyr^{682}-phosphorylated APP βCTF. *Sci Adv.* 2023;9(30):eadg1925.

Imbimbo BP, Balducci C, Ippati S, Watling M. Initial failures of anti-tau antibodies in Alzheimer's disease are reminiscent of the amyloid-β story. *Neural Regen Res.* 2023;18(1):117–118.

Ising C, Venegas C, Zhang S, et al. NLRP3 inflammasome activation drives tau pathology. *Nature.* 2019;575(7784):669–673.

Isop LM, Neculau AE, Necula RD, Kakucs C, Moga MA, Dima L. Metformin: The winding path from understanding its molecular mechanisms to proving therapeutic benefits in neurodegenerative disorders. *Pharmaceuticals (Basel).* 2023;16(12):1714.

Jadhav D, Saraswat N, Vyawahare N, Shirode D. Targeting the molecular web of Alzheimer's disease: Unveiling pathways for effective pharmacotherapy. *Egypt J Neurol Psychiatry Neurosurg.* 2024;60(1):7.

Jorfi M, Park J, Hall CK, Lin CJ, Chen M, von Maydell D, Kruskop JM, Kang B, Choi Y, Prokopenko D, Irimia D, Kim DY, Tanzi RE. Infiltrating CD8+ T cells exacerbate Alzheimer's disease pathology in a 3D human neuroimmune axis model. *Nat Neurosci.* 2023;26(9):1489–1504.

Kabir MT, Uddin MS, Mamun AA, et al. Combination drug therapy for the management of Alzheimer's disease. *Int J Mol Sci.* 2020;21(9):3272.

Kamatham PT, Shukla R, Khatri DK, Vora LK. Pathogenesis, diagnostics, and therapeutics for Alzheimer's disease: Breaking the memory barrier. *Ageing Res Rev.* 2024;101:102481.

Kargbo RB. PROTAC compounds targeting α-synuclein protein for treating neurogenerative disorders: Alzheimer's and Parkinson's diseases. *ACS Med Chem Lett.* 2020;11(6):1086–1087.

Kumar N, Kumar V, Anand P, Kumar V, Ranjan Dwivedi A, Kumar V. Advancements in the development of multi-target directed ligands for the treatment of Alzheimer's disease. *Bioorg Med Chem.* 2022;61:116742.

Kumar S, Reddy AP, Yin X, Reddy PH. Novel MicroRNA-455-3p and its protective effects against abnormal APP processing and amyloid beta toxicity in Alzheimer's disease. *Biochim Biophys Acta Mol Basis Dis.* 2019;1865(9):2428–2440.

Kurkinen M, Fułek M, Fułek K, Beszłej JA, Kurpas D, Leszek J, et al. The amyloid cascade hypothesis in Alzheimer's disease: Should we change our thinking? *Biomolecules.* 2023;13:453.

Kwak KA, Lee SP, Yang JY, Park YS. Current perspectives regarding stem cell-based therapy for Alzheimer's disease. *Stem Cells Int.* 2018;2018:6392986.

Lee JH, McBrayer MK, Wolfe DM, Haslett LJ, Kumar A, Sato Y, Lie PP, Mohan P, Coffey EE, Kompella U, Mitchell CH, Lloyd-Evans E, Nixon RA. Presenilin 1 maintains Lysosomal Ca(2+) Homeostasis via TRPML1 by regulating vATPase-mediated lysosome acidification. *Cell Rep.* 2015;12(9):1430–1444.

Leinenga G, Koh WK, Götz J. A comparative study of the effects of Aducanumab and scanning ultrasound on amyloid plaques and behavior in the APP23 mouse model of Alzheimer disease. *Alzheimers Res Ther.* 2021;13(1):76.

Leng F, Edison P. Neuroinflammation and microglial activation in Alzheimer disease: Where do we go from here?. *Nat Rev Neurol.* 2021;17(3):157–172.

Liang M, Gu L, Zhang H, et al. Design, synthesis, and bioactivity of novel bifunctional small molecules for Alzheimer's disease. *ACS Omega.* 2022;7(30):26308–26315.

Litus EA, Kazakov AS, Deryusheva EI, et al. Ibuprofen favors binding of amyloid-β peptide to its depot, serum albumin. *Int J Mol Sci.* 2022;23(11):6168.

Lo CH, Zeng J. Defective lysosomal acidification: A new prognostic marker and therapeutic target for neurodegenerative diseases. *Transl Neurodegener.* 2023;12(1):29

Loh JS, Mak WQ, Tan LKS, Ng CX, Chan HH, Yeow SH, Foo JB, Ong YS, How CW, Khaw, KY. Microbiota-gut-brain axis and its therapeutic applications in neurodegenerative diseases. *Signal Transduct Target Ther.* 2024;9(1):37.

Luo JE, Li YM. Turning the tide on Alzheimer's disease: Modulation of γ-secretase. *Cell Biosci.* 2022;12(1):2.

Macfarlane S, Grimmer T, Teo K, et al. Blarcamesine for the treatment of early Alzheimer's disease: Results from the ANAVEX2-73-AD-004 Phase IIB/III trial. *J Prev Alzheimers Dis.* 2025;12(1):100016.

Machado A, Ferreira D, Grothe MJ, Eyjolfsdottir H, Almqvist PM, Cavallin L, Lind G, Linderoth B, Seiger Å, Teipel S, Wahlberg LU, Wahlund LO, Westman E, Eriksdotter M, et al. Alzheimer's disease neuroimaging initiative. The cholinergic system in sub types of Alzheimer's disease: An in vivo longitudinal MRI study. *Alzheimers Res Ther.* 2020;12:51.

Manzano S, Agüera L, Aguilar M, Olazarán J. A review on tramiprosate (homotaurine) in Alzheimer's disease and other neurocognitive disorders. *Front Neurol.* 2020;11:614.

McDaid J, Mustaly-Kalimi S, Stutzmann GE. Ca^{2}+ Dyshomeostasis disrupts neuronal and synaptic function in Alzheimer's disease. *Cells.* 2020;9(12):2655.

Merighi S, Travagli A, Nigro M, Pasquini S, Cappello M, Contri C, Varani K, Vincenzi F, Borea PA, Gessi S. Caffeine for prevention of Alzheimer's disease: Is the A_{2A} adenosine receptor its target? *Biomolecules.* 2023; 13(6):967.

Mintun MA, Lo AC, Duggan Evans C, et al. Donanemab in early Alzheimer's disease. *N Engl J Med.* 2021;384(18):1691–1704.

Mishra S, Knupp A, Szabo MP, Williams CA, Kinoshita C, Hailey DW, Wang Y, Andersen OM, Young JE. The Alzheimer's gene SORL1 is a regulator of endosomal traffic and recycling in human neurons. *Cell Mol Life Sci.* 2022;79(3):162.

Misra A, Chakrabarti SS, Gambhir IS. New genetic players in late-onset Alzheimer's disease: Findings of genome-wide association studies. *Indian J Med Res.* 2018;148(2):135–144.

Misrani A, Tabassum S, Yang L. Mitochondrial dysfunction and oxidative stress in Alzheimer's disease. *Front Aging Neurosci.* 2021;13:617588.

Moreta MP-G, Burgos-Alonso N, Torrecilla M, Marco-Contelles J, Bruzos-Cidón C. Efficacy of acetylcholinesterase inhibitors on cognitive function in Alzheimer's disease. Review of reviews. *Biomedicines.* 2021; 9(11):1689.

Mouofo EN, Spires-Jones TL. Reeling from news that reelin defends the brain against Alzheimer's. *Cell Rep Med.* 2023;4(7):101111.

Mustaly-Kalimi S, Gallegos W, Marr RA, Gilman-Sachs A, Peterson DA, Sekler I, Stutzmann GE. Protein mishandling and impaired lysosomal proteolysis generated through calcium dysregulation in Alzheimer's disease. *Proc Natl Acad Sci U S A.* 2022;119(49):e2211999119.

Nehra G, Bauer B, Hartz AMS. Blood-brain barrier leakage in Alzheimer's disease: From discovery to clinical relevance. *Pharmacol Ther.* 2022;234:108119.

Nowell J, Blunt E, Gupta D, Edison P. Antidiabetic agents as a novel treatment for Alzheimer's and Parkinson's disease. *Ageing Res Rev.* 2023;89:101979.

Ogunsuyi OB, Aro OP, Oboh G, Olagoke OC. Curcumin improves the ability of donepezil to ameliorate memory impairment in *Drosophila melanogaster*: Involvement of cholinergic and cnc/Nrf2-redox systems. *Drug Chem Toxicol.* 2023;46(5):1035–1043.

Penke B, Szucs M, Bogar F. Oligomerization and conformational change turn monomeric β-amyloid and tau proteins toxic: Their role in Alzheimer's pathogenesis. *Molecules.* 2020;25:1659.

Punyakoti P, Behl T, Sehgal A, et al. Postulating the possible cellular signalling mechanisms of antibody drug conjugates in Alzheimer's disease. *Cell Signal.* 2023;102:110539.

Qu W, Li L. Microglial TREM2 at the intersection of brain aging and Alzheimer's disease. *Neuroscientist.* 2023;29(3):302–316.

Querfurth H, Lee HK. Mammalian/mechanistic target of rapamycin (mTOR) complexes in neurodegeneration. *Mol Neurodegener.* 2021;16(1):44.

Rodriguez-Soacha D, Cañon A, Sánchez LT, et al. Assessment of the Empact of IGC-AD1 on hepatic biomarkers in Alzheimer's disease: Insights from a phase 1 trial. *Alzheimers Dement.* 2025;20(Suppl 8):e095744.

Rofo F, Meier SR, Metzendorf NG, et al. A brain-targeting bispecific-multivalent antibody clears soluble amyloid-beta aggregates in Alzheimer's disease mice. *Neurotherapeutics.* 2022;19(5):1588–1602.

Simic G, Babic Leko M, Wray S, Harrington C, Delalle I, Jovanov-Milošević N, Bažadona D, Buee L, De Silva R, Di Giovanni G, et al. Tau protein hyperphosphorylation and aggregation in Alzheimer's disease and other tauopathies, and possible neuroprotective strategies. *Biomolecules.* 2016;6:6.

Singh B, Day CM, Abdella S, Garg S. Alzheimer's disease current therapies, novel drug delivery systems and future directions for better disease management. *J Control Release.* 2024;367:402–424.

Song C, Shi J, Zhang P, et al. Immunotherapy for Alzheimer's disease: Targeting β-amyloid and beyond. *Transl Neurodegener.* 2022;11(1):18.

Sun J, Xu M, Ru J, et al. Small molecule-mediated targeting of microRNAs for drug discovery: Experiments, computational techniques, and disease implications. *Eur J Med Chem.* 2023;257:115500.

Tampi RR, Forester BP, Agronin M. Aducanumab: Evidence from clinical trial data and controversies. *Drugs Context.* 2021;10: 2021-7-3.

Tang BL. Enhancing α-secretase processing for Alzheimer's disease-A view on SFRP1. *Brain Sci.* 2020;10(2):122.

Tonegawa-Kuji R, Hou Y, Hu B, et al. Efficacy and safety of passive immunotherapies targeting amyloid beta in Alzheimer's disease: A systematic review and meta-analysis. *PLoS Med.* 2025;22(3):e1004568.

Ulku I, Leung R, Herre F, Walther L, Shobo A, Saftig P, Hancock MA, Liebsch F, Multhaup G. Inhibition of BACE1 affected both its Aβ producing and degrading activities and increased Aβ42 and Aβ40 levels at high-level BACE1 expression. *J Biol Chem.* 2024;300(8):107510.

van Dyck CH, Swanson CJ, Aisen P, et al. Lecanemab in early Alzheimer's disease. *N Engl J Med.* 2023;388(1):9–21.

Vasic V, Barth K, Schmidt MHH. Neurodegeneration and neuro-regeneration-Alzheimer's disease and stem cell therapy. *Int J Mol Sci.* 2019;20(17):4272.

Vitek GE, Decourt B, Sabbagh MN. Lecanemab (BAN2401): An anti-beta-amyloid monoclonal antibody for the treatment of Alzheimer disease. *Expert Opin Investig Drugs.* 2023;32(2):89–94.

Wang C, Fan L, Khawaja RR, Liu B, Zhan L, Kodama L, Chin M, Li Y, Le D, Zhou Y, Condello C, Grinberg LT, Seeley WW, Miller BL, Mok SA, Gestwicki JE, Cuervo AM, Luo W, Gan L. Microglial NF-κB drives tau spreading and toxicity in a mouse model of tauopathy. *Nat Commun.* 2022;13(1):1969.

Wang R, Reddy PH. Role of glutamate and NMDA receptors in Alzheimer's disease. *J Alzheimers Dis.* 2017;57(4):1041–1048. doi:10.3233/JAD-160763

Wang W, Zhao F, Ma X, Perry G, Zhu X. Mitochondria dysfunction in the pathogenesis of Alzheimer's disease: Recent advances. *Mol Neurodegener.* 2020;15(1):30.

Wang W, Zhou Q, Jiang T, et al. A novel small-molecule PROTAC selectively promotes tau clearance to improve cognitive functions in Alzheimer-like models. *Theranostics.* 2021;11(11):5279–5295.

Wang ZH, Xiang J, Liu X, Yu SP, Manfredsson FP, Sandoval IM, Wu S, Wang JZ, Ye K. Deficiency in BDNF/TrkB neurotrophic activity stimulates δ-secretase by upregulating C/EBPβ in Alzheimer's disease. *Cell Rep.* 2019;28(3):655–669.e5.

Wiessner C, Wiederhold KH, Tissot AC, et al. The second-generation active Aβ immunotherapy CAD106 reduces amyloid accumulation in APP transgenic mice while minimizing potential side effects. *J Neurosci.* 2011;31(25):9323–9331.

Winkler EA, Nishida Y, Sagare AP, Rege SV, Bell RD, Perlmutter D, Sengillo JD, Hillman S, Kong P, Nelson AR, Sullivan JS, Zhao Z, Meiselman HJ, Wendy RB, Soto J, Abel ED, Makshanoff J, Zuniga E, De Vivo DC, Zlokovic BV. GLUT1 reductions exacerbate Alzheimer's disease vasculo-neuronal dysfunction and degeneration. *Nat Neurosci.* 2015;18(4):521–530.

Wischik CM, Bentham P, Gauthier S, Miller S, Kook K, Schelter BO. Oral Tau aggregation inhibitor for Alzheimer's Disease: Design, progress and basis for selection of the 16 mg/day dose in a Phase 3, randomized, placebo-controlled trial of hydromethylthionine mesylate. *J Prev Alzheimers Dis.* 2022;9(4):780–790.

Wong-Guerra M, Calfio C, Maccioni RB, Rojo LE. Revisiting the neuroinflammation hypothesis in Alzheimer's disease: A focus on the druggability of current targets. *Front Pharmacol.* 2023;14: 1161850.

Wu T, Lin D, Cheng Y, Jiang S, Riaz MW, Fu N, Mou C, Ye M, Zheng Y, et al. Amyloid cascade hypothesis for the treatment of Alzheimer's disease: Progress and challenges. *Aging Dis.* 2022;13:1745–58.

Xhima K, Markham-Coultes K, Hahn Kofoed R, Saragovi HU, Hynynen K, Aubert I. Ultrasound delivery of a TrkA agonist confers neuroprotection to Alzheimer-associated pathologies. *Brain*. 2022;145(8): 2806–2822.

Xiang J, Wang ZH, Ahn EH, Liu X, Yu SP, Manfredsson FP, Sandoval IM, Ju G, Wu S, Ye K. Delta-secretase-cleaved Tau antagonizes TrkB neurotrophic signalings, mediating Alzheimer's disease pathologies. *Proc Natl Acad Sci U S A*. 2019;116(18):9094–9102.

Xiao S, Chan P, Wang T, et al. A 36-week multicenter, randomized, double-blind, placebo-controlled, parallel-group, phase 3 clinical trial of sodium oligomannate for mild-to-moderate Alzheimer's dementia. *Alzheimers Res Ther*. 2021;13(1):62.

Yang LG, March ZM, Stephenson RA, Narayan PS. Apolipoprotein E in lipid metabolism and neurodegenerative disease. *Trends Endocrinol Metab*. 2023;34(8):430–445.

Yu W, Jin H, Huang Y. Mitochondria-associated membranes (MAMs): A potential therapeutic target for treating Alzheimer's disease. *Clin Sci (Lond)*. 2021;135(1):109–126.

Zhang H, Wei W, Zhao M, Ma L, Jiang X, Pei H, Cao Y, Li H. Interaction between Aβ and Tau in the pathogenesis of Alzheimer's disease. *Int J Biol Sci*. 2021;17(9):2181–2192.

Zhang J, Zhang Y, Wang J, Xia Y, Zhang J, Chen L. Recent advances in Alzheimer's disease: Mechanisms, clinical trials and new drug development strategies. *Signal Transduct Target Ther*. 2024;9(1):211.

Zhao K, Liu J, Sun T, Zeng L, Cai Z, Li Z, Liu R. The miR-25802/KLF4/NF-κB signaling axis regulates microglia-mediated neuroinflammation in Alzheimer's disease. *Brain Behav Immun*. 2024;118:31–48.

Zhong L, Sheng X, Wang W, Li Y, Zhuo R, Wang K, Zhang L, Hu DD, Hong Y, Chen L, Rao H, Li T, Chen M, Lin Z, Zhang YW, Wang X, Yan XX, Chen X, Bu G, Chen XF. TREM2 receptor protects against complement-mediated synaptic loss by binding to complement C1q during neurodegeneration. *Immunity*. 2023;56(8):1794–1808.e8.

3 Parkinson's Disease
Unravelling the Mechanisms, Pathogenesis, Genetic Basis, and Emerging Therapies

Nibha Singh, Yogesh Sharma, Bhaskar Sharma, Srishti Jain, and Prabhakar Singh

3.1 INTRODUCTION

The progressive degeneration of dopaminergic neurons in the substantia nigra is a key feature of Parkinson's disease (PD), a neurodegenerative disorder characterized by motor impairments such as bradykinesia, stiffness, postural instability, and resting tremor [Goetz et al, 2008; Poewe & Mahlknecht, 2009; Prashanth & Roy, 2018]. The disease also involves non-motor signs of the illness can include autonomic dysfunction and cognitive decline [Maruyama & Yanagisawa, 2006]. The main motor symptoms of PD are attributed to the dysfunction of the nigrostriatal dopamine system [Maruyama & Yanagisawa, 2006]. This dopaminergic depletion is linked to tremors and stiffness, whereas disturbances in the brain processes that link sensory input to motor output cause akinesia and bradykinesia [Maruyama & Yanagisawa, 2006]. Posture and variations in walking patterns are connected to lesions in the basal ganglia, pedunculopontine nucleus, and frontal lobe, which are involved in gait control [Maruyama & Yanagisawa, 2006]. People with PD may need extra training to master and remember motor skills, and they may depend more on a preliminary control system for learning and controlling their movements [Olson et al, 2019].

Although PD has a long history, English surgeon James Parkinson was the first to formally characterize it in the early 19th century. In "An Essay on the Shaking Palsy," published in 1817, Parkinson described the signs of the condition, including postural instability, trembling, and the progressive development of motor dysfunction [Pearce, 1989]. He referred to the condition as "paralysis agitans," which was later renamed "Parkinson's disease" by French neurologist Jean-Martin Charcot, who further defined its clinical features and distinguished it from similar disorders. Throughout the 20th century, significant advancements were made in understanding PD's pathology, including the identification of dopaminergic neuron loss in the substantia nigra, the role of Lewy bodies, and α-synuclein aggregation. PD is recognized as a complex disorder with both motor and non-motor manifestations. It is the second most common neurodegenerative disorder after Alzheimer's disease, affecting over 10 million people worldwide. The incidence of PD increases with age, with around 1% of people over 60 years old affected. However, a small percentage of cases (5–10%) occur before the age of 50, often linked to genetic mutations [Przedborski, 2017]. Geographical variations in PD prevalence may be influenced by environmental factors, genetic predisposition, and diagnostic practices. The societal impact of PD is profound, affecting patients, their families, caregivers, and healthcare systems. The progressive nature of PD leads to increasing disability, which severely impacts daily activities and independence. Non-motor symptoms, such as cognitive decline, depression, anxiety, and sleep disturbances, further compound the disease burden [Dorsey et al, 2018]. The economic impact is also significant, encompassing direct medical expenses for treatment and care, as well as indirect costs related to lost efficiency and the need for caregiving.

DOI: 10.1201/9781003500025-3

With these compelling challenges, a deep understanding of PD is essential to meet such dilemmas in the diagnosis and treatment [Bloem et al, 2021]. Management, given the effects of dopaminergic medications on motor symptoms, symptomatic relief is seen with treatment; however, there has been no evidence to suggest that any therapy halts disease progression. Long-term use of these medications can also have severe side effects, such as dyskinesias [Aradi & Hauser, 2020]. This chapter aims to explore the intricacies of PD by delving into its pathogenesis, including the role of alpha-synuclein (α-syn) aggregation, mitochondrial impairment, and neuroinflammation.

3.2 CLINICAL FEATURES OF PARKINSON'S DISEASE

3.2.1 Motor Symptoms

PD primarily affects movement, with tremor often being the first symptom, typically presenting as a unilateral resting tremor that decreases with voluntary movement. Bradykinesia, or slowness of movement, significantly impairs daily activities, while rigidity leads to stiffness and decreased range of motion. Postural instability increases the risk of falls and severely limits mobility, often requiring assistive devices. This can affect any part of the body, leading to stiffness and decreased range of motion. Rigidity is often described as a "cogwheel" when combined with tremor, creating a ratchet-like feel during joint movement [Bloem et al, 2021]. This symptom significantly affects mobility and independence, frequently requiring the use of assistive devices such as walkers or canes [Armstrong & Okun, 2020]. The impairment in balance and coordination associated with postural instability can severely limit a PD patient's ability to perform everyday activities and can lead to a decline in overall quality of life.

3.2.2 Non-Motor Symptoms

Cognitive impairment is a common symptom in PD and can range from mild cognitive decline to severe dementia. This impairment affects various cognitive functions, including executive functions, memory, attention, and visuospatial skills [Emre, 2003]. Mild cognitive impairment (MCI) is prevalent in the early stages of PD and can progress to Parkinson's disease dementia (PDD) as the disease advances [Weintraub & Burn, 2011]. The cognitive impairment observed in PD is linked to several underlying factors, such as Lewy body pathology, cholinergic deficits, and other neurodegenerative processes occurring in the brain. Autonomic dysfunction impacts various bodily systems, leading to issues like orthostatic hypotension and urinary incontinence [Zheng et al, 2023]. Disruptions in sleep patterns, such as difficulty falling asleep and REM sleep behavior abnormalities, affect up to 90% of patients, exacerbating the disease's overall burden. These symptoms can arise directly from the neurochemical changes in the brain caused by the disease or as a reaction to the diagnosis and progression of PD (Figure 3.1). Unfortunately, depression and anxiety in PD are often not properly identified or treated, emphasizing the need for comprehensive management strategies to address these issues. Common sleep disorders include insomnia, excessive daytime drowsiness, leg discomfort with the urge to move, and abnormalities in REM sleep behavior [Hoehn & Yahr, 1967]. These disturbances can result from neurodegeneration in sleep-regulating brain regions, medication side effects, or the motor symptoms of PD themselves.

3.3 MECHANISMS OF PARKINSON'S DISEASE

A pivotal structure for movement regulation in the midbrain is the substantia nigra. It has two parts: the pars compacta, which contains dopaminergic neurons and is the primary source of the neurotransmitter dopamine for several basal ganglia circuits, and the pars reticulata, which is the output stage of the basal ganglia system and contains mostly GABAergic (gamma-aminobutyric acid) neurons [Przedborski, 2017]. PD involves a gradual degeneration not only of the neurons in the

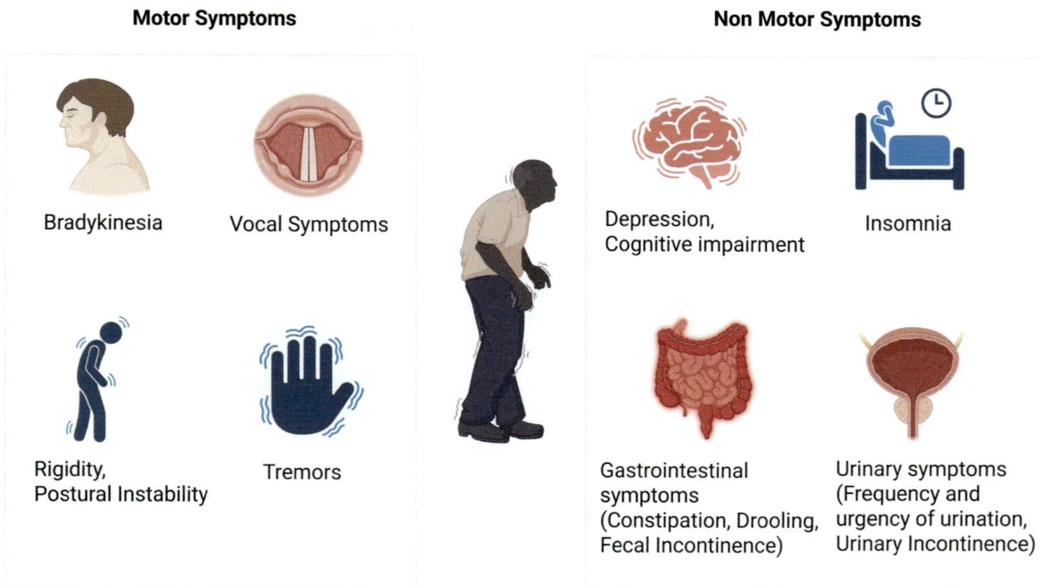

FIGURE 3.1 Symptoms of Parkinson's disease.

pars compacta but also the outputs through the pars reticulata that form the closed loop of circuitry around the basal ganglia that feeds back to the cortex. Lewy bodies are abnormal protein deposits that primarily consist of the protein α-syn and are one of the main signs of PD and other neurological diseases that are grouped as synucleinopathies [Dong-Chen et al, 2023; Grassi et al, 2018]. The presence of Lewy bodies in the brain is strongly associated with the development and progression of PD, as they are found in the substantia nigra and other brain regions affected by the disease. α-syn, a synaptic protein that is normally involved in neurotransmitter release and synaptic function, can misfold and aggregate, leading to the formation of Lewy bodies [Grassi et al, 2018], which contributes to neuronal dysfunction and death. Lewy bodies have been shown to induce mitochondrial damage and mitophagy, further exacerbating neuronal degeneration [Grassi et al, 2018].

3.3.1 Neurochemical Pathways

Dopamine deficiency, particularly within the nigrostriatal pathway, is closely linked to neurodegenerative diseases like PD. In PD, the gradual loss of dopamine-producing neurons in the substantia nigra leads to reduced dopamine levels in the striatum, disrupting the basal ganglia circuits that are essential for motor control [Sveinbjornsdottir, 2016]. The basal ganglia, which include the striatum, globus pallidus (external and internal), subthalamic nucleus, and substantia nigra, modulate motor commands initiated by the cerebral cortex [Kowal et al, 2013]. Dopamine loss in these circuits creates an imbalance between the direct (movement-exciting) and indirect (movement-inhibiting) cascades, resulting in bradykinesia, rigidity, tremor, and postural instability. Dopamine depletion leads to reduced activation of D1 and D2 receptors, affecting the direct and indirect pathways, respectively. This excess results in hyperactivity of the subthalamic nucleus and Gpi, which increases inhibition to the thalamus (and ultimately motor cortex), just as reduced striatal output caused by cell loss decreases activation of these pathways, leading to some but not all hypokinetic symptoms seen in PD [Kalia et al, 2015]. Consequently, hyperactivity in the subthalamic nucleus and globus pallidus internus increases inhibition of the thalamus and, ultimately, the motor cortex, contributing to the hypokinetic symptoms of PD. While other neurotransmitters like acetylcholine and serotonin also have a major part in the course of PD, dopamine is the main neurotransmitter involved.

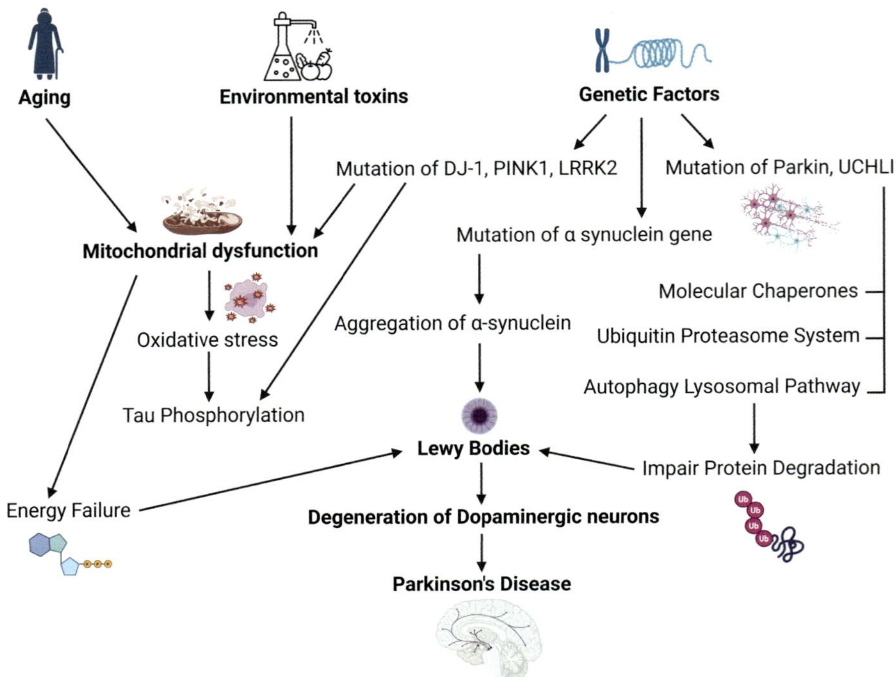

FIGURE 3.2 Molecular mechanisms in Parkinson's disease.

The cholinergic system within the striatum, which often counteracts dopamine's effects, becomes more active when dopamine is depleted, further contributing to PD's motor symptoms. In PD, both motor and non-motor symptoms arise as the disease impacts regions where serotonin is released (Figure 3.2). Serotonin modulates the release of dopamine and the sensitivity of dopamine receptors, contributing to the pathology through serotonergic dysfunction [Hirsch et al, 2016].

3.4 PATHOGENESIS OF PARKINSON'S DISEASE

PD is A complex neurodegenerative condition with multiple etiology that includes both genetic and environmental elements. Several environmental agents have been implicated as potential risk factors, including industrial chemicals, pesticides, solvents, and metals. In addition to environmental factors, epigenetic modifications, such as DNA methylation, have also been associated with PD [Brundin, 2002]. Alterations in DNA methylation patterns at specific genes like PARK7 and SLC17A6 have been observed in PD, suggesting that epigenetic dysregulation might be involved in the pathogenesis [Kochmanski et al, 2022].

The protein α-synuclein, central to PD pathogenesis, is encoded by the SNCA gene. The point mutations in the SNCA gene, such as A53T or E46K, and the duplication/triplication of genes lead to abnormal α-syn aggregation into Lewy bodies, one of the pathognomonic hallmarks of PD [Lim & Klein, 2024]. These mutations induce a toxic gain-of-function effect that interferes with the normal function of neurons, thus leading to neurodegeneration, particularly of dopaminergic neurons of the substantia nigra. The SNCA gene was the first gene identified to have an association with familial PD, and its role underlines the general importance of protein misfolding in the disease [Jankovic, 2008].

LRRK2 mutations, in particular the G2019S mutation, increase kinase activity, and it is that which is hypothesized to contribute to PD pathogenesis via numerous means involving altered autophagy, mitochondrial dysfunction, and cytoskeletal dynamics. PARK2 is responsible for autosomal recessive juvenile parkinsonism and encodes the E3-ubiquitin-ligase Parkin [Gibb & Lees, 1988]. In this case, mutations cause the dysfunction of PARK2; that is, disturbance of the ubiquitin–proteasome system and impairment of mitochondrial quality control. Indeed, the role played by Parkin in promoting damaged mitochondria degradation through mitophagy is important for cell homeostasis [McAvoy & Kawamata, 2019]. These accumulating, mutated Parkin-induced dysfunctional mitochondria increase the level of oxidative stress and neuronal death, particularly dopaminergic neurons.

3.4.1 Epigenetic Modifications in Parkinson's Disease

Epigenetic modifications in gene expression are a very important step in regulation without actually affecting the DNA sequence. Although these modifications can also mediate environmental factors on gene expression in PD, they have been associated with the cause and progression of the condition. This chapter examines three main epigenetic mechanisms—DNA methylation, histone modifications, and non-coding RNAs—in PD pathogenesis (Figure 3.3). The process of adding a methyl group to cytosine residues at their 5′ position is known as DNA methylation, typically at CpG dinucleotides [Armstrong & Okun, 2020]. This frequently leads to gene silencing at its promoter regions by either obstructing the binding of transcription factors or recruiting several methyl-binding proteins that enhance chromatin condensation.

Some reports describe that the promoter region of the SNCA gene may be hypermethylated, hence explaining its reduced expression in several brain areas, although hypomethylation in other areas might lead to overexpression and aggregation [Bogdanov et al, 2008]. Several studies have reported gene SNCA to be differentially methylated in patients with PD compared with controls, an

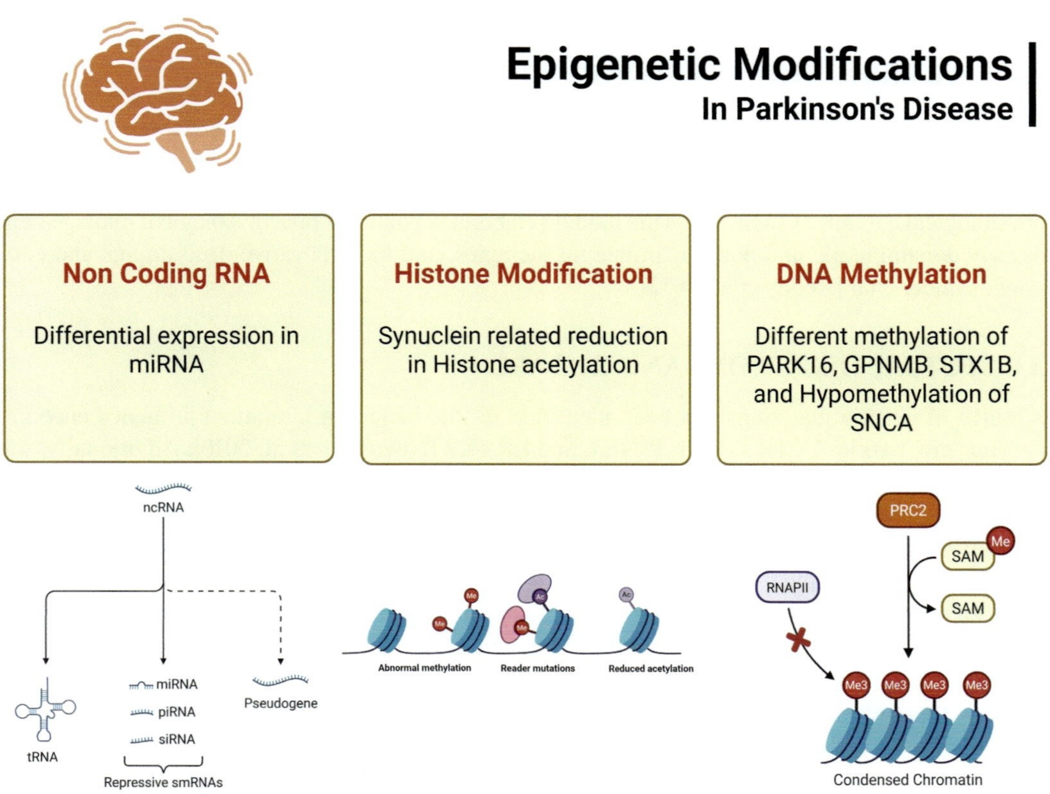

FIGURE 3.3 Epigenetic modification in Parkinson's disease.

indication of a role in the heterogeneity of the disease. Hypomethylation of SNCA in dopaminergic neurons might be associated with increased α-syn expression, promoting its pathologic aggregation into Lewy bodies [Lin et al, 2022]. Histone modifications are post-translational alterations of histone proteins around which DNA is wound and therefore involve modulation of chromatin structure and gene expression. In the case of PD, changes in histone acetylation have been shown to underlie the misexpression of genes that regulate neuronal survival and stress responses. Preclinical studies have shown that HDAC inhibitors can be beneficial by up-regulating neuroprotective genes.

Histone methylation may lead to the activation or repression of genes, based on the quantity of methyl groups bonded and the particular amino acid residue altered. Abnormal histone methylation patterns at the genes related to inflammation, apoptosis, and synaptic function have been reported in PD [Zhao et al, 2020]. By base-pairing with complementary sequences on mRNAs, single-stranded RNA molecules alter gene expression, typically leading to mRNA breakdown or translational suppression. miRNA expression, especially its dysregulation, has been gaining increasing recognition in the potential role of controlling genes related to apoptosis, autophagy, and synaptic plasticity in the pathogenesis of PD [Arrázola et al, 2015]. LncRNAs are long single-stranded RNA transcripts that can regulate gene expression through multiple mechanisms, which include chromatin remodeling, transcriptional interference, and post-transcriptional regulation. Several lncRNAs have been implicated in PD, where they are known to alter the expression of multiple genes that are associated with neuroprotection and neurodegeneration. For instance, one recent study found that the lncRNA HOTAIR can alter histone modulations at the SNCA locus and, hence, may exert functional effects on α-syn expression. In PD, some circRNAs are differentially expressed and therefore might have effects on the function of their targeting miRNAs and downstream gene expression. Specifically, circSLC8A1 can sequester miR-128, up-regulating the expression of BACE1, a gene involved in the manufacture of amyloid-β, which could contribute to neurodegeneration processes in PD.

3.5 EMERGING THERAPIES FOR PARKINSON'S DISEASE

The conventional palliative treatments of PD are guided toward symptom alleviation, particularly motor signs, through dopaminergic therapies. Innovative therapies are those that aspire to furnish a wider approach to management with involvement in different levels of engagement in PD pathophysiological cascades (Table 3.1). This module elaborates on recent pharmacological management, namely dopaminergic and non-dopaminergic therapies, and the innovative drug targets and compounds under trial [Reddy et al, 2012].

3.6 FUTURE DIRECTIONS AND RESEARCH

A range of genetic mutations has been identified in PD, including alterations in genes encoding α-synuclein, parkin, UCHL1, DJ-1, PINK1, and LRRK2 [Nuyteman et al, 2010]. Additionally, several biochemical factors have a pivotal function in the pathology, such as free radicals, mitochondrial complex 1 deficiencies, excitotoxicity, and inflammation [Olanow et al, 2003]. Advances in neuroimaging have greatly contributed to early detection. Techniques like MRI, proton magnetic resonance spectroscopy, transcranial B-mode sonography, SPECT, and PET are increasingly utilized for identifying biomarkers linked to PD [Saeed et al, 2020]. Treating PD has shifted toward personalized medicine, which takes into factors like age, lifestyle, and genetic predisposition. For instance, carriers of the LRRK2 gene mutation could benefit from targeted treatments using LRRK2 inhibitors [Titova & Chaudhuri, 2017]. Personalized medicine aims to predict an individual's susceptibility to developing PD and tailor treatments accordingly, thus improving outcomes and managing symptoms more effectively. There are many developments underway that promise to revolutionize the treatment of PD. These include immunotherapies, drug repurposing, and treatments that target non-dopaminergic systems. Regenerative approaches, such as the use of stem

TABLE 3.1
Different Emerging Therapies for Parkinson's Disease, Highlighting Their Mechanisms, Examples, and Relevant Notes

Therapy Type	Mechanism of Action	Examples
Dopaminergic Therapies	**Levodopa**: Crosses the blood-brain barrier, converted to dopamine, and replenishes dopamine in the striatum.	Levodopa
	Dopamine Agonists: Directly activate dopamine receptors (D2-like family).	Pramipexole, Ropinirole, Rotigotine
Non-Dopaminergic Therapies	**MAO-B Inhibitors**: Inhibit dopamine breakdown, increase dopamine availability, and have neuroprotective effects.	Selegiline, Rasagiline
	COMT Inhibitors: Prevent peripheral breakdown of levodopa, and stabilize dopamine levels in the brain.	Entacapone, Opicapone
Immunotherapy & Vaccines	Target alpha-synuclein, enhance clearance of pathogenic aggregates, and slow disease progression.	Prasinezumab, Vaccines
Neuroprotective Agents	Reduce oxidative stress and neuroinflammation, target mitochondrial function.	Coenzyme Q10, NSAIDs, Creatine, Mitochondrial-targeted antioxidants
Neurotrophic Factors	Support survival and regeneration of neurons, delivered via gene therapy or infusion.	GDNF, BDNF

cells, and advanced surgical options like deep brain stimulation (DBS) are also under investigation. Furthermore, non-invasive therapies are showing the potential to reshape PD management [Stoker & Barker, 2020]. PD is multifaceted, with motor and non-motor symptoms that extend into cognitive, psychological, and social domains. Multidisciplinary care teams are essential for addressing complexities and bringing together neurologists, physiotherapists, psychologists, and other specialists to ensure comprehensive treatment [Radder et al, 2020]. Global initiatives such as Aligning Science Across Parkinson's (ASAP), the Global Parkinson's Genetics Program (GP2), the International Parkinson's Disease Genomics Consortium (IPDGC), and the Genetic Epidemiology of Parkinson's Disease (GEoPD) have emerged, fostering collaborative research and accelerating discoveries that may lead to more effective treatments.

3.7 CONCLUSION

Neurodegenerative disorders such as PD are mostly untreatable due to the irreversible death of specific neurons in affected individuals. Several factors contribute to the development of PD, including age, genetics, and environmental influences. Growing research has highlighted a strong link between genetic predispositions and environmental triggers in PD-related neurodegeneration. Environmental factors and PD-related gene mutations appear to function in interconnected or overlapping pathways, suggesting shared molecular mechanisms. In recent years, significant advancements have been made in understanding PD's pathogenesis, particularly concerning genetic mutations, mitochondrial dysfunction, oxidative stress, and neuroinflammation. Notable genetic research has identified mutations in genes such as α-synuclein, LRRK2, PINK1, and DJ-1, which are pivotal in PD development. Although this chapter explores biological pathways, the primary challenge ahead lies in identifying effective therapies that target the dysregulated mechanisms driving neurodegenerative diseases. Future research on PD should focus on advancements in personalized medicine, incorporating genetic predispositions and lifestyle factors, and hold promise for developing targeted therapies, such as LRRK2 inhibitors, that could revolutionize treatment strategies. Additionally, exploring regenerative approaches, like stem cell therapy, and refining non-invasive techniques, such as advanced neuroimaging and DBS, will be critical in enhancing early diagnosis and improving patient outcomes.

ABBREVIATIONS

PD	Parkinson's Disease
DA	Dopamine
CNS	Central Nervous System
SNpc	Substantia Nigra pars compacta
α-syn	Alpha-synuclein
UPS	Ubiquitin–Proteasome System
ROS	Reactive Oxygen Species
NO	Nitric Oxide
NF-κB	Nuclear Factor kappa-light-chain-enhancer of activated B cells
LPS	Lipopolysaccharide
MPP$^+$	1-Methyl-4-phenyl pyridinium
MPTP	1-Methyl-4-phenyl-1,2,3,6-tetrahydropyridine
MAO-B	Monoamine Oxidase-B
COMT	Catechol-O-methyltransferase
VMAT2	Vesicular Monoamine Transporter 2
PET	Positron Emission Tomography
MRI	Magnetic Resonance Imaging
LD	Levodopa
DBS	Deep Brain Stimulation
FDA	Food and Drug Administration
GDNF	Glial cell line-derived Neurotrophic Factor
BDNF	Brain-Derived Neurotrophic Factor
iPSC	Induced Pluripotent Stem Cells
TH	Tyrosine Hydroxylase
IL-1β	Interleukin 1 beta
IL-6	Interleukin 6
TNF-α	Tumor Necrosis Factor-alpha
BBB	Blood–Brain Barrier
AAV	Adeno-Associated Virus
ATP	Adenosine Triphosphate

REFERENCES

Aradi SD, Hauser RA. Medical management and prevention of motor complications in Parkinson's disease. Neurotherapeutics. 2020;17(4):1339–65.

Armstrong MJ, Okun MS. Diagnosis and treatment of Parkinson disease: a review. JAMA. 2020;323(6):548–60.

Arrázola MS, Silva-Alvarez C, Inestrosa NC. How the Wnt signaling pathway protects from neurodegeneration: the mitochondrial scenario. Frontiers in Cellular Neuroscience. 2015;9:166.

Bloem BR, Okun MS, Klein C. Parkinson's disease. The Lancet. 2021;397(10291):2284–303.

Bogdanov M, Matson WR, Wang L, Matson T, Saunders-Pullman R, Bressman SS, Beal MF. Metabolomic profiling to develop blood biomarkers for Parkinson's disease. Brain. 2008;131(2):389–96.

Brundin P. GDNF treatment in Parkinson's disease: time for controlled clinical trials? Brain. 2002;125(10):2149–51.

Dong-Chen X, Yong C, Yang X, Chen-Yu S, Li-Hua P. Signaling pathways in Parkinson's disease: molecular mechanisms and therapeutic interventions. Signal Transduction and Targeted Therapy. 2023;8(1):73.

Dorsey ER, Elbaz A, Nichols E, Abbasi N, Abd-Allah F, Abdelalim A, Adsuar JC, Ansha MG, Brayne C, Choi JY, Collado-Mateo D. Global, regional, and national Burden of Parkinson's Disease, 1990–2016: a systematic analysis for the Global Burden of Disease Study 2016. The Lancet Neurology. 2018;17(11):939–53.
Emre M. Dementia associated with Parkinson's disease. The Lancet Neurology. 2003;2(4):229–37.
Gibb WR, Lees A. The relevance of the Lewy body to the pathogenesis of idiopathic Parkinson's disease. Journal of Neurology, Neurosurgery & Psychiatry. 1988;51(6):745–52.
Goetz CG, Tilley BC, Shaftman SR, Stebbins GT, Fahn S, Martinez-Martin P, Poewe W, Sampaio C, Stern MB, Dodel R, Dubois B. Movement Disorder Society-sponsored revision of the Unified Parkinson's Disease Rating Scale (MDS-UPDRS): scale presentation and clinimetric testing results. Movement Disorders: Official Journal of the Movement Disorder Society. 2008;23(15):2129–70.
Grassi D, Howard S, Zhou M, Diaz-Perez N, Urban NT, Guerrero-Given D, Kamasawa N, Volpicelli-Daley LA, LoGrasso P, Lasmézas CI. Identification of a highly neurotoxic α-synuclein species inducing mitochondrial damage and mitophagy in Parkinson's disease. Proceedings of the National Academy of Sciences. 2018;115(11):E2634–43.
Hirsch L, Jette N, Frolkis A, Steeves T, Pringsheim T. The incidence of Parkinson's disease: a systematic review and meta-analysis. Neuroepidemiology. 2016;46(4):292–300.
Hoehn MM, Yahr MD. Parkinsonism: onset, progression, and mortality. Neurology. 1967;17(5):427–42.
Jankovic J. Parkinson's disease: clinical features and diagnosis. Journal of Neurology, Neurosurgery & Psychiatry. 2008;79(4):368–76.
Kalia LV, Lang AE, Hazrati LN, Fujioka S, Wszolek ZK, Dickson DW, Ross OA, Van Deerlin VM, Trojanowski JQ, Hurtig HI, Alcalay RN. Clinical correlations with Lewy body pathology in LRRK2-related Parkinson disease. JAMA Neurology. 2015;72(1):100–5.
Kochmanski J, Kuhn NC, Bernstein AI. Parkinson's disease-associated, sex-specific changes in DNA methylation at PARK7 (DJ-1), SLC17A6 (VGLUT2), PTPRN2 (IA-2β), and NR4A2 (NURR1) in cortical neurons. NPJ Parkinson's Disease. 2022;8(1):120.
Kowal SL, Dall TM, Chakrabarti R, Storm MV, Jain A. The current and projected economic burden of Parkinson's disease in the United States. Movement Disorders. 2013;28(3):311–8.
Lim SY, Klein C. Parkinson's disease is predominantly a genetic disease. Journal of Parkinson's Disease. 2024;1:1–6.
Lin KJ, Chen SD, Lin KL, Liou CW, Lan MY, Chuang YC, Wang PW, Lee JJ, Wang FS, Lin HY, Lin TK. Iron brain menace: the involvement of ferroptosis in Parkinson disease. Cells. 2022;11(23):3829.
Maruyama T, Yanagisawa N. Cognitive impact on freezing of gait in Parkinson's disease. Parkinsonism & Related Disorders. 2006;12(Suppl 2):S77–82.
McAvoy K, Kawamata H. Glial mitochondrial function and dysfunction in health and neurodegeneration. Molecular and Cellular Neuroscience. 2019;101:103417.
Nuytemans K, Theuns J, Cruts M, Van Broeckhoven C. Genetic etiology of Parkinson disease associated with mutations in the SNCA, PARK2, PINK1, PARK7, and LRRK2 genes: a mutation update. Human Mutation. 2010;31(7):763–80.
Olanow CW, Schapira AH, Agid Y. Neurodegeneration and prospects for neuroprotection and rescue in Parkinson's disease. Annals of Neurology. 2003;53(Suppl 3):S1–2.
Olson M, Lockhart TE, Lieberman A. Motor learning deficits in Parkinson's disease (PD) and their effect on training response in gait and balance: a narrative review. Frontiers in Neurology. 2019;10:62.
Pearce JM. Aspects of the history of Parkinson's disease. Journal of Neurology, Neurosurgery, and Psychiatry. 1989;52(Suppl):6.
Poewe W, Mahlknecht P. The clinical progression of Parkinson's disease. Parkinsonism & Related Disorders. 2009;15(Suppl 4):S28–32.
Prashanth R, Roy SD. Early detection of Parkinson's disease through patient questionnaire and predictive modelling. International Journal of Medical Informatics. 2018;119:75–87.
Przedborski S. The two-century journey of Parkinson disease research. Nature Reviews Neuroscience. 2017;18(4):251–9.
Radder DLM, Nonnekes J, van Nimwegen M, Eggers C, Abbruzzese G, Alves G, Browner N, Chaudhuri KR, Ebersbach G, Ferreira JJ, Fleisher JE, Fletcher P, Frazzitta G, Giladi N, Guttman M, Iansek R, Khandhar S, Klucken J, Lafontaine AL, Marras C, Nutt J, Okun MS, Parashos SA, Munneke M, Bloem BR. Recommendations for the Organization of Multidisciplinary Clinical Care Teams in Parkinson's Disease. Journal of Parkinson's Disease. 2020;10(3):1087–98.
Reddy P, Martinez-Martin P, Rizos A, Martin A, Faye GC, Forgacs I, Odin P, Antonini A, Chaudhuri KR. Intrajejunal levodopa versus conventional therapy in Parkinson disease: motor and nonmotor effects. Clinical Neuropharmacology. 2012;35(5):205–7.

Saeed U, Lang AE, Masellis M. Neuroimaging advances in Parkinson's disease and atypical parkinsonian syndromes. Frontiers in Neurology. 2020;11:572976.

Stoker TB, Barker RA. Recent developments in the treatment of Parkinson's Disease. F1000Research. 2020;9:F1000 Faculty Rev-862.

Sveinbjornsdottir S. The clinical symptoms of Parkinson's disease. Journal of Neurochemistry. 2016;139:318–24.

Titova N, Chaudhuri KR. Personalized medicine in Parkinson's disease: time to be precise. Movement Disorders. 2017;32(8):1147–54.

Weintraub D, Burn DJ. Parkinson's disease: the quintessential neuropsychiatric disorder. Movement Disorders. 2011;26(6):1022–31.

Zhao Y, Qin L, Pan H, Liu Z, Jiang L, He Y, Zeng Q, Zhou X, Zhou X, Zhou Y, Fang Z. The role of genetics in Parkinson's disease: a large cohort study in Chinese mainland population. Brain. 2020;143(7):2220–34.

Zheng Z, Zhu Z, Zhou C, Cao L, Zhao G. Burden of Parkinson Disease in China, 1990–2019: findings from the 2019 Global Burden of Disease Study. Neuroepidemiology. 2023;57(1):51–64.

4 The Brain's Response to Neurological Disorders
Structural and Functional Changes

SK Venkata Manjari and Pragya Komal

4.1 INTRODUCTION

The most comprehensive study reflects that the burden of neurological conditions affects 43% of the world's population as per the daily-adjusted life years (DALY) reports (Huang et al. 2023). Neurological disorders are characterized by different behavioral symptoms that vary depending on the specific conditions, neuropathology, and the affected brain region (Kim et al. 2014; Agnihotri and Aruoma 2020; Andhale et al. 2022; Galoppin et al. 2022; Manjari et al. 2022). This can manifest in numerous ways, reflecting the intricate relationship between brain structure and function (Bachiller et al. 2018). For instance, Alzheimer's disease (AD) primarily affects memory and cognition due to the progressive degeneration of cortical and hippocampal neurons (Agnihotri and Aruoma 2020). In contrast, Parkinson's disease (PD) predominantly impacts motor control, as it leads to the degeneration of dopaminergic neurons in the substantia nigra, a brain region essential for the regulation of movement which may further lead to posture instability (Amende et al. 2005; Agnihotri and Aruoma 2020). On the other hand, multiple sclerosis (MS) is characterized by the demyelination of nerve fibers which can affect various sensory, motor, and cognitive functions (Dörr and Paul 2015). Huntington's disease (HD) is another significant neurological disorder that is characterized by selective neurodegeneration of neurons in the basal ganglia which is crucial for coordinating movement and regulating various cognitive functions (Andhale et al. 2022; Manjari et al. 2023). Schizophrenia is another neurological disorder that is characterized by the disruption of thought processes, perception, emotional responsiveness, and social interactions (Chelini et al. 2018; Cui et al. 2021).

Each of these conditions disrupts the brain's normal operations in different ways, but they share common themes of neuronal dysfunction (Wang et al. 2023a, 2023b). The brain's ability to process information, control movement, and regulate emotions relies on the integrity of its neural networks and synaptic connections (Šimić et al. 2021; Appelbaum et al. 2023; Braine and Georges 2023). When these networks are compromised, the resulting impairments can be severe and multifaceted, affecting not only the individual's cognitive and physical capabilities but also their emotional and social well-being (Amende et al. 2005; Chelini et al. 2018; Agnihotri and Aruoma 2020; Andhale et al. 2022; Wang et al. 2023a).

Moreover, these functional consequences often interact with each other, creating a complex web of symptoms that can significantly impact an individual's daily life. For example, cognitive decline in AD can affect a person's ability to carry out daily activities, while motor symptoms in PD can lead to difficulties with mobility and self-care. The interplay between these various symptoms underscores the need for a comprehensive understanding of how neurological disorders impact the brain.

Despite the damage caused by neurological disorders, the brain has a remarkable ability to adapt through a process known as neuroplasticity. Neuroplasticity refers to the brain's capacity to reorganize itself by forming new neural connections in response to damage or altered conditions

DOI: 10.1201/9781003500025-4

(Abdulghani et al. 2023). This adaptive response can help compensate for lost functions and sometimes improve outcomes, even in the presence of ongoing disease.

However, neuroplasticity is not always beneficial. In some cases, the brain's attempts to reorganize can lead to maladaptive changes that exacerbate symptoms. For example, in chronic neurological conditions, the brain's compensatory mechanisms may result in abnormal neural activity or altered connectivity patterns that contribute to further cognitive or motor impairments.

Understanding the profound effects of neurological disorders on brain structure and function is crucial for advancing both our scientific knowledge and clinical practice. By examining how these conditions disrupt normal brain operations, we can gain insights into the mechanisms underlying their symptoms and develop more effective strategies for diagnosis and treatment. This chapter sets the stage for a detailed exploration of these impacts, offering a foundation for subsequent discussions on the pathophysiology of neurological disorders and potential therapeutic approaches.

4.2 STRUCTURAL CHANGES IN NEUROLOGICAL DISORDERS

4.2.1 Alzheimer's Disease

AD is the most prevalent form of dementia, characterized by a it progressive decline in cognitive abilities, including memory, reasoning, and behavior. It poses significant challenges to individuals, caregivers, and healthcare systems worldwide. AD is a progressive neurodegenerative disease caused by the accumulation of amyloid-β plaques (Aβ) in the cortical and hippocampal regions leading to the massive loss of neurons (Breijyeh and Karaman 2020). At present, there are around 416 million AD patients and this number may increase drastically in the coming years (Gustavsson et al. 2023). At present, there is no cure for AD, although the available treatments just improve or rectify the symptoms of AD.

In the initial stages of the disease, the connections among the neurons in the parts of the brain involved in memory, including the cortex and hippocampus get disrupted (DeTure and Dickson 2019; Rao et al. 2022). It later affects the areas in the cerebral cortex responsible for reasoning and social behavior (DeTure and Dickson 2019; Ávila-Villanueva et al. 2022). This might be caused due to the presence of aberrant structures, senile plaques, and neurofibrillary tangles (Ávila-Villanueva et al. 2022). As the disease progresses, Aβ plaques appear during the development of the disease by following a specific pattern (Chen et al. 2017; Hampel et al. 2021). This pattern starts at the neocortex and continues to the hippocampus, basal ganglia midbrain, and cerebellum (Hampel et al. 2021). The process of degeneration starts way before the diagnosis of the disease (Breijyeh and Karaman 2020). Later on, it can be found by the presence of Tau protein aggregates which lead to the propagation of neuronal death in different pathways (Tolkovsky and Spillantini 2021; Chen and Yu 2023). In this way, two different pathologies, i.e., the aggregation of Aβ plaques and Tau proteins with different structural changes (Bloom 2014; Sengupta and Kayed 2022). As the disease progresses, both of the pathologies overlap. In this way, accumulation of protein aggregates in the different regions of brain lead to the degeneration of neurons (Bloom 2014; Sengupta and Kayed 2022).

The diagnosis of disease can be carried out by noninvasive methods like magnetic resonance imaging (MRI). The AD brain often has at least moderate cortical atrophy that is most marked in multimodal association cortices and limbic lobe structures. There is increasing recognition of atrophy in posterior cortical areas in AD, most notably the precuneus and posterior cingulate gyrus, driven in part by functional imaging studies. This can be determined by the enlargement in the frontal and temporal horns of the lateral ventricles and decreased brain weight in the affected individuals. This analysis is further followed by the microscopic examination of multiple brain regions.

4.2.2 Parkinson's Disease

PD is a progressive neurodegenerative disorder that impairs movement and cognition, with frequent neuropsychiatric complications (DeMaagd and Philip 2015). The pathophysiology of the disease is the loss of dopaminergic neurons in the substatia nigra (SN), resulting in the dysfunction and structural modification of basal ganglia-thalamocortical circuits modulated by dopamine (Galvan et al. 2015; DeMaagd and Philip 2015; Sayyaed et al. 2023; Yang et al. 2024). Several studies have reported that the total volume of the subcortical nuclei mainly caudate and putamen were changed in patients with PD (Garg et al. 2015; He et al. 2020; Saeed et al. 2020; Charroud and Turella 2021; Jeong et al. 2022; Zhou et al. 2024). Acetylcholine (ACh), norepinephrine (NE), and serotonin may have a role in nonmotor features of PD (Buddhala et al. 2015; Prakash et al. 2016; Ye 2022). As the disease progresses by basal ganglia is the most effected brain area in PD (Caligiore et al. 2016; McGregor and Nelson 2019). Studies have determined that PD patients have reduced fractional anisotropy within the SN and increased mean and radial diffusivity within the SN and globus palladius (Nagae et al. 2016; Shim and Baek 2021; Bergsland et al. 2021; Welton et al. 2023). In PD patients, loss of attenuation and length of the dendritic spines of medium-sized spiny neurons located in the striatum has been reported (Prakash et al. 2016; Zhang et al. 2017).

Similarly, in PD patients, atrophy of the brain was observed in many cortical and subcortical areas, which contributes in a decrease in the volume of the brain (Prakash et al. 2016; Yang et al. 2024). Interestingly, it was reported that the volume of the frontal lobe, temporoparietal lobe junction, parietal lobe, insula, anterior cingulate cortex, basal ganglia, and thalamus increased in PD patients determining the loss of those neurons (Prakash et al. 2016; Filippi et al. 2020; Pieperhoff et al. 2022; Onofrj et al. 2023). This shows that there are neuroanatomical changes in the major brain structures related to cognition in PD.

4.2.3 Multiple Sclerosis

MS is an inflammatory demyelinating autoimmune disease of the central nervous system (CNS) (Galoppin et al. 2022). As per the 2020 census, 35.9/1,00,000 population are affected by MS (Walton et al. 2020). Myelin, the protective sheath that surrounds nerve fibers (axons), is crucial for the rapid transmission of electrical signals. In MS, the immune system erroneously attacks oligodendrocytes, the cells responsible for producing and maintaining myelin (Ghasemi et al. 2017; López-Muguruza and Matute 2023). This immune-mediated damage leads to the loss of myelin, resulting in the formation of lesions or plaques, which are the hallmark of MS (Wingerchuk et al. 2001; Popescu et al. 2013; Lassmann 2019). These lesions predominantly occur in the white matter regions of the brain, such as the periventricular areas, the corpus callosum, optic nerves, and the spinal cord (Kidd et al. 1999; Filippi et al. 2019; Wildner et al. 2020). The extent and location of these plaques can vary, correlating with the diverse clinical symptoms of MS (Lassmann 2018). Over time, the process of demyelination not only disrupts nerve signal transmission but also contributes to long-term neurological deficits (Bakirtzis et al. 2023; López-Muguruza and Matute 2023).

Although MS has traditionally been considered a white matter disease, gray matter is also significantly affected (Ontaneda et al. 2021). Gray matter demyelination occurs in the cerebral cortex, deep gray matter structures like the thalamus, and the spinal cord (Klaver et al. 2013). Cortical demyelination, particularly in the subpial regions, is common and can lead to cognitive dysfunction, such as memory loss and impaired executive function (Calabrese et al. 2009; DeLuca et al. 2014). Thalamic atrophy is another critical feature of MS, with studies showing that the extent of thalamic volume loss correlates strongly with both physical and cognitive disability (Azevedo et al. 2018; Sandroff et al. 2022). The mechanisms leading to gray matter atrophy include retrograde degeneration from white matter lesions, direct immune-mediated attacks, and the loss of synaptic connections (Lie et al. 2022). The involvement of gray matter is increasingly recognized as a major contributor to the clinical symptoms and disease progression in MS (Ontaneda et al. 2021).

The spinal cord is a frequent target of MS, with demyelination and atrophy occurring in various segments, particularly in the cervical and thoracic regions (Marrodan et al. 2020; Kreiter et al. 2024). Spinal cord lesions are a major contributor to the motor and sensory deficits seen in MS, including spasticity, weakness, and sensory loss (Marrodan et al. 2020). These lesions are often associated with more severe clinical symptoms, and their presence is used to confirm the diagnosis of MS (Wildner et al. 2020). Over time, chronic spinal cord damage leads to atrophy, further impairing the transmission of signals between the brain and the rest of the body (Alizadeh et al. 2019; Hu et al. 2023). The degree of spinal cord atrophy is closely linked to the severity of physical disability, making it an important marker of disease progression (Ruggieri et al. 2021; Hu et al. 2023).

Cortical demyelination is a significant but less well-known aspect of MS pathology (Popescu et al. 2013). Unlike white matter lesions, which are easily detected with conventional MRI, cortical lesions are more challenging to visualize (Kidd et al. 1999; Calabrese et al. 2009). These lesions, particularly those affecting the subpial regions of the cortex, are associated with cognitive deficits in MS patients (Calabrese et al. 2009; Magliozzi et al. 2018). Cortical demyelination disrupts neural networks that are essential for higher cognitive functions, such as attention, memory, and problem-solving (Damasceno 2012; Filley and Fields 2016; Margoni et al. 2023). This disruption can lead to cognitive impairment, which affects a substantial proportion of MS patients (DeLuca et al. 2014). The degree of cognitive decline is often related to the extent and distribution of cortical demyelination, highlighting the importance of cortical involvement in the overall disease process (DeLuca et al. 2014; Magliozzi et al. 2018).

The CNS has an inherent ability to repair damaged myelin through a process known as remyelination, primarily mediated by oligodendrocyte precursor cells (OPCs) (Keough and Yong 2013; López-Muguruza and Matute 2023). In the early stages of MS, remyelination can occur, leading to partial recovery of nerve function (Chari 2007). However, as the disease progresses, remyelination becomes less effective (Neumann et al. 2019). Factors such as chronic inflammation, the presence of inhibitory molecules, and the depletion of OPCs contribute to the failure of remyelination in chronic MS lesions (Gruchot et al. 2019; Maciak et al. 2023; López-Muguruza and Matute 2023). The inability to effectively remyelinate leads to the progressive accumulation of neurological deficits and contributes to the transition to progressive forms of the disease (Gruchot et al. 2019). Understanding the mechanisms that inhibit remyelination is crucial for developing therapies that can promote repair and halt the progression of MS.

The structural changes in MS are complex and multifaceted, involving both white and gray matter and affecting various regions of the CNS. These changes underlie the diverse clinical manifestations of the disease and contribute to its progressive nature. While demyelination and axonal damage are central features of MS, the involvement of gray matter and the failure of remyelination are increasingly recognized as critical factors in the disease's progression. Understanding these structural changes is essential for developing targeted therapies that can slow or halt the progression of MS and improve the quality of life for patients. Continued research into the pathophysiology of MS will provide further insights into the mechanisms driving these structural alterations and pave the way for more effective treatments.

4.2.4 Huntington's Disease

HD is a neurodegenerative disorder characterized by progressive structural changes in the brain, primarily affecting the basal ganglia, particularly the caudate nucleus and putamen, which are part of the striatum (Manjari et al. 2022; Andhale et al. 2022). These regions are crucial for motor control, cognitive functions, and emotional regulation, and their degeneration leads to the hallmark symptoms of HD, including involuntary movements (chorea), cognitive decline, and psychiatric disturbances (Molnar et al. 2022; Manjari et al. 2022; Jiang et al. 2023). The earliest and most severe structural change in HD is the significant atrophy of the striatum, where medium spiny neurons, the primary neuronal type in this region, undergo selective and progressive degeneration

(Morigaki and Goto 2017; Manjari et al. 2022). This neuronal loss is accompanied by a decrease in overall brain volume, which can be detected using imaging techniques such as MRI (Kinnunen et al. 2021).

As the disease progresses, the cortical regions of the brain also undergo atrophy, particularly in the frontal and temporal lobes (Montoya et al. 2006; Johnson et al. 2021). This cortical degeneration correlates with the cognitive decline and psychiatric symptoms observed in HD patients (Montoya et al. 2006; Johnson et al. 2021). Additionally, the white matter tracts connecting these cortical areas and the basal ganglia show signs of degeneration, disrupting the communication between different brain regions and further exacerbating the clinical symptoms (Blumenstock and Dudanova 2020; Tan et al. 2022).

At the cellular level, HD is marked by the accumulation of mutant huntingtin protein (mHTT) aggregates, which are toxic to neurons (Sari 2011; Manjari et al. 2022; Miguez et al. 2023). These protein aggregates are found not only in the striatum but also in other brain regions, contributing to widespread neuronal dysfunction (Caron et al. 2020; Barron et al. 2021; Miguez et al. 2023). The presence of these aggregates triggers a cascade of pathological events, including mitochondrial dysfunction, impaired energy metabolism, oxidative stress, and disruption of normal cellular processes, all of which contribute to the structural changes seen in HD (Manjari et al. 2022; Wang et al. 2023a; Chen et al. 2023; Manjari et al. 2023).

Furthermore, the ventricles, fluid-filled spaces in the brain, become enlarged (a condition known as ventriculomegaly) as the surrounding brain tissue atrophies, further indicating the extent of neurodegeneration (Todd et al. 2018). The degree of striatal atrophy and ventricular enlargement are often used as biomarkers to assess disease severity and progression in HD patients.

4.2.5 Schizophrenia

Schizophrenia is a complex psychiatric disorder characterized by a range of cognitive, emotional, and behavioral symptoms (Cui et al. 2021). One of the key aspects of schizophrenia is its impact on brain structure, which has been extensively studied using neuroimaging techniques (Dabiri et al. 2022; Howes et al. 2023). Structural changes observed in individuals with schizophrenia include alterations in brain volume, cortical thickness, and the integrity of white matter pathways (Karlsgodt et al. 2010; Dabiri et al. 2022; Adamu et al. 2023).

A prominent structural feature of schizophrenia is the reduction in overall brain volume (Karlsgodt et al. 2010). Studies using MRI have consistently shown that individuals with schizophrenia exhibit decreased gray matter volume in several key brain regions (Dabiri et al. 2022; Howes et al. 2023; Adamu et al. 2023). This reduction is particularly pronounced in the frontal and temporal lobes, which are critical for cognitive processes such as reasoning, planning, and auditory processing (Lesh et al. 2011; Friedman and Robbins 2022). The decrease in gray matter volume is often associated with enlargements of the lateral and third ventricles, indicating a loss of surrounding brain tissue (Horga et al. 2011; Vita et al. 2012; Adamu et al. 2023).

In addition to volume reductions, schizophrenia is associated with abnormalities in cortical thickness (Sun et al. 2023). Research has shown that individuals with schizophrenia often have thinner cortical regions in the prefrontal cortex and temporal lobes (Kaur et al. 2020; Howes et al. 2023). These areas are involved in higher-order cognitive functions, including executive control and sensory processing (Jung et al. 2017). Thinning of the cortex in these regions may contribute to the cognitive deficits observed in schizophrenia, such as impaired working memory and attention (Alkan et al. 2021; McCutcheon et al. 2023).

Structural changes in white matter are also evident in schizophrenia (Karlsgodt et al. 2010). White matter tracts, which facilitate communication between different brain regions, show reduced integrity in individuals with the disorder (Karlsgodt et al. 2010; Mamah et al. 2018). Diffusion tensor imaging (DTI) studies have revealed abnormalities in the organization and connectivity of major white matter pathways, such as the arcuate fasciculus, which connects the frontal and temporal lobes (Tae et al. 2018; Li et al. 2022a). Reduced white matter integrity is thought to reflect disruptions in neural connectivity, which may underlie some of the cognitive and functional impairments seen in schizophrenia (Karlsgodt et al. 2010; Liu et al. 2013).

The hippocampus and amygdala, structures involved in memory and emotion processing, also exhibit structural changes in schizophrenia (Karlsgodt et al. 2010). Studies have reported reduced hippocampal volume and alterations in the shape and function of the amygdala (Zheng et al. 2019; Perez-Rando et al. 2023). These changes are thought to contribute to the mnemonic and affective disturbances commonly experienced by individuals with schizophrenia.

Many of the structural changes observed in schizophrenia are believed to arise from neurodevelopmental abnormalities (Karlsgodt et al. 2010; Luvsannyam et al. 2022). Evidence suggests that these changes may occur early in life and progress over time, potentially due to a combination of genetic predispositions and environmental factors (Robinson and Bergen 2021; McCutcheon et al. 2023). Structural abnormalities may reflect disruptions in normal brain maturation processes, leading to the manifestation of schizophrenia later in life (Karlsgodt et al. 2010; McCutcheon et al. 2023).

These structural abnormalities provide valuable insights into the neurobiological underpinnings of the disorder and highlight the complex interplay between genetic, developmental, and environmental factors in its etiology. Understanding these structural changes is crucial for developing targeted interventions and improving our approach to the treatment and management of schizophrenia.

4.3 FUNCTIONAL CHANGES IN NEUROLOGICAL DISORDERS

4.3.1 Alzheimer's Disease

One of the earliest and most critical functional changes in AD is synaptic dysfunction, which precedes neuronal death. Amyloid β (Aβ) peptide accumulation, particularly soluble oligomers, interferes with synaptic function by disrupting neurotransmitter release and activation (Damasceno 2012; Bloom 2014; Hampel et al. 2021; Ávila-Villanueva et al. 2022; Chen and Yu 2023). This leads to impaired synaptic plasticity, which is essential for learning and memory. This is strongly correlated with cognitive decline and memory impairment.

AD can be characterized by neurochemical alterations that play a critical role in its pathophysiology (Teleanu et al. 2022). These neurochemical alterations contribute to cognitive decline, memory impairment, and other neuropsychiatric symptoms observed in AD patients (Rao et al. 2022; Ávila-Villanueva et al. 2022). In AD, there is a marked loss of cholinergic neurons in the forebrain which leads to the significant reduction of ACh levels in the cortex and hippocampus (Ferreira-Vieira et al. 2016; Chen et al. 2022). This leads to the disruption of cholinergic signaling, impairing synaptic plasticity and leading to deficits in attention, learning, and memory. In AD, dysregulation of glutamatergic signaling, particularly through overactivation of N-methyl-D-aspartate (NMDA) receptors leads to excitotoxicity (Wang and Reddy 2017). This process involves excessive calcium influx into neurons, triggering a cascade of events that result in oxidative stress, mitochondrial dysfunction, and ultimately neuronal death (Calvo-Rodriguez et al. 2020; Verma et al. 2022; Clemente-Suárez et al. 2023). This excitotoxicity contributes to synaptic loss and neurodegeneration, exacerbating the cognitive decline and memory impairment in AD patients (Wang and Reddy 2017; Verma et al. 2022). The serotonergic system, which influences mood, behavior, and cognitive function, is also affected in AD as serotonin (5-HT) modulates synaptic plasticity and has neuroprotective effects (Rodríguez et al. 2012; Švob Štrac et al. 2016). The serotonergic deficits in AD are associated with neuropsychiatric symptoms such as depression, anxiety, aggression, and sleep disturbances (Heilman and Nadeau 2022; Pierson et al. 2024). Along with this, dopaminergic system gets implicated in AD in the most pronounced brain regions like the prefrontal cortex and striatum (Sala et al. 2021; Chaudhary et al. 2022). Gamma-aminobutyric acid (GABA) is the main the main inhibitory neurotransmitter in the brain which has a key role in maintaining the balance of excitatory and inhibitory signals. AD is associate with alterations in GABAergic neurotransmission, including changes in GABA receptor expression and function (Gravielle 2021). These alterations can disrupt the excitatory/inhibitory balance, contributing to neuronal hyperexcitability and cognitive impairment (Scaduto et al. 2023). Also, in AD, there is significant degeneration of the locus coeruleus, leading to reduced levels of NE which is involved in attention, arousal, stress responses, and neuroprotective properties (Gutiérrez et al. 2022; Evans et al. 2022). This degeneration is one of the earliest changes in AD and may contribute to the disease's progression (DeTure and Dickson 2019). This implicated that NE deficits in AD also play a major role in attention deficits and cognitive decline (Evans et al. 2022). This implies that neurochemical changes in AD are central to its pathophysiology, driving the cognitive decline and neuropsychiatric symptoms.

The main neuropathology involved in AD is the formation of Amyloid-β (Aβ) peptides. Their accumulation in the form of plaques is the major hallmark of the disease (Sengupta and Kayed 2022). Aβ disrupts various neurotransmitter systems, including cholinergic, glutamatergic, and GABAergic pathways, leading to widespread synaptic and neuronal dysfunction (Sharma et al. 2021; Meftah and Gan 2023). Tau protein also stabilizes microtubules in neurons. In AD, tau becomes hyperphosphorylated and forms neurofibrillary tangles, which disrupt cellular function. Abnormal tau can interfere with axonal transport, affecting neurotransmitter release and synaptic functioning (Alonso et al. 2018; Rawat et al. 2022).

Among the various pathological mechanisms implicated in AD, oxidative stress plays a significant role in the disease's progression (Agnihotri and Aruoma 2020). Oxidative damage results from an imbalance between the production of reactive oxygen species (ROS) and the brain's ability to detoxify these reactive intermediates (Lin et al. 2020). Mitochondria are the primary source of ROS in cells (Calvo-Rodriguez et al. 2020). They generate ROS as by-products of the electron transport chain (ETC) during ATP production. In AD, mitochondrial dysfunction leads to an increased production of ROS. This occurs due to an increased production of ROS. This occurs due to impaired electron transport, decreased efficiency of oxidative phosphorylation, and mitochondrial DNA mutations. In addition to this, Aβ aggregates on the mitochondrial membrane impair complex IV of the ETC, leading to increased ROS production (Chen et al. 2023). Along with this, Aβ aggregates interact with redox-active metals such as iron and copper, catalyzing the formation of free radicles, and leading to neuronal damage, synaptic dysfunction, and AD propagation (Olufunmilayo et al. 2023; Wang et al. 2023a; Liu et al. 2023). Due to the generation of Aβ plaques and Tau tangles, brain resident immune cells and microglial cells get activated and produce ROS and reactive nitrogen species, leading to chronic neuroinflammation (Zhang et al. 2023a, 2023b; Chen and Yu 2023). In addition to ROS production, the Aβ aggregates lead to lipid peroxidation, protein oxidation, and DNA and RNA damage due to excess ROS production (Olufunmilayo et al. 2023; Sharma and Kim 2023; Houldsworth 2024).

Mitochondrial dysfunction is increasingly recognized as a central feature in the pathogenesis of AD (Calvo-Rodriguez et al. 2020). In AD, mitochondrial functions are compromised, leading to a cascade of cellular disturbances contributing to the disease's progression (Agnihotri and Aruoma 2020). As mitochondrial dysfunction and oxidative stress are interrelated, additionally, calcium dysregulation also takes place due to the enhanced intracellular calcium levels triggering the opening of the mitochondrial permeability transition pore (mPTP), leading to the release of pro-apoptotic factors (Matuz-Mares et al. 2022; Olufunmilayo et al. 2023; De Nicolo et al. 2023; Zong et al. 2024). Along with these agents, such as peroxisome proliferator-activated receptor gamma coactivator 1-alpha (PGC-1α) activators have been investigated for their potential to enhance mitochondrial biogenesis and improve cellular energy metabolism, which gets modulated in AD (Wareski et al. 2009; Zhang et al. 2019).

4.3.2 Parkinson's Disease

PD is a progressive neurodegenerative disorder primarily characterized by motor symptoms such as tremors, bradykinesia, rigidity, and postural instability (Filippi et al. 2020). However, PD also involves significant non-motor symptoms, including cognitive decline, mood disorders, and autonomic dysfunction (McGregor and Nelson 2019; Filippi et al. 2020). These symptoms arise due to the dysfunction of neurotransmitter systems and neurotrophic factors, particularly in the brain's dopaminergic pathways. In PD, there is a significant loss of dopaminergic neurons in the Substantia nigra pars compacta (SNpC), leading to a marked reduction in striatal dopamine levels (Maiti et al. 2017; Ramesh and Arachchige 2023). This depletion disrupts the balance between excitatory and inhibitory signals within the basal ganglia circuitry, resulting in the characteristic motor symptoms of PD as shown in Figure 4.1 (McGregor and Nelson 2019; Ramesh and Arachchige 2023). Dopamine dysfunction not only causes motor impairments but also contributes to non-motor symptoms such as depression, apathy, and cognitive dysfunction (Ramesh and Arachchige 2023). Similarly, Serotonin dysfunction in PD leads to the degeneration of serotonin-producing neurons and the formation of Lewy bodies in the raphe nuclei (Vegas-Suarez et al. 2019; Miquel-Rio et al. 2023). This leads to a decrease in serotonin levels in various brain regions, including the cortex and limbic system, thereby contributing to non-motor symptoms, such as depression, anxiety, and sleep disturbances (Švob Štrac et al. 2016). As the disease progresses, the locus coeruleus, which is a primary source of NE, undergoes significant degeneration. This is linked to autonomic dysfunction such as orthostatic hypotension and mood disorders (Weinshenker 2018; Zhou et al. 2021a). In PD, there is a disruption in cholinergic transmission, partly due to the loss of cholinergic neurons and the imbalance between dopaminergic and cholinergic systems (Cai et al. 2021; Pasquini et al. 2021; Björklund and Barker 2024). The degeneration of the pedunculopontine nucleus, which plays a role in gait and posture control, also affects cholinergic signaling (French and Muthusamy 2018).

Along with this, neurotrophic dysfunction also takes place in PD. Reduced levels of brain-derived neurotrophic factor (BDNF) which plays a key role in survival, growth, and differentiation have been observed in the brains of individuals with PD (Bathina and Das 2015). This reduction is

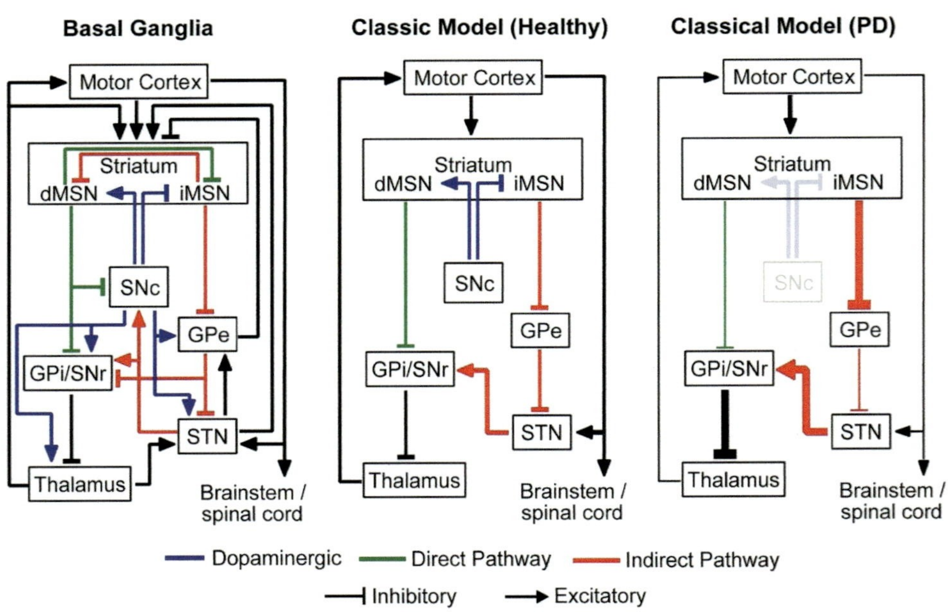

FIGURE 4.1 The classical model (centre) simplifies basal ganglia connectivity (left) by highlighting the role of dopamine on direct and indirect pathway activity and motor output. In the healthy condition (centre), dopamine (blue) from the SNc to the striatum activates the direct pathway (green) and inhibits the indirect pathway (red) MSNs. This effect decreases GPi output, releasing inhibition on the thalamus and cortex and promoting movement. In the Parkinsonian condition (right), loss of SNc dopamine causes hypoactivity of the direct pathway and hyperactivity of the indirect pathway that leads to excessive GPi output. As a result, over-inhibition of the thalamus and cortex leads to a suppression of movement. Adapted from McGregor and Nelson (2019), originally based on DeLong (1990). Licence number: 6262431303598.

particularly evident in the substantia nigra and striatum, regions that are critically affected by PD. Glial cell-derived neurotrophic factor (GDNF) is another neurotrophin that supports the survival and function of dopaminergic neurons. In PD, the expression of GDNF is diminished, leading to a lack of neurotrophic support for the surviving dopaminergic neurons (Allen et al. 2013; Duarte Azevedo et al. 2020). The reduced availability of GDNF exacerbates neuronal loss in the substantia nigra and contributes to the progression of motor symptoms (Allen et al. 2013; Duarte Azevedo et al. 2020). Along with this, alterations in nerve growth factor (NGF) signaling which is involved in survival and maintenance of neurons have been observed in PD which in turn contributes with cholinergic dysfunction and cognitive decline (Allen et al. 2013).

A hallmark of PD pathology is the formation of protein aggregates, particularly involving the protein α-synuclein (α-Syn), which aggregates into Lewy bodies (Calabresi et al. 2023). The accumulation of these protein aggregates is closely linked to mitochondrial dysfunction and oxidative damage, which play central roles in the progression of the disease (Han et al. 2020; Calabresi et al. 2023). α-Syn is a small, soluble protein predominantly expressed in the brain that is involved in synaptic vesicle trafficking and neurotransmitter release (Calabresi et al. 2023). In PD, α-Syn undergoes misfolding and aggregation into oligomers, protofibrils, and eventually fibrils that form Lewy bodies (Han et al. 2020; Calabresi et al. 2023). These Lewy bodies are intracellular inclusions found in the neurons of PD patients composed of aggregates of α-synuclein, along with other proteins such as ubiquitin, parkin, and neurofilaments (Giráldez-Pérez et al. 2014; Srinivasan et al. 2021). Recent evidence suggests that α-Syn aggregates can act as seeds, promoting the misfolding and aggregation of native α-Syn in a prion-like manner (Jan et al. 2021; Calabresi et al. 2023; Negi et al. 2024). This seeding can spread from neuron to neuron, contributing to the progressive nature of PD. Aggregated forms of α-Syn can disrupt mitochondrial membranes, leading to impaired mitochondrial dynamics, including fission, fusion, and biogenesis. This disruption contributes to mitochondrial fragmentation and loss of function (Rocha et al. 2018; Lin et al. 2019; Thorne and Tumbarello 2022). Excessive fission and reduced fusion lead to mitochondrial fragmentation, which is a common feature in PD (Gao et al. 2022). α-Syn aggregates have been shown to inhibit the activity of mitochondrial complex I, a critical component of the ETC, which leads to a reduction in ATP production and an increase in electron leakage, leading to the elevated ROS (Rocha et al. 2018; Lin et al. 2019; Gao et al. 2022). In addition to this, α-Syn aggregates can localize to mitochondria and interact with mtDNA, causing mutations and deletions, which further leads to a decline in mitochondrial function, including reduced ATP production and increased ROS generation (Lin et al. 2019; Thorne and Tumbarello 2022). This suggests that α-Syn aggregates lead to protein disruption, thereby leading to oxidative damage in PD.

Also, α-Syn is associated with oxidative stress. In PD, the auto-oxidation of dopamine generates ROS, which can further drive the aggregation of α-Syn (Thorne and Tumbarello 2022). Iron can also bind to α-Syn, promoting its aggregation and enhancing oxidative damage. Activated microglia in PD produce pro-inflammatory cytokines and ROS, contributing to a toxic environment that promotes protein aggregation and neuronal death (Ma et al. 2021; Yi et al. 2022; Gao et al. 2023). Generation of ROS leads to lipid peroxidation, protein oxidation, and DNA damage contributing to the loss of dopaminergic neurons in PD (Dias et al. 2013; Weng et al. 2018). In PD, there is often a reduction in the levels of key antioxidants, such as glutathione. This decrease leaves neurons more vulnerable to oxidative stress, accelerating the progression of the disease (Olufunmilayo et al. 2023; Houldsworth 2024). This infers that the formation of protein aggregates, particularly involving α-Syn, plays a critical role in the pathogenesis of PD by driving mitochondrial dysfunction and oxidative damage.

4.3.3 Multiple Sclerosis

MS induces a wide array of functional changes across the central nervous system (CNS), driven by the complex interplay of demyelination, axonal damage, and neuroinflammation (Ghasemi et al. 2017; Papiri et al. 2023). These changes manifest in diverse neurological dysfunctions that impact motor, sensory, cognitive, and autonomic functions (Margoni et al. 2023). One of the primary functional consequences of MS is impaired signal conduction in affected nerves due to the loss of myelin, which leads to a variety of motor symptoms such as muscle weakness, spasticity, and coordination difficulties (Papiri et al. 2023; Margoni et al. 2023). These motor impairments are often assessed through clinical scales like the Expanded Disability Status Scale (EDSS), and their severity is linked to the extent of white matter lesion load, which can be visualized using MRI (Gajofatto et al. 2013; Reilmann et al. 2013; Kenyon et al. 2023).

Cognitive dysfunction is another significant aspect of MS, with up to 70% of patients experiencing deficits in memory, attention, and executive functions (Oreja-Guevara et al. 2019). These cognitive changes correlate with the presence of cortical and subcortical gray matter atrophy, particularly in the thalamus and hippocampus. Biomarkers such as neurofilament light chain (NfL) in cerebrospinal fluid (CSF) and blood have been identified as indicators of axonal damage and are associated with both cognitive decline and overall disease progression (Schultz et al. 2020; Zhang et al. 2022a). High levels of NfL are linked to active neurodegeneration and can predict future disability.

Additionally, the disruption of synaptic plasticity due to demyelination and neuronal loss in MS contributes to cognitive and motor deficits (Bellingacci et al. 2021). This is evidenced by decreased levels of BDNF, a key molecule involved in neuroplasticity, which is found to be reduced in MS patients and is associated with poorer cognitive outcomes (Bathina and Das 2015; Miranda et al. 2019).

Autonomic dysfunctions are also prevalent in MS, with bladder, bowel, and sexual dysfunctions being common (Pintér et al. 2015). These issues are often related to lesions in the spinal cord and brainstem, regions critical for autonomic control. Biomarkers such as increased levels of oxidative stress markers (e.g., malondialdehyde) and inflammatory cytokines (e.g., interleukin-6 [IL-6], tumor necrosis factor-alpha [TNF-α]) in CSF and blood have been associated with the severity of autonomic symptoms, suggesting that ongoing inflammation and oxidative damage play a significant role in these dysfunctions (Zhang et al. 2020, 2023b; Qu et al. 2023; Burgetova et al. 2023).

Moreover, MS patients frequently suffer from chronic fatigue, a multifaceted symptom that may arise from mitochondrial dysfunction and the resultant energy deficits within affected neurons (Chen et al. 2023). Elevated levels of lactate in the CSF, indicating impaired oxidative phosphorylation, have been observed in MS patients and correlate with the severity of fatigue (Albanese et al. 2016).

This infers that the functional changes in MS are multifaceted and deeply interwoven with the underlying pathophysiological processes of the disease. The identification of specific biomarkers such as NfL, BDNF, oxidative stress markers, and inflammatory cytokines offers insights into the mechanisms driving these dysfunctions and provides potential targets for monitoring disease progression and therapeutic intervention. Understanding these functional changes and their biomarkers is crucial for developing more effective treatments and improving the quality of life for MS patients.

4.3.4 Huntington's Disease

HD induces profound functional changes in the brain, driven by a complex interplay of neurotrophin dysfunction, neurotransmitter imbalances, neuroinflammation, oxidative stress, and mitochondrial dysfunction (Maity et al. 2022; Manjari et al. 2022, 2023). These factors collectively contribute to the disease's hallmark symptoms, including motor impairments, cognitive decline, and psychiatric disturbances.

Neurotrophins, such as BDNF and NGF are essential for neuronal survival, growth, and synaptic plasticity (Zuccato and Cattaneo 2007; Zhou et al. 2021b; Manjari et al. 2022; Speidell et al. 2023). The loss of BDNF in HD is largely attributed to the toxic effects of mHTT aggregates, which interfere with intracellular signaling pathways essential for BDNF expression (Zuccato et al. 2001; Zuccato and Cattaneo 2007). These aggregates cause cellular stress and damage, leading to decreased BDNF production and impaired delivery to neurons. The reduction in BDNF availability exacerbates the vulnerability of medium spiny neurons in the striatum, a region critical for motor control and cognitive function (Blumenstock and Dudanova 2020; Li et al. 2022b; Speidell et al. 2023). The disruption of neurotrophic support not only accelerates neuronal degeneration but also impairs cognitive and motor functions, highlighting the importance of neurotrophins in maintaining brain health and function (Bathina and Das 2015; Speidell et al. 2023).

As, HD is characterized by selective neurodegeneration of medium spiny neurons in the striatum, a key component of the basal ganglia, which leads to profound alterations in neurotransmitter systems (Garret et al. 2018; Blumenstock and Dudanova 2020). One of the most notable changes is the dysregulation of dopamine, GABA, and glutamate, which are critical for maintaining motor control and cognitive function (Teleanu et al. 2022). The decrease in GABA levels disrupts the delicate balance between excitatory and inhibitory signals in the brain, leading to hyperexcitability and impaired motor control (Hsu et al. 2018; Teleanu et al. 2022). This imbalance manifests as the characteristic motor symptoms of HD, including chorea (involuntary, irregular movements) and dystonia (sustained muscle contractions). The loss of GABAergic neurons in the striatum leads to a compensatory increase in dopaminergic activity, as dopamine-producing neurons in the substantia nigra attempt to counterbalance the reduced inhibitory input (Teleanu et al. 2022). This hyperdopaminergic state contributes to the motor disturbances seen in HD, including dyskinesias and altered motor coordination. Furthermore, the glutamatergic system, which involves excitatory neurotransmission, is also disrupted in HD (Hsu et al. 2018; Andersen et al. 2021). The accumulation of mHTT affects glutamate receptors and transporters, leading to excitotoxicity. These imbalances contribute to the diverse range of symptoms observed in HD, from motor impairments and cognitive decline to psychiatric disturbances (Jimenez-Sanchez et al. 2017; Verma et al. 2022; Tong et al. 2024). Understanding these dysfunctions is crucial for developing targeted treatments aimed at restoring neurotransmitter balance and improving patient outcomes (Teleanu et al. 2022; Appelbaum et al. 2023; Clemente-Suárez et al. 2023).

The cholinergic system, which involves the neurotransmitter ACh, is crucial for various cognitive functions, including memory, attention, and learning. In HD, there is a marked degeneration of cholinergic neurons, particularly those located in the basal forebrain, which results in increased activity of acetylcholinesterase in critical brain regions such as the cortex and striatum (Manjari et al. 2023). This cholinergic deficit is closely linked to the cognitive decline observed in HD patients. The loss of cholinergic input impairs synaptic transmission and plasticity, leading to difficulties in learning and memory formation.

Neuroinflammation is a central feature of HD and significantly contributes to the progression and severity of the disorder (Hsiao et al. 2013; Manjari et al. 2023). This inflammatory response is characterized by the activation of glial cells, including microglia and astrocytes, which play a critical role in maintaining brain homeostasis and responding to neuronal damage. In HD, the accumulation of mHTT triggers chronic neuroinflammation, which exacerbates neuronal death (Sari 2011). In HD, microglia which are resident immune cells of the CNS become activated in response to the presence of mHTT aggregates and neuronal cell death (Yang et al. 2017; Gao et al. 2023). Activated microglia release a range of pro-inflammatory cytokines, such as TNF-α, IL-1β, and IL-6 (Qin et al. 2023; Manjari et al. 2023). These cytokines contribute to further neuronal injury by promoting oxidative stress, disrupting neuronal function, and amplifying local inflammation. In HD, astrocytes also become reactive and undergo functional changes in response to mHTT-induced stress. This reactivity includes altered expression of various inflammatory mediators and a reduction in neuroprotective factors (Franklin et al. 2021; Gao et al. 2023). Neuroinflammation in HD is closely linked to oxidative stress leading to the production of ROS and oxidative stress markers. This oxidative stress further damages neuronal cells and amplifies the inflammatory response, creating a vicious cycle of inflammation and neuronal injury. The accumulation of ROS can damage cellular lipids, proteins, and DNA, contributing to neuronal dysfunction and death.

In HD, the accumulation of mHTT in neurons leads to increased production of ROS (Tong et al. 2024). mHTT aggregates disrupt normal cellular functions and promote the generation of free radicals (Olufunmilayo et al. 2023). ROS, including superoxide anions, hydrogen peroxide, and hydroxyl radicals, can damage a wide range of cellular components, including lipids, proteins, and nucleic acids impairs neuronal function and contributes to cell death. Under normal conditions, the brain utilizes various antioxidant defense mechanisms to neutralize ROS and mitigate oxidative damage. These mechanisms include antioxidant enzymes such as superoxide dismutase (SOD), catalase, and glutathione peroxidase. In HD, the capacity of these antioxidant defenses is overwhelmed by the excessive production of ROS (Manjari et al. 2022, 2023). The efficiency of these antioxidant systems is compromised, leading to a decrease in their protective effects and an increase in oxidative damage. The oxidative damage induced by ROS affects various cellular processes, including lipid peroxidation, protein oxidation, and DNA damage. This oxidative damage leads to significant neuronal injury and contributes to the motor, cognitive, and psychiatric symptoms of the disease.

In HD, mitochondrial dysfunction contributes to neuronal damage through a variety of mechanisms, including impaired energy production, increased oxidative stress, and disrupted mitochondrial dynamics (Maity et al. 2022; Clemente-Suárez et al. 2023). In HD, the accumulation of mHTT disrupts mitochondrial function, leading to a significant reduction in ATP production. This energy deficit impairs various cellular processes, including neurotransmitter synthesis, ion homeostasis, and protein degradation, ultimately compromising neuronal function and survival (Olufunmilayo et al. 2023; Clemente-Suárez et al. 2023). Disrupted mitochondrial dynamics, which include alterations in mitochondrial morphology and function, are another hallmark of HD. These disruptions include changes in the levels of mitochondrial fusion proteins (e.g., OPA1, MFN1, MFN2) and fission proteins (e.g., DRP1) (Reddy 2014; Jurcau and Jurcau 2022; Dai et al. 2023). Altered levels of mitochondrial trafficking proteins, such as Miro 1 and Milton impair the mitochondrial movement which is a significant feature of mitochondrial dysfunction in HD (Glater et al. 2006; Jurcau and Jurcau 2022; Aspenström 2024).

This infers that the functional changes in HD are driven by a complex interaction of neurotrophin dysfunction, neurotransmitter imbalances, neuroinflammation, oxidative stress, and mitochondrial dysfunction. These pathological processes collectively disrupt normal neuronal function, leading to the characteristic motor, cognitive, and psychiatric symptoms of HD. Understanding these changes is crucial for developing targeted therapies aimed at slowing disease progression and improving patient outcomes.

4.3.5 Schizophrenia

Schizophrenia is a severe and chronic mental disorder characterized by a complex interplay of genetic, neurobiological, and environmental factors. Receptor dysfunction is a critical feature of schizophrenia, significantly influencing the disorder's clinical manifestations (Chelini et al. 2018; Cui et al. 2021). Dopaminergic receptor dysregulation, particularly involving increased D2 receptor binding in the mesolimbic system, is associated with positive symptoms like hallucinations and delusions, as evidenced by elevated dopamine receptor availability in positron emission tomography (PET) imaging studies (Yang and Tsai 2017; Malik et al. 2023; McCutcheon et al. 2023). This heightened receptor activity correlates with the severity of psychotic symptoms. In contrast, reduced D1 receptor activity in the prefrontal cortex is linked to cognitive impairments, with PET studies showing diminished D1 receptor availability in this brain area (Malik et al. 2023). Additionally, glutamate receptor dysfunction, particularly at NMDA receptors, contributes to cognitive deficits and negative symptoms with magnetic resonance spectroscopy (MRS) revealing decreased glutamate levels in affected brain regions (Kruse and Bustillo 2022; McCutcheon et al. 2023). Serotonin receptor alterations, especially in the 5-HT2A subtype, impact mood and cognition, with PET imaging and CSF analyses showing altered serotonin receptor binding (Sumiyoshi et al. 2014).

Along with this, the cholinergic system gets disrupted in schizophrenia, which is characterized by decreased levels of ACh and alteration of cholinergic receptors (Björklund and Barker 2024). Studies have identified binding to muscarinic and nicotinic ACh receptors, indicating a decline in cholinergic neurotransmission. The muscarinic M1 receptor, which is implicated in cognitive processes, shows reduced availability and binding in schizophrenia patients (Bakker et al. 2018; McCutcheon et al. 2023). Additionally, nicotinic receptors, which are important for attention and cognitive function, also demonstrate decreased binding and altered function. These cholinergic impairments contribute to cognitive deficits and attention disturbances observed in schizophrenia, underscoring the importance of the cholinergic system in understanding and potentially targeting therapeutic strategies for the disorder (Sarter et al. 2012; Eickhoff et al. 2022).

In schizophrenia, neurotrophins such as BDNF and NGF play critical roles in neuronal health, synaptic plasticity, and cognitive function (Autry and Monteggia 2012; Nieto et al. 2013; Miranda et al. 2019; Turkmen et al. 2021; Zhang et al. 2022a). Alterations in these neurotrophins have been implicated in the disorder's pathology. BDNF, which supports neuronal survival and synaptic plasticity, is notably reduced in individuals with schizophrenia. Reduced BDNF is associated with cognitive deficits and structural brain abnormalities, such as decreased hippocampal and prefrontal cortex volumes (Miranda et al. 2019). NGF, another neurotrophin involved in neurogenesis and neuronal differentiation, also shows altered levels in schizophrenia, although its role is less studied than BDNF (Turkmen et al. 2021). The reduced levels of these neurotrophins highlight disruptions in neuroplasticity and neuronal support, which contribute to the cognitive impairments and brain structural changes observed in schizophrenia.

Oxidative stress is a key factor in the pathophysiology of schizophrenia, characterized by an imbalance between ROS production and the brain's antioxidant defenses (Madireddy and Madireddy 2020; Ermakov et al. 2021; Murray et al. 2021). This imbalance leads to significant oxidative damage to cellular macromolecules, contributing to neuronal dysfunction and disease progression. Biomarkers of oxidative stress in schizophrenia include elevated levels of malondialdehyde (MDA) and 4-hydroxynonenal (4-HNE), which are lipid peroxidation products. These biomarkers are found in increased concentrations in the blood and CSF of patients with schizophrenia, indicating substantial oxidative damage to lipids (Romano et al. 2017; Guidara et al. 2020). Additionally, oxidative damage to proteins, such as increased carbonylation, and DNA, with markers like 8-hydroxydeoxyguanosine (8-OHdG), are also elevated (Goh et al. 2021). Furthermore, a reduction in key antioxidant defenses, including decreased glutathione levels and altered activity of antioxidant enzymes such as SOD and catalase, has been documented (Buosi et al. 2021; Djordjević et al. 2022). These findings underscore the role of oxidative stress in the neuronal damage observed in schizophrenia.

Mitochondrial damage is increasingly recognized as another significant aspect of schizophrenia, impacting cellular energy production, metabolism, and overall neuronal function (Rajasekaran et al. 2015; Roberts 2021). In schizophrenia, mitochondrial dysfunction manifests as reduced mitochondrial ATP production, impaired mitochondrial DNA (mtDNA) integrity, and altered mitochondrial dynamics (Roberts 2021; Bulduk et al. 2024). This is estimated by decreased mtDNA copy number and increased levels of mtDNA mutations, which are detectable in blood and postmortem brain tissue. Additionally, the presence of elevated mitochondrial ROS reflects oxidative stress within mitochondria, further contributing to cellular damage (Madireddy and Madireddy 2020; Murray et al. 2021). Studies also show reduced activity of key mitochondrial enzymes, such as cytochrome c oxidase (complex IV), which is essential for the ETC and ATP synthesis (Ermakov et al. 2021; Roberts 2021). Increased levels of mitochondrial-derived damage-associated molecular patterns (DAMPs) in CSF indicate heightened mitochondrial stress and dysfunction (Morris and Berk 2015; Wilkins et al. 2017; Lin et al. 2022). These biomarkers highlight the role of mitochondrial damage in the pathophysiology of schizophrenia, providing insights into potential therapeutic strategies aimed at restoring mitochondrial function and improving neuronal health.

Lastly, neuroinflammation is a significant factor in the pathophysiology of schizophrenia, contributing to the disorder's symptomatology and progression (Vallée 2022; Chaves et al. 2024). This inflammatory response is characterized by elevated levels of pro-inflammatory cytokines, chemokines, and acute-phase proteins, which are indicative of ongoing neuroimmune activation. There is increased concentrations of C-reactive protein (CRP) and IL-6 in blood and CSF highlight systemic and central inflammation (Gallego et al. 2018; Campana et al. 2022; Warren et al. 2024). Elevated levels of TNF-α and IL-1β have also been reported, reflecting the involvement of key inflammatory pathways (Zhu et al. 2018, 2020; Dawidowski et al. 2021). These neuroinflammatory changes are linked to neuronal damage, synaptic dysfunction, and alterations in brain structure and function, underscoring the importance of inflammation in schizophrenia's pathophysiology.

4.4 DIAGNOSTIC AND RESEARCH METHODS

Diagnosing and researching neurological diseases such as AD, PD, MS, HD, and schizophrenia require a multifaceted approach that integrates clinical assessments, neuroimaging techniques, biomarker analysis, and genetic testing. Clinical assessments are foundational, with cognitive tests like the Mini-Mental State Examination (MMSE) commonly used in Alzheimer's to evaluate memory and cognitive function, while the Unified Parkinson's Disease Rating Scale (UPDRS) is critical for assessing motor symptoms in PD (Ivey et al. 2012; Arevalo-Rodriguez et al. 2015).

4.4.1 Imaging Techniques

Imaging techniques play a critical role in the diagnosis and research of neurological diseases such as AD, PD, MS, HD, and schizophrenia. MRI is one of the most widely used methods, offering detailed images of brain structure. In AD, MRI can detect early hippocampal atrophy, a hallmark of the disease, and monitor cortical thinning and white matter changes over time (Johnson et al. 2012; Chandra et al. 2019). Functional MRI (fMRI) extends this utility by capturing real-time brain activity, which is particularly useful in understanding the cognitive deficits associated with schizophrenia and mapping brain function alterations in response to tasks or stimuli (Gur and Gur 2010; Dabiri et al. 2022).

PET and single photon emission computed tomography (SPECT) are pivotal in assessing the molecular and metabolic aspects of these diseases. In Alzheimer's, PET scans are employed to visualize amyloid plaques and tau tangles, key pathological features, using radiotracers like Florbetapir and Flortaucipir (James et al. 2021; Matsuda et al. 2022). For PD, PET and SPECT imaging of dopamine transporters using tracers such as Fluorodopa (FDOPA) or Ioflupane (DaTSCAN) is crucial for assessing the integrity of dopaminergic neurons in the basal ganglia, correlating with motor symptom severity (Cummings et al. 2011; Akdemir et al. 2021; O'Shea et al. 2024). DTI, a specialized form of MRI, is used in MS to detect microstructural changes in white matter tracts, revealing the extent of demyelination and axonal damage that characterizes the disease (Bao et al. 2022; Lopez-Soley et al. 2023).

In HD, MRI and PET scans are used to observe atrophy in the caudate nucleus and other basal ganglia structures, which correlates with motor and cognitive decline (Niccolini and Politis 2014; Negi et al. 2014; Pagano et al. 2016).

MRS is another valuable technique, allowing for the non-invasive measurement of biochemical changes in the brain, such as reductions in N-acetylaspartate (NAA), which indicate neuronal loss or dysfunction in several of these diseases (Hnilicová et al. 2020; Gudmundson et al. 2023). Fluorodeoxyglucose PET (FDG-PET) is particularly useful in Alzheimer's for measuring glucose metabolism, often showing hypometabolism in the parietal and temporal lobes, which precedes clinical symptoms (Marcus et al. 2014; Ou et al. 2019).

Together, these imaging techniques provide a comprehensive toolkit for diagnosing neurological diseases, tracking disease progression, and guiding treatment decisions by visualizing both the structural and functional changes occurring within the brain.

4.4.2 Biomarkers

The diagnosis of neurological diseases such as AD's, PD's, MS, HD, and schizophrenia increasingly relies on specific biomarkers and advanced diagnostic techniques.

4.4.2.1 Alzheimer's Disease

Diagnostic biomarkers for AD have significantly advanced in recent years, offering more accurate and earlier detection of the disease. The primary biomarkers are typically categorized into those that reflect the core pathological features of AD like amyloid-beta plaques, tau tangles, and neurodegeneration (Bloom 2014; Chen and Yu 2023). Amyloid-beta (Aβ) biomarkers are crucial for early diagnosis. These include decreased levels of Aβ42 and the Aβ42/Aβ40 ratio in CSF, as well as amyloid PET imaging, which can visualize amyloid plaque deposition in the brain (Constantinides et al. 2023; Andersson et al. 2023). The presence of amyloid plaques is considered a hallmark of Alzheimer's and can be detected years before symptoms appear. Tau biomarkers are another critical component. Hyperphosphorylated tau protein (p-tau) and total tau (t-tau) levels in CSF are elevated in AD patients, reflecting the formation of neurofibrillary tangles (Bloom 2014; Alonso et al. 2018; Rocha et al. 2018). Recently, specific forms of p-tau, such as p-tau181, p-tau217, and p-tau231, have gained attention for their ability to predict AD with high accuracy (Leuzy et al. 2021; Ashton et al. 2022). Neurodegeneration markers include NfL, which can be measured in both CSF and blood. Elevated NfL levels correlate with neuronal damage and disease progression, making it a valuable marker for tracking the course of AD (Mielke et al. 2019; Hu et al. 2024). Plasma levels of Aβ42/Aβ40, p-tau181, and NfL are being integrated into diagnostic frameworks and have shown promising results in identifying AD at various stages (Leuzy et al. 2021; Ashton et al. 2022). These biomarkers are increasingly being used in combination to improve diagnostic accuracy, allowing for earlier detection, better differentiation from other types of dementia, and more precise monitoring of disease progression.

4.4.2.2 Parkinson's Disease

Diagnostic biomarkers for PD have advanced, aiming to improve early detection, differentiate PD from other neurodegenerative disorders, and monitor disease progression. The primary focus of these biomarkers is on the pathological hallmark of PD, which is the aggregation of α-synuclein protein, alongside indicators of dopaminergic neuron loss and neuroinflammation. α-Synuclein is a central biomarker in PD. Misfolded α-Synuclein aggregates into Lewy bodies and Lewy neurites, which are characteristic of the disease. These pathological forms of α-synuclein can be detected in CSF, blood plasma, and even in peripheral tissues like skin and submandibular gland biopsies (Chahine et al. 2020; Kharel et al. 2022). Recent developments include the use of seed amplification assays (RT-QuIC), which can detect minute amounts of misfolded α-synuclein with high sensitivity and specificity (Srivastava et al. 2022; Grossauer et al. 2023). Neuroinflammation and oxidative stress are also areas of interest in PD biomarker research. Elevated levels of inflammatory cytokines, such as IL-6 and TNF-α, have been observed in the blood and CSF of PD patients, reflecting the ongoing neuroinflammatory processes (Qu et al. 2023). Additionally, markers of oxidative stress, such as 8-OHdG, are being studied for their role in PD pathology (Gmitterová et al. 2009). NfL is emerging as a general marker of neurodegeneration and has shown promise in PD. Elevated NfL levels in CSF and blood correlate with disease severity and progression, making it a useful tool for tracking the course of the disease. Lastly, genetic biomarkers are important for identifying individuals at risk for PD, particularly in familial cases. Mutations in genes such as α-Synuclein, LRRK2, PARKIN, and PINK1 are associated with an increased risk of developing PD and can be detected through genetic testing (Nuytemans et al. 2010). These biomarkers, when used in combination, offer a more comprehensive approach to diagnosing PD, enabling earlier detection, differentiation from other parkinsonian syndromes, and better monitoring of disease progression.

4.4.2.3 Multiple Sclerosis

Diagnostic biomarkers for MS are critical for confirming the diagnosis, assessing disease activity, predicting progression, and monitoring response to treatment. MS is characterized by immune-mediated demyelination and neurodegeneration within the central nervous system (CNS), and the biomarkers reflect these pathological processes. Oligoclonal bands (OCBs) in CSF are among the most established biomarkers for MS (Graner et al. 2020). OCBs represent a pattern of immunoglobulin production in the CNS and are present in approximately 95% of MS patients (Karrenbauer et al. 2021). The presence of OCBs, particularly when absent in serum but present in CSF, strongly supports an MS diagnosis and indicates intrathecal antibody production. NfL has emerged as a promising biomarker for neurodegeneration in MS. Myelin oligodendrocyte glycoprotein (MOG) antibodies and aquaporin-4 (AQP4) antibodies are also important, though primarily for differential diagnosis. MOG antibodies are associated with a distinct demyelinating disorder known as MOG antibody-associated disease (MOGAD), while AQP4 antibodies are indicative of neuromyelitis optica spectrum disorder (NMOSD). Testing for these antibodies helps differentiate MS from these other demyelinating diseases, which can present with similar clinical features (Martinez-Hernandez et al. 2015; Alkabie and Budhram 2022). Chitinase-3-like-1 (CHI3L1) is another biomarker gaining attention in MS. Elevated levels of CHI3L1 in CSF are associated with a more aggressive disease course and greater disability progression (Talaat et al. 2023). These biomarkers, particularly when used in combination, enable a more accurate and early diagnosis of MS, differentiation from other neurological disorders, and ongoing assessment of disease activity and progression. This comprehensive biomarker profile is essential for guiding treatment decisions and improving patient outcomes in MS.

4.4.2.4 Huntington's Disease

Diagnostic biomarkers for HD are essential for confirming diagnosis, assessing disease progression, and differentiating HD from other neurodegenerative disorders. The most definitive biomarker is genetic testing, which identifies the expanded Cytosine-Adenine-Guanine (CAG) repeat in the HTT gene. This mutation leads to the production of a mHTT, which is responsible for the characteristic neuronal damage in HD. Genetic testing is crucial for diagnosing HD, especially in asymptomatic individuals with a family history of the disease (Nopoulos 2016). NfL, a marker of neurodegeneration, is increasingly being studied in HD. Elevated levels of NfL in CSF and blood are associated with neuronal damage and can provide insights into disease activity and progression. This marker is useful for monitoring changes over time and assessing the efficacy of therapeutic interventions. Biomarkers of neuronal dysfunction and damage, such as mHTT levels in CSF, are under investigation. Although not yet widely used clinically, these biomarkers offer the potential for better understanding disease mechanisms and monitoring therapeutic responses. Together, these biomarkers—genetic, imaging, and biochemical—provide a comprehensive approach to diagnosing HD, assessing its progression, and guiding treatment strategies.

4.4.2.5 Schizophrenia

Diagnostic biomarkers for schizophrenia are an area of active research, aiming to enhance early diagnosis, differentiate schizophrenia from other psychiatric and neurological disorders, and monitor disease progression. While no definitive biomarkers are yet established for clinical use, several promising candidates are being investigated. CSF biomarkers are also being studied. Elevated levels of certain proteins, such as NfL and glutamate, have been associated with neurodegeneration and disease severity in schizophrenia. These biomarkers might help in understanding disease mechanisms and monitoring treatment effects. Blood-based biomarkers are gaining interest due to their potential for non-invasive testing. Research has identified alterations in inflammatory markers, such as IL-6 and CRP, which may reflect the neuroinflammatory processes associated with schizophrenia (Chase et al. 2016; Hartwig et al. 2017; Chumakov et al. 2022). Additionally, changes in levels of oxidative stress markers, like malondialdehyde (MDA), are being explored as potential indicators of cellular damage in schizophrenia. Genetic biomarkers involve identifying specific gene variants associated with schizophrenia, such as those in the Disrupted in Schizophrenia 1 (DISC1), catechol-O-methyltransferase (COMT), and BDNF genes (Dong et al. 2016; Harari et al. 2017; Năstase et al. 2022). Genetic risk scores and polygenic risk scores based on multiple genetic variants are used to estimate the likelihood of developing the disorder, though they are not yet used for diagnostic purposes. Epigenetic modifications and microRNA profiles are emerging areas of research. Changes in DNA methylation patterns and specific microRNAs have been observed in individuals with schizophrenia, suggesting that these epigenetic and regulatory elements may play a role in disease pathogenesis and progression. Ongoing research aims to validate these biomarkers and integrate them into clinical settings to provide a more accurate and individualized approach to managing schizophrenia.

4.5 THERAPEUTIC APPROACHES AND FUTURE DIRECTIONS

Therapeutic approaches for neurological diseases such as AD's, PD's, MS, HD, and schizophrenia are diverse and continually evolving, reflecting the complexity and varied nature of these disorders. These approaches range from symptomatic treatments aimed at improving quality of life to disease-modifying therapies (DMTs) designed to alter the course of the disease.

4.5.1 Current Therapies

4.5.1.1 Alzheimer's Disease

Current therapies for AD are designed to address symptoms, slow disease progression, and, in some cases, modify the underlying pathology. Symptomatic treatments primarily focus on improving cognitive function and quality of life. Cholinesterase inhibitors, such as donepezil, rivastigmine, and galantamine, are commonly prescribed to enhance cholinergic neurotransmission by inhibiting the breakdown of ACh, a neurotransmitter involved in memory and cognition (Li et al. 2019; Marucci et al. 2021). These drugs can provide modest improvements in cognitive symptoms and daily functioning in the early to moderate stages of AD. Memantine, an NMDA receptor antagonist, is used to manage moderate to severe AD. It works by regulating glutamate activity, which is thought to be involved in neuronal damage when present in excessive amounts. Memantine can help improve cognitive function and reduce symptoms of agitation and aggression (Liu et al. 2019; Siddiqui et al. 2023). In recent years, the focus has expanded to include DMTs aimed at targeting the underlying pathology of AD. One notable development is the advent of amyloid-beta-targeted therapies. Aducanumab, a monoclonal antibody, received FDA approval as the first therapy intended to reduce amyloid-beta plaques in the brain, which are thought to be a key driver of disease progression (Synnott et al. 2021; Rahman et al. 2023). However, its approval has been controversial due to debates over its clinical efficacy and the interpretation of clinical trial data. Other amyloid-targeting agents, such as lecanemab and donanemab, are in various stages of clinical trials and show promise in reducing amyloid burden and potentially slowing cognitive decline (Volloch and Rits-Volloch 2023; Espay et al. 2024a). Lifestyle and multi-modal interventions are also gaining traction. Research into combining pharmacological treatments with lifestyle interventions such as cognitive training, physical exercise, and dietary changes aims to provide a holistic approach to managing AD. These interventions may help enhance cognitive function and overall quality of life, potentially working synergistically with pharmacological therapies.

4.5.1.2 Parkinson's Disease

Current therapies for PD focus on managing symptoms, improving quality of life, and slowing disease progression. These therapies include pharmacological treatments, surgical interventions, and supportive care strategies. Pharmacological treatments are the cornerstone of PD management. The most commonly used medication is levodopa, which is converted into dopamine in the brain, helping to alleviate motor symptoms such as tremors, stiffness, and bradykinesia (slowness of movement) (Tambasco et al. 2018). Levodopa is often combined with carbidopa (as in Sinemet) to prevent its premature conversion to dopamine outside the brain and reduce side effects (Espay et al. 2024a). Dopamine agonists such as pramipexole, ropinirole, and rotigotine mimic the effects of dopamine by stimulating dopamine receptors in the brain (Contin et al. 2019; Ruan et al. 2021; Ferraiolo and

Hermans 2023). These are used both as initial treatments and as adjuncts to levodopa, particularly in younger patients to delay the onset of levodopa-related motor complications. Monoamine oxidase-B (MAO-B) inhibitors like selegiline and rasagiline help to prolong the action of dopamine by inhibiting its breakdown (Peretz et al. 2016; Cereda et al. 2017). These can provide mild symptomatic relief and are often used in early-stage PD or in combination with other medications. COMT inhibitors such as entacapone and tolcapone work by blocking the enzyme that breaks down dopamine, thereby prolonging the effects of levodopa (Artusi et al. 2021; Song et al. 2021). These are typically used in conjunction with levodopa to manage motor fluctuations and "wearing-off" symptoms. Anticholinergics like trihexyphenidyl and benztropine are used to manage tremors and rigidity, particularly in younger patients with early-stage PD (Barrett et al. 2021; Jost et al. 2023). However, their use is generally limited due to potential side effects such as cognitive impairment. Amantadine is an antiviral medication with dopaminergic and anticholinergic properties (Rascol et al. 2021). It is used to manage motor symptoms and dyskinesias (involuntary movements) associated with long-term levodopa therapy. Surgical interventions are considered for patients with advanced PD who do not respond well to medication. Deep brain stimulation (DBS) involves implanting electrodes into specific brain regions (such as the subthalamic nucleus or globus pallidus interna) to modulate abnormal brain activity. DBS can help reduce motor symptoms and medication requirements, though it does not halt disease progression. Supportive care is an integral part of PD management and includes physical therapy to improve mobility, strength, and balance; occupational therapy to assist with daily activities; and speech therapy to address speech and swallowing difficulties. Additionally, lifestyle modifications such as regular exercise, a balanced diet, and mental health support are important for managing symptoms and enhancing overall well-being. Overall, while current therapies significantly improve the quality of life for many patients with PD, ongoing research aims to develop disease-modifying treatments and address the unmet needs of those with advanced disease.

4.5.1.3 Multiple Sclerosis

Current therapies for MS encompass a range of strategies aimed at managing symptoms, reducing disease activity, and slowing progression. DMTs are central to treatment and include several classes of medications. Interferon-beta and glatiramer acetate are among the first-line options, working to modulate the immune response and reduce inflammation. Oral DMTs such as fingolimod, dimethyl fumarate, and teriflunomide offer more convenience and help manage disease activity by modulating immune cell function or reducing inflammation (Kalincik et al. 2019; Lorscheider et al. 2021). Sphingosine-1-phosphate receptor modulators like ozanimod and siponimod also target immune cell trafficking (Kish 2018; Swallow et al. 2023). B-cell depleting therapies such as ocrelizumab and rituximab focus on reducing the number of B-cells involved in the autoimmune process, while alemtuzumab targets a broader range of immune cells (Katsavos and Coles 2018; Roos et al. 2023). Cladribine, a newer oral agent, selectively targets lymphocytes to modify immune response (Nabizadeh et al. 2023). For managing acute relapses, corticosteroids are commonly used to reduce inflammation, and plasma exchange may be considered for severe cases. Symptomatic treatments address specific issues such as spasticity, pain, and bladder dysfunction, using medications like muscle relaxants, anticholinergics, and pain management options. Comprehensive care involving physical, occupational, and speech therapy, as well as psychological support, is crucial for addressing functional impairments and enhancing quality of life.

4.5.1.4 Huntington's Disease

Current therapies for HD are primarily focused on managing symptoms and improving quality of life, as there is no cure for the condition. Tetrabenazine and deutetrabenazine are specifically used to manage chorea, the involuntary and erratic movements characteristic of the disease (Rodrigues et al. 2017). These drugs work by depleting neurotransmitters such as dopamine, which can help reduce the severity of chorea. For mood disturbances and psychiatric symptoms, such as depression and anxiety, patients may be prescribed antidepressants (e.g., selective serotonin reuptake inhibitors like fluoxetine) and antipsychotics (e.g., quetiapine or olanzapine) (Andriessen et al. 2022). These medications help manage mood swings, irritability, and other psychiatric issues associated with HD. Cognitive and behavioral interventions are used to help patients manage cognitive decline, as well as to cope with the emotional and psychological aspects of the disease. These therapies can include cognitive exercises, counseling, and support groups. Physical therapy helps maintain mobility and balance, while occupational therapy focuses on daily living skills and adaptive techniques to improve function and independence. Speech therapy addresses difficulties with communication and swallowing, which can be affected as the disease progresses. Due to the potential for weight loss and difficulties with swallowing, nutritional support is important. Dietitians may recommend high-calorie diets or feeding tubes if necessary to ensure adequate nutrition. While these treatments help manage symptoms, research continues into more targeted therapies that may slow or alter the disease course.

4.5.1.5 Schizophrenia

Current therapies for schizophrenia aim to address the disorder's complex symptoms and enhance the overall quality of life for affected individuals. The primary treatment modality is antipsychotic medications, which are classified into two categories: first-generation (typical) and second-generation (atypical) antipsychotics. First-generation antipsychotics, such as haloperidol and chlorpromazine, primarily target dopamine receptors to alleviate symptoms like delusions and hallucinations but can cause side effects such as movement disorders, including tardive dyskinesia (Leucht et al. 2008). Second-generation antipsychotics, including risperidone, olanzapine, quetiapine, and aripiprazole, offer a broader spectrum of action on various neurotransmitter systems, aiming to manage a wider array of symptoms with a generally lower risk of extrapyramidal side effects (Huhn et al. 2019; Kim et al. 2021; Sampogna et al. 2023). Alongside medication, psychosocial interventions play a crucial role; cognitive-behavioral therapy (CBT) helps individuals address distressing symptoms and develop coping strategies, while social skills training and family therapy improve interpersonal interactions and support systems. Rehabilitation programs focus on enhancing daily living skills, vocational training, and social integration, aiding individuals in achieving greater independence and stability. Community support services, such as case management, supported housing, and assertive community treatment (ACT), provide continuous care to manage the complexities of the disorder and minimize hospitalizations. Long-acting injectable antipsychotics are also utilized to improve medication adherence and prevent relapse. Additionally, addressing co-occurring conditions, such as substance abuse or physical health issues, through integrated treatment approaches is essential for comprehensive care. Ongoing research aims to develop novel antipsychotic medications, neuromodulation techniques, and personalized medicine approaches to improve treatment outcomes further and tailor interventions to individual needs.

4.5.2 Emerging Therapies

4.5.2.1 Alzheimer's Disease

Emerging therapies for AD represent a dynamic and promising frontier in the quest to more effectively treat and ultimately cure the disease. These therapies encompass a range of innovative approaches targeting the underlying pathophysiology of AD, with several key areas of development showing potential. Tau-targeting therapies are also under investigation. Tau protein aggregation forms neurofibrillary tangles, which are another hallmark of AD pathology (Chen and Yu 2023). Clinical trials are exploring monoclonal antibodies and small molecules aimed at reducing tau aggregation or promoting tau clearance. In addition to these targeted approaches, combination therapies are being explored to enhance treatment efficacy. These might include combining cholinesterase inhibitors with memantine or integrating lifestyle interventions, such as cognitive training and physical exercise, which may have adjunctive benefits (Li et al. 2019). Gene therapy and gene editing represent cutting-edge strategies aimed at directly modifying genetic factors involved in AD. Techniques such as CRISPR/Cas9 are being explored to target and correct genetic mutations associated with familial AD (De Plano et al. 2022). Additionally, research into gene silencing approaches, like RNA interference and antisense oligonucleotides (ASOs), seeks to reduce the expression of harmful proteins or enhance the expression of protective genes (Chen et al. 2013). Neuroprotection and regenerative approaches are another area of active research. These include therapies designed to protect neurons from damage or promote neuronal repair and regeneration. Agents that modulate neuroinflammation, oxidative stress, and mitochondrial function are being investigated for their potential to slow disease progression and support neuronal health. For instance, drugs targeting inflammatory pathways, such as those inhibiting microglial activation, and compounds aimed at reducing oxidative damage are under clinical evaluation. Immunotherapies, beyond amyloid and tau targeting, are exploring broader immune system modulation. These include vaccines designed to stimulate the immune system to clear amyloid-beta or tau aggregates, as well as agents that regulate the immune response to reduce neuroinflammation and support neuronal health. These emerging therapies reflect a broader shift toward personalized and disease-modifying treatments, offering hope for more effective management of AD and potentially transformative changes in how the disease is approached in the future.

4.5.2.2 Parkinson's Disease

Emerging therapies for PD are focused on addressing the disease's underlying pathology, improving symptom management, and potentially altering the course of the disease. These innovative approaches include advanced pharmacological treatments, gene and cell therapies, and novel surgical techniques. Gene therapy is a burgeoning field in Parkinson's research. The goal is to modify or replace defective genes associated with PD or deliver genes that can produce therapeutic proteins. For instance, gene delivery of neurotrophic factors, such as GDNF, aims to support the survival and function of dopaminergic neurons (Chmielarz and Saarma 2020; Palasz et al. 2020). Clinical trials are investigating the delivery of these factors directly into the brain to potentially slow disease progression and alleviate symptoms. Cell-based therapies are also showing promise. These approaches involve transplanting dopamine-producing neurons or stem cells into the brain to replace the lost neurons in PD. Dopaminergic neuronal transplantation involves using stem cells or fetal tissue to generate new dopamine-producing neurons, intending to restore dopaminergic function in the brain. Recent advancements in induced pluripotent stem cell (iPSC) technology are enabling researchers to generate patient-specific neurons for transplantation, reducing the risk of immune rejection and providing personalized treatment options (Stoddard-Bennett and Reijo Pera 2019). DMTs are being explored to slow or halt the progression of PD. Alpha-synuclein-targeted therapies aim to address one of the key pathological features of Parkinson's—misfolded alpha-synuclein aggregates. Strategies include the use of monoclonal antibodies to neutralize misfolded alpha-synuclein or promote its clearance. Other approaches involve small molecules or vaccines designed to target and reduce alpha-synuclein accumulation. Novel pharmacological agents are under investigation to improve symptom management and reduce side effects. For example, new dopamine agonists and enzyme inhibitors are being developed to enhance dopamine receptor stimulation or inhibit dopamine breakdown more effectively. Additionally, adenosine A2A receptor antagonists are being studied for their potential to modulate neurotransmitter systems and improve motor and non-motor symptoms in PD. Neuroprotective strategies are another exciting area of research. These include drugs and compounds that aim to protect neurons from damage, reduce oxidative stress, or modulate inflammation. For instance, mitochondrial-targeted antioxidants and anti-inflammatory agents are being evaluated for their potential to slow neuronal loss and improve disease outcomes. Advanced neuromodulation techniques, such as focused ultrasound, are being explored as non-invasive alternatives to DBS (Groiss et al. 2009). Overall, these emerging therapies represent a significant advancement in the quest to better understand, manage, and potentially cure PD.

As research progresses, these innovations hold the potential to transform the treatment landscape and improve the quality of life for individuals affected by this challenging neurodegenerative disorder.

4.5.2.3 Multiple Sclerosis

Emerging therapies for MS are advancing rapidly, offering innovative approaches to potentially modify disease progression and enhance treatment efficacy. Stem cell-based therapies are a significant area of exploration, with autologous hematopoietic stem cell transplantation (AHSCT) being investigated for its ability to reset the immune system and potentially halt disease activity (Mohammadi et al. 2021). This approach involves collecting a patient's stem cells, administering high-dose chemotherapy to ablate the existing immune system, and then reintroducing the stem cells to reconstitute a new immune system. Mesenchymal stem cell (MSC) therapies are also under investigation for their potential to promote the repair and regeneration of damaged tissues in MS (Islam et al. 2023). Gene therapy and gene editing techniques are promising avenues for targeting the underlying mechanisms of MS. Approaches like CRISPR/Cas9 aim to correct genetic mutations associated with MS or modify immune cell function to reduce autoimmunity (Lee et al. 2022). Research is exploring the delivery of genes that encode neuroprotective factors or regulate immune responses directly into the central nervous system. Remyelination therapies are focused on repairing the damaged myelin sheath, a key feature of MS pathology (Harlow et al. 2015). OPC transplantation and the use of growth factors such as B-cell activating factor (BAFF) and neurotrophic factors aim to stimulate the regeneration of myelin (Ben-Hur 2011; Mangale et al. 2019). Neuroprotective agents are another promising area, focusing on protecting neurons from damage and reducing neuroinflammation. Nanomedicine and targeted delivery systems are being explored to enhance drug delivery to the central nervous system. Nanoparticles and liposomal formulations aim to improve the precision of drug delivery, reduce systemic side effects, and enhance therapeutic efficacy. These emerging therapies represent a significant advancement in the management of MS, offering hope for more effective treatments that not only alleviate symptoms but also address the underlying disease mechanisms and improve long-term outcomes for patients.

4.5.2.4 Huntington's Disease

Emerging therapies for HD are increasingly focused on innovative approaches to address the underlying genetic causes and potentially alter the disease's progression. ASOs represent a leading therapeutic strategy. These molecules are designed to target and degrade the mRNA produced by the mutant HTT gene, thereby reducing the levels of the toxic huntingtin protein (Lemprière 2019). Clinical trials, such as those involving Ionis-HTTRx (tofersen), are investigating the safety and efficacy of this approach in reducing disease symptoms and slowing progression (Tabrizi et al. 2019). Technologies like CRISPR/Cas9 are being explored for their potential to directly correct the genetic mutations responsible for HD (Seo et al. 2023). By targeting and modifying the specific DNA sequences involved, researchers hope to correct or eliminate the mutation that causes the disease, potentially offering a cure. Researchers are developing small molecules that can modulate the pathways involved in the aggregation and processing of the mHTT. These compounds aim to reduce the toxic effects of protein aggregates and protect neuronal function. Novel compounds are being tested for their ability to protect neurons from damage and promote neuronal survival. These agents could help mitigate the neurodegenerative effects of HD and preserve cognitive and motor functions. Advances in stem cell technology offer the potential for replacing damaged neurons or delivering therapeutic agents directly to the affected brain regions (Liang et al. 2023; Conner et al. 2023). Researchers are exploring ways to use stem cells to regenerate brain tissue or provide a delivery mechanism for other therapies. Emerging therapies are also investigating how modulating the immune system might impact HD. This includes research into how inflammatory responses contribute to the disease and whether targeting these responses could provide therapeutic benefits. These emerging therapies represent a promising shift toward more targeted and potentially disease-modifying treatments, although they are still in various stages of development and clinical testing.

4.5.2.5 Schizophrenia

Emerging therapies for schizophrenia are focusing on novel approaches that target the underlying mechanisms of the disorder and aim to improve treatment outcomes. Research is ongoing to develop new antipsychotic drugs with improved efficacy and fewer side effects. This includes drugs targeting neurotransmitter systems beyond dopamine, such as glutamate, which may help address symptoms that are less responsive to current treatments. Medications that modulate the glutamatergic system, such as NMDA receptor modulators, are being investigated (Hanson et al. 2024). These agents aim to address cognitive and negative symptoms of schizophrenia that are not well-managed by traditional antipsychotics. There is growing interest in the role of the immune system in schizophrenia. Therapies that target inflammatory processes or modulate immune responses are being explored, as emerging evidence suggests that inflammation may play a role in the pathophysiology of the disorder (Fond et al. 2020). Drugs and interventions aimed at improving cognitive function are under development. This includes agents designed to enhance cognitive deficits that are common in schizophrenia, such as impaired attention, memory, and executive function. Advanced neuromodulation methods, such as transcranial magnetic stimulation (TMS) and transcranial direct current stimulation (tDCS), are being studied for their potential to improve symptoms by modulating brain activity non-invasively (Cole et al. 2015; Cheng et al. 2020). Advances in genomics and pharmacogenomics are paving the way for personalized treatment approaches. By analyzing genetic and biochemical profiles, researchers hope to tailor therapies to individual patients, improving efficacy and reducing adverse effects. Experimental research into the use of psychedelics, such as psilocybin, is underway to explore their potential in treating symptoms of schizophrenia, particularly in conjunction with psychotherapy. These emerging therapies represent a significant advancement in the field, aiming to address the limitations of current treatments and offer new hope for managing schizophrenia more effectively.

4.6 CONCLUSION

In conclusion, the brain's response to neurological disorders involves complex structural and functional changes that reflect the impact of these conditions on neural networks and overall brain health. Structural alterations, such as atrophy, lesions, and changes in brain connectivity, can significantly affect cognitive, motor, and emotional functions. Functional changes, including altered neurotransmitter activity, disrupted neural circuits, and changes in brain activity patterns, further contribute to the clinical manifestations of neurological disorders. Understanding these changes is crucial for developing targeted therapies and interventions aimed at mitigating the effects of these disorders. Advances in neuroimaging and neurophysiological techniques continue to provide deeper insights into how neurological disorders affect the brain, enabling more effective diagnostic and therapeutic strategies. By elucidating the mechanisms underlying these structural and functional changes, researchers and clinicians can work toward improving patient outcomes and enhancing the quality of life for individuals affected by neurological conditions.

REFERENCES

Abdulghani A, Poghosyan M, Mehren A, et al. (2023) Neuroplasticity to autophagy cross-talk in a therapeutic effect of physical exercises and irisin in ADHD. Front Mol Neurosci 15:997054. https://doi.org/10.3389/fnmol.2022.997054

Adamu MJ, Qiang L, Nyatega CO, et al. (2023) Unraveling the pathophysiology of schizophrenia: Insights from structural magnetic resonance imaging studies. Front Psychiatry 14. https://doi.org/10.3389/fpsyt.2023.1188603

Agnihotri A, Aruoma OI (2020) Alzheimer's disease and Parkinson's disease: A nutritional toxicology perspective of the impact of oxidative stress, mitochondrial dysfunction, nutrigenomics and environmental chemicals. J Am College Nutr 39:16–27. https://doi.org/10.1080/07315724.2019.1683379

Akdemir ÜÖ, Bora TA, Atayl LÖ (2021) Dopamine transporter SPECT imaging in Parkinson's disease and Parkinsonian disorders. Turk J Med Sci 51:400–410. https://doi.org/10.3906/sag-2008-253

Albanese M, Zagaglia S, Landi D, et al. (2016) Cerebrospinal fluid lactate is associated with multiple sclerosis disease progression. J Neuroinflammation 13:36. https://doi.org/10.1186/s12974-016-0502-1

Alizadeh A, Dyck SM, Karimi-Abdolrezaee S (2019) Traumatic spinal cord injury: An overview of pathophysiology, models and acute injury mechanisms. Front Neurol 10(282). https://doi.org/10.3389/fneur.2019.00282

Alkabie S, Budhram A (2022) Testing for antibodies against aquaporin-4 and myelin oligodendrocyte glycoprotein in the diagnosis of patients with suspected autoimmune myelopathy. Front Neurol 13:912050. https://doi.org/10.3389/fneur.2022.912050

Alkan E, Davies G, Evans SL (2021) Cognitive impairment in schizophrenia: Relationships with cortical thickness in fronto-temporal regions, and dissociability from symptom severity. NPJ Schizophr 7:20. https://doi.org/10.1038/s41537-021-00149-0

Allen SJ, Watson JJ, Shoemark DK, et al. (2013) GDNF, NGF and BDNF as therapeutic options for neurodegeneration. Pharmacol Ther 138:155–175. https://doi.org/10.1016/j.pharmthera.2013.01.004

Alonso AD, Cohen LS, Corbo C, et al. (2018) Hyperphosphorylation of tau associates with changes in its function beyond microtubule stability. Front Cell Neurosci 12. https://doi.org/10.3389/fncel.2018.00338

Amende I, Kale A, McCue S, et al. (2005) Gait dynamics in mouse models of Parkinson's disease and Huntington's disease. J NeuroEngineering Rehabil 2:20. https://doi.org/10.1186/1743-0003-2-20

Andersen JV, Markussen KH, Jakobsen E, et al. (2021) Glutamate metabolism and recycling at the excitatory synapse in health and neurodegeneration. Neuropharmacology 196:108719. https://doi.org/10.1016/j.neuropharm.2021.108719

Andersson E, Schultz N, Saito T, et al. (2023) Cerebral Aβ deposition precedes reduced cerebrospinal fluid and serum Aβ42/Aβ40 ratios in the AppNL–F/NL–F knock-in mouse model of Alzheimer's disease. Alzheimers Res Ther 15:64. https://doi.org/10.1186/s13195-023-01196-8

Andhale R, Shrivastava D, Andhale R, Shrivastava D (2022) Huntington's disease: A clinical review. Cureus 14(8):e28484. https://doi.org/10.7759/cureus.28484

Andriessen RL, Oosterloo M, Hollands A, et al. (2022) Psychotropic medication use in Huntington's disease: A retrospective cohort study. Parkinsonism Relat Disord 105:69–74. https://doi.org/10.1016/j.parkreldis.2022.11.004

Appelbaum LG, Shenasa MA, Stolz L, Daskalakis Z (2023) Synaptic plasticity and mental health: Methods, challenges and opportunities. Neuropsychopharmacol 48:113–120. https://doi.org/10.1038/s41386-022-01370-w

Arevalo-Rodriguez I, Smailagic N, Roqué i Figuls M, et al. (2015) Mini-Mental State Examination (MMSE) for the detection of Alzheimer's disease and other dementias in people with mild cognitive impairment (MCI). Cochrane Database Syst Rev 2015:CD010783. https://doi.org/10.1002/14651858.CD010783.pub2

Artusi CA, Sarro L, Imbalzano G, et al. (2021) Safety and efficacy of tolcapone in Parkinson's disease: Systematic review. Eur J Clin Pharmacol 77:817–829. https://doi.org/10.1007/s00228-020-03081-x

Ashton NJ, Janelidze S, Mattsson-Carlgren N, et al. (2022) Differential roles of Aβ42/40, p-tau231 and p-tau217 for Alzheimer's trial selection and disease monitoring. Nat Med 28:2555–2562. https://doi.org/10.1038/s41591-022-02074-w

Aspenström P (2024) Miro GTPases at the crossroads of cytoskeletal dynamics and mitochondrial trafficking. Cells 13:647. https://doi.org/10.3390/cells13070647

Autry AE, Monteggia LM (2012) Brain-derived neurotrophic factor and neuropsychiatric disorders. Pharmacol Rev 64:238–258. https://doi.org/10.1124/pr.111.005108

Ávila-Villanueva M, Marcos Dolado A, Gómez-Ramírez J, Fernández-Blázquez M (2022) Brain structural and functional changes in cognitive impairment due to Alzheimer's disease. Front Psychol 13. https://doi.org/10.3389/fpsyg.2022.886619

Azevedo CJ, Cen SY, Khadka S, et al. (2018) Thalamic atrophy in multiple sclerosis: A magnetic resonance imaging marker of neurodegeneration throughout disease. Ann Neurol 83:223–234. https://doi.org/10.1002/ana.25150

Bachiller S, Jiménez-Ferrer I, Paulus A, et al. (2018) Microglia in neurological diseases: A road map to brain-disease dependent-inflammatory response. Front Cell Neurosci 12:488.

Bakirtzis C, Lima M, De Lorenzo SS, et al. (2023) Secondary central nervous system demyelinating disorders in the elderly: A narrative review. Healthcare 11:2126. https://doi.org/10.3390/healthcare11152126

Bakker G, Vingerhoets C, Boucherie D et al. (2018) Relationship between muscarinic M1 receptor binding and cognition in medication-free subjects with psychosis. Neuroimage Clin 18:713–719. https://doi.org/10.1016/j.nicl.2018.02.030

Bao J, Tu H, Li Y, et al. (2022) Diffusion tensor imaging revealed microstructural changes in normal-appearing white matter regions in relapsing–remitting multiple sclerosis. Front Neurosci 16:837452. https://doi.org/10.3389/fnins.2022.837452

Barrett MJ, Sargent L, Nawaz H, et al. (2021) Antimuscarinic anticholinergic medications in Parkinson disease: To prescribe or deprescribe? Movem Disord Clin Prac 8:1181–1188. https://doi.org/10.1002/mdc3.13347

Barron JC, Hurley EP, Parsons MP (2021) Huntingtin and the synapse. Front Cell Neurosci 15:689332. https://doi.org/10.3389/fncel.2021.689332

Bathina S, Das UN (2015) Brain-derived neurotrophic factor and its clinical implications. Arch Med Sci 11:1164–1178. https://doi.org/10.5114/aoms.2015.56342

Bellingacci L, Mancini A, Gaetani L, et al. (2021) Synaptic dysfunction in multiple sclerosis: A red thread from inflammation to network disconnection. Int J Mol Sci 22:9753. https://doi.org/10.3390/ijms22189753

Ben-Hur T (2011) Cell therapy for multiple sclerosis. Neurotherapeutics 8:625. https://doi.org/10.1007/s13311-011-0073-x

Bergsland N, Pelizzari L, Laganá MM, et al. (2021) Automated assessment of the Substantia Nigra Pars Compacta in Parkinson's disease: A diffusion tensor imaging study. J Personal Med 11:1235. https://doi.org/10.3390/jpm11111235

Björklund A, Barker RA (2024) The basal forebrain cholinergic system as target for cell replacement therapy in Parkinson's disease. Brain 147:1937–1952. https://doi.org/10.1093/brain/awae026

Bloom GS (2014) Amyloid-β and tau: The trigger and bullet in Alzheimer disease pathogenesis. JAMA Neurol 71:505–508. https://doi.org/10.1001/jamaneurol.2013.5847

Blumenstock S, Dudanova I (2020) Cortical and striatal circuits in Huntington's disease. Front Neurosci 14:82. https://doi.org/10.3389/fnins.2020.00082

Braine A, Georges F (2023) Emotion in action: When emotions meet motor circuits. Neurosci Biobehav Rev 155:105475. https://doi.org/10.1016/j.neubiorev.2023.105475

Breijyeh Z, Karaman R (2020) Comprehensive review on Alzheimer's disease: Causes and treatment. Molecules 25:5789. https://doi.org/10.3390/molecules25245789

Buddhala C, Loftin SK, Kuley BM, et al. (2015) Dopaminergic, serotonergic, and noradrenergic deficits in Parkinson disease. Ann Clin Transl Neurol 2:949–959. https://doi.org/10.1002/acn3.246

Bulduk BK, Tortajada J, Valiente-Pallejà A, et al. (2024) High number of mitochondrial DNA alterations in postmortem brain tissue of patients with schizophrenia compared to healthy controls. Psychiat Res 337:115928. https://doi.org/10.1016/j.psychres.2024.115928

Buosi P, Borghi FA, Lopes AM, et al. (2021) Oxidative stress biomarkers in treatment-responsive and treatment-resistant schizophrenia patients. Trends Psychiat Psychother 43:278–285. https://doi.org/10.47626/2237-6089-2020-0078

Burgetova A, Dusek P, Uher T, et al. (2023) CSF markers of oxidative stress are associated with brain atrophy and iron accumulation in a 2-year longitudinal cohort of early MS. Int J Mol Sci 24:10048. https://doi.org/10.3390/ijms241210048

Cai Y, Nielsen BE, Boxer EE, et al. (2021) Loss of nigral excitation of cholinergic interneurons contributes to parkinsonian motor impairments. Neuron 109:1137–1149.e5. https://doi.org/10.1016/j.neuron.2021.01.028

Calabrese M, Agosta F, Rinaldi F, et al. (2009) Cortical lesions and atrophy associated with cognitive impairment in relapsing-remitting multiple sclerosis. Arch Neurol 66:1144–1150. https://doi.org/10.1001/archneurol.2009.174

Calabresi P, Mechelli A, Natale G, et al. (2023) Alpha-synuclein in Parkinson's disease and other synucleinopathies: From overt neurodegeneration back to early synaptic dysfunction. Cell Death Dis 14:1–16. https://doi.org/10.1038/s41419-023-05672-9

Caligiore D, Helmich RC, Hallett M, et al. (2016) Parkinson's disease as a system-level disorder. NPJ Parkinson's Dis 2:1–9. https://doi.org/10.1038/npjparkd.2016.25

Calvo-Rodriguez M, Hou SS, Snyder AC, et al. (2020) Increased mitochondrial calcium levels associated with neuronal death in a mouse model of Alzheimer's disease. Nat Commun 11:2146. https://doi.org/10.1038/s41467-020-16074-2

Campana M, Strauß J, Münz S, et al. (2022) Cerebrospinal fluid pathologies in schizophrenia-spectrum disorder-A retrospective chart review. Schizophr Bull 48:47–55. https://doi.org/10.1093/schbul/sbab105

Caron NS, Wright GE, Hayden MR (2020) Huntington disease. University of Washington, Seattle.

Cereda E, Cilia R, Canesi M, et al. (2017) Efficacy of rasagiline and selegiline in Parkinson's disease: A head-to-head 3-year retrospective case–control study. J Neurol 264:1254–1263. https://doi.org/10.1007/s00415-017-8523-y

Chahine LM, Beach TG, Brumm MC, et al. (2020) In vivo distribution of α-synuclein in multiple tissues and biofluids in Parkinson disease. Neurology 95:e1267–e1284. https://doi.org/10.1212/WNL.0000000000010404

Chandra A, Dervenoulas G, Politis M (2019) Magnetic resonance imaging in Alzheimer's disease and mild cognitive impairment. J Neurol 266:1293–1302. https://doi.org/10.1007/s00415-018-9016-3

Chari DM (2007) Remyelination in multiple sclerosis. Int Rev Neurobiol 79:589–620. https://doi.org/10.1016/S0074-7742(07)79026-8

Charroud C, Turella L (2021) Subcortical grey matter changes associated with motor symptoms evaluated by the Unified Parkinson's disease Rating Scale (part III): A longitudinal study in Parkinson's disease. Neuroimage Clin 31:102745. https://doi.org/10.1016/j.nicl.2021.102745

Chase KA, Cone JJ, Rosen C, Sharma RP (2016) The value of interleukin 6 as a peripheral diagnostic marker in schizophrenia. BMC Psychiatry 16:152. https://doi.org/10.1186/s12888-016-0866-x

Chaudhary S, Zhornitsky S, Chao HH, et al. (2022) Emotion processing dysfunction in Alzheimer's disease: An overview of behavioral findings, systems neural correlates, and underlying neural biology. Am J Alzheimers Dis Other Demen 37:15333175221082834. https://doi.org/10.1177/15333175221082834

Chaves C, Dursun SM, Tusconi M, Hallak JEC (2024) Neuroinflammation and schizophrenia – Is there a link? Front Psychiatry 15, 1356975. https://doi.org/10.3389/fpsyt.2024.1356975

Chelini G, Pantazopoulos H, Durning P, Berretta S (2018) The tetrapartite synapse: A key concept in the pathophysiology of schizophrenia. European Psychiatry 50:60–69. https://doi.org/10.1016/j.eurpsy.2018.02.003

Chen G, Xu T, Yan Y, et al. (2017) Amyloid beta: Structure, biology and structure-based therapeutic development. Acta Pharmacol Sin 38:1205–1235. https://doi.org/10.1038/aps.2017.28

Chen S, Ge X, Chen Y, et al. (2013) Advances with RNA interference in Alzheimer's disease research. Drug Des Devel Ther 7:117–125. https://doi.org/10.2147/DDDT.S40229

Chen W, Zhao H, Li Y (2023) Mitochondrial dynamics in health and disease: Mechanisms and potential targets. Sig Transduct Target Ther 8:1–25. https://doi.org/10.1038/s41392-023-01547-9

Chen Y, Yu Y (2023) Tau and neuroinflammation in Alzheimer's disease: Interplay mechanisms and clinical translation. J Neuroinflammation 20:165. https://doi.org/10.1186/s12974-023-02853-3

Chen Z-R, Huang J-B, Yang S-L, Hong F-F (2022) Role of cholinergic signaling in Alzheimer's disease. Molecules 27(1816). https://doi.org/10.3390/molecules27061816

Cheng PWC, Louie LLC, Wong YL, et al. (2020) The effects of transcranial direct current stimulation (tDCS) on clinical symptoms in schizophrenia: A systematic review and meta-analysis. Asian J Psychiatr 53:102392. https://doi.org/10.1016/j.ajp.2020.102392

Chmielarz P, Saarma M (2020) Neurotrophic factors for disease-modifying treatments of Parkinson's disease: Gaps between basic science and clinical studies. Pharmacol Rep 72:1195–1217. https://doi.org/10.1007/s43440-020-00120-3

Chumakov E, Dorofeikova M, Tsyrenova K, Petrova N (2022) A cross-sectional study on associations between BDNF, CRP, IL-6 and clinical symptoms, cognitive and personal performance in patients with paranoid schizophrenia. Front Psychiatry 13(943869). https://doi.org/10.3389/fpsyt.2022.943869

Clemente-Suárez VJ, Redondo-Flórez L, Beltrán-Velasco AI, et al. (2023) Mitochondria and brain disease: A comprehensive review of pathological mechanisms and therapeutic opportunities. Biomedicines 11:2488. https://doi.org/10.3390/biomedicines11092488

Cole JC, Green Bernacki C, Helmer A, et al. (2015) Efficacy of transcranial magnetic stimulation (TMS) in the treatment of schizophrenia: A review of the literature to Date. Innov Clin Neurosci 12:12–19.

Conner L, Srinageshwar B, Bakke J, et al. (2023) Advances in stem cell and other therapies for Huntington's disease: An update. Brain Res Bull 199:110673. https://doi.org/10.1016/j.brainresbull.2023.110673

Constantinides VC, Paraskevas GP, Boufidou F, et al. (2023) CSF Aβ42 and Aβ42/Aβ40 ratio in Alzheimer's disease and frontotemporal dementias. Diagnostics (Basel) 13:783. https://doi.org/10.3390/diagnostics13040783

Contin M, Lopane G, Mohamed S, et al. (2019) Clinical pharmacokinetics of pramipexole, ropinirole and rotigotine in patients with Parkinson's disease. Parkinsonism Relat Disord 61:111–117. https://doi.org/10.1016/j.parkreldis.2018.11.007

Cui X, McGrath JJ, Burne THJ, Eyles DW (2021) Vitamin D and schizophrenia: 20 years on. Mol Psychiatry 26:2708–2720. https://doi.org/10.1038/s41380-021-01025-0

Cummings JL, Henchcliffe C, Schaier S, et al. (2011) The role of dopaminergic imaging in patients with symptoms of dopaminergic system neurodegeneration. Brain 134:3146–3166. https://doi.org/10.1093/brain/awr177

Dabiri M, Dehghani Firouzabadi F, Yang K, et al. (2022) Neuroimaging in schizophrenia: A review article. Front Neurosci 16(1042814). https://doi.org/10.3389/fnins.2022.1042814

Dai Y, Wang H, Lian A, et al. (2023) A comprehensive perspective of Huntington's disease and mitochondrial dysfunction. Mitochondrion 70:8–19. https://doi.org/10.1016/j.mito.2023.03.001

Damasceno BP (2012) Relationship between cortical microinfarcts and cognitive impairment in Alzheimer's disease. Dement Neuropsychol 6:131–136. https://doi.org/10.1590/S1980-57642012DN06030004

Dawidowski B, Górniak A, Podwalski P, et al. (2021) The role of cytokines in the pathogenesis of schizophrenia. J Clin Med 10:3849. https://doi.org/10.3390/jcm10173849

DeLuca GC, Yates RL, Beale H, Morrow SA (2014) Cognitive impairment in multiple sclerosis: Clinical, radiologic and pathologic insights. Brain Pathol 25:79–98. https://doi.org/10.1111/bpa.12220

DeMaagd G, Philip A (2015) Parkinson's disease and its management. Pharm Ther 40:504–532.

De Nicolo B, Cataldi-Stagetti E, Diquigiovanni C, Bonora E (2023) Calcium and reactive oxygen species signaling interplays in cardiac physiology and pathologies. Antioxidants (Basel) 12:353. https://doi.org/10.3390/antiox12020353

De Plano LM, Calabrese G, Conoci S, et al. (2022) Applications of CRISPR-Cas9 in Alzheimer's disease and related disorders. Int J Mol Sci 23:8714. https://doi.org/10.3390/ijms23158714

DeTure MA, Dickson DW (2019) The neuropathological diagnosis of Alzheimer's disease. Mol Neurodegener 14:32. https://doi.org/10.1186/s13024-019-0333-5

Dias V, Junn E, Mouradian MM (2013) The role of oxidative stress in Parkinson's disease. J Parkinsons Dis 3:461–491. https://doi.org/10.3233/JPD-130230

Djordjević VV, Kostić J, Krivokapić Ž, et al. (2022) Decreased activity of erythrocyte catalase and glutathione peroxidase in patients with schizophrenia. Medicina (Kaunas) 58:1491. https://doi.org/10.3390/medicina58101491

Dong Z, Sun X, Pan C, et al. (2016) Association of DISC1, BDNF, and COMT polymorphisms with exploratory eye movement of schizophrenia in a Chinese Han population. Psychiatr Genet 26:258–265. https://doi.org/10.1097/YPG.0000000000000138

Dörr J, Paul F (2015) The transition from first-line to second-line therapy in multiple sclerosis. Curr Treat Options Neurol 17:354. https://doi.org/10.1007/s11940-015-0354-5

Duarte Azevedo M, Sander S, Tenenbaum L (2020) GDNF, a neuron-derived factor upregulated in glial cells during disease. J Clin Med 9:456. https://doi.org/10.3390/jcm9020456

Eickhoff S, Franzen L, Korda A, et al. (2022) The basal forebrain cholinergic nuclei and their relevance to schizophrenia and other psychotic disorders. Front Psychiatry 13. https://doi.org/10.3389/fpsyt.2022.909961

Ermakov EA, Dmitrieva EM, Parshukova DA, et al. (2021) Oxidative stress-related mechanisms in schizophrenia pathogenesis and new treatment perspectives. Oxid Med Cell Longev 2021:8881770. https://doi.org/10.1155/2021/8881770

Espay AJ, Kepp KP, Herrup K (2024a) Lecanemab and donanemab as therapies for Alzheimer's disease: An illustrated perspective on the data. eNeuro 11. https://doi.org/10.1523/ENEURO.0319-23.2024

Evans AK, Defensor E, Shamloo M (2022) Selective vulnerability of the locus coeruleus noradrenergic system and its role in modulation of neuroinflammation, cognition, and neurodegeneration. Front Pharmacol 13. https://doi.org/10.3389/fphar.2022.1030609

Ferraiolo M, Hermans E (2023) The complex molecular pharmacology of the dopamine D2 receptor: Implications for pramipexole, ropinirole, and rotigotine. Pharmacol Ther 245:108392. https://doi.org/10.1016/j.pharmthera.2023.108392

Ferreira-Vieira TH, Guimaraes IM, Silva FR, Ribeiro FM (2016) Alzheimer's disease: Targeting the cholinergic system. Curr Neuropharmacol 14:101–115. https://doi.org/10.2174/1570159X13666150716165726

Filippi M, Preziosa P, Banwell BL, et al. (2019) Assessment of lesions on magnetic resonance imaging in multiple sclerosis: Practical guidelines. Brain 142:1858–1875. https://doi.org/10.1093/brain/awz144

Filippi M, Sarasso E, Piramide N, et al. (2020) Progressive brain atrophy and clinical evolution in Parkinson's disease. Neuroimage Clin 28:102374. https://doi.org/10.1016/j.nicl.2020.102374

Filley CM, Fields RD (2016) White matter and cognition: Making the connection. J Neurophysiol 116:2093–2104. https://doi.org/10.1152/jn.00221.2016

Fond G, Lançon C, Korchia T, et al. (2020) The role of inflammation in the treatment of schizophrenia. Front Psychiatry 11:160. https://doi.org/10.3389/fpsyt.2020.00160

Franklin H, Clarke BE, Patani R (2021) Astrocytes and microglia in neurodegenerative diseases: Lessons from human in vitro models. Prog Neurobiol 200:101973. https://doi.org/10.1016/j.pneurobio.2020.101973

French IT, Muthusamy KA (2018) A review of the pedunculopontine nucleus in Parkinson's disease. Front Aging Neurosci 10:99. https://doi.org/10.3389/fnagi.2018.00099

Friedman NP, Robbins TW (2022) The role of prefrontal cortex in cognitive control and executive function. Neuropsychopharmacol 47:72–89. https://doi.org/10.1038/s41386-021-01132-0

Gajofatto A, Calabrese M, Benedetti MD, Monaco S (2013) Clinical, MRI, and CSF markers of disability progression in multiple sclerosis. Dis Markers 35:484959. https://doi.org/10.1155/2013/484959

Gallego JA, Blanco EA, Husain-Krautter S, et al. (2018) Cytokines in cerebrospinal fluid of patients with schizophrenia spectrum disorders: New data and an updated meta-analysis. Schizophr Res 202:64–71. https://doi.org/10.1016/j.schres.2018.07.019

Galoppin M, Kari S, Soldati S, et al. (2022) Full spectrum of vitamin D immunomodulation in multiple sclerosis: Mechanisms and therapeutic implications. Brain Communications 4:fcac171. https://doi.org/10.1093/braincomms/fcac171

Galvan A, Devergnas A, Wichmann T (2015) Alterations in neuronal activity in basal ganglia-thalamocortical circuits in the parkinsonian state. Front Neuroanat 9(5). https://doi.org/10.3389/fnana.2015.00005

Gao C, Jiang J, Tan Y, Chen S (2023) Microglia in neurodegenerative diseases: Mechanism and potential therapeutic targets. Sig Transduct Target Ther 8:1–37. https://doi.org/10.1038/s41392-023-01588-0

Gao X-Y, Yang T, Gu Y, Sun X-H (2022) Mitochondrial dysfunction in Parkinson's disease: From mechanistic insights to therapy. Front Aging Neurosci 14:885500. https://doi.org/10.3389/fnagi.2022.885500

Garg A, Appel-Cresswell S, Popuri K, et al. (2015) Morphological alterations in the caudate, putamen, pallidum, and thalamus in Parkinson's disease. Front Neurosci 9. https://doi.org/10.3389/fnins.2015.00101

Garret M, Du Z, Chazalon M, et al. (2018) Alteration of GABAergic neurotransmission in Huntington's disease. CNS Neurosci Ther 24:292–300. https://doi.org/10.1111/cns.12826

Ghasemi N, Razavi S, Nikzad E (2017) Multiple sclerosis: Pathogenesis, symptoms, diagnoses and cell-based therapy. Cell J 19:1–10.

Giráldez-Pérez RM, Antolín-Vallespín M, Muñoz MD, Sánchez-Capelo A (2014) Models of α-synuclein aggregation in Parkinson's disease. Acta Neuropathol Commun 2:176. https://doi.org/10.1186/s40478-014-0176-9

Glater E, Megeath L, Stowers R, Schwarz T (2006) Axonal transport of mitochondria requires Milton to recruit kinesin heavy chain and is light chain independent. J Cell Biol 173:545–57. https://doi.org/10.1083/jcb.200601067

Gmitterová K, Heinemann U, Gawinecka J, et al. (2009) 8-OHdG in cerebrospinal fluid as a marker of oxidative stress in various neurodegenerative diseases. Neurodegener Dis 6:263–269. https://doi.org/10.1159/000237221

Goh XX, Tang PY, Tee SF (2021) 8-hydroxy-2'-deoxyguanosine and reactive oxygen species as biomarkers of oxidative stress in mental illnesses: A meta-analysis. Psychiatry Investig 18:603–618. https://doi.org/10.30773/pi.2020.0417

Graner M, Pointon T, Manton S, et al. (2020) Oligoclonal IgG antibodies in multiple sclerosis target patient-specific peptides. PLoS One 15:e0228883. https://doi.org/10.1371/journal.pone.0228883

Gravielle MC (2021) Regulation of GABAA receptors induced by the activation of L-type voltage-gated calcium channels. Membranes 11:486. https://doi.org/10.3390/membranes11070486

Groiss SJ, Wojtecki L, Südmeyer M, Schnitzler A (2009) Deep brain stimulation in Parkinson's disease. Ther Adv Neurol Disord 2:20–28. https://doi.org/10.1177/1756285609339382

Grossauer A, Hemicker G, Krismer F, et al. (2023) α-Synuclein seed amplification assays in the diagnosis of synucleinopathies using cerebrospinal fluid—a systematic review and meta-analysis. Mov Disord Clin Pract 10:737–747. https://doi.org/10.1002/mdc3.13710

Gruchot J, Weyers V, Göttle P et al. (2019) The molecular basis for remyelination failure in multiple sclerosis. Cells 8:825. https://doi.org/10.3390/cells8080825

Gudmundson AT, Koo A, Virovka A, et al. (2023) Meta-analysis and open-source database for in vivo brain magnetic resonance spectroscopy in health and disease. bioRxiv 2023.02.10.528046. https://doi.org/10.1101/2023.02.10.528046

Guidara W, Messedi M, Naifar M, et al. (2020) Predictive value of oxidative stress biomarkers in drug-free patients with schizophrenia and schizo-affective disorder. Psychiatry Research 293:113467. https://doi.org/10.1016/j.psychres.2020.113467

Gur RE, Gur RC (2010) Functional magnetic resonance imaging in schizophrenia. Dialogues Clin Neurosci 12:333–343.

Gustavsson A, Norton N, Fast T, et al. (2023) Global estimates on the number of persons across the Alzheimer's disease continuum. Alzheimers Dement 19:658–670. https://doi.org/10.1002/alz.12694

Gutiérrez IL, Dello Russo C, Novellino F, et al. (2022) Noradrenaline in Alzheimer's disease: A new potential therapeutic target. Int J Mol Sci 23:6143. https://doi.org/10.3390/ijms23116143

Hampel H, Hardy J, Blennow K, et al. (2021) The amyloid-β pathway in Alzheimer's disease. Mol Psychiatry 26:5481–5503. https://doi.org/10.1038/s41380-021-01249-0

Han D, Zheng W, Wang X, Chen Z (2020) Proteostasis of α-synuclein and its role in the pathogenesis of Parkinson's disease. Front Cell Neurosci 14:45. https://doi.org/10.3389/fncel.2020.00045

Hanson JE, Yuan H, Perszyk RE, et al. (2024) Therapeutic potential of N-methyl-D-aspartate receptor modulators in psychiatry. Neuropsychopharmacol 49:51–66. https://doi.org/10.1038/s41386-023-01614-3

Harari JH, Díaz-Caneja CM, Janssen J, et al. (2017) The association between gene variants and longitudinal structural brain changes in psychosis: A systematic review of longitudinal neuroimaging genetics studies. NPJ Schizophr 3:1–12. https://doi.org/10.1038/s41537-017-0036-2

Harlow DE, Honce JM, Miravalle AA (2015) Remyelination therapy in multiple sclerosis. Front Neurol 6:257. https://doi.org/10.3389/fneur.2015.00257

Hartwig FP, Borges MC, Horta BL, et al. (2017) Inflammatory biomarkers and risk of schizophrenia: A 2-Sample Mendelian Randomization Study. JAMA Psychiatry 74:1226–1233. https://doi.org/10.1001/jamapsychiatry.2017.3191

He H, Liang L, Tang T, et al. (2020) Progressive brain changes in Parkinson's disease: A meta-analysis of structural magnetic resonance imaging studies. Brain Res 1740:146847. https://doi.org/10.1016/j.brainres.2020.146847

Heilman KM, Nadeau SE (2022) Emotional and neuropsychiatric disorders associated with Alzheimer's disease. Neurotherapeutics 19:99–116. https://doi.org/10.1007/s13311-021-01172-w

Hnilicová P, Štrbák O, Kolisek M, et al. (2020) Current methods of magnetic resonance for noninvasive assessment of molecular aspects of pathoetiology in multiple sclerosis. Int J Mol Sci 21:6117. https://doi.org/10.3390/ijms21176117

Horga G, Bernacer J, Dusi N, et al. (2011) Correlations between ventricular enlargement and gray and white matter volumes of cortex, thalamus, striatum, and internal capsule in schizophrenia. Eur Arch Psychiatry Clin Neurosci 261:467–476. https://doi.org/10.1007/s00406-011-0202-x

Houldsworth A (2024) Role of oxidative stress in neurodegenerative disorders: A review of reactive oxygen species and prevention by antioxidants. Brain Commun 6:fcad356. https://doi.org/10.1093/braincomms/fcad356

Howes OD, Cummings C, Chapman GE, Shatalina E (2023) Neuroimaging in schizophrenia: An overview of findings and their implications for synaptic changes. Neuropsychopharmacol 48:151–167. https://doi.org/10.1038/s41386-022-01426-x

Hsiao H-Y, Chen Y-C, Chen H-M, et al (2013) A critical role of astrocyte-mediated nuclear factor-κB-dependent inflammation in Huntington's disease. Hum Mol Genet 22:1826–1842. https://doi.org/10.1093/hmg/ddt036

Hsu Y-T, Chang Y-G, Chern Y (2018) Insights into GABAAergic system alteration in Huntington's disease. Open Biol 8:180165. https://doi.org/10.1098/rsob.180165

Hu Q, Shi M, Li Y, Zhao X (2024) Elevated plasma neurofilament light was associated with multi-modal neuroimaging features in Alzheimer's disease signature regions and predicted future tau deposition. Res Sq rs.3.rs–3946421. https://doi.org/10.21203/rs.3.rs-3946421/v1

Hu X, Xu W, Ren Y, et al. (2023) Spinal cord injury: Molecular mechanisms and therapeutic interventions. Sig Transduct Target Ther 8:1–28. https://doi.org/10.1038/s41392-023-01477-6

Huang Y, Li Y, Pan H, Han L (2023) Global, regional, and national burden of neurological disorders in 204 countries and territories worldwide. J Glob Health 13:04160. https://doi.org/10.7189/jogh.13.04160

Huhn M, Nikolakopoulou A, Schneider-Thoma J, et al. (2019) Comparative efficacy and tolerability of 32 oral antipsychotics for the acute treatment of adults with multi-episode schizophrenia: A systematic review and network meta-analysis. Lancet 394:939–951. https://doi.org/10.1016/S0140-6736(19)31135-3

Islam MA, Alam SS, Kundu S, et al. (2023) Mesenchymal stem cell therapy in multiple sclerosis: A systematic review and meta-analysis. J Clin Med 12:6311. https://doi.org/10.3390/jcm12196311

Ivey FM, Katzel LI, Sorkin JD, et al. (2012) The Unified Parkinson's Disease Rating Scale as a predictor of peak aerobic capacity and ambulatory function. J Rehabil Res Dev 49:1269–1276.

James OG, Linares AR, Hellegers C, et al. (2021) Evaluating Alzheimer disease with flortaucipir and florbetapir PET: A clinical case series. Clin Nucl Med 46:605–608. https://doi.org/10.1097/RLU.0000000000003493

Jan A, Gonçalves NP, Vaegter CB, et al. (2021) The prion-like spreading of alpha-synuclein in Parkinson's disease: Update on models and hypotheses. Int J Mol Sci 22:8338. https://doi.org/10.3390/ijms22158338

Jeong SH, Lee E-C, Chung SJ, et al. (2022) Local striatal volume and motor reserve in drug-naïve Parkinson's disease. NPJ Parkinsons Dis 8:1–9. https://doi.org/10.1038/s41531-022-00429-1

Jiang A, Handley RR, Lehnert K, Snell RG (2023) From pathogenesis to therapeutics: A review of 150 years of Huntington's disease research. International Journal of Molecular Sciences 24:13021. https://doi.org/10.3390/ijms241613021

Jimenez-Sanchez M, Licitra F, Underwood BR, Rubinsztein DC (2017) Huntington's disease: Mechanisms of pathogenesis and therapeutic strategies. Cold Spring Harb Perspect Med 7:a024240. https://doi.org/10.1101/cshperspect.a024240

Johnson EB, Ziegler G, Penny W, et al. (2021) Dynamics of cortical degeneration over a decade in Huntington's disease. Biol Psychiatry 89:807–816. https://doi.org/10.1016/j.biopsych.2020.11.009

Johnson KA, Fox NC, Sperling RA, Klunk WE (2012) Brain imaging in Alzheimer disease. Cold Spring Harb Perspect Med 2:a006213. https://doi.org/10.1101/cshperspect.a006213

Jost ST, Kaldenbach M-A, Antonini A, et al. (2023) Levodopa dose equivalency in Parkinson's disease: Updated systematic review and proposals. Movement Dis 38:1236–1252. https://doi.org/10.1002/mds.29410

Jung J, Cloutman LL, Binney RJ, Lambon Ralph MA (2017) The structural connectivity of higher order association cortices reflects human functional brain networks. Cortex 97:221–239. https://doi.org/10.1016/j.cortex.2016.08.011

Jurcau A, Jurcau CM (2022) Mitochondria in Huntington's disease: Implications in pathogenesis and mitochondrial-targeted therapeutic strategies. Neural Regen Res 18:1472–1477. https://doi.org/10.4103/1673-5374.360289

Kalincik T, Kubala Havrdova E, Horakova D, et al. (2019) Comparison of fingolimod, dimethyl fumarate and teriflunomide for multiple sclerosis. J Neurol Neurosurg Psychiat 90:458–468. https://doi.org/10.1136/jnnp-2018-319831

Karlsgodt KH, Sun D, Cannon TD (2010) Structural and functional brain abnormalities in schizophrenia. Curr Dir Psychol Sci 19:226–231. https://doi.org/10.1177/0963721410377601

Karrenbauer VD, Bedri SK, Hillert J, Manouchehrinia A (2021) Cerebrospinal fluid oligoclonal immunoglobulin gamma bands and long-term disability progression in multiple sclerosis: A retrospective cohort study. Sci Rep 11:14987. https://doi.org/10.1038/s41598-021-94423-x

Katsavos S, Coles A (2018) Alemtuzumab as treatment for multiple sclerosis. Cold Spring Harb Perspect Med 8:a032029. https://doi.org/10.1101/cshperspect.a032029

Kaur A, Basavanagowda DM, Rathod B, et al. (2020) Structural and functional alterations of the temporal lobe in schizophrenia: A literature review. Cureus 12:e11177. https://doi.org/10.7759/cureus.11177

Kenyon KH, Boonstra F, Noffs G, et al. (2023) An update on the measurement of motor cerebellar dysfunction in multiple sclerosis. Cerebellum 22:761–775. https://doi.org/10.1007/s12311-022-01435-y

Keough MB, Yong VW (2013) Remyelination therapy for multiple sclerosis. Neurotherapeutics 10:44–54. https://doi.org/10.1007/s13311-012-0152-7

Kharel S, Ojha R, Bist A, et al. (2022) Salivary alpha-synuclein as a potential fluid biomarker in Parkinson's disease: A systematic review and meta-analysis. Aging Med 5:53–62. https://doi.org/10.1002/agm2.12192

Kidd D, Barkhof F, McConnell R, et al. (1999) Cortical lesions in multiple sclerosis. Brain 122:17–26. https://doi.org/10.1093/brain/122.1.17

Kim DD, Barr AM, Lian L, et al. (2021) Efficacy and tolerability of aripiprazole versus D2 antagonists in the early course of schizophrenia: A systematic review and meta-analysis. NPJ Schizophr 7:1–9. https://doi.org/10.1038/s41537-021-00158-z

Kim HK, Andreazza AC, Yeung PY, et al. (2014) Oxidation and nitration in dopaminergic areas of the prefrontal cortex from patients with bipolar disorder and schizophrenia. J Psychiatry Neurosci 39:276–285. https://doi.org/10.1503/jpn.130155

Kinnunen KM, Schwarz AJ, Turner EC, et al. (2021) Volumetric MRI-based biomarkers in Huntington's disease: An evidentiary review. Front Neurol 12. https://doi.org/10.3389/fneur.2021.712555

Kish T (2018) Promising multiple sclerosis agents in late-stage development. Pharm Ther 43:750–772.

Klaver R, De Vries HE, Schenk GJ, Geurts JJG (2013) Grey matter damage in multiple sclerosis. Prion 7:66–75. https://doi.org/10.4161/pri.23499

Kreiter D, Postma AA, Hupperts R, Gerlach O (2024) Hallmarks of spinal cord pathology in multiple sclerosis. J Neurol Sci 456:122846. https://doi.org/10.1016/j.jns.2023.122846

Kruse AO, Bustillo JR (2022) Glutamatergic dysfunction in Schizophrenia. Transl Psychiat 12:1–13. https://doi.org/10.1038/s41398-022-02253-w

Lassmann H (2018) Multiple sclerosis pathology. Cold Spring Harb Perspect Med 8:a028936. https://doi.org/10.1101/cshperspect.a028936

Lassmann H (2019) Pathogenic mechanisms associated with different clinical courses of multiple sclerosis. Front Immunol 9. https://doi.org/10.3389/fimmu.2018.03116

Lee MH, Shin JI, Yang JW, et al. (2022) Genome editing using CRISPR-Cas9 and autoimmune diseases: A comprehensive review. Int J Mol Sci 23:1337. https://doi.org/10.3390/ijms23031337

Lemprière S (2019) Antisense oligonucleotide reduces mutant protein in patients with HD. Nat Rev Neurol 15:435–435. https://doi.org/10.1038/s41582-019-0211-3

Lesh TA, Niendam TA, Minzenberg MJ, Carter CS (2011) Cognitive control deficits in schizophrenia: Mechanisms and meaning. Neuropsychopharmacol 36:316–338. https://doi.org/10.1038/npp.2010.156

Leucht C, Kitzmantel M, Chua L, et al. (2008) Haloperidol versus chlorpromazine for treatment of schizophrenia. Schizophr Bull 34:813–815. https://doi.org/10.1093/schbul/sbn087

Leuzy A, Janelidze S, Mattsson-Carlgren N, et al. (2021) Comparing the clinical utility and diagnostic performance of CSF P-Tau181, P-Tau217, and P-Tau231 assays. Neurology 97:e1681–e1694. https://doi.org/10.1212/WNL.0000000000012727

Li D-D, Zhang Y-H, Zhang W, Zhao P (2019) Meta-analysis of randomized controlled trials on the efficacy and safety of donepezil, galantamine, rivastigmine, and memantine for the treatment of Alzheimer's disease. Front Neurosci 13. https://doi.org/10.3389/fnins.2019.00472

Li M, Wang Y, Tachibana M, et al. (2022a) Atypical structural connectivity of language networks in autism spectrum disorder: A meta-analysis of diffusion tensor imaging studies. Autism Res 15:1585–1602. https://doi.org/10.1002/aur.2789

Li Y, Li F, Qin D, et al. (2022b) The role of brain derived neurotrophic factor in central nervous system. Front Aging Neurosci 14. https://doi.org/10.3389/fnagi.2022.986443

Liang X-S, Sun Z-W, Thomas AM, Li S (2023) Mesenchymal stem cell therapy for Huntington disease: A meta-analysis. Stem Cells Int 2023:1109967. https://doi.org/10.1155/2023/1109967

Lie IA, Weeda MM, Mattiesing RM, et al. (2022) Relationship between white matter lesions and gray matter atrophy in multiple sclerosis. Neurology 98:e1562–e1573. https://doi.org/10.1212/WNL.0000000000200006

Lin C-I, Chang Y-C, Kao N-J, et al. (2020) 1,25(OH)2D3 alleviates Aβ(25-35)-induced tau hyperphosphorylation, excessive reactive oxygen species, and apoptosis through interplay with glial cell line-derived neurotrophic factor signaling in SH-SY5Y cells. Int J Mol Sci 21:4215. https://doi.org/10.3390/ijms21124215

Lin K-J, Lin K-L, Chen S-D, et al. (2019) The overcrowded crossroads: Mitochondria, alpha-synuclein, and the endo-lysosomal system interaction in Parkinson's disease. Int J Mol Sci 20:5312. https://doi.org/10.3390/ijms20215312

Lin M, Liu N, Qin Z, Wang Y (2022) Mitochondrial-derived damage-associated molecular patterns amplify neuroinflammation in neurodegenerative diseases. Acta Pharmacol Sin 43:2439–2447. https://doi.org/10.1038/s41401-022-00879-6

Liu G, Yang C, Wang X, et al. (2023) Oxygen metabolism abnormality and Alzheimer's disease: An update. Redox Biol 68:102955. https://doi.org/10.1016/j.redox.2023.102955

Liu J, Chang L, Song Y, et al. (2019) The role of NMDA receptors in Alzheimer's disease. Front Neurosci 13:43. https://doi.org/10.3389/fnins.2019.00043

Liu X, Lai Y, Wang X, et al. (2013) Reduced white matter integrity and cognitive deficit in never-medicated chronic schizophrenia: A diffusion tensor study using TBSS. Behav Brain Res 252:157–163. https://doi.org/10.1016/j.bbr.2013.05.061

López-Muguruza E, Matute C (2023) Alterations of oligodendrocyte and myelin energy metabolism in multiple sclerosis. Int J Mol Sci 24:12912. https://doi.org/10.3390/ijms241612912

Lopez-Soley E, Martinez-Heras E, Solana E, et al. (2023) Diffusion tensor imaging metrics associated with future disability in multiple sclerosis. Sci Rep 13:3565. https://doi.org/10.1038/s41598-023-30502-5

Lorscheider J, Benkert P, Lienert C, et al. (2021) Comparative analysis of dimethyl fumarate and fingolimod in relapsing–remitting multiple sclerosis. J Neurol 268:941–949. https://doi.org/10.1007/s00415-020-10226-6

Luvsannyam E, Jain MS, Pormento MKL, et al. (2022) Neurobiology of schizophrenia: A comprehensive review. Cureus 14:e23959. https://doi.org/10.7759/cureus.23959

Ma L, Gholam Azad M, Dharmasivam M, et al. (2021) Parkinson's disease: Alterations in iron and redox biology as a key to unlock therapeutic strategies. Redox Biol 41:101896. https://doi.org/10.1016/j.redox.2021.101896

Maciak K, Dziedzic A, Saluk J (2023) Remyelination in multiple sclerosis from the miRNA perspective. Front Mol Neurosci 16:1199313. https://doi.org/10.3389/fnmol.2023.1199313

Madireddy S, Madireddy S (2020) Regulation of reactive oxygen species-mediated damage in the pathogenesis of schizophrenia. Brain Sci 10:742. https://doi.org/10.3390/brainsci10100742

Magliozzi R, Reynolds R, Calabrese M (2018) MRI of cortical lesions and its use in studying their role in MS pathogenesis and disease course. Brain Pathol 28:735–742. https://doi.org/10.1111/bpa.12642

Maiti P, Manna J, Dunbar GL (2017) Current understanding of the molecular mechanisms in Parkinson's disease: Targets for potential treatments. Trans Neurodegener 6:28. https://doi.org/10.1186/s40035-017-0099-z

Maity S, Komal P, Kumar V, et al. (2022) Impact of ER stress and ER-mitochondrial crosstalk in Huntington's disease. Int J Mol Sci 23:780. https://doi.org/10.3390/ijms23020780

Malik JA, Yaseen Z, Thotapalli L et al. (2023) Understanding translational research in schizophrenia: A novel insight into animal models. Mol Biol Rep 50:3767–3785. https://doi.org/10.1007/s11033-023-08241-7

Mamah D, Ji A, Rutlin J, Shimony JS (2018) White matter integrity in schizophrenia and bipolar disorder: Tract- and voxel-based analyses of diffusion data from the Connectom scanner. Neuroimage Clin 21:101649. https://doi.org/10.1016/j.nicl.2018.101649

Mangale V, McIntyre LL, Walsh CM, et al. (2019) Promoting remyelination through cell transplantation therapies in a model of viral-induced neurodegenerative disease. Dev Dyn 248:43–52. https://doi.org/10.1002/dvdy.24658

Manjari S, Abraham SM, Poornima R, et al. (2023) Unprecedented effect of vitamin D3 on T-cell receptor beta subunit and alpha7 nicotinic acetylcholine receptor expression in a 3-nitropropionic acid induced mouse model of Huntington's disease. IBRO Neurosci Rep 15:116–125. https://doi.org/10.1016/j.ibneur.2023.07.001

Manjari SKV, Maity S, Poornima R, et al. (2022) Restorative action of vitamin D3 on motor dysfunction through enhancement of neurotrophins and antioxidant expression in the striatum. Neuroscience 492:67–81. https://doi.org/10.1016/j.neuroscience.2022.03.039

Marcus C, Mena E, Subramaniam RM (2014) Brain PET in the diagnosis of Alzheimer's disease. Clin Nucl Med 39:e413–e426. https://doi.org/10.1097/RLU.0000000000000547

Margoni M, Preziosa P, Rocca MA, Filippi M (2023) Depressive symptoms, anxiety and cognitive impairment: Emerging evidence in multiple sclerosis. Transl Psychiat 13:1–13. https://doi.org/10.1038/s41398-023-02555-7

Marrodan M, Gaitán MI, Correale J (2020) Spinal cord involvement in MS and other demyelinating diseases. Biomedicines 8:130. https://doi.org/10.3390/biomedicines8050130

Martinez-Hernandez E, Sepulveda M, Rostásy K, et al. (2015) Antibodies to aquaporin 4, myelin-oligodendrocyte glycoprotein, and the Glycine receptor α1 subunit in patients with isolated optic neuritis. JAMA Neurol 72:187–193. https://doi.org/10.1001/jamaneurol.2014.3602

Marucci G, Buccioni M, Ben DD, et al. (2021) Efficacy of acetylcholinesterase inhibitors in Alzheimer's disease. Neuropharmacology 190:108352. https://doi.org/10.1016/j.neuropharm.2020.108352

Matsuda H, Okita K, Motoi Y, et al. (2022) Clinical impact of amyloid PET using 18F-florbetapir in patients with cognitive impairment and suspected Alzheimer's disease: A multicenter study. Ann Nucl Med 36:1039–1049. https://doi.org/10.1007/s12149-022-01792-y

Matuz-Mares D, González-Andrade M, Araiza-Villanueva MG, et al. (2022) Mitochondrial calcium: Effects of its imbalance in disease. Antioxidants (Basel) 11:801. https://doi.org/10.3390/antiox11050801

McCutcheon RA, Keefe RSE, McGuire PK (2023) Cognitive impairment in schizophrenia: Aetiology, pathophysiology, and treatment. Mol Psychiatry 28:1902–1918. https://doi.org/10.1038/s41380-023-01949-9

McGregor MM, Nelson AB (2019) Circuit mechanisms of Parkinson's disease. Neuron 101:1042–1056. https://doi.org/10.1016/j.neuron.2019.03.004

Meftah S, Gan J (2023) Alzheimer's disease as a synaptopathy: Evidence for dysfunction of synapses during disease progression. Front Synaptic Neurosci 15:1129036. https://doi.org/10.3389/fnsyn.2023.1129036

Mielke MM, Syrjanen JA, Blennow K, et al. (2019) Plasma and CSF neurofilament light. Neurology 93:e252–e260. https://doi.org/10.1212/WNL.0000000000007767

Miguez A, Gomis C, Vila C, et al. (2023) Soluble mutant huntingtin drives early human pathogenesis in Huntington's disease. Cell Mol Life Sci 80:238. https://doi.org/10.1007/s00018-023-04882-w

Miquel-Rio L, Sarriés-Serrano U, Pavia-Collado R, et al. (2023) The role of α-synuclein in the regulation of serotonin system: Physiological and pathological features. Biomedicines 11(541). https://doi.org/10.3390/biomedicines11020541

Miranda M, Morici JF, Zanoni MB, Bekinschtein P (2019) Brain-derived neurotrophic factor: A key molecule for memory in the healthy and the pathological brain. Front Cell Neurosci 13(363). https://doi.org/10.3389/fncel.2019.00363

Mohammadi R, Aryan A, Omrani MD, et al. (2021) Autologous hematopoietic stem cell transplantation (AHSCT): An evolving treatment avenue in multiple sclerosis. Biol Targ Ther 15:53. https://doi.org/10.2147/BTT.S267277

Molnar MJ, Molnar V, Fedor M, et al. (2022) Improving mood and cognitive symptoms in Huntington's disease with Cariprazine treatment. Front Psychiatry 12. https://doi.org/10.3389/fpsyt.2021.825532

Montoya A, Price BH, Menear M, Lepage M (2006) Brain imaging and cognitive dysfunctions in Huntington's disease. J Psychiatry Neurosci 31:21–29.

Morigaki R, Goto S (2017) Striatal vulnerability in Huntington's disease: Neuroprotection versus neurotoxicity. Brain Sci 7:63. https://doi.org/10.3390/brainsci7060063

Morris G, Berk M (2015) The many roads to mitochondrial dysfunction in neuroimmune and neuropsychiatric disorders. BMC Med 13:68. https://doi.org/10.1186/s12916-015-0310-y

Murray AJ, Rogers JC, Katshu MZUH, et al. (2021) Oxidative stress and the pathophysiology and symptom profile of schizophrenia spectrum disorders. Front Psychiatry 12(703452). https://doi.org/10.3389/fpsyt.2021.703452

Nabizadeh F, Mohamadi M, Rahmani S, et al. (2023) Safety and efficacy of cladribine in multiple sclerosis: A systematic review and meta-analysis. Neurol Sci 1–13. https://doi.org/10.1007/s10072-023-06794-w

Nagae LM, Honce JM, Tanabe J, et al. (2016) Microstructural changes within the basal ganglia differ between Parkinson disease subtypes. Front Neuroanat 10(17). https://doi.org/10.3389/fnana.2016.00017

Năstase MG, Vlaicu I, Trifu SC (2022) Genetic polymorphism and neuroanatomical changes in schizophrenia. Rom J Morphol Embryol 63:307–322. https://doi.org/10.47162/RJME.63.2.03

Negi RS, Manchanda KL, Sanga S (2014) Imaging of Huntington's disease. Med J Armed Forces India 70:386–388. https://doi.org/10.1016/j.mjafi.2012.08.002

Negi S, Khurana N, Duggal N (2024) The misfolding mystery: α-synuclein and the pathogenesis of Parkinson's disease. Neurochem Int 177:105760. https://doi.org/10.1016/j.neuint.2024.105760

Neumann B, Segel M, Chalut KJ, Franklin RJ (2019) Remyelination and ageing: Reversing the ravages of time. Mult Scler 25:1835–1841. https://doi.org/10.1177/1352458519884006

Niccolini F, Politis M (2014) Neuroimaging in Huntington's disease. World J Radiol 6:301–312. https://doi.org/10.4329/wjr.v6.i6.301

Nieto R, Kukuljan M, Silva H (2013) BDNF and schizophrenia: From neurodevelopment to neuronal plasticity, learning, and memory. Front Psychiatry 4. https://doi.org/10.3389/fpsyt.2013.00045

Nopoulos PC (2016) Huntington disease: A single-gene degenerative disorder of the striatum. Dialogues Clin Neurosci 18:91–98.

Nuytemans K, Theuns J, Cruts M, Van Broeckhoven C (2010) Genetic etiology of Parkinson disease associated with mutations in the SNCA, PARK2, PINK1, PARK7, and LRRK2 genes: A mutation update. Hum Mutat 31:763–780. https://doi.org/10.1002/humu.21277

O'Shea DM, Arkhipenko A, Galasko D, et al. (2024) Practical use of DAT SPECT imaging in diagnosing dementia with Lewy bodies: A US perspective of current guidelines and future directions. Front Neurol 15. https://doi.org/10.3389/fneur.2024.1395413

Olufunmilayo EO, Gerke-Duncan MB, Holsinger RMD (2023) Oxidative stress and antioxidants in neurodegenerative disorders. Antioxidants 12(517). https://doi.org/10.3390/antiox12020517

Onofrj M, Russo M, Delli Pizzi S, et al. (2023) The central role of the Thalamus in psychosis, lessons from neurodegenerative diseases and psychedelics. Transl Psychiatry 13:1–14. https://doi.org/10.1038/s41398-023-02691-0

Ontaneda D, Raza PC, Mahajan KR, et al. (2021) Deep grey matter injury in multiple sclerosis: A NAIMS consensus statement. Brain 144:1974–1984. https://doi.org/10.1093/brain/awab132

Oreja-Guevara C, Ayuso Blanco T, Brieva Ruiz L, et al. (2019) Cognitive dysfunctions and assessments in multiple sclerosis. Front Neurol 10:581. https://doi.org/10.3389/fneur.2019.00581

Ou Y-N, Xu W, Li J-Q, et al. (2019) FDG-PET as an independent biomarker for Alzheimer's biological diagnosis: A longitudinal study. Alzheimers Res Ther 11:57. https://doi.org/10.1186/s13195-019-0512-1

Pagano G, Niccolini F, Politis M (2016) Current status of PET imaging in Huntington's disease. Eur J Nucl Med Mol Imaging 43:1171–1182. https://doi.org/10.1007/s00259-016-3324-6

Palasz E, Wysocka A, Gasiorowska A, et al. (2020) BDNF as a promising therapeutic agent in Parkinson's disease. Int J Mol Sci 21:1170. https://doi.org/10.3390/ijms21031170

Papiri G, D'Andreamatteo G, Cacchiò G, et al. (2023) Multiple sclerosis: Inflammatory and neuroglial aspects. Curr Issues Mol Biol 45:1443–1470. https://doi.org/10.3390/cimb45020094

Pasquini J, Brooks DJ, Pavese N (2021) The cholinergic brain in Parkinson's disease. Mov Disord Clin Pract 8:1012–1026. https://doi.org/10.1002/mdc3.13319

Peretz C, Segev H, Rozani V, et al. (2016) Comparison of Selegiline and Rasagiline therapies in Parkinson disease: A real-life study. Clin Neuropharmacol 39:227–231. https://doi.org/10.1097/WNF.0000000000000167

Perez-Rando M, Penades-Gomiz C, Martinez-Marin P, et al. (2023) Volume alterations of the hippocampus and amygdala in patients with schizophrenia and persistent auditory hallucinations. Rev Psiquiatr Salud Ment 18(4):241–249. https://doi.org/10.1016/j.rpsm.2023.05.002

Pieperhoff P, Südmeyer M, Dinkelbach L, et al. (2022) Regional changes of brain structure during progression of idiopathic Parkinson's disease – A longitudinal study using deformation based morphometry. Cortex 151:188–210. https://doi.org/10.1016/j.cortex.2022.03.009

Pierson SR, Kolling LJ, James TD, et al. (2024) Serotonergic dysfunction may mediate the relationship between alcohol consumption and Alzheimer's disease. Pharmacol Res 203:107171. https://doi.org/10.1016/j.phrs.2024.107171

Pintér A, Cseh D, Sárközi A, et al. (2015) Autonomic dysregulation in multiple sclerosis. Int J Mol Sci 16:16920–16952. https://doi.org/10.3390/ijms160816920

Popescu BFGh, Pirko I, Lucchinetti CF (2013) Pathology of multiple sclerosis: Where do we stand? Continuum (Minneap Minn) 19:901–921. https://doi.org/10.1212/01.CON.0000433291.23091.65

Prakash KG, Bannur BM, Chavan MD, et al. (2016) Neuroanatomical changes in Parkinson's disease in relation to cognition: An update. J Adv Pharm Technol Res 7:123–126. https://doi.org/10.4103/2231-4040.191416

Qin J, Ma Z, Chen X, Shu S (2023) Microglia activation in central nervous system disorders: A review of recent mechanistic investigations and development efforts. Front Neurol 14:1103416. https://doi.org/10.3389/fneur.2023.1103416

Qu Y, Li J, Qin Q, et al. (2023) A systematic review and meta-analysis of inflammatory biomarkers in Parkinson's disease. NPJ Parkinsons Dis 9:18. https://doi.org/10.1038/s41531-023-00449-5

Rahman A, Hossen MA, Chowdhury MFI, et al. (2023) Aducanumab for the treatment of Alzheimer's disease: A systematic review. Psychogeriatrics 23:512–522. https://doi.org/10.1111/psyg.12944

Rajasekaran A, Venkatasubramanian G, Berk M, Debnath M (2015) Mitochondrial dysfunction in schizophrenia: Pathways, mechanisms and implications. Neurosci Biobehav Rev 48:10–21. https://doi.org/10.1016/j.neubiorev.2014.11.005

Ramesh S, Arachchige ASPM (2023) Depletion of dopamine in Parkinson's disease and relevant therapeutic options: A review of the literature. AIMS Neurosci 10:200–231. https://doi.org/10.3934/Neuroscience.2023017

Rao YL, Ganaraja B, Murlimanju BV, et al. (2022) Hippocampus and its involvement in Alzheimer's disease: A review. 3 Biotech 12:55. https://doi.org/10.1007/s13205-022-03123-4

Rascol O, Fabbri M, Poewe W (2021) Amantadine in the treatment of Parkinson's disease and other movement disorders. Lancet Neurol 20:1048–1056. https://doi.org/10.1016/S1474-4422(21)00249-0

Rawat P, Sehar U, Bisht J, et al. (2022) Phosphorylated Tau in Alzheimer's disease and other tauopathies. Int J Mol Sci 23:12841. https://doi.org/10.3390/ijms232112841

Reddy PH (2014) Increased mitochondrial fission and neuronal dysfunction in Huntington's disease: Implications for molecular inhibitors of excessive mitochondrial fission. Drug Discov Today 19:951–955. https://doi.org/10.1016/j.drudis.2014.03.020

Reilmann R, Holtbernd F, Bachmann R, et al. (2013) Grasping multiple sclerosis: Do quantitative motor assessments provide a link between structure and function? J Neurol 260:407–414. https://doi.org/10.1007/s00415-012-6639-7

Roberts RC (2021) Mitochondrial dysfunction in schizophrenia: With a focus on postmortem studies. Mitochondrion 56:91–101. https://doi.org/10.1016/j.mito.2020.11.009

Robinson N, Bergen SE (2021) Environmental risk factors for schizophrenia and bipolar disorder and their relationship to genetic risk: Current knowledge and future directions. Front Genet 12. https://doi.org/10.3389/fgene.2021.686666

Rocha EM, De Miranda B, Sanders LH (2018) Alpha-synuclein: Pathology, mitochondrial dysfunction and neuroinflammation in Parkinson's disease. Neurobiol Dis 109:249–257. https://doi.org/10.1016/j.nbd.2017.04.004

Rodrigues FB, Duarte GS, Costa J, et al. (2017) Tetrabenazine versus deutetrabenazine for Huntington's disease: Twins or distant cousins? Mov Disord Clin Pract 4:582–585. https://doi.org/10.1002/mdc3.12483

Rodríguez JJ, Noristani HN, Verkhratsky A (2012) The serotonergic system in ageing and Alzheimer's disease. Prog Neurobiol 99:15–41. https://doi.org/10.1016/j.pneurobio.2012.06.010

Romano A, Serviddio G, Calcagnini S, et al. (2017) Linking lipid peroxidation and neuropsychiatric disorders: Focus on 4-hydroxy-2-nonenal. Free Rad Biol Med 111:281–293. https://doi.org/10.1016/j.freeradbiomed.2016.12.046

Roos I, Hughes S, McDonnell G et al. (2023) Rituximab vs Ocrelizumab in relapsing-remitting multiple sclerosis. JAMA Neurol 80:789–797. https://doi.org/10.1001/jamaneurol.2023.1625

Ruan X, Lin F, Wu D, et al. (2021) Comparative efficacy and safety of dopamine agonists in advanced Parkinson's disease with motor fluctuations: A systematic review and network meta-analysis of double-blind randomized controlled trials. Front Neurosci 15. https://doi.org/10.3389/fnins.2021.728083

Ruggieri S, Petracca M, De Giglio L, et al. (2021) A matter of atrophy: Differential impact of brain and spine damage on disability worsening in multiple sclerosis. J Neurol 268:4698–4706. https://doi.org/10.1007/s00415-021-10576-9

Saeed U, Lang AE, Masellis M (2020) Neuroimaging advances in Parkinson's disease and atypical Parkinsonian syndromes. Front Neurol 11. https://doi.org/10.3389/fneur.2020.572976

Sala A, Caminiti SP, Presotto L, et al. (2021) In vivo human molecular neuroimaging of dopaminergic vulnerability along the Alzheimer's disease phases. Alzheimers Res Ther 13:187. https://doi.org/10.1186/s13195-021-00925-1

Sampogna G, Di Vincenzo M, Giuliani L, et al. (2023) A systematic review on the effectiveness of antipsychotic drugs on the quality of life of patients with schizophrenia. Brain Sci 13:1577. https://doi.org/10.3390/brainsci13111577

Sandroff BM, Motl RW, Román CAF, et al. (2022) Thalamic atrophy moderates associations among aerobic fitness, cognitive processing speed, and walking endurance in persons with multiple sclerosis. J Neurol 269:5531–5540. https://doi.org/10.1007/s00415-022-11205-9

Sari Y (2011) Huntington's disease: From mutant huntingtin protein to neurotrophic factor therapy. Int J Biomed Sci 7:89–100.

Sarter M, Lustig C, Taylor SF (2012) Cholinergic contributions to the cognitive symptoms of schizophrenia and the viability of cholinergic treatments. Neuropharmacology 62:1544–1553. https://doi.org/10.1016/j.neuropharm.2010.12.001

Sayyaed A, Saraswat N, Vyawahare N, Kulkarni A (2023) A detailed review of pathophysiology, epidemiology, cellular and molecular pathways involved in the development and prognosis of Parkinson's disease with insights into screening models. Bull Natl Res Cent 47:70. https://doi.org/10.1186/s42269-023-01047-4

Scaduto P, Lauterborn JC, Cox CD, et al. (2023) Functional excitatory to inhibitory synaptic imbalance and loss of cognitive performance in people with Alzheimer's disease neuropathologic change. Acta Neuropathol 145:303–324. https://doi.org/10.1007/s00401-022-02526-0

Schultz SA, Strain JF, Adedokun A, et al. (2020) Serum neurofilament light chain levels are associated with white matter integrity in autosomal dominant Alzheimer's disease. Neurobiol Dis 142:104960. https://doi.org/10.1016/j.nbd.2020.104960

Sengupta U, Kayed R (2022) Amyloid β, Tau, and α-Synuclein aggregates in the pathogenesis, prognosis, and therapeutics for neurodegenerative diseases. Prog Neurobiol 214:102270. https://doi.org/10.1016/j.pneurobio.2022.102270

Seo JH, Shin JH, Lee J, et al. (2023) DNA double-strand break-free CRISPR interference delays Huntington's disease progression in mice. Commun Biol 6:1–12. https://doi.org/10.1038/s42003-023-04829-8

Sharma C, Kim SR (2023) Oxidative stress: Culprit or consequence in Alzheimer's amyloidopathy. Neural Regen Res 18:1948. https://doi.org/10.4103/1673-5374.367843

Sharma K, Pradhan S, Duffy LK, et al. (2021) Role of receptors in relation to plaques and tangles in Alzheimer's disease pathology. Int J Mol Sci 22:12987. https://doi.org/10.3390/ijms222312987

Shim J-H, Baek H-M (2021) Diffusion measure changes of substantia nigra subregions and the ventral tegmental area in newly diagnosed Parkinson's disease. Exp Neurobiol 30:365–373. https://doi.org/10.5607/en21025

Siddiqui AJ, Badraoui R, Jahan S, et al. (2023) Targeting NMDA receptor in Alzheimer's disease: Identifying novel inhibitors using computational approaches. Front Pharmacol 14. https://doi.org/10.3389/fphar.2023.1208968

Šimić G, Tkalčić M, Vukić V, et al. (2021) Understanding emotions: Origins and roles of the Amygdala. Biomolecules 11:823. https://doi.org/10.3390/biom11060823

Song Z, Zhang J, Xue T, et al. (2021) Different Catechol-O-Methyl transferase inhibitors in Parkinson's disease: A Bayesian network meta-analysis. Front Neurol 12. https://doi.org/10.3389/fneur.2021.707723

Speidell A, Bin Abid N, Yano H (2023) Brain-derived neurotrophic factor dysregulation as an essential pathological feature in Huntington's disease: Mechanisms and potential therapeutics. Biomedicines 11:2275. https://doi.org/10.3390/biomedicines11082275

Srinivasan E, Chandrasekhar G, Chandrasekar P, et al. (2021) Alpha-synuclein aggregation in Parkinson's disease. Front Med (Lausanne) 8:736978. https://doi.org/10.3389/fmed.2021.736978

Srivastava A, Alam P, Caughey B (2022) RT-QuIC and related assays for detecting and quantifying prion-like pathological seeds of α-synuclein. Biomolecules 12:576. https://doi.org/10.3390/biom12040576

Stoddard-Bennett T, Reijo Pera R (2019) Treatment of Parkinson's disease through personalized medicine and induced pluripotent stem cells. Cells 8:26. https://doi.org/10.3390/cells8010026

Sumiyoshi T, Kunugi H, Nakagome K (2014) Serotonin and dopamine receptors in motivational and cognitive disturbances of schizophrenia. Front Neurosci 8. https://doi.org/10.3389/fnins.2014.00395

Sun S, Xiao S, Guo Z, et al. (2023) Meta-analysis of cortical thickness reduction in adult schizophrenia. J Psychiatry Neurosci 48:E461–E470. https://doi.org/10.1503/jpn.230081

Švob Štrac D, Pivac N, Mück-Šeler D (2016) The serotonergic system and cognitive function. Transl Neurosci 7:35–49. https://doi.org/10.1515/tnsci-2016-0007

Swallow E, Pham T, Patterson-Lomba O, et al. (2023) Comparative efficacy and safety of ozanimod and ponesimod for relapsing multiple sclerosis: A matching-adjusted indirect comparison. Mult Scler Relat Disord 71:104551. https://doi.org/10.1016/j.msard.2023.104551

Synnott PG, Whittington MD, Lin GA, et al. (2021) The effectiveness and value of aducanumab for Alzheimer's disease. J Manag Care Spec Pharm 27:10.18553/jmcp.2021.27.11.1613. https://doi.org/10.18553/jmcp.2021.27.11.1613

Tabrizi SJ, Leavitt BR, Landwehrmeyer GB, et al. (2019) Targeting huntingtin expression in patients with Huntington's disease. N Engl J Med 380:2307–2316. https://doi.org/10.1056/NEJMoa1900907

Tae W-S, Ham B-J, Pyun S-B, et al. (2018) Current clinical applications of diffusion-tensor imaging in neurological disorders. J Clin Neurol 14:129–140. https://doi.org/10.3988/jcn.2018.14.2.129

Talaat F, Abdelatty S, Ragaie C, Dahshan A (2023) Chitinase-3-like 1-protein in CSF: A novel biomarker for progression in patients with multiple sclerosis. Neurol Sci 44:3243–3252. https://doi.org/10.1007/s10072-023-06764-2

Tambasco N, Romoli M, Calabresi P (2018) Levodopa in Parkinson's disease: Current status and future developments. Curr Neuropharmacol 16:1239–1252. https://doi.org/10.2174/1570159X15666170510143821

Tan B, Shishegar R, Oldham S, et al. (2022) Investigating longitudinal changes to frontal cortico-striatal tracts in Huntington's disease: The IMAGE-HD study. Brain Imaging Behav 16:2457–2466. https://doi.org/10.1007/s11682-022-00699-6

Teleanu RI, Niculescu A-G, Roza E, et al. (2022) Neurotransmitters—Key factors in neurological and neurodegenerative disorders of the Central nervous system. Int J Mol Sci 23:5954. https://doi.org/10.3390/ijms23115954

Thorne NJ, Tumbarello DA (2022) The relationship of alpha-synuclein to mitochondrial dynamics and quality control. Front Mol Neurosci 15:947191. https://doi.org/10.3389/fnmol.2022.947191

Todd KL, Brighton T, Norton ES, et al. (2018) Ventricular and periventricular anomalies in the aging and cognitively impaired brain. Front Aging Neurosci 9:445. https://doi.org/10.3389/fnagi.2017.00445

Tolkovsky AM, Spillantini MG (2021) Tau aggregation and its relation to selected forms of neuronal cell death. Essays Biochem 65:847–857. https://doi.org/10.1042/EBC20210030

Tong H, Yang T, Xu S, et al. (2024) Huntington's disease: Complex pathogenesis and therapeutic strategies. Int J Mol Sci 25:3845. https://doi.org/10.3390/ijms25073845

Turkmen BA, Yazici E, Erdogan DG, et al. (2021) BDNF, GDNF, NGF and Klotho levels and neurocognitive functions in acute term of schizophrenia. BMC Psychiatry 21:562. https://doi.org/10.1186/s12888-021-03578-4

Vallée A (2022) Neuroinflammation in schizophrenia: The key role of the WNT/β-catenin pathway. Int J Mol Sci 23:2810. https://doi.org/10.3390/ijms23052810

Vegas-Suarez S, Paredes-Rodriguez E, Aristieta A, et al. (2019) Chapter nine – Dysfunction of serotonergic neurons in Parkinson's disease and dyskinesia. In Sharma HS, Sharma A (eds) International review of neurobiology. Academic Press, pp. 259–279.

Verma M, Lizama BN, Chu CT (2022) Excitotoxicity, calcium and mitochondria: A triad in synaptic neurodegeneration. Trans Neurodegener 11:3. https://doi.org/10.1186/s40035-021-00278-7

Vita A, De Peri L, Deste G, Sacchetti E (2012) Progressive loss of cortical gray matter in schizophrenia: A meta-analysis and meta-regression of longitudinal MRI studies. Transl Psychiatry 2:e190. https://doi.org/10.1038/tp.2012.116

Volloch V, Rits-Volloch S (2023) Effect of Lecanemab and Donanemab in early Alzheimer's disease: Mechanistic interpretation in the Amyloid Cascade Hypothesis 2.0 perspective. J Alzheimers Dis 93:1277–1284. https://doi.org/10.3233/JAD-230164

Walton C, King R, Rechtman L, et al. (2020) Rising prevalence of multiple sclerosis worldwide: Insights from the Atlas of MS, third edition. Mult Scler 26:1816–1821. https://doi.org/10.1177/1352458520970841

Wang J, Fu J, Zhao Y, et al. (2023a) Iron and targeted iron therapy in Alzheimer's disease. Int J Mol Sci 24:16353. https://doi.org/10.3390/ijms242216353

Wang R, Reddy PH (2017) Role of glutamate and NMDA receptors in Alzheimer's disease. J Alzheimers Dis 57:1041–1048. https://doi.org/10.3233/JAD-160763

Wang W, Li Y, Meng X (2023b) Vitamin D and neurodegenerative diseases. Heliyon 9:e12877. https://doi.org/10.1016/j.heliyon.2023.e12877

Wareski P, Vaarmann A, Choubey V, et al. (2009) PGC-1{alpha} and PGC-1{beta} regulate mitochondrial density in neurons. J Biol Chem 284:21379–21385. https://doi.org/10.1074/jbc.M109.018911

Warren, N, O'Gorman C, Horgan I, et al. (2024) Inflammatory cerebrospinal fluid markers in schizophrenia spectrum disorders: A systematic review and meta-analysis of 69 studies with 5710 participants. Schizophr Res 266:24–31. https://doi.org/10.1016/j.schres.2024.02.001

Weinshenker D (2018) Long road to ruin: Noradrenergic dysfunction in neurodegenerative disease. Trends Neurosci 41:211–223. https://doi.org/10.1016/j.tins.2018.01.010

Welton T, Hartono S, Shih Y-C, et al. (2023) Microstructure of brain nuclei in early Parkinson's disease: Longitudinal diffusion kurtosis imaging. JPD 13:233–242. https://doi.org/10.3233/JPD-225095

Weng M, Xie X, Liu C, et al. (2018) The sources of reactive oxygen species and its possible role in the pathogenesis of Parkinson's disease. Parkinsons Dis 2018:9163040. https://doi.org/10.1155/2018/9163040

Wildner P, Stasiołek M, Matysiak M (2020) Differential diagnosis of multiple sclerosis and other inflammatory CNS diseases. Multiple Sclerosis Relat Dis 37:101452. https://doi.org/10.1016/j.msard.2019.101452

Wilkins HM, Weidling IW, Ji Y, Swerdlow RH (2017) Mitochondria-derived damage-associated molecular patterns in neurodegeneration. Front Immunol 8:508. https://doi.org/10.3389/fimmu.2017.00508

Wingerchuk DM, Lucchinetti CF, Noseworthy JH (2001) Multiple sclerosis: Current pathophysiological concepts. Lab Invest 81:263–281. https://doi.org/10.1038/labinvest.3780235

Yang AC, Tsai S-J (2017) New targets for schizophrenia treatment beyond the dopamine hypothesis. Int J Mol Sci 18:1689. https://doi.org/10.3390/ijms18081689

Yang H-M, Yang S, Huang S-S, et al. (2017) Microglial activation in the pathogenesis of Huntington's disease. Front Aging Neurosci 9:193. https://doi.org/10.3389/fnagi.2017.00193

Yang W, Bai X, Guan X, et al. (2024) The longitudinal volumetric and shape changes of subcortical nuclei in Parkinson's disease. Sci Rep 14:7494. https://doi.org/10.1038/s41598-024-58187-4

Ye Z (2022) Mapping neuromodulatory systems in Parkinson's disease: Lessons learned beyond dopamine. Curr Med 1:15. https://doi.org/10.1007/s44194-022-00015-w

Yi S, Wang L, Wang H, et al. (2022) Pathogenesis of α-synuclein in Parkinson's disease: From a neuron-glia crosstalk perspective. Int J Mol Sci 23:14753. https://doi.org/10.3390/ijms232314753

Zhang J, Cheng H, Liu W, et al. (2022a) Neurofilament light chain in cerebrospinal fluid or blood as a biomarker for mild cognitive impairment. Medicine (Baltimore) 101:e28932. https://doi.org/10.1097/MD.0000000000028932

Zhang Q, Lei Y-H, Zhou J-P, et al. (2019) Role of PGC-1α in mitochondrial quality control in neurodegenerative diseases. Neurochem Res 44:2031–2043. https://doi.org/10.1007/s11064-019-02858-6

Zhang S-Y, Gui L-N, Liu Y-Y, et al. (2020) Oxidative stress marker aberrations in multiple sclerosis: A meta-analysis study. Front Neurosci 14:823. https://doi.org/10.3389/fnins.2020.00823

Zhang W, Xiao D, Mao Q, Xia H (2023a) Role of neuroinflammation in neurodegeneration development. Sig Transduct Target Ther 8:1–32. https://doi.org/10.1038/s41392-023-01486-5

Zhang X, Liu S, Zhan F, et al. (2017) The effects of medium spiny neuron morphologcial changes on basal ganglia network under external electric field: A computational modeling study. Front Comput Neurosci 11. https://doi.org/10.3389/fncom.2017.00091

Zhang Z, Zhao L, Zhou X, et al. (2023b) Role of inflammation, immunity, and oxidative stress in hypertension: New insights and potential therapeutic targets. Front Immunol 13:1098725. https://doi.org/10.3389/fimmu.2022.1098725

Zheng F, Li C, Zhang D, et al. (2019) Study on the sub-regions volume of hippocampus and amygdala in schizophrenia. Quant Imaging Med Surg 9:1025–1036. https://doi.org/10.21037/qims.2019.05.21

Zhou C, Guo T, Bai X, et al. (2021a) Locus coeruleus degeneration is associated with disorganized functional topology in Parkinson's disease. Neuroimage Clin 32:102873. https://doi.org/10.1016/j.nicl.2021.102873

Zhou W, Tang M, Sun L, et al. (2024) Subcortical structure alteration in patients with drug-induced parkinsonism: Evidence from neuroimaging. IBRO Neurosci Rep 16:436–442. https://doi.org/10.1016/j.ibneur.2024.03.001

Zhou Z, Zhong S, Zhang R, et al. (2021b) Functional analysis of brain derived neurotrophic factor (BDNF) in Huntington's disease. Aging (Albany NY) 13:6103–6114. https://doi.org/10.18632/aging.202603

Zhu F, Zhang L, Liu F, et al. (2018) Altered serum tumor necrosis factor and interleukin-1β in first-episode drug-naive and chronic schizophrenia. Front Neurosci 12. https://doi.org/10.3389/fnins.2018.00296

Zhu S, Zhao L, Fan Y, et al. (2020) Interaction between TNF-α and oxidative stress status in first-episode drug-naïve schizophrenia. Psychoneuroendocrinology 114:104595. https://doi.org/10.1016/j.psyneuen.2020.104595

Zong Y, Li H, Liao P, et al. (2024) Mitochondrial dysfunction: Mechanisms and advances in therapy. Sig Transduct Target Ther 9:1–29. https://doi.org/10.1038/s41392-024-01839-8

Zuccato C, Cattaneo E (2007) Role of brain-derived neurotrophic factor in Huntington's disease. Prog Neurobiol 81:294–330. https://doi.org/10.1016/j.pneurobio.2007.01.003

Zuccato C, Ciammola A, Rigamonti D, et al. (2001) Loss of huntingtin-mediated BDNF gene transcription in Huntington's disease. Science 293:493–498. https://doi.org/10.1126/science.1059581

5 Epilepsy
From Seizures to Understanding the Neural Circuitry

Harsha, Kapil Suchal, Soumya S. Pattnaik, Vinay Shankar Dubey, Deepak Rathi, and Prabhakar Singh

5.1 INTRODUCTION: EPILEPSY AS A DISORDER OF NEURAL CIRCUIT

Epilepsy is a persistent neurological condition marked by recurring seizures. It affects around 50 million people worldwide, ranking as one of the most frequent neurological disorders. Nearly 80% of people with epilepsy reside in low or middle-income countries, posing a substantial burden on both individuals and society (Srivastava et al., 2021).

Seizures are defined as transient symptoms and signs due to abnormal, excessive, or simultaneous neuronal activity of a population of neuronal cells in the brain. Epilepsy is defined as a chronic disorder of the brain characterized by an enduring disposition toward recurrent unprovoked seizures and by the neurobiological, cognitive, psychological, and social consequences of this condition. The diagnosis of epilepsy requires at least two unprovoked seizures occurring >24 h apart. A syndrome is defined as a characteristic seizure associated with abnormal investigations that occur together in a recognizable pattern. It usually includes more than one type of epilepsy.

Understanding the microscopic mechanisms underlying seizures is crucial for effective epilepsy treatment, although the etiology of epilepsy is notably complex. It encompasses a vast array of modifications in neural circuits, which span from molecular alterations to network-level changes (Laxpati et al., 2014). Neural circuits are intricate connections formed by neurons of diverse properties and functions within the brain (Lerner et al., 2016; Zhang et al., 2021). Neurons with distinct properties and functions establish complex connections in various forms, constituting neural circuits and neural networks at different levels (Malezieux et al., 2023). These activities take various forms, such as series, parallel, feedforward, and feedback (positive or negative) (EyeWirers et al., 2016). In recent years, the emergence of advanced techniques, such as viral tracing, optogenetics, and chemogenetics, has facilitated a deeper understanding of the intricate neural networks implicated in epilepsy (Lv et al., 2022; Xiao et al., 2022). The application of these techniques substantially enriches our comprehension of the disorder's pathophysiology and has been instrumental in the identification of new therapeutic targets.

5.2 SEIZURE CLASSIFICATION AND CLINICAL PRESENTATION

The International League Against Epilepsy (ILAE) operational classification of seizure types was published in 2017 (Table 5.1). In 2023, the ILAE's Executive Committee appointed a working group assigned to assess the real-world application of the 2017 seizure classification and recommend adjustments while preserving the fundamental framework of the 2017 classification. The basic organization of the 2017 classification evolved from the original 1981 version through subsequent modifications. The primary objective remains the establishment of a common language and standardized definitions for clinical practice. Emphasizing flexibility, the classification aims to accommodate diverse clinical settings, including resource-limited areas and highly specialized centers.

DOI: 10.1201/9781003500025-5

TABLE 5.1
ILAE 2017 Seizure Classification

Main Class (Onset)	Sub-class/Awareness	Motor	Non-motor	Special Type
Focal-onset	*Aware* or *Impaired awareness*	Automatisms, atonic, clonic, epileptic spasms, hyperkinetic, myoclonic, tonic	Autonomic, cognitive, emotional, sensory, behavior arrest	Focal → bilateral tonic-clonic ("secondarily generalized")
Generalized-onset	(Not subclassified by awareness)	Tonic-clonic, clonic, tonic, myoclonic, myoclonic-tonic-clonic, myoclonic-atonic, atonic, epileptic spasms	Absence seizures: typical, atypical, myoclonic, eyelid myoclonic	
Unknown-onset	Use available semiology	Same potential features as focal or generalized	Same as above	If evolving to bilateral tonic-clonic, noted as such
Unclassified	When onset and features cannot be confidently assigned	–	–	Reserved for seizures lacking sufficient information

Simultaneously, it aims to provide a clear and robust framework for implementation in research databases and clinical trials (Beniczky et al., 2025).

5.2.1 ILAE 2025 Operational Seizure Classification

Four main seizure classes: Focal, generalized, unknown (whether focal or generalized), and unclassified.
Total of 21 seizure types, reduced from 63 in the 2017 framework
Six key updates:

1. Removed "onset" from class names.
2. Distinguished *classifiers* (biological) vs. *descriptors* (semiological features).
3. Replaced *awareness* with *consciousness* (awareness + responsiveness).
4. Shifted from "motor vs non-motor" to *observable vs. non-observable manifestations.*
5. Semiological description now follows *chronological sequence.*
6. Recognized *epileptic negative myoclonus* as a seizure type.

5.3 CELLULAR AND SYNAPTIC MECHANISMS OF EPILEPTOGENESIS

Epileptogenesis, the process leading to the development of epilepsy, is a complex phenomenon that involves cellular and synaptic mechanisms, including dysfunctions in ion channels, alterations in synaptic plasticity, and disruptions to the excitatory-inhibitory balance. Understanding the interplay of these factors is crucial for elucidating the mechanisms of seizure initiation and propagation.

5.3.1 Ion Channel Dysfunction

Dysfunction of ion channels is a critical aspect of the pathophysiology of epilepsy. Mutations in genes encoding various ion channels—specifically voltage-gated sodium channels (VGSCs), calcium channels, and potassium channels—are recognized as underlying causes of different epileptic syndromes. Evidence indicates that approximately 25% of genetic epilepsy cases are attributable to mutations in ion channel genes (). For example, mutations in the SCN1A gene, which encodes a sodium channel, are associated with Dravet syndrome and other severe childhood epilepsies, leading to significant network hyperexcitability primarily due to altered ion current dynamics (Klassen et al., 2011).

In addition to inherited mutations, de novo mutations in ion channels are frequently found in genetic forms of epilepsy, complicating diagnosis and personalized risk assessment for affected individuals. Functional studies demonstrate that these mutations can disrupt normal cellular processes, resulting in hyperexcitability within neuronal circuits due to either gain-of-function or loss-of-function effects. Potassium channels, such as KCNQ2 and KCNQ3, are essential for maintaining resting membrane potential and repolarizing action potentials; mutations in these channels can increase neuronal excitability, thus contributing to seizure susceptibility.

Furthermore, changes in ion channel expression and function occur after acute neurological insults, such as traumatic brain injury or status epilepticus (SE), suggesting a progressive alteration in neuronal circuit excitability. Studies using animal models reveal that alterations in ion channel expression during the latent period after initial seizures enhance neuronal hyperexcitability and may facilitate the eventual onset of spontaneous seizures (Jung et al., 2007).

5.3.2 Synaptic Plasticity Alterations

The role of synaptic plasticity in epileptogenesis is fundamental, as it affects circuit functionality and the alteration of synaptic strength following seizures. Proper neuronal communication relies on a delicate balance of excitation and inhibition, mediated through long-term potentiation (LTP) and long-term depression (LTD) mechanisms. Following a seizure, there is often a heightened excitatory synaptic transmission that may predispose the network to further seizure activity.

Research indicates that even subtle injuries or disruptions can result in enhanced synaptic efficacy due to synaptic plasticity mechanisms operating under Hebbian learning principles. For example, experiments in animal models have shown that after trauma, there is an increase in excitatory postsynaptic potentials (EPSPs) and modifications to dendritic spines, mirroring changes seen with LTP. These alterations can create a state of hyperexcitability in neuronal circuits, making seizure activity more likely.

Moreover, recent findings suggest that AMPA receptors and NMDA receptors undergo significant changes in their expression and function after seizures. Specifically, AMPAR trafficking to the postsynaptic membrane following neonatal seizures enhances synaptic strength, which is pivotal for the long-term consequences of such events. Consequently, an imbalance favoring excitation over inhibition develops, contributing greatly to the emergence of chronic epilepsy characterized by recurrent seizures.

5.3.3 Excitatory-Inhibitory Balance in Seizure Initiation

The excitatory-inhibitory (E-I) balance is central to maintaining normal brain function, and its disruption is a hallmark of epileptogenesis. In a functional network, excitatory neurons primarily release neurotransmitters like glutamate, while inhibitory interneurons release gamma-aminobutyric acid (GABA). This equilibrium regulates neuronal excitability and prevents excessive synchronous firing characteristic of seizures.

Research indicates that alterations in GABAergic signaling—often influenced by ion channel dysfunction—can diminish inhibitory control over excitatory networks. For instance, decreased efficacy of GABA receptors is observed in various epilepsy models, resulting in reduced inhibitory postsynaptic currents (IPSCs) that increase the overall excitatory drive. This effect is particularly evident in temporal lobe epilepsy, where synaptic reorganization often accompanies a relative loss of inhibitory interneuron function and an increase in excitatory connectivity.

The interaction between disrupted ion channel function and altered synaptic plasticity highlights the critical role these mechanisms play in fostering a pro-epileptogenic environment. Indeed, models of acquired epilepsy suggest that the loss of chronic inhibition—whether through environmental changes, excitotoxicity, or genetic predisposition—can heighten seizure susceptibility by enhancing synaptic strength and reducing inhibition (Leite et al., 2005). Thus, the dynamics of channelopathies combined with synaptic changes can activate feedback loops that worsen the E-I imbalance, facilitating both the development and escalation of seizure frequency and severity.

5.3.4 Integrated View of Mechanisms

In summary, the mechanisms underlying epileptogenesis are intricately interconnected, encompassing ion channel dysfunction, changes in synaptic plasticity, and subsequent disruptions to excitatory-inhibitory balance. These processes foster an environment conducive for seizures, where heightened cellular excitability and synaptic strengthening culminate in a self-perpetuating cycle of hyperexcitability.

Insights from genetic studies underscore the importance of comprehending both the individual contributions of channel mutations and their interactions with synaptic mechanisms in establishing epilepsy (Li et al., 2021). Given the complexity of these relationships, a multifaceted approach is essential for unraveling the molecular, cellular, and network-level contributions to epilepsy, paving the way for novel therapeutic strategies targeting these mechanisms effectively.

As ongoing research continues to illuminate the diverse genetic, synaptic, and cellular interactions involved in epilepsy, valuable information is unveiled that can inform future interventions aimed at restoring a healthier balance of excitatory and inhibitory signaling in the brain.

The synaptic and molecular processes behind epileptogenesis are depicted in Figure 5.1.

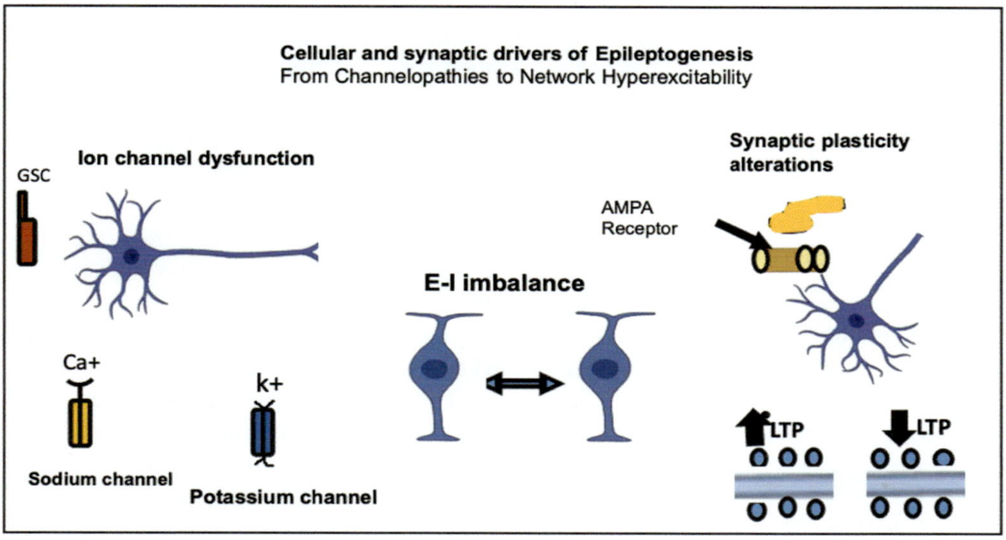

FIGURE 5.1 Cellular and synaptic drivers of epileptogenesis: From channelopathies to network hyperexcitability. This outline shows the synaptic and molecular processes behind epileptogenesis. An imbalance between excitatory and inhibitory transmission (E–I imbalance) and aberrant neuronal firing are caused by dysfunction of voltage-gated sodium, potassium, and calcium channels. Hyperexcitable networks are a result of altered synaptic plasticity, which includes aberrant LTP and defective LTD. Depolarization is exacerbated, and normal inhibitory control is disrupted by alterations in AMPA and GABA receptor expression. The biological underpinning for seizure formation and propagation is created when these cellular and synaptic abnormalities work together to create a self-reinforcing state of network hyperexcitability.

5.4 NETWORK LEVEL REORGANIZATION AND SEIZURE PROPAGATION

Epilepsy isn't limited to single neurons firing abnormally—it involves complex networks of interconnected brain regions. Epileptic activity, initiating from a local source, spreads to wider brain regions by specific network nodes and pathways, which leads to an epileptic seizure. Spencer recognized the epilepsy network as a whole system, which is composed of a set of bilaterally distributed cortical and subcortical brain structures and regions that are functionally and anatomically connected. In this network, the activity of any one part can affect the activity of the other parts. Previously, a large number of functional imaging studies confirmed that bilaterally distributed brain networks were involved in epileptic seizures of Childhood Absence Epilepsy, mainly including the medial frontal lobe (FL), parietal lobe (PL), anterior/posterior cingulate cortex (ACC/PCC), thalamus, and ascending brainstem. Some studies also reported the involution of temporal lobe (TL) and occipital lobe (OL). These brain network regions constitute an important network pathway, namely cortical-thalamic circuit, highlighting the important role of cortex and thalamus in the occurrence of epilepsy. However, whether the cortical or subcortical areas are dominant is still unknown. Early experiments were in favor of cortical theory and laid the foundation for the cortico-reticular concept, which stressed that the thalamus/reticular formation and cortex constituted the epilepsy network and were responsible for generating SWDs.

The concept of functional connectivity (FC) was gradually introduced to further study how the various brain regions in the brain network are connected and their related functions. FC is defined as the correlation between remote neurophysiological events, which is used to determine whether there is a functional correlation between these events.

The following framework outlines how seizures manifest, propagate, and can be modulated at a systems level.

5.4.1 Fundamentals of Brain Networks

- Brain networks consist of *nodes* (neurons or brain regions) and *edges* (synapses or white-matter tracts), analyzed using *graph theory* tools such as network density, clustering coefficient, path length, and efficiency.
- Healthy brains typically exhibit *small-world* organization: high local clustering and short global paths, facilitating efficient communication and synchronization.

5.4.2 Epileptic Network Dynamics

- *Partial (focal) seizures* typically begin within an *epileptogenic zone (EZ network)*—a group of highly synchronized nodes generating fast discharges. This network may include lesions and exhibit seizure onset rhythms and desynchronization events.
- As seizures evolve, a broader *propagation network* emerges, involving cortical and subcortical regions (e.g., thalamus), with slower, more synchronous activity triggering clinical signs.

5.4.3 Temporal Network Phases During Seizures

Phase	Network Characteristics
Interictal	Altered baseline connectivity; elevated clustering, longer paths, hub fragility in focal epilepsies (especially TLE)
Ictal onset	EZ nodes fire synchronously; network segregates into tightly connected modules (increased assortativity), isolating focus from surrounding tissue
Propagation	Recruitment of new nodes across cortex and thalamus; network expands and synchronizes over large scales
Termination	Network reverts toward higher segregation; synchronous modules break down facilitating seizure end

5.4.4 Micro- vs. Macro-Circuit Contributions

- *Microcircuits*: Abnormal inhibitory/excitatory balance at neuron level—often involving GABAergic interneuron deficits and glutamate hyperactivation, notably in hippocampus and thalamus.
- *Macrocircuits*: Changes in structural and functional connectivity among large brain regions (e.g., hippocampus, subiculum, entorhinal cortex, thalamus, raphe) alter seizure susceptibility and spread.

5.4.5 Clinical Implications: Modulating Epileptic Networks

- *Neuromodulation techniques* leverage network dynamics:
  - *Vagal nerve stimulation (VNS)* modulates cortical-limbic connectivity, reconfiguring salience networks and increasing regional homogeneity.
  - *Deep brain stimulation (DBS)* targeting the anterior nucleus of the thalamus (ANT) influences limbic and cortical circuits; responsiveness is linked to network desynchronization patterns.
  - *Responsive neurostimulation (RNS)* adapts stimulation based on detected ictal patterns; effective RNS patients show specific connectivity patterns linking medial prefrontal, cingulate, and pulvinar–precuneus hubs.

The Network reorganization and seizure propagation pathways in epilepsy are depicted in Figure 5.2.

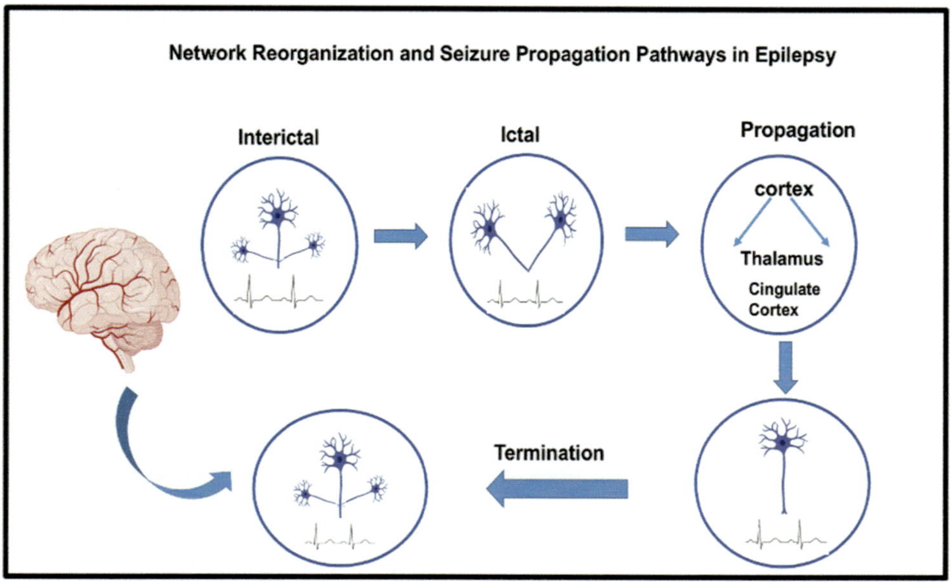

FIGURE 5.2 Network reorganization and seizure propagation pathways in epilepsy. Diagrammatic illustration showing the dynamic rearrangement of neural networks during the interictal, ictal propagation, and termination phases of seizure activity. Starting in cortical areas, epileptic discharges propagate through interrelated subcortical structures such as the thalamus and cingulate cortex. The spread of seizures is aided by the thalamocortical network, a vital relay system that synchronizes and amplifies excitatory neural activity. Loss of synchronization and restoration of inhibitory tone are factors in seizure cessation during the terminal phase. The intricacy of network-level changes and the part cortical–subcortical communication plays in epileptogenesis and seizure dynamics are highlighted in this picture.

5.5 NEURODEVELOPMENTAL VS. ACQUIRED EPILEPSIES: STRUCTURAL AND CIRCUIT DIFFERENCES

Neurodevelopmental and acquired epilepsy present markedly different etiological backgrounds and temporal dynamics, leading to substantial distinctions in the underlying neural structures, circuit functions, and treatment strategies. Etiology and timing critically influence neural development, brain plasticity, and therapeutic responses, reshaping the understanding of epilepsy management. This synthesis reviews the role of etiology and timing in shaping structural and functional characteristics of the brain, particularly concerning neurodevelopmental and acquired epilepsy.

5.5.1 Etiology and Its Role in Neural Development

Neurodevelopmental epilepsies often arise from congenital abnormalities or genetic mutations occurring during critical periods of brain development. These conditions, such as infantile spasms and Dravet syndrome, are frequently associated with mutations in ion channel genes or other genes involved in synaptic functioning, indicating a strong genetic predisposition (Anney, 2014; Perucca & Scheffer, 2020). Conversely, acquired epilepsies develop due to external insults or stressors, including stroke, traumatic brain injury, or infections, often leading to secondary changes in established neural circuits (Pijet et al., 2024; Reddy et al., 2016; Vezzani et al., 2015). For example, ischemic events can result in a latent period before the onset of seizures, suggesting a complex interaction between the initial injury and subsequent neural changes (Pijet et al., 2024; Zheng et al., 2013).

The timing of the insult is paramount, as the brain undergoes significant structural changes during development. Early insults during critical periods can lead to alterations in neurogenesis and synaptic plasticity, setting the stage for abnormal circuit formation and increased susceptibility to seizures later in life (Dahl-Hansen et al., 2018; Vezzani et al., 2015). In neurodevelopmental disorders, disruptions in neural circuits can manifest as altered connectivity and function, consistently linked with specific cognitive and behavioral outcomes, which can be less predictable in acquired epilepsies. For example, studies have shown that the structural brains of children with epilepsy differ significantly from controls, with marked deviations in gray and white matter correlating with different syndromes (Stevelink et al., 2023; Zhang et al., 2019). This structural and functional perturbation often signifies a less adaptable brain capable of recovery, potentially reducing the effectiveness of conventional therapeutic strategies (Diessen et al., 2014; Pauletti et al., 2019).

5.5.2 Structural and Circuit Differences

Neurodevelopmental and acquired epilepsy are characterized by distinct structural and functional brain alterations. In the former, specific regions often demonstrate marked abnormalities at a very young age, evident in volumetric reductions, which may be indicative of sensorimotor dysfunction. For instance, temporal lobe epilepsy has frequently been associated with reductions in hippocampal volume attributed to various genetic conditions resulting in structural malformations (Haast et al., 2022; Memarian et al., 2013). Similarly, findings from the ENIGMA study reveal common neuroanatomical signatures across various epilepsy types but highlight significant abnormalities unique to developmental epilepsies, emphasizing a potential shared pathway influenced by the timing of etiological factors (Stevelink et al., 2023; Whelan et al., 2018).

In acquired epilepsy, structural changes arise from neurotoxic processes, inflammation, or injury, leading to neurodegeneration. Studies involving patients post-stroke have shown specific patterns of gray matter atrophy contextualized within the network topology of seizure propagation. Here, the volume reductions and connectivity alterations tend to concentrate in specific regions, which is markedly different from the widespread alterations observed in developmental forms

(Diessen et al., 2014; Reddy et al., 2016). Moreover, immune responses, particularly involving the neuroinflammatory cascade, are foundational to the progression from an initial insult to the development of chronic epilepsy; elevated levels of inflammatory biomarkers have been consistently linked to reduced neuroplasticity and susceptibility to seizure development post-injury (Pauletti et al., 2019; Vezzani et al., 2015).

5.5.3 The Impact of Time on Treatment Approaches

The timing of seizures significantly impacts treatment modalities for epilepsy. In neurodevelopmental forms, the use of antiseizure medications (ASMs) can modulate abnormal cortical excitability; however, interventions must be carefully timed to align with developmental milestones to prevent altering or stunting neural growth. Certain medications like sodium valproate have been associated with cognitive side effects, emphasizing the need for individualized treatment regimens that consider developmental stages (Pauletti et al., 2019; Tondelli et al., 2016).

In contrast, acquired epilepsies often require a multifaceted treatment approach addressing not just seizure control but also underlying brain health and restoration. Recent advances emphasize using ASMs alongside neuroprotective strategies aimed at mitigating secondary damage and promoting recovery after the initial injury (Pauletti et al., 2019; Reddy et al., 2016). Ongoing research indicates that treatments targeting oxidative stress can ameliorate outcomes in acquired epilepsy through restorations of neural integrity, suggesting that benefits gained from early intervention may significantly improve patient quality of life and reduce the incidence of subsequent seizures (Pauletti et al., 2019; Pijet et al., 2024).

5.5.4 Brain Plasticity and Epilepsy

Brain plasticity—its capacity for functional and structural reorganization—varies significantly between neurodevelopmental and acquired contexts. In neurodevelopmental epilepsy, the plasticity of the developing brain can facilitate some recovery from structural abnormalities; when appropriate therapeutic interventions are applied, they may enhance synaptic repair or reorganization. However, treatment responses are often inconsistent and may result in varying degrees of success depending on the timing and severity of the initial developmental disruption (Dahl-Hansen et al., 2018; Perucca & Scheffer, 2020).

On the other hand, in acquired epilepsies, plasticity is often maladaptive following injury, leading to the establishment of hyperexcitable networks that exacerbate seizure activity. Mechanisms that encourage rearrangement in neural pathways post-injury are influenced by both the type of acquired insult and the resultant pathological changes. Research indicates that repeated seizures can lead to additional neural reorganization, resulting in a vicious cycle of further seizure activity and structural injury (Pauletti et al., 2019; Pijet et al., 2024). Therapeutic focus thus not only includes managing acute seizures but also may involve implementing neuroprotective therapies capable of fostering beneficial plasticity while curtailing detrimental reorganization in the chronic phase post-injury (Pijet et al., 2024; Wang et al., 2023).

5.6 GENETIC AND MOLECULAR FOUNDATIONS OF EPILEPSY

Epilepsy is a heterogeneous neurological disorder characterized by recurrent seizures resulting from excessive neuronal excitability. The genetic and molecular underpinnings of epilepsy have emerged as a focal point of research, elucidating the mechanisms underlying various seizure phenotypes. Specifically, channelopathies, regulatory mutations, and changes in gene expression contribute significantly to the pathophysiology of epilepsy. This chapter aims to explore these components, detailing the complex interactions that lead to the manifestation of different seizure types.

5.6.1 Channelopathies in Epilepsy

Channelopathies, or disorders arising from dysfunctional ion channels, have been extensively linked to epilepsy. Mutations in ion channel genes can significantly alter neuronal excitability, predisposing individuals to seizures. The study by Subaran and Greenberg (2014) highlighted the pivotal role of ion channel genes in epilepsy, noting that early investigations identified numerous monogenic forms of epilepsy linked to such mutations.

One of the most well-studied ion channel genes is SCN1A, which encodes a sodium channel that is integral to action potential generation in neurons. Mutations in this gene have been identified in various epilepsy syndromes, particularly in Dravet syndrome, characterized by early-onset seizures and developmental delays (Klassen et al., 2011). Further, research by Klassen et al. revealed that benign variants in ion channel genes could complicate personal risk assessment in epilepsy, indicating the genetic complexity involved in these disorders (Klassen et al., 2011).

Another critical ion channel contributing to seizure phenotypes is the KCNMA1 gene, which encodes large-conductance calcium-activated potassium channels (BK channels). Mutations in KCNMA1 have been associated with a spectrum of epilepsy syndromes, particularly generalized epilepsy with absences, as discussed by Dong et al. (2022). The alteration in the function of these channels can lead to persistent membrane depolarization, fostering seizure susceptibility.

Moreover, mutations in HCN channels have been implicated in familial forms of epilepsy. Chang et al. suggested that variations in HCN channel genes may enhance neuronal excitability and thereby contribute to the onset of seizures via mechanisms referred to as "h-channelopathies" (Chang et al., 2019). These variations encompass structural aberrations and transcriptional and regulatory deficits that may significantly alter channel function.

5.6.2 Mechanisms of Channelopathy

The functional consequences of channel mutations extend beyond simple loss of channel activity. Research has shown that certain mutations can yield gain-of-function phenotypes, wherein the altered channels contribute positively to excitability rather than diminishing it. For example, Zybura et al. discussed how pharmacological modulation of SCN8A mutations demonstrates varied responses, indicating differential regulatory properties of channel mutations in seizure pathology (Zybura et al., 2022).

The gain-of-function mechanism suggests that, beyond traditionally understood loss-of-function channelopathies, the excitatory drive can stem from altered channel kinetics and voltage-dependence, leading to heightened neuronal excitability and facilitating seizure genesis. Such insights are essential when developing targeted therapies that account for the specific mutation profiles in individual patients.

5.6.3 Regulatory Mutations: Beyond Channelopathies

In addition to channelopathies, regulatory mutations in single genes can profoundly impact neuronal function and development, fostering epileptic phenomena. For instance, the mTOR signaling pathway plays a vital role in regulating cellular growth, proliferation, and neuronal differentiation. Dysregulation of this pathway due to mutations in genes such as DEPDC5 or NPRL3 leads to malformations of cortical development and subsequent epilepsy associated with focal cortical dysplasia (Iffland et al., 2019; Li et al., 2021).

The role of GATOR1 complex mutations associated with mTOR signaling was extensively delineated by Korenke et al. who noted that these mutations cause disruptions in translational regulation, leading to hyperactivation of mTORC1 and resulting in cortical malformations that relate directly to epileptogenic processes (Korenke et al., 2016). This highlights the importance of understanding broader regulatory networks contributing to epilepsy rather than focusing solely on direct channel function.

Notably, STXBP1, a gene crucial for neurotransmitter release, illustrates how regulatory mutations can severely impact neuronal function. Truncating mutations in STXBP1 lead to severe infantile epilepsy and are critical in the convergence of various epilepsy disorders (Hormozdiari et al., 2014). The role of such regulatory proteins underscores the multifaceted nature of epilepsy, wherein mutations may contribute both directly and indirectly to seizure generation.

5.6.4 Impacts on Gene Expression and Epileptic Phenotypes

Gene expression changes governed by mutations and regulatory elements also significantly affect seizure phenotypes. For example, alterations in gene expression patterns within interneuron populations have been shown to diminish inhibitory drive, allowing for excessive excitatory activity and increased seizure susceptibility. This phenomenon has been linked to transcription factors like CHD2, which, when mutated, can skew the development and function of cortical interneurons (Meganathan et al., 2017).

Additionally, the regulatory landscape of genes involved in GABAergic signaling, such as those coding for GABAA receptors, illustrates how gene expression changes must be incorporated into the understanding of epilepsy. The research conducted by Durisic et al. demonstrated the effectiveness of SAHA (Vorinostat) in correcting synaptic deficiencies caused by mutations in GABAA receptor subunits, showcasing how pharmacological interventions can be refined by understanding the underlying regulatory mutations and their effects on gene expression (Durisic et al., 2018).

The interplay between mutations in regulatory genes and their subsequent impact on neurodevelopment and circuit formation is crucial for understanding the etiology of epilepsy. Misregulation of transcription factors and downstream effects on target gene expression can create vulnerabilities in neural circuits that predispose them to hyperexcitability (Lewis & Chetkovich, 2011).

5.6.5 Epigenetics and Epilepsy

Emerging research into epigenetic modifications is beginning to illuminate how environmental factors can exacerbate or initiate seizure disorders alongside genetic predispositions. It has been postulated that the interplay between genetic mutations, epigenetic mechanisms, and environmental triggers could comprehensively explain the diversity of epilepsy phenotypes observed in the clinic. For example, factors such as stress, metabolic disturbances, or neuroinflammatory responses can lead to changes in epigenetic marks, altering gene expression profiles and potentially unleashing a latent predisposition to epilepsy (Yao et al., 2021).

Given the complexity surrounding gene expression regulation, it becomes imperative to assess how these epigenetic layers can be considered in the framework of epilepsy treatment and management. Approaches targeting epigenetic modulation show promise, not only in the treatment of epilepsy but also in paving the way for developing individualized therapeutics based on a patient's specific genetic and epigenetic backgrounds.

5.7 NEUROIMAGING AND ELECTROPHYSIOLOGICAL TOOLS FOR CIRCUIT MAPPING

Epilepsy is a complex neurological disorder characterized by spontaneous and recurrent seizures resulting from abnormal electrical activity in the brain. Establishing a clear localization of the EZ—the area of the brain responsible for generating seizures—is imperative for effective treatment, particularly surgical intervention. The development of innovative neuroimaging and electrophysiological tools has bolstered our understanding of brain networks involved in epilepsy, enhancing the identification and characterization of the EZ. This chapter discusses prominent techniques employed in mapping brain circuits in epilepsy, exploring the strengths, weaknesses, and clinical applications of electroencephalography (EEG), magnetoencephalography (MEG), functional magnetic resonance imaging (fMRI), positron emission tomography (PET), and stereo-electroencephalography (sEEG).

5.7.1 EEG and Its Role in Localizing Epileptogenic Zones

EEG offers a high temporal resolution, capturing electrical brain activity with millisecond precision. This technique records electrical potentials at the scalp through electrodes, allowing for the detection of interictal epileptiform discharges (IEDs) and ictal activities that signify epileptic events (Tamilia et al., 2021). EEG has long been a cornerstone in the presurgical evaluation of epilepsy, primarily due to its availability and cost-effectiveness.

The sensitivity of EEG to detect IEDs varies across different brain regions, with notable limitations in detecting activity from deep structures such as the hippocampus and insular cortex (Shultz et al., 2014). Advances in high-density EEG, characterized by an increased number of electrodes, have improved spatial resolution and localization of seizure foci. Studies such as those conducted by Michel and Brunet have demonstrated that combining high-density EEG recordings with sophisticated source localization algorithms provides three-dimensional images of electrical activity, enhancing our understanding of epileptic dynamics (Michel & Brunet, 2019).

Despite its advantages, EEG is not without drawbacks. It is susceptible to noise and artifacts from physiological sources, such as muscle activity and eye movements, which can impact interpretability. Moreover, classic EEG may miss critical foci located in regions with complex morphologies (Chowdhury et al., 2015). These limitations underscore the necessity of multimodal approaches to enhance localization accuracy and depth assessment.

5.7.2 Magnetoencephalography (MEG): A Complementary Approach

MEG provides a non-invasive method for studying neuronal activity by measuring magnetic fields generated by electrical currents in the brain. One advantage of MEG over EEG is its superior spatial resolution due to the minimal distortion of magnetic fields by surrounding tissues, making it particularly useful for localizing deep brain structures (Neymotin et al., 2019).

Recent studies have highlighted the synergistic potential of combining EEG and MEG in clinical settings. Chowdhury et al. emphasize that the fusion of EEG and MEG data improves the localization of epileptic foci, even in cases with low signal-to-noise ratios (SNR) (Chowdhury et al., 2015). This integrative approach enables a more comprehensive evaluation of electrophysiological data, ultimately guiding surgical decisions.

MEG can map brain activity during resting states and while patients engage in various tasks, providing insights into network behavior in epilepsy. For instance, Neymotin et al. developed the Human Neocortical Neurosolver (HNN) software, which interprets MEG/EEG data at the microcircuit level, allowing for predictions about transient oscillations relevant to epilepsy (Neymotin et al., 2019). Such strategies are vital for understanding the functional architecture of brain networks affected by epileptogenesis.

5.7.3 The Role of Functional Magnetic Resonance Imaging (fMRI)

fMRI measures brain activity indirectly through changes in blood flow and oxygenation levels, providing insights into regional brain function over time. Unlike EEG and MEG, fMRI has excellent spatial resolution, allowing for precise anatomical localization (Wang et al., 2024).

Recent applications in epilepsy include resting-state fMRI, which provides valuable information about network connectivity even in the absence of overt seizure activity. It allows the investigation of functional connectivity patterns between brain regions that host epileptic networks, assisting in the identification of the EZ among individuals with non-lesional epilepsy (Piastra et al., 2020). However, fMRI is limited by its temporal resolution, making it less suitable for capturing rapid oscillatory events associated with seizures. Consequently, its most effective application lies in conjunction with other modalities, offering complementary data to facilitate surgical planning (Tamilia et al., 2021).

5.7.4 Positron Emission Tomography (PET) in Epilepsy Research

PET imaging utilizes radioactive tracers to evaluate metabolic activity within the brain, providing critical insights into the pathophysiological processes associated with epilepsy. The development of new radiotracers targeting various neurotransmitter systems adds depth to epilepsy research. For instance, alterations in glucose metabolism can serve as biomarkers for the EZ in drug-resistant patients (Shultz et al., 2014).

Notably, PET can be instrumental in localizing the EZ in patients with MRI-negative focal epilepsy by assessing regional metabolic patterns. The investigation conducted by Wang et al. demonstrates the utility of multimodal approaches involving PET alongside other neuroimaging techniques in patients with non-lesional epilepsy (Wang et al., 2024). However, PET's relatively low temporal resolution limits its utility in capturing transient events characteristic of seizures, highlighting its most effective application in assessing chronic pathology over acute seizure dynamics.

5.7.5 Stereo-Electroencephalography (sEEG): Invasive Precision Mapping

sEEG involves implanting depth electrodes in the brain to directly record electrophysiological activity from specific brain regions. This technique is considered the gold standard for localizing the EZ, particularly in patients with drug-resistant epilepsy (DRE) and complex morphologies. By allowing for the collection of both local field potentials and single-unit recordings, sEEG provides unparalleled insights into the dynamics of seizure initiation and propagation (Klink et al., 2017).

sEEG has demonstrated exceptional sensitivity in identifying small and subtle structures implicated in seizure generation, yielding highly localized data on seizure onset zones. Current research indicates that including sEEG data in decision-making processes significantly improves surgical outcomes, resulting in higher rates of seizure freedom following operative interventions (Tamilia et al., 2021). Conversely, the invasiveness of sEEG increases associated risks, which must be weighed against its diagnostic benefits.

5.7.6 Multimodal Imaging: A Comprehensive Approach

The integration of various neuroimaging and electrophysiological modalities—often referred to as multimodal imaging—has emerged as a powerful strategy for epilepsy research and clinical application. Each modality provides unique strengths and perspectives on brain function and pathophysiology. For instance, combining EEG with fMRI can harness the high temporal resolution of EEG to identify seizure onsets while using the spatial resolution of fMRI to localize activity, thereby yielding a more comprehensive view of the epileptogenic network (Nissen et al., 2016).

Chowdhury et al.'s study on the reproducibility of EEG-MEG fusion highlights the significance of multimodal approaches in epilepsy presurgical evaluation, emphasizing that combining the insights obtained through different modalities ultimately enhances the cognitive and clinical understanding of the disease (Chowdhury et al., 2017). The future of epilepsy management will benefit from such integrative strategies, as they can accommodate individual differences across patients and provide tailored interventions.

5.7.7 Investigating Network Behavior and Dynamics

Analyzing the dynamics of brain networks is crucial for understanding the mechanisms underlying epilepsy and its manifestations. Recent advances in utilizing network theory allow for the exploration of functional connectivity patterns, shedding light on how disparate brain regions interact in the context of seizure activity. Studies, such as those by Nissen et al., indicate that applying network-based approaches to interictal MEG data can reveal the hub status of local brain areas, aiding in EZ identification (Nissen et al., 2016).

Furthermore, electroencephalographic coherence and connectivity analyses have provided insights into neural synchrony and desynchronization patterns associated with seizure generation. By delineating connectivity changes during seizure events or resting states, researchers are now able to characterize the dynamical behavior of cortical networks involved in epilepsy.

5.8 GLIAL AND NEUROIMMUNE CONTRIBUTIONS TO EPILEPTOGENESIS

Epileptogenesis is a multifaceted process characterized by the development of epilepsy following an initial brain insult or abnormal neural activity. Recent research has demonstrated that glial cells, specifically astrocytes and microglia, play critical roles in this process through their involvement in neuroinflammation and neuronal modulation. These glial cells not only provide structural and metabolic support to neurons but also actively participate in immune responses and signaling pathways that can promote the initiation and spread of seizures. This chapter will explore the contributions of astrocytes, microglia, and inflammatory signaling in epileptogenesis, highlighting their potential as targets for therapeutic intervention.

5.8.1 Glial Cells: Key Players in Epileptogenesis

5.8.1.1 Astrocytes

Astrocytes, the most abundant glial cells in the central nervous system (CNS), are involved in maintaining homeostasis, modulating synaptic transmission, and responding to CNS injuries. Their activation during pathological processes typically leads to a phenomenon known as reactive astrogliosis, which is characterized by hypertrophy, proliferation, and altered gene expression (Sofroniew & Vinters, 2009). In the context of epilepsy, astrocytes have been shown to play dual roles: protective and detrimental.

When activated, astrocytes release various neurotrophic factors that can protect neurons and support recovery following injury. Conversely, excessive activation of astrocytes can lead to the release of pro-inflammatory cytokines and contribute to neuroinflammation (Gibbons et al., 2013; Sano et al., 2019). For instance, following SE, astrocytes can produce interleukin-1 beta (IL-1β), which has been implicated in promoting epileptogenesis (Vinet et al., 2016). Elevated IL-1β levels can enhance glutamate release and reduce the efficacy of GABAergic signaling, ultimately leading to increased neuronal excitability (Vezzani & Granata, 2005).

Recent studies, such as those by Gibbons et al., highlight significant changes in astrocyte function following SE, suggesting that understanding astrocytic behavior during this time could reveal critical insights into the mechanisms underlying the transition from acute injury to chronic epilepsy (Gibbons et al., 2013). Furthermore, astrocytes are crucial for maintaining the blood–brain barrier (BBB), and their dysfunction can lead to increased permeability, contributing to neuroinflammation and seizure susceptibility (Klein et al., 2017).

5.8.1.2 Microglia

Microglia serve as the resident immune cells of the CNS and are essential for maintaining homeostasis and responding to injury or disease. These cells act as first responders to brain insults, undergoing activation that typically leads to a pro-inflammatory state characterized by the release of cytokines and chemokines (Maroso et al., 2011; Peng et al., 2020). The role of microglia in epilepsy has gained attention, especially concerning their timing and effects on seizure initiation and propagation.

Research by Sano et al. has indicated that microglia are initially activated following SE, and this early activation is critical for stimulating both microglial and astrocytic responses that ultimately contribute to epileptogenesis (Sano et al., 2019). Importantly, microglial activation is linked with the expression of IL-1β, which can have significant downstream effects on astrocytes and neuronal cells, further enhancing seizure pathways (Maroso et al., 2011). Furthermore, the work published by Vinet et al. shows that microglial inflammation peaks immediately following SE, indicating the potentially acute interplay between microglial activation and early inflammatory responses (Vinet et al., 2016).

Unlike astrocytes, which predominantly display a protective retrograde effect, activated microglia can produce neurotoxic factors that exacerbate neuronal injury during the epileptogenic process. For example, microglial secretion of reactive oxygen species (ROS) and matrix metalloproteinases can lead to neuronal death and subsequent circuit reorganization that favors hyperexcitability (Peng et al., 2020).

5.8.2 Inflammatory Signaling Pathways

The relationship between glial cell activation and inflammatory signaling pathways active in the brain during epileptogenesis is crucial. Inflammatory cytokines, such as tumor necrosis factor-alpha (TNF-α) and IL-1β, act to enhance neuronal excitability and modify synaptic transmission, promoting seizure activity (Maroso et al., 2011; Yang et al., 2016). These cytokines can signal through various pathways, including the Janus kinase/signal transducer and activator of transcription (JAK/STAT) pathway, which has emerged as an essential mechanism regulating glial activation during epileptogenesis (Han et al., 2018, 2020).

Notably, the hypoxia-inducible factor 1-alpha (HIF-1α) pathway is also implicated in mediating inflammatory responses in epilepsy, influencing expression levels of TNF-α and other pro-inflammatory mediators (Yang et al., 2016). The interplay between neuroinflammatory signaling and adenosine signaling pathways—cascades often associated with neuronal excitability—is vital, as adenosine typically serves a protective function against seizures but can be dysregulated during chronic inflammation (Liu et al., 2024).

Another critical aspect of inflammatory signaling involves microRNAs (miRNAs), which have been recently identified as significant regulators of glial cell responses and cytokine expression in the context of epilepsy. For instance, miR-16-1's involvement in modulating astrocytic inflammatory responses highlights the complexity of neuroimmune interactions in epileptogenesis (Xu et al., 2021; Yao et al., 2017).

5.8.3 The Interaction Between Glial Cells and Seizures

5.8.3.1 Seizure Initiation

The initiation of seizures is a complex interplay involving aberrant neuronal activity and glial responses. Evidence suggests that reactive astrocytes can facilitate seizure spread by releasing excess neurotransmitters such as glutamate, which can lead to excitotoxic events. This process often accompanies the release of inflammatory cytokines that contribute to further neuronal depolarization and excitability (Gibbons et al., 2013; Jones & Ali, 2021).

Research focusing on the temporally distinct activation of microglia and astrocytes suggests a collaborative mechanism in seizure progression. Reactive microglia respond rapidly to the initial seizures, laying the groundwork for astrocytic activation to solidify neuronal hyperexcitability (Jones & Ali, 2021). This sequenced activation is critical, as microglial responses may dictate the subsequent astrocytic behavior, thereby influencing the overall progression toward uncontrolled seizure activity.

5.8.3.2 Seizure Spread and Network Effects

Once seizures are initiated, the spread of this hyperexcitable state typically occurs via interconnected neuronal networks. Microglial activation can exacerbate this spread by promoting inflammatory signaling, which can rewire synaptic connections and alter network dynamics, predisposing regions of the brain to further seizure activity (Peng et al., 2020; Yan et al., 2023).

Astrocytes contribute to this network behavior by forming functional networks that can synchronize or desynchronize neuronal firing patterns, depending on their activation state. The complexity of astrocytic signaling—particularly calcium signaling through inositol trisphosphate receptors (IP3R)—can significantly influence neuronal network behavior and seizure susceptibility (Jones & Ali, 2021; Sofroniew & Vinters, 2009). This underscores the bidirectional communication between neurons and glial cells; a dysregulation in either can contribute to establishing a more seizure-prone environment.

Figure 5.3 illustrates the role of proinflammatory mediators and glial cell activation induced by neuroinflammation in promoting epileptogenesis.

5.8.4 Targeting Glial Responses for Therapeutic Interventions

Given the critical roles that astrocytes and microglia play in both seizure initiation and propagation, targeting their responses may present new avenues for treatment. Modulating inflammatory signaling pathways with pharmacological agents may mitigate the harmful effects of glial activation while preserving their protective roles. Conventional anti-inflammatory drugs have shown promise

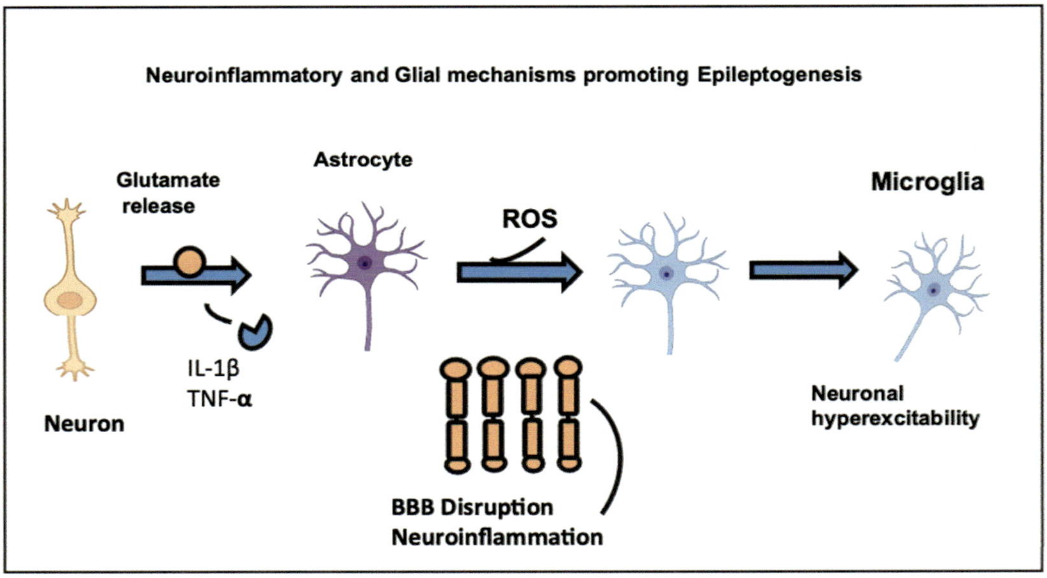

FIGURE 5.3 Neuroinflammatory and glial mechanisms promoting epileptogenesis. The illustration showing how proinflammatory mediators and glial cell activation caused by neuroinflammation aid in epileptogenesis. ROS and cytokines including IL-1β and TNF-α are secreted by microglia and astrocytes, which can disrupt the BBB and cause excitotoxicity. These mediators promote neuronal hyperexcitability by increasing glutamate release and inhibiting its reuptake. Excitatory signaling is maintained by long-term glial and inflammatory pathway activation, which strengthens epileptogenic circuits.

in preclinical models of epilepsy, but further refinement and understanding of specific pathways are necessary for successful translation to clinical practice.

Currently, researchers are exploring gene therapy and targeted molecular strategies aimed at modulating glial cell activity and associated inflammatory responses during the early stages of epileptogenesis. For example, interventions targeting the JAK/STAT pathway or the expression of specific miRNAs linked to glial activation could yield substantial benefits by restoring normal glial function and thereby limiting the seizure-generating capacity of affected networks (Han et al., 2018; Klein et al., 2017).

5.9 COGNITIVE AND BEHAVIORAL DIMENSIONS OF EPILEPSY

Cognitive impairment and behavioral disturbances are devastating co-morbidities of epilepsy. In some patients, these co-morbidities may be of greater consequence than the epilepsy itself. There is increasing recognition that cognitive and behavioral co-morbidities can be both chronic, primarily due to the underlying etiology of the epilepsy, and in dynamic evolution because of recurrent seizures or interictal spikes. With both chronic and dynamic effects, the type and severity of the impairment are related to the maturational stage of the brain at the time epilepsy appears. Several morphological changes can occur with epilepsy, including cell loss, synaptic reorganization, and changes in neurogenesis. Seizures can also result in physiological alterations in excitatory and inhibitory currents, as well as alterations in the temporal coding of information and impaired single-cell firing patterns. In parallel with these morphological and physiological changes, rats subjected to seizures show considerable cognitive and behavioral deficits. Likewise, interictal spikes can result in cognitive impairment, and in the developing animal, impaired neurogenesis and cell loss. Epilepsy is a disorder that affects neuronal networks, and the cognitive and behavioral deficits related to epilepsy are due to the pathological interactions between many brain components. Recent studies have suggested that pathological alterations in oscillatory patterns have an important role in the cognitive and behavioral co-morbidities of epilepsy.

Among the co-morbidities associated with epilepsy, cognitive and behavioral abnormalities are the most common and severe. Mental retardation, learning disabilities, memory impairment, attention deficit hyperactivity disorder, autism, anxiety, and conduct disorders are greatly over-represented in individuals with epilepsy, and the consequences of such co-morbidities greatly diminish the quality of life. Indeed, many people with epilepsy, and their families, consider the cognitive and behavioral consequences of seizures to be at least as troubling as the seizures themselves. Cognitive disorders can be found in a wide range of seizure disorders including temporal and frontal lobe epilepsies, primarily generalized idiopathic epilepsies and epileptic encephalopathies. While our understanding of the mechanisms responsible for epilepsy-related cognitive and behavioral problems lags far behind our knowledge of the mechanism of epilepsy, increased attention has recently been directed to investigating these co-morbidities.

When considering the pathophysiological mechanisms of the cognitive and behavioral consequences of epilepsy, it is helpful to distinguish between impairments that are permanent, and those that are dynamic, i.e., progressive or transient. Permanent deficits originate from a large number of etiologies in epilepsy: trauma, hypoxia-ischemia insults, genetic disorders, mesial temporal sclerosis as a result of SE, and malformations of cortical development. In addition to causing seizures, these disorders may result in cognitive and behavioral disturbances, with the severity of such disturbances related to the severity of the etiology. While the cognitive and behavioral issues associated with these brain insults may evolve due to maturational and aging changes in the brain, they are relatively fixed and remain attributable to the underlying brain disorder. Only if the underlying cause of epilepsy is remedied do the cognitive and behavioral deficits improve. This aspect of cognitive and behavioral impairments in the context of epilepsy is particularly under-investigated. The second type of impairment is dynamic in the sense that the deficits are either happening in stages or transiently affecting the patients. These cognitive and behavioral deficits can occur as a result of the seizures themselves, interictal epileptiform abnormalities, or antiepileptic drug therapy. Although the dynamic impairments can be observed in the absence of the permanent ones, they often contribute together in affecting the patient's quality of life (Kleen et al., 2012).

5.10 PSYCHIATRIC COMORBIDITIES IN EPILEPSY: A CLINICAL SIDEBAR

Comorbid psychiatric disorders are common in epilepsy patients. Patients with epilepsy are prone to psychosis, depression, personality disorders, hyposexuality, and other behavioral disorders. Comorbid psychiatric disorders are particularly common in temporal lobe epilepsy or complex partial seizures. Though psychiatric comorbidity is common in epilepsy, it is under-recognized and under-treated, both in adult and pediatric patients in specialty health care centers as well as in community-based health care centers. Early recognition and management of psychiatric disorders in patients with epilepsy is extremely important because it improves the quality of life and aids in better seizure control.

People with epilepsy have a 2–5 times increased risk of developing any psychiatric disorder, and 1 in 3 patients with epilepsy have a lifetime psychiatric diagnosis. Psychiatric comorbidities represent a poor prognostic marker as they have been associated with a poor response to treatment (drugs and surgery), increased morbidity, and mortality. Validated screening instruments are available for mood and anxiety disorders in adults as well as attention-deficit hyperactivity disorder in children with epilepsy (Mula et al., 2021).

Psychiatric problems could be due to shared neurobiological mechanisms, a consequence of having epilepsy, or simply due to the unfortunate occurrence of two conditions in the same individual. Nevertheless, the management of these people can be challenging regardless of the causes. For this reason, physicians must identify comorbid psychiatric disorders and incorporate them into the comprehensive individual management. Accordingly, a multidisciplinary approach is required. This entails the need for health care professionals in other disciplines, such as psychiatrists, clinical psychologists, neuropsychologists, psychiatric nurses, and social workers, to ensure that the individual receives the best management.

In epilepsy, treatment and prognosis are dependent on the accurate definition of the epilepsy syndrome. It would then be reasonable to apply the same model for psychiatric comorbidities (e.g., psychiatric disorders being more common in TLE compared with generalized syndromes). The relative contribution of syndrome-specific variables is still unclear, and clinical evidence suggests that psychiatric comorbidities do not necessarily respect such borders. It is now evident that psychiatric comorbidities have to be considered when informing people about the long-term prognosis of the epilepsy itself.

In people with epilepsy, psychiatric comorbidities have been historically associated with poor quality of life, but there are now data suggesting their role as a prognostic indicator. Psychiatric comorbidities are associated with a high risk of side effects, especially cognitive complaints and psychiatric side effects. Psychiatric comorbidities, particularly depression, represent an important cause of cognitive complaints.

Psychiatric comorbidities are associated with a 4-fold increased risk of drug resistance in focal (Nogueira et al., 2017) and generalized epilepsies. The impact of psychiatric comorbidities in terms of seizure outcome and psychiatric outcome, in epilepsy surgery, is complex and yet to be established. Some studies have found a lower probability of achieving seizure freedom after temporal lobectomy (Bell et al., 2017). The same holds for psychiatric outcomes, as some studies showed an increased risk of recurrence of depression or anxiety during the first year after surgery, whereas other studies showed long-term improvement.

Psychiatric comorbidities are associated with premature mortality in epilepsy (Fazel et al., 2013). This may be due to a variety of reasons, including increased risk of substance or alcohol abuse, increased risk of injury, poor adherence to medications, and increased suicide rates. Data from a population-based study of over 57,000 people in Sweden showed that females with epilepsy and

TABLE 5.2
Prevalence and Odds Ratio of Psychiatric Disorders in People with Epilepsy

Psychiatric Disorder	Pooled Prevalence in People with Epilepsy (PWE)	Odds Ratio vs. General Population
Mood disorders (e.g., depression)	~35 %; Major Depressive Disorder ≈ 24%	~2.5
Anxiety disorders (e.g., GAD, PTSD)	~25.6%; PTSD ~14.2%, GAD ~11.1%	~2.1
Psychotic disorders (including psychosis, schizophrenia)	~5.7% psychosis; schizophrenia ~1.7%	~4
Bipolar disorder	~6.2%	OR ~3.1
Obsessive-compulsive disorder	~3.8%	OR ~2.7

psychiatric comorbidities had a 5-fold increased risk of sudden unexpected death in epilepsy compared with those without such comorbidities.

Last, it is evident that psychiatric comorbidities increase the global burden of epilepsy from a public health perspective, with increased health costs. People with epilepsy and psychiatric disorders have high health resource utilization, including increased emergency department admissions and outpatient visits (Table 5.2). Data from a US nationwide study assessing almost 400,000 hospital admissions showed that psychiatric comorbidities, depression, and psychosis, in particular, increase length of stay and inpatient costs for people with epilepsy (Patel et al., 2017).

5.11 EXPERIMENTAL MODELS AND TRANSLATIONAL TOOLS

The quest to understand and treat epilepsy has driven significant advancements in experimental models and translational research tools over the years. Epileptic disorders are characterized by diverse seizure types and complex pathophysiologies, leading to the necessity of multifaceted approaches for studying seizure mechanisms and therapeutic strategies. This chapter delves into the utilization of rodent models, human neuronal cultures, and optogenetic methodologies to decode the circuits responsible for seizure generation and propagation.

5.11.1 Rodent Models of Epilepsy

Rodent models have been indispensable in epilepsy research due to their genetic, biochemical, and anatomical similarities to humans. These models enable the exploration of specific mechanisms of seizure generation, propagation, and the underlying neurobiological alterations that accompany epilepsy. Notable models include the pilocarpine and kainic acid models, which induce SE and lead to the subsequent development of chronic epilepsy, closely mirroring the human condition (Chen et al., 2024).

5.11.1.1 Induction of Seizures

In these models, intrahippocampal injections of chemicals such as pilocarpine can induce seizures that evolve into a chronic state, reflective of temporal lobe epilepsy in humans. The use of these models allows researchers to study the cellular and molecular changes that occur post-injury, including glial activation, neuronal death, and synaptic reorganization (Shen et al., 2025).

Moreover, genetic models, such as the Scn1a mutant mice, provide insights into the effects of specific genetic mutations on seizure susceptibility. These models help illuminate the pathways modulated by genetic factors and their roles in influencing seizure thresholds and comorbidities (Mattis et al., 2022). Experimental precision allows for the examination of how distinct circuit alterations contribute to diverse epileptic phenotypes, leading to better-targeted therapeutic interventions.

5.11.1.2 Advantages and Limitations

The major advantages of rodent models include their ability to undergo extensive genetic manipulation and their relatively short lifespan, allowing for rapid assessment of long-term effects of treatments. However, one limitation is the potential discrepancy between rodent-specific seizure mechanisms and human epilepsy, necessitating careful caution in translational studies (Krook-Magnuson et al., 2015). In addition, ethical considerations regarding animal research have prompted the exploration of other models, including human neuronal cultures.

5.11.2 Human Neuronal Cultures

Human neuronal cultures provide an invaluable resource for studying the cellular and molecular dynamics underlying epilepsy. These cultures can be derived from pluripotent stem cells (PSCs), providing a human-specific context that is critically relevant given the species differences observed in rodent models (Ghosh et al., 2023).

5.11.2.1 Derivation of Neuronal Cultures

Neuronal cultures can be generated from neural progenitor cells differentiated from PSCs or from human brain tissue resected during epilepsy surgeries. These cultures retain the genetic and phenotypic characteristics of the patient, enabling the study of patient-specific cellular mechanisms (Soper et al., 2016). Detailed analysis of these cultures illuminates the role of aberrant synaptic transmission, ion channel dysfunction, and glial interactions in seizure activities.

5.11.2.2 Advantages and Challenges

Human neuronal cultures facilitate high-throughput drug screening and personalized approaches to epilepsy treatment. By evaluating the response of human neurons to various pharmacological agents, researchers can obtain clinically relevant data that may better predict outcomes in human patients. However, these models are limited by their inherent complexity and the challenge of maintaining long-term cultures with stable phenotypes.

5.11.3 Optogenetics in Decoding Seizure Circuits

Optogenetics is a groundbreaking technique that enables precise control of neuronal circuits using light-sensitive proteins. This methodology has been increasingly integrated into epilepsy research to dissect the mechanisms underlying seizure initiation and propagation. By using optogenetic tools, researchers can manipulate specific neuronal populations, dissecting the role of various circuits in epilepsy.

5.11.3.1 Functional Dissection of Circuits

Optogenetics allows for the selective excitation or inhibition of genetically targeted neurons, providing insights into their causal roles in seizure dynamics. For example, by activating inhibitory interneurons using channelrhodopsins, researchers can suppress seizure activity and examine the effects on seizure propagation pathways (Khoshkhoo et al., 2017; Tung et al., 2018). Moreover, different neuronal populations can be targeted simultaneously, enabling complex manipulations essential for studying multi-region interactions that are critical during seizures (Weitz et al., 2015).

The application of optogenetics alongside other technologies, such as fMRI, has provided a novel way to investigate the functional connectivity and activation patterns of neuronal networks during seizure events (Sorokin et al., 2017). A recent study emphasized the use of optogenetics combined with video and calcium imaging to correlate behavioral changes with neural activity, elucidating how different circuits contribute to seizure-like behaviors (Shen et al., 2025).

5.11.3.2 Innovations and Limitations

Innovative optogenetic tools, including chemogenetic methods, have emerged, allowing for the manipulation of neuronal activity in a flexible and non-invasive manner (Tung et al., 2018). Despite its advantages, optogenetics faces challenges related to the depth of light penetration in biological tissues and potential unintended activation of neurons outside the target population. Nonetheless, advancements in fiber-optic technologies and light delivery systems continue to enhance the applicability of optogenetic approaches in complex seizure networks.

5.11.4 Integrative Approaches: Combining Models and Techniques

To decipher the intricate circuits involved in epilepsy comprehensively, integrative approaches that meld rodent models, human neuronal cultures, and optogenetic methods can bring forward a fuller understanding of seizure mechanisms and therapeutic actions. The integration of behavioral, cellular, and network-level studies enhances our capacity to elucidate the complexity of seizure dynamics.

For instance, researchers have utilized rodent models of epilepsy alongside in vitro human neuronal cultures to compare cellular responses to specific treatments, assessing the translational potential of therapeutic candidates (Chen et al., 2024). Concurrently, optogenetic manipulations can reveal the impacts of targeted interventions on seizure propagation across different network architectures, thereby providing insights into circuit-specific vulnerabilities in epilepsy (Paz & Huguenard, 2015).

5.11.4.1 Implications for Therapeutic Development

Such integrative approaches might yield groundbreaking advances in treatment strategies for epilepsy management. Understanding how distinct circuits contribute to seizure dynamics enables the design of targeted therapies that could modulate specific neuronal populations, ultimately leading to more effective interventions with reduced side effects (He et al., 2023).

Progress in the fields of neural decoding and artificial intelligence can further enhance the efficacy of treatment approaches through personalized medicine. Machine learning algorithms are increasingly being employed to analyze data gathered from optogenetic experiments, allowing for rapid identification of optimal intervention strategies for individual patients based on real-time monitoring of seizure signals (He et al., 2023).

5.12 CURRENT AND EMERGING PHARMACOLOGICAL THERAPIES

Epilepsy is a prevalent neurological disorder that affects approximately 65 million individuals globally. While approximately 65–70% of patients can achieve seizure control through pharmacological interventions, a significant proportion—estimated at around 30%—suffers from DRE (French & Perucca, 2020; Kalilani et al., 2018). This chapter discusses the current landscape of antiseizure medications (ASMs), the mechanisms by which they operate, the challenges posed by DRE, and emerging therapeutic targets aimed at improving treatment efficacy.

5.12.1 Mechanism-Based Antiseizure Medications

ASMs are typically categorized based on their mechanisms of action, which include modulation of ion channels, enhancement of inhibitory neurotransmission, inhibition of excitatory neurotransmission, and modification of neurotransmitter release (Silva et al., 2023).

5.12.1.1 Modulation of Ion Channels

Many ASMs exert their effects by targeting voltage-gated ion channels responsible for neuronal excitability. For instance, *sodium channel blockers*, such as carbamazepine and lamotrigine, inhibit the sustained high-frequency firing of action potentials observed in epileptic neurons (Perucca, 2019). These drugs stabilize inactivated channels, thereby reducing the likelihood of seizure initiation.

Calcium channel modulators represent another class of ASMs, with drugs like ethosuximide specifically targeting T-type calcium channels implicated in absence seizures. By inhibiting these channels, ethosuximide limits excitatory neurotransmitter release and diminishes neuronal excitability (Silva et al., 2023).

Potassium channel openers, such as retigabine, enhance the outward potassium current, promoting neuronal hyperpolarization and inhibiting excitability. While initially showing promise, retigabine has faced withdrawal from the market due to safety concerns (Talevi, 2022).

5.12.1.2 Enhancement of Inhibitory Neurotransmission

The potentiation of GABAergic neurotransmission is a primary mechanism for numerous ASMs. Drugs like benzodiazepines and barbiturates enhance the binding of GABA to GABA_A receptors, increasing chloride influx and hyperpolarizing the neuron. This action decreases the likelihood of the neuron reaching the threshold for action potential generation (Sills & Rogawski, 2020).

New approaches are also exploring therapeutic potentials through modulation of GABAergic pathways. For instance, the *GABA analogs*, such as gabapentin and pregabalin, while originally designed to target calcium channels, also exhibit GABA-enhancing properties, although the latter's precise mechanisms remain incompletely understood (Ghosh et al., 2023; Talevi, 2022).

5.12.1.3 Inhibition of Excitatory Neurotransmission

Conversely, some ASMs aim to inhibit excitatory neurotransmission. *Vigabatrin*, for instance, irreversibly inhibits the enzyme GABA transaminase, leading to increased GABA levels and subsequently reduced excitatory signaling. However, its side effect profile has limited its use in clinical practice (Talevi, 2022).

Another promising area of research includes targeting glutamate receptors. Experimental drugs that modulate the activity of AMPA and NMDA receptors aim to reduce glutamate's excitatory effects on the brain (Waris et al., 2024). Recent findings indicate that targeting the glutamatergic system, specifically within hyperexcitable circuits, may lead to new therapeutic avenues (Ravizza & Vezzani, 2018).

5.12.2 Challenges of Drug-Resistant Epilepsy

Despite a diverse arsenal of ASMs, the challenge of DRE remains formidable, with many patients failing to achieve sustained seizure control. Literature indicates that resistance is multifactorial—stemming from genetic, pharmacokinetic, and pharmacodynamic factors that render traditional treatments ineffective (Kalilani et al., 2018; Löscher et al., 2020; Perucca, 2019).

5.12.2.1 Genetic Factors

Recent advances in genomics and bioinformatics have opened avenues for dissecting the genetic underpinnings of DRE. Genetic predispositions can affect drug metabolism, receptor subtypes, and the expression levels of target proteins (Kalilani et al., 2018). For example, polymorphisms in genes related to drug transport or hepatic enzymes can significantly influence drug efficacy and safety profiles (Hu et al., 2023). Consequently, personalized medicine—tailoring ASMs to the individual genetic makeup of patients—holds promise for enhancing treatment outcomes.

5.12.2.2 Pharmacokinetic and Pharmacodynamic Considerations

The pharmacokinetics of ASMs, influenced by their absorption, distribution, metabolism, and excretion, play a critical role in their effectiveness. In drug-resistant patients, abnormal pharmacokinetic profiles may lead to subtherapeutic drug levels, thereby necessitating higher doses or alternative therapies (Silva et al., 2023). Additionally, drug interactions, particularly common in patients who may be on polytherapy regimens, can further complicate pharmacodynamics, leading to non-optimal therapeutic outcomes.

5.12.3 Emerging Therapeutic Targets

Given the persistent challenges associated with DRE and the limitations of current pharmacological approaches, significant research has turned toward novel therapeutic targets. These approaches aim to address the underlying mechanisms of epilepsy rather than merely suppressing symptoms.

5.12.4 Neuroinflammatory Pathways

Recent studies have increasingly recognized the role of neuroinflammation in epileptogenesis and seizure exacerbation. Targeting inflammatory cytokines, such as IL-1β and TNF-α, has presented a promising therapeutic avenue. Drugs that inhibit neuroinflammatory pathways and modulate the immune response may prevent, or at least reduce the frequency of seizures, particularly in patients with inflammatory or autoimmune components to their epilepsy (Hu et al., 2023; Ravizza & Vezzani, 2018; Silva et al., 2023).

5.12.5 Epigenetic Modifications

Investigative efforts into epigenetic therapies highlight the potential of modifying gene expression related to seizure susceptibility. Drugs that influence chromatin remodeling and transcription factor activity carry the potential for disease modification. Targeting dysregulated transcription factors associated with epilepsy could help restore normal gene expression profiles in at-risk populations (Perucca, 2021; Zeibich et al., 2025).

5.12.6 Adenosine-Targeted Therapies

The adenosinergic system's interactions with seizure activity offer a new promising therapeutic target. Agents that selectively enhance adenosine A1 receptor (A1R) signaling have shown potential antiepileptic activity by promoting inhibitory neurotransmission and dampening excitatory responses (Klaft et al., 2020). Optogenetic approaches to manipulate A1R signaling in neural circuits have enhanced our understanding of this pathway, revealing its role in modulating seizure susceptibility (Klaft et al., 2020; Spanoghe et al., 2024).

5.12.7 Combination Therapies

The complexity of epilepsy's pathophysiology suggests that a singular therapeutic approach may often be insufficient. As a result, combination therapies that employ existing ASMs alongside novel therapeutic agents are gaining traction. This poly-pharmacological approach holds the potential to target multiple seizure pathways simultaneously, increasing overall therapeutic effectiveness while potentially reducing the likelihood of drug resistance (Ghosh et al., 2023; Riva et al., 2021).

5.13 NEUROMODULATION AND SURGICAL APPROACHES TO CIRCUIT DISRUPTION

Epilepsy is a multifaceted disorder that can significantly impair the quality of life due to recurrent seizures. While a substantial number of patients can achieve seizure control through antiseizure medications (ASMs), approximately 30% of individuals suffer from DRE, a condition that poses a considerable challenge for effective management (French & Perucca, 2020). Recent advances in neuromodulation and surgical interventions offer promising alternative strategies for disrupting pathological neural circuits responsible for seizure activity. This chapter provides an in-depth overview of neuromodulation techniques, such as DBS, vagus nerve stimulation (VNS), and RNS, along with the intricacies of respective surgeries guided by high-resolution circuit mapping.

5.13.1 Neuromodulation Techniques

5.13.1.1 Vagus Nerve Stimulation (VNS)

VNS has been employed for more than two decades as a therapeutic option for treatment-resistant epilepsy. It involves the implantation of a pulse generator that delivers regular electrical impulses to the vagus nerve, thereby modulating neuronal excitability and neurotransmitter release in the brain. The therapeutic efficacy of VNS has been demonstrated in numerous studies; a meta-analysis of 74 clinical studies encompassing over 3,300 patients indicated that VNS provided a greater than 50% reduction in seizure frequency for many patients after one year of therapy.

The mechanism by which VNS exerts its effects remains multifaceted. The stimulation of afferent fibers of the vagus nerve influences various brain regions implicated in seizure development, including the locus coeruleus, the thalamus, and potentially the cortex. Moreover, VNS has neuroprotective effects, which may contribute to its antiepileptic efficacy. Research suggests that VNS can enhance GABAergic transmission while simultaneously inhibiting excitatory neurotransmission, thus promoting a more balanced network state conducive to seizure control.

5.13.1.2 Deep Brain Stimulation (DBS)

DBS is another neuromodulatory technique that has garnered interest in the treatment of epilepsy, particularly for patients with focal epilepsy who are unresponsive to medication. This approach involves the implantation of electrodes within specific brain areas, such as the anterior nucleus of the thalamus, with stimulation delivered via an implanted pulse generator. DBS has been shown to alter abnormal firing patterns within cortical and subcortical structures, thereby deterring seizure initiation and propagation.

The mode of action of DBS is thought to involve modulation of the disrupted thalamocortical networks associated with seizure generation. Its efficacy has been established through clinical trials, with studies revealing notable reductions in seizure frequency and improvements in quality of life following stimulation (Vezzani et al., 2015). Importantly, the refinement of electrode placement through advanced imaging techniques has allowed clinicians to better target the specific circuitry involved in the patient's epilepsy, further enhancing treatment efficacy (Li et al., 2021).

5.13.1.3 Responsive Neurostimulation (RNS)

The RNS system offers a distinct approach by delivering localized electrical stimulation in response to the detection of abnormal electrical activity, essentially forming a closed-loop system. The RNS device is implanted within the skull, with leads positioned on or adjacent to the EZ (Maroso et al., 2011). This targeted stimulation aims to prevent seizure onset before they fully develop, enabling timely intervention to avert the seizure.

Clinical trials have demonstrated that RNS can significantly reduce seizure frequency in patients with focal epilepsy. In patients with DRE, studies have shown that RNS therapy can lead to up to a 50% reduction in seizure frequency after one year of therapy, with sustained effects over an extended follow-up. The RNS device's ability to provide tailored stimulation based on individual seizure patterns represents a substantial advancement in the field of epilepsy management.

5.13.2 Surgical Approaches: Resective Surgery

For patients with focal epilepsy where the EZ can be clearly identified, resective surgery remains a cornerstone of treatment. Traditionally, surgical interventions focused on resecting the seizure foci; recent advances in imaging and circuit mapping technologies have refined these surgical approaches.

5.13.3 High-Resolution Circuit Mapping

High-resolution circuit mapping techniques, such as intraoperative electrocorticography (ECoG) and advanced imaging modalities like fMRI and diffusion tensor imaging (DTI), play a pivotal role in accurately localizing the EZ. These techniques can reveal the connectivity and network dynamics of brain regions, thus informing surgical decision-making.

Intraoperative ECoG provides real-time recording of cortical electrical activity during surgery, allowing for precise identification of critical functional areas that must be preserved while excising the epileptogenic hemisphere or lobe (Zheng et al., 2013). Additionally, recent advancements have integrated machine learning tools with imaging data, enhancing the predictive accuracy of seizure localization and surgical outcomes.

The interplay between imaging and electrical recordings can provide a comprehensive understanding of seizure circuitry. This synergy can guide the identification of networks involved in seizure generation, leading to more effective surgical strategies and reduced postoperative complications.

5.13.4 Outcomes of Resective Surgery

Surgical outcomes can be highly favorable for patients with DRE when the resected tissue contains the seizure-generating focus. Studies have reported seizure freedom rates ranging from 50% to 80% following resective surgeries. The outcomes are nuanced, as factors such as age, presence of structural lesions, and duration of epilepsy prior to surgery can significantly influence results.

Epilepsy surgery not only aims to reduce seizures but also endeavors to improve cognitive function and enhance the overall quality of life. Postoperative assessments reflect improvements in cognitive domain functioning, and patients report enhanced psychosocial well-being, thereby supporting surgical strategies as an essential component of epilepsy treatment.

5.13.5 Future Directions in Neuromodulation and Surgical Approaches

The therapeutic landscape for epilepsy is rapidly evolving, with ongoing research aimed at refining existing techniques and developing new interventions. This section addresses current priorities and potential innovations in the field.

5.13.6 Precision Medicine Approaches

The next frontier in epilepsy treatment lies within precision medicine, where specific patient characteristics, including genetic profiles and specific anatomical features, will inform the therapeutic strategy. Biomarker development for identifying patients who are likely to respond to specific neuromodulatory therapies or surgical interventions may significantly enhance treatment outcomes and reduce pharmacoresistance.

5.13.7 Hybrid Technologies

Furthermore, integrating neuromodulation techniques with coordinated therapies such as ASMs or gene therapies can optimize outcome strategies, potentially addressing the multifactorial etiology of epilepsy. For instance, combining RNS with targeted drug delivery systems or pharmacological agents that modulate inflammatory pathways has shown promise in preclinical studies. Such hybrid approaches may lead to a more comprehensive strategy for controlling seizures.

5.13.8 Longitudinal Monitoring and Adaptation

As technologies continue to advance, longitudinal monitoring of neural circuits via implantable devices may become commonplace. This enables ongoing assessment and real-time data collection on seizure activity, allowing for adaptive modulation of stimulation patterns that align with the patient's evolving condition. The incorporation of machine learning algorithms and artificial intelligence can revolutionize these adaptive systems, making personalized epilepsy management possible.

5.14 PERSONALIZED NETWORK EPILEPSY CARE

Personalized epilepsy care leverages individual brain network architecture and dynamics-derived from neuroimaging and electrophysiology—to tailor diagnostics, surgery, stimulation, and pharmacotherapy.

5.14.1 Individualized Structural and Functional Modeling

- *Diffusion MRI and connectomics* are used to construct patient-specific network models that predict seizure propagation pathways and surgical outcomes with ~90% accuracy when combined with clinical data.
- *Dynamic systems modeling* of seizure networks (from EEG or sEEG) helps simulate seizure initiation, spread, and targetable nodes for intervention.

5.14.2 Guiding Treatment through Network Insights

- Modeling allows prediction of surgical effectiveness, aiding clinicians in selecting optimal targets while minimizing unnecessary resection.
- *Network-guided neuromodulation*: RNS and DBS can be personalized based on connectivity architecture and dynamic biomarkers of seizure onset.

5.14.3 Network Pharmacology and Systems Biology

- Multi-target drug combos (e.g., levetiracetam + topiramate), selected via network pharmacology, showed synergy and prevented epilepsy progression in preclinical models.
- Integrating genomic, proteomic, and metabolomic data into network models enables precision selection of pharmacotherapy tailored to individual disease pathways.

5.14.4 Clinical Integration across the Care Continuum

5.14.4.1 Pre-Surgical Planning

- Network mapping refines localization of EZs, distinguishing focal cases versus diffuse epileptic networks and informing surgical or neuromodulatory candidacy phases.

5.14.4.2 Post-Treatment Monitoring

- Serial network analyses (pre- and post-intervention) help evaluate network normalization, residual epileptic connectivity, and guide decisions on drug tapering or further intervention.

The field of epilepsy is rapidly moving beyond the concept of a single seizure focus to view it as a network disorder. Multimodal imaging and electrophysiological data, integrated into patient-specific structural and functional brain models—sometimes called "virtual brains"—enable dynamic simulations of seizure initiation, spread, and treatment effects. Studies consistently show that higher structure–function coupling in preoperative data predicts better surgical outcomes, with predictive accuracies approaching 85% or more.

Further advances using graph neural networks on intracranial EEG (sEEG) deliver >90% accuracy in predicting seizure freedom and key network nodes relevant for treatment planning. These insights are transforming therapeutic strategies—informing surgical planning, targeted neuromodulation (e.g., DBS, RNS), and network-guided pharmacology—all tailored to an individual's epileptic network.

5.15 CONCLUSION: EPILEPSY AS A WINDOW INTO BRAIN NETWORK DYSFUNCTION

Epilepsy is not merely a neurological disorder defined by the occasional occurrence of seizures; instead, it represents a profound disruption in the functional architecture of the brain, illuminating the intricate dynamics of neural circuitry. Understanding epilepsy offers unique insights into brain network dysfunction, facilitating potential innovations in both research and clinical care. As we progress into a new era of brain research rooted in circuit neuroscience, future investigations must focus on elucidating the mechanisms underlying epileptogenesis, the interplay between genetic and environmental factors, and the therapeutic implications derived from these findings. This chapter synthesizes current knowledge and outlines future directions that may enhance epilepsy management through translational research efforts.

5.15.1 Epilepsy as an Illustrative Model of Brain Dysfunction

5.15.1.1 Neurodevelopmental and Acquired Epilepsies

Epilepsy manifests in two primary forms: neurodevelopmental and acquired. Neurodevelopmental epilepsies often have genetic underpinnings, differentiating them from acquired epilepsies, which typically arise following traumatic brain injuries, strokes, or infections. Each form exhibits unique structural and functional characteristics that elucidate how diverse pathological insults lead to similar outcomes: recurrent seizures.

Neurodevelopmental epilepsy typically results from congenital abnormalities that disrupt early brain development and synaptic integrity. This disruption can lead to a failing excitatory-inhibitory balance and aberrant circuit formation, as evidenced in conditions like Dravet syndrome, where mutations in ion channel genes lead to increased neuronal excitability. On the other hand, acquired epilepsy often involves secondary structural changes following initial insults to the brain; this is reflected in various studies that highlight the relationship between acquired brain injuries and subsequent epileptogenesis.

5.15.1.2 The Role of Circuit Dysfunction

From the synaptic level to the macro-architecture of the brain, epilepsy signifies a breakdown in the delicate balance necessary for normative functioning. Studies have demonstrated that network connectivity, as illustrated through techniques like fMRI and DTI, changes demonstrably post-insult, leading to maladaptive neuroplasticity that perpetuates seizure activity. This highlights the crucial discourse surrounding synaptic plasticity, as increased excitability driven by repeated seizures alters synaptic efficacy, which can foster further hyperexcitability, thereby reinforcing the self-perpetuating cycle of epilepsy.

Understanding epilepsy enables researchers and clinicians to address the mechanisms linked to seizure development, contributing to a bigger narrative surrounding brain health and cognitive functioning. The modeling of epilepsy can illustrate broader principles of network function, recruitment, and resilience, considerably enhancing our grasp of brain dynamics.

5.15.1.3 Future Directions in Research

The future landscape of epilepsy research is poised for transformative changes, particularly with advancements in technology and methodologies that illuminate underlying circuit dynamics. Below are focal areas that warrant further exploration.

5.15.1.4 Precision Medicine Approaches

Precision medicine, which tailors treatment based on individual genetic and clinical profiles, holds promise in addressing the complexities of epilepsy management. Identifying genetic mutations and metabolic pathways associated with specific epilepsy types can lead to the development of targeted therapies. For example, understanding mechanisms related to mutations in the *DEPDC5* gene could open avenues for tailored interventions utilizing mTOR inhibitors such as everolimus, demonstrating anti-seizure efficacy in syndromes like tuberous sclerosis.

Improving genetic screening and biomarker identification will also facilitate earlier diagnoses and prompt treatment adjustments, optimizing outcomes for individuals with epilepsy. Developing databases that aggregate genetic and clinical data will enhance our understanding of genotype-phenotype relationships, further aiding in the personalization of therapies.

5.15.1.5 Neuroinflammation and Immune Modulation

In recent years, an increasing body of evidence has highlighted the role of neuroinflammation in the pathophysiology of epilepsy. Research should focus on the underlying mechanisms by which inflammatory pathways contribute to seizure susceptibility and circuit changes. Identifying specific inflammatory mediators and their roles in excitatory-inhibitory balance disruption can serve as potential therapeutic targets.

Recent trials developing immune-modulating therapies have shown promise. Drugs that effectively inhibit neuroinflammatory pathways or cytokine release hold potential to reduce seizure frequency and improve overall brain health. This could reveal new avenues for intervention, particularly in patients with established epileptogenesis related to inflammatory brain conditions.

5.15.1.6 Neuromodulation Techniques

The growing field of neuromodulation offers innovative strategies for epilepsy treatment. Technologies like RNS and DBS are gaining traction as adjunctive therapies for DRE. Future research should focus on optimizing device designs and electrode placements, leveraging the insights gleaned from circuit mapping (Chen et al., 2024).

Emerging modalities, such as optogenetics, provide high specificity in modulating neural networks and could elucidate the effects of excitatory versus inhibitory stimulation on seizure control. Such experimental frameworks will inform precision targeting of neuromodulatory therapies to restore balance in aberrant circuits.

5.15.1.7 Clinical Implications and Translatability

As foundational research leads to an increased understanding of the cellular and synaptic mechanisms behind epilepsy, clinical ramifications will grow. Several areas present exciting opportunities for translational applications that directly benefit patient care.

5.15.1.8 Multidisciplinary Treatment Approaches

Advances in understanding the complex interplay of genetic, structural, and biochemical factors underlying seizures can guide multidisciplinary approaches in managing epilepsy. Collaboration across neurology, genetics, psychiatry, and psychosocial care is essential to holistically address the needs of epilepsy patients, particularly those with comorbid conditions, such as psychiatric disorders, which enhance the treatment burden.

Utilizing technology to enhance patient engagement and support, including telemedicine consultations and mobile apps for seizure tracking, can also improve adherence to treatment regimens and facilitate timely interventions.

5.15.1.9 Addressing the Epilepsy Treatment Gap

A critical focus for clinical care is addressing the existing treatment gap that afflicts many individuals with epilepsy. Prior work has highlighted barriers to care, including stigmas associated with epilepsy, socioeconomic factors, and geographic limitations affecting access to healthcare resources. Efforts to enhance the delivery of care in underserved populations through community health initiatives must remain a priority.

Implementing strategies to deliver care directly to patients, such as healthcare visits in home settings and strengthening community support networks, can significantly impact treatment accessibility, ultimately providing needed support for optimal management of the condition.

5.15.1.10 Surveillance and Sudden Unexpected Death in Epilepsy (SUDEP)

As understanding of the intricacies of epilepsy improves, raising awareness about SUDEP is crucial. Research indicates that patients with epilepsy face markedly increased risks of sudden death, emphasizing the necessity for routine risk assessments and prevention strategies in clinical practice. Creating standardized protocols for monitoring patients with epilepsy, particularly in at-risk populations, can serve to directly address this alarming outcome.

5.16 CONCLUSION

Epilepsy serves as a unique window into brain network dysfunction, revealing critical insights into the underlying mechanisms governing neuronal activity and synaptic alterations. Understanding the cellular, molecular, and structural components of epilepsy is fundamental to informing future research and clinical practices. As we move forward into a new wave of advancements fueled by technology and interdisciplinary collaborations, there is hope for refining treatment strategies that are both effective and personalized. By striving to address the multifactorial nature of epilepsy, researchers and clinicians can optimize strategies that significantly improve patients' quality of life and overall health outcomes.

REFERENCES

Anney, R. (2014). Genetic determinants of common epilepsies: A meta-analysis of genome-wide association studies. The Lancet Neurology, 13(9), 893–903. https://doi.org/10.1016/s1474-4422(14)70171-1

Bell, G.S., De Tisi, J., Gonzalez-Fraile, J.C., Peacock, J.L., McEvoy, A.W., & Harkness, W.F., et al. (2017). Factors affecting seizure outcome after epilepsy surgery: An observational series. J Neurol Neurosurg Psychiatry, 88(11), 933–940.

Beniczky, S., Trinka, E., Wirrell, E., Abdulla, F., Al Baradie, R., & Alonso Vanegas, M. et al. (2025). Updated classification of epileptic seizures: Position paper of the international league against epilepsy. Epilepsia, 66(6), 1804–1823.

Chang, X., Wang, J., Jiang, H., Shi, L., & Xie, J. (2019). Hyperpolarization-activated cyclic nucleotide-gated channels: An emerging role in neurodegenerative diseases. Frontiers in Molecular Neuroscience, 12. https://doi.org/10.3389/fnmol.2019.00141

Chen, Y., Litt, B., Vitale, F., & Takano, H. (2024). On-demand seizures facilitate rapid screening of therapeutics for epilepsy. https://doi.org/10.1101/2024.08.26.609726

Chowdhury, R., Pellegrino, G., Aydın, Ü., Lina, J., Dubeau, F., Kobayashi, E., ... & Grova, C. (2017). Reproducibility OFEEG-MEG fusion source analysis of interictal spikes: Relevance in presurgical evaluation of epilepsy. Human Brain Mapping, 39(2), 880–901. https://doi.org/10.1002/hbm.23889

Chowdhury, R., Zerouali, Y., Hedrich, T., Heers, M., Kobayashi, E., Lina, J., ... & Grova, C. (2015). MEG–EEG information fusion and electromagnetic source imaging: From theory to clinical application in epilepsy. Brain Topography, 28(6), 785–812. https://doi.org/10.1007/s10548-015-0437-3

Dahl-Hansen, E., Koht, J., & Syvertsen, M. (2018). Epilepsy at different ages—etiologies in a Norwegian population. Epilepsia Open, 4(1), 176–181. https://doi.org/10.1002/epi4.12292

Diessen, E., Zweiphenning, W., Jansen, F., Stam, C., Braun, K., & Otte, W. (2014). Brain network organization in focal epilepsy: A systematic review and meta-analysis. PLOS One, 9(12), e114606. https://doi.org/10.1371/journal.pone.0114606

Dong, P., Zhang, Y., Hunanyan, A., Mikati, M., Cui, J., & Yang, H. (2022). Neuronal mechanism of a bk channelopathy in absence epilepsy and dyskinesia. Proceedings of the National Academy of Sciences, 119(12). https://doi.org/10.1073/pnas.2200140119

Durisic, N., Keramidas, A., Dixon, C., & Lynch, J. (2018). SAHA (vorinostat) corrects inhibitory synaptic deficits caused by missense epilepsy mutations to the GABAA receptor γ2 subunit. Frontiers in Molecular Neuroscience, 11. https://doi.org/10.3389/fnmol.2018.00089

EyeWirers, Sharpee TO, Destexhe A, Kawato M, Sekulić V, Skinner FK, et al. 25th annual computational neuroscience meeting: CNS-2016. BMC Neuroscience [Internet]. 2016 Aug [cited 2025 Jul 23];17(S1). Available from: https://bmcneurosci.biomedcentral.com/articles/10.1186/s12868-016-0283-6

Fazel, S., Wolf, A., Långström, N., Newton, C.R., & Lichtenstein, P. (2013). Premature mortality in epilepsy and the role of psychiatric comorbidity: A total population study. The Lancet, 382(9905), 1646–1654.

French, J., & Perucca, E. (2020). Time to start calling things by their own names? The case for antiseizure medicines. Epilepsy Currents, 20(2), 69–72. https://doi.org/10.1177/1535759720905516

Ghosh, S., Sinha, J., Ghosh, S., Sharma, H., Bhaskar, R., & Narayanan, K. (2023). A comprehensive review of emerging trends and innovative therapies in epilepsy management. Brain Sciences, 13(9), 1305. https://doi.org/10.3390/brainsci13091305

Gibbons, M., Smeal, R., Takahashi, D., Vargas, J., & Wilcox, K. (2013). Contributions of astrocytes to epileptogenesis following status epilepticus: Opportunities for preventive therapy? Neurochemistry International, 63(7), 660–669. https://doi.org/10.1016/j.neuint.2012.12.008

Haast, R., Testud, B., Makhalova, J., Dary, H., Cabane, A., Troter, A., ... & Guye, M. (2022). Multi-scale structural alterations of the thalamus and basal ganglia in focal epilepsy as demonstrated by 7t MRI. https://doi.org/10.1101/2022.11.01.514655

Han, C., Ge, M., Liu, Y., Zhao, X., Wang, K., Chen, N., ... & Meng, F. (2018). LncRNA h19 contributes to hippocampal glial cell activation via JAK/STAT signaling in a rat model of temporal lobe epilepsy. Journal of Neuroinflammation, 15(1). https://doi.org/10.1186/s12974-018-1139-z

Han, C., Liu, Y., Guo, C., Du, T., Jiang, Y., Wang, K., ... & Zhang, J. (2020). The lncRNA H19 binding to let-7b promotes hippocampal glial cell activation and epileptic seizures by targeting Stat3 in a rat model of temporal lobe epilepsy. Cell Proliferation, 53(8). https://doi.org/10.1111/cpr.12856

He, Z., Zheng, J., Duan, J., Jin, Z., Huang, Z., Wu, S., & Xiong, Z. (2023). Long-term reliable neural decoding based on flexible implantable microelectronics and machine learning for seizure prediction application. *bioRxiv*. https://doi.org/10.1101/2023.03.08.531452

Hormozdiari, F., Penn, O., Borenstein, E., & Eichler, E. (2014). The discovery of integrated gene networks for autism and related disorders. Genome Research, 25(1), 142–154. https://doi.org/10.1101/gr.178855.114

Hu, T., Zhang, J., Wang, J., Sha, L., Xia, Y., Ortyl, T., ... & Chen, L. (2023). Advances in epilepsy: Mechanisms, clinical trials, and drug therapies. Journal of Medicinal Chemistry, 66(7), 4434–4467. https://doi.org/10.1021/acs.jmedchem.2c01975

Iffland, P., Carson, V., Bordey, A., & Crino, P. (2019). Gatoropathies: The role of amino acid regulatory gene mutations in epilepsy and cortical malformations. Epilepsia, 60(11), 2163–2173. https://doi.org/10.1111/epi.16370

Jones, N., & Ali, I. (2021). Glial cell collaboration in space and time contributes to epileptogenesis. Epilepsy Currents, 21(6), 452–454. https://doi.org/10.1177/15357597211041199

Kalilani, L., Sun, X., Pelgrims, B., Noack-Rink, M., & Villanueva, V. (2018). The epidemiology of drug-resistant epilepsy: A systematic review and meta-analysis. Epilepsia, 59(12), 2179–2193. https://doi.org/10.1111/epi.14596

Khoshkhoo, S., Vogt, D., & Sohal, V. (2017). Dynamic, cell-type-specific roles for GABAergic interneurons in a mouse model of optogenetically inducible seizures. Neuron, 93(2), 291–298. https://doi.org/10.1016/j.neuron.2016.11.043

Klaft, Z., Duerrwald, L., Gerevich, Z., & Dulla, C. (2020). The adenosine A1 receptor agonist wag 994 suppresses acute kainic acid-induced status epilepticus in vivo. Neuropharmacology, 176, 108213. https://doi.org/10.1016/j.neuropharm.2020.108213

Klassen, T., Davis, C., Goldman, A., Burgess, D., Chen, T., Wheeler, D., ... & Noebels, J. (2011). Exome sequencing of ion channel genes reveals complex profiles confounding personal risk assessment in epilepsy. Cell, 145(7), 1036–1048. https://doi.org/10.1016/j.cell.2011.05.025

Klein, P., Dingledine, R., Aronica, E., Bernard, C., Blümcke, I., Boison, D., ... & Löscher, W. (2017). Commonalities in epileptogenic processes from different acute brain insults: Do they translate?. Epilepsia, 59(1), 37–66. https://doi.org/10.1111/epi.13965

Kleen, J.K., Scott, R.C., Lenck-Santini, P.-P., & Holmes, G.L. (2012). *Jaspers Basic Mechanisms of the Epilepsies* (4th ed.). NCBI.

Klink, N., Rosmalen, F., Nenonen, J., Burnos, S., Helle, L., Taulu, S., ... & Hillebrand, A. (2017). Automatic detection and visualisation of meg ripple oscillations in epilepsy. Neuroimage Clinical, 15, 689–701. https://doi.org/10.1016/j.nicl.2017.06.024

Korenke, G., Eggert, M., Thiele, H., Nürnberg, P., Sander, T., & Steinlein, O. (2016). Nocturnal frontal lobe epilepsy caused by a mutation in the gatorl complex gene NPRL3. Epilepsia, 57(3). https://doi.org/10.1111/epi.13307

Krook-Magnuson, E., Armstrong, C., Bui, A., Lew, S., Oijala, M., & Soltész, I. (2015). In vivo evaluation of the dentate gate theory in epilepsy. The Journal of Physiology, 593(10), 2379–2388. https://doi.org/10.1113/jp270056

Laxpati, N.G., Kasoff, W.S., & Gross, R.E. (2014). Deep brain stimulation for the treatment of epilepsy: Circuits, targets, and trials. Neurotherapeutics, 11(3), 508–26.

Lerner, T.N., Ye, L., & Deisseroth, K. (2016). Communication in neural circuits: Tools, opportunities, and challenges. Cell, 164(6), 1136–50.

Lewis, A., & Chetkovich, D. (2011). HCN channels in behavior and neurological disease: Too hyper or not active enough? Molecular and Cellular Neuroscience, 46(2), 357–367. https://doi.org/10.1016/j.mcn.2010.11.007

Li, Y., Zhao, X., Wang, S., Xu, K., Zhao, X., Huang, S., ... & Zhu, S. (2021). A novel loss-of-function mutation in the NPRL3 gene identified in Chinese familial focal epilepsy with variable foci. Frontiers in Genetics, 12. https://doi.org/10.3389/fgene.2021.766354

Liu, S., Luo, Z., Li, F., Zhang, L., Xie, M., Yang, J., ... & Xu, Z. (2024). The JNK signaling pathway regulates seizures through ENT1 in pilocarpine-induced epilepsy rat model. CNS Neuroscience & Therapeutics, 30(12). https://doi.org/10.1111/cns.70190

Löscher, W., Potschka, H., Sisodiya, S., & Vezzani, A. (2020). Drug resistance in epilepsy: Clinical impact, potential mechanisms, and new innovative treatment options. Pharmacological Reviews, 72(3), 606–638. https://doi.org/10.1124/pr.120.019539

Lv, Q.Y., Chen, M.M., Li, Y., Yu, Y., & Liao, H. (2022). Brain circuit dysfunction in specific symptoms of depression. European Journal Neuroscience, 55(9–10), 2393–2403.

Malezieux, M., Klein, A.S., & Gogolla, N. (2023). Neural circuits for emotion. Annual Review of Neuroscience, Jul 10;46(1), 211–231.

Maroso, M., Balosso, S., Ravizza, T., Liu, J., Bianchi, M., & Vezzani, A. (2011). Interleukin-1 type 1 receptor/toll-like receptor signalling in epilepsy: The importance of IL-1beta and high-mobility group box 1. Journal of Internal Medicine, 270(4), 319–326. https://doi.org/10.1111/j.1365-2796.2011.02431.x

Mattis, J., Somarowthu, A., Goff, K., Jiang, E., Yom, J., Sotuyo, N., ... & Goldberg, E. (2022). Corticohippocampal circuit dysfunction in a mouse model of Dravet syndrome. Elife, 11. https://doi.org/10.7554/elife.69293

Meganathan, K., Lewis, E., Gontarz, P., Liu, S., Stanley, E., Elefanty, A., ... & Kroll, K. (2017). Regulatory networks specifying cortical interneurons from human embryonic stem cells reveal roles for CHD2 in interneuron development. Proceedings of the National Academy of Sciences, 114(52). https://doi.org/10.1073/pnas.1712365115

Memarian, N., Thompson, P., Engel, J., & Staba, R. (2013). Quantitative analysis of structural neuroimaging of mesial temporal lobe epilepsy. Imaging in Medicine, 5(3), 219–235.

Michel, C., & Brunet, D. (2019). EEG source imaging: A practical review of the analysis steps. Frontiers in Neurology, 10. https://doi.org/10.3389/fneur.2019.00325

Mula, M., Kanner, A.M., Jetté, N., & Sander, J.W. (2021). Psychiatric Comorbidities in People With Epilepsy. Neurology Clinical Practice [Internet]. 2021 Apr [cited 2025 Jul 25];11(2). Available from: https://www.neurology.org/doi/10.1212/CPJ.0000000000000874

Neymotin, S., Daniels, D., Caldwell, B., McDougal, R., Carnevale, N., Jas, M., … & Jones, S. (2019). Human neocortical neurosolver (HNN): A new software tool for interpreting the cellular and network origin of human MEG/EEG data. https://doi.org/10.1101/740597

Nissen, I., Stam, C., Reijneveld, J., Straaten, E., Hendriks, E., Baayen, J., … & Hillebrand, A. (2016). Identifying the epileptogenic zone in interictal resting-state MEG source-space networks. Epilepsia, 58(1), 137–148. https://doi.org/10.1111/epi.13622

Nogueira, M.H., Yasuda, C.L., Coan, A.C., Kanner, A.M., & Cendes, F. (2017). Concurrent mood and anxiety disorders are associated with pharmacoresistant seizures in patients with MTLE. Epilepsia, 58(7), 1268–1276.

Patel, R.S., Elmaadawi, A., Mansuri, Z., Kaur, M., Shah, K., & Nasr, S. (2017). Psychiatric comorbidities and outcomes in epilepsy patients: An insight from a nationwide inpatient analysis in the United States. Cureus [Internet]. Sep 14 [cited 2025 Jul 25]; Available from: http://www.cureus.com/articles/8654-psychiatric-comorbidities-and-outcomes-in-epilepsy-patients-an-insight-from-a-nationwide-inpatient-analysis-in-the-united-states.

Pauletti, A., Terrone, G., Shekh-Ahmad, T., Salamone, A., Ravizza, T., Rizzi, M., … & Vezzani, A. (2019). Targeting oxidative stress improves disease outcomes in a rat model of acquired epilepsy. Brain, 142(7), e39–e39. https://doi.org/10.1093/brain/awz130

Paz, J., & Huguenard, J. (2015). Optogenetics and epilepsy: Past, present and future. Epilepsy Currents, 15(1), 34–38. https://doi.org/10.5698/1535-7597-15.1.34

Peng, X., Li, C., Yu, W., Liu, S., Cong, Y., Fan, G., … & Qi, S. (2020). Propofol attenuates hypoxia-induced inflammation in bv2 microglia by inhibiting oxidative stress and NF-κb/HIF-1α signaling. Biomed Research International, 2020(1). https://doi.org/10.1155/2020/8978704

Perucca, E. (2019). Antiepileptic drugs: Evolution of our knowledge and changes in drug trials. Epileptic Disorders, 21(4), 319–329. https://doi.org/10.1684/epd.2019.1083

Perucca, E. (2021). The pharmacological treatment of epilepsy: Recent advances and future perspectives. Acta Epileptologica, 3(1). https://doi.org/10.1186/s42494-021-00055-z

Perucca, P., & Scheffer, I. (2020). Genetic contributions to acquired epilepsies. Epilepsy Currents, 21(1), 5–13. https://doi.org/10.1177/1535759720954254

Piastra, M., Nüßing, A., Vorwerk, J., Clerc, M., Engwer, C., & Wolters, C. (2020). A comprehensive study on electroencephalography and magnetoencephalography sensitivity to cortical and subcortical sources. Human Brain Mapping, 42(4), 978–992. https://doi.org/10.1002/hbm.25272

Pijet, B., Kostrzewska-Księżyk, A., Pijet-Kucicka, M., & Kaczmarek, L. (2024). Matrix metalloproteinase-9 contributes to epilepsy development after ischemic stroke in mice. International Journal of Molecular Sciences, 25(2), 896. https://doi.org/10.3390/ijms25020896

Ravizza, T., & Vezzani, A. (2018). Pharmacological targeting of brain inflammation in epilepsy: Therapeutic perspectives from experimental and clinical studies. Epilepsia Open, 3(S2), 133–142. https://doi.org/10.1002/epi4.12242

Reddy, D., Bhimani, A., Kuruba, R., Park, M., & Sohrabji, F. (2016). Prospects of modeling poststroke epileptogenesis. Journal of Neuroscience Research, 95(4), 1000–1016. https://doi.org/10.1002/jnr.23836

Riva, A., Golda, A., Balagura, G., Amadori, E., Vari, M., Piccolo, G., … & Striano, P. (2021). New trends and most promising therapeutic strategies for epilepsy treatment. Frontiers in Neurology, 12. https://doi.org/10.3389/fneur.2021.753753

Sano, F., Shigetomi, E., Shinozaki, Y., Tsuzukiyama, H., Saito, K., Mikoshiba, K., & Koizumi, S. (2019). Reactive astrocyte-driven epileptogenesis is induced by microglia initially activated following status epilepticus. https://doi.org/10.1101/806398

Shen, Y., Thomas, J., Chen, X., Zelidon, J., Hahn, A., Sathyanesan, A., … & Gu, B. (2025). Behavior decoding delineates seizure microfeatures and associated sudden death risks in mice. https://doi.org/10.1101/2025.01.13.632867.

Shultz, S., O'Brien, T., Stefanidou, M., & Kuzniecky, R. (2014). Neuroimaging the epileptogenic process. Neurotherapeutics, 11(2), 347–357. https://doi.org/10.1007/s13311-014-0258-1

Sills, G., & Rogawski, M. (2020). Mechanisms of action of currently used antiseizure drugs. Neuropharmacology, 168, 107966. https://doi.org/10.1016/j.neuropharm.2020.107966

Silva, R., Colom, H., Bicker, J., Almeida, A., Silva, A., Sales, F., … & Fortuna, A. (2023). Population pharmacokinetic analysis of Perampanel in Portuguese patients diagnosed with refractory epilepsy. Pharmaceutics, 15(6), 1704. https://doi.org/10.3390/pharmaceutics15061704

Sofroniew, M., & Vinters, H. (2009). Astrocytes: Biology and pathology. Acta Neuropathologica, 119(1), 7–35. https://doi.org/10.1007/s00401-009-0619-8

Soper, C., Wicker, E., Kulick, C., N'Gouemo, P., & Forcelli, P. (2016). Optogenetic activation of superior colliculus neurons suppresses seizures originating in diverse brain networks. Neurobiology of Disease, 87, 102–115. https://doi.org/10.1016/j.nbd.2015.12.012

Sorokin, J., Davidson, T., Fréchette, É, Abramian, A., Deisseroth, K., Huguenard, J., … & Paz, J. (2017). Bidirectional control of generalized epilepsy networks via rapid real-time switching of firing mode. Neuron, 93(1), 194–210. https://doi.org/10.1016/j.neuron.2016.11.026

Spanoghe, J., Boon, P., Vergaelen, M., Colvenaer, M., Mariman, T., Vonck, K., … & Raedt, R. (2024). Photopharmacological activation of adenosine a1receptor signaling suppresses seizures in a mouse model for temporal lobe epilepsy. https://doi.org/10.1101/2024.07.04.602052

Srivastava, K., Agarwal, E., Rajadhyaksha, S., Deshmukh, R., Kulkarni, V., & Doke, P., et al. (2021). Significant treatment gap and co-morbidities identified in an epidemiological survey of pediatric epilepsy in rural suburbs of India. Seizure, 91, 417–424.

Stevelink, R., Koeleman, B., & Sisodiya, S. (2023). Distinct genetic basis of common epilepsies and structural magnetic resonance imaging measures. Epilepsia, 64(5). https://doi.org/10.1111/epi.17529

Subaran, R., & Greenberg, D. (2014). The genetics of common epilepsy disorders: Lessons learned from the channelopathy era. Current Genetic Medicine Reports, 2(3), 190–200. https://doi.org/10.1007/s40142-014-0040-z

Talevi, A. (2022). Antiseizure medication discovery: Recent and future paradigm shifts. Epilepsia Open, 7(S1). https://doi.org/10.1002/epi4.12581

Tamilia, E., Matarrese, M., Ntolkeras, G., Grant, P., Madsen, J., Stufflebeam, S., … & Papadelis, C. (2021). Noninvasive mapping of ripple onset predicts outcome in epilepsy surgery. Annals of Neurology, 89(5), 911–925. https://doi.org/10.1002/ana.26066

Tondelli, M., Vaudano, A., Ruggieri, A., & Meletti, S. (2016). Cortical and subcortical brain alterations in juvenile absence epilepsy. Neuroimage Clinical, 12, 306–311. https://doi.org/10.1016/j.nicl.2016.07.007

Tung, J., Shiu, F., Ding, K., & Gross, R. (2018). Chemically activated luminopsins allow optogenetic inhibition of distributed nodes in an epileptic network for non-invasive and multi-site suppression of seizure activity. Neurobiology of Disease, 109, 1–10. https://doi.org/10.1016/j.nbd.2017.09.007

Vezzani, A., Fujinami, R., White, H., Preux, P., Blümcke, I., Sander, J., … & Löscher, W. (2015). Infections, inflammation and epilepsy. Acta Neuropathologica, 131(2), 211–234. https://doi.org/10.1007/s00401-015-1481-5

Vezzani, A., & Granata, T. (2005). Brain inflammation in epilepsy: Experimental and clinical evidence. Epilepsia, 46(11), 1724–1743. https://doi.org/10.1111/j.1528-1167.2005.00298.x

Vinet, J., Vainchtein, I., Spano, C., Giordano, C., Bordini, D., Curia, G., … & Biagini, G. (2016). Microglia are less pro-inflammatory than myeloid infiltrates in the hippocampus of mice exposed to status epilepticus. Glia, 64(8), 1350–1362. https://doi.org/10.1002/glia.23008

Wang, M., Cheng, X., Shi, Q., Xu, B., Hou, X., Zhao, H., … & Ding, Z. (2023). Brain diffusion tensor imaging reveals altered connections and networks in epilepsy patients. Frontiers in Human Neuroscience, 17. https://doi.org/10.3389/fnhum.2023.1142408

Wang, W., Huang, Q., Zhou, Q., Han, J., Zhang, X., Li, L., … & Wang, Y. (2024). Multimodal non-invasive evaluation in MRI-negative epilepsy patients. Epilepsia Open, 9(2), 765–775. https://doi.org/10.1002/epi4.12896

Waris, A., Asim, M., Ullah, A., & Alhumaydhi, F. (2024). Various pharmacological agents in the pipeline against intractable epilepsy. Archiv Der Pharmazie, 357(9). https://doi.org/10.1002/ardp.202400229

Weitz, A., Fang, Z., Lee, H., Fisher, R., Smith, W., Choy, M., … & Lee, J. (2015). Optogenetic fMRI reveals distinct, frequency-dependent networks recruited by dorsal and intermediate hippocampus stimulations. Neuroimage, 107, 229–241. https://doi.org/10.1016/j.neuroimage.2014.10.039

Whelan, C., Altmann, A., Botía, J., Jahanshad, N., Hibar, D., Absil, J., … & Sisodiya, S. (2018). Structural brain abnormalities in the common epilepsies assessed in a worldwide enigma study. Brain, 141(2), 391–408. https://doi.org/10.1093/brain/awx341.

Xiao, W., Jiao, Z.L., Senol, E., Yao, J., Zhao, M., & Zhao, Z.D., et al. (2022). Neural circuit control of innate behaviors. Science China Life Sciences, 65(3), 466–499.

Xu, J., Sun, M., Li, X., Huang, L., Gao, Z., Gao, J., … & Xie, A. (2021). MicroRNA expression profiling after recurrent febrile seizures in rat and emerging role of MIR-148A-3P/SYNJ1 axis. Scientific Reports, 11(1). https://doi.org/10.1038/s41598-020-79543-0

Yan, B., Sha, R., & Yu, H. (2023). Interaction between neuroimmune activation mechanisms in epilepsy: The S100B signaling network and brain-gut-bio-axis. Theoretical and Natural Science, 3(1), 505–512. https://doi.org/10.54254/2753-8818/3/20220334

Yang, J., He, F., Meng, Q., Sun, Y., Wang, W., & Wang, C. (2016). Inhibiting HIF-1α decreases expression of TNF-α and caspase-3 in specific brain regions exposed kainic acid-induced status epilepticus. Cellular Physiology and Biochemistry, 38(1), 75–82. https://doi.org/10.1159/000438610

Yao, R., Zhou, Y., Tang, J., Li, N., Yu, T., He, Y., … & Wang, J. (2021). Genetic diagnosis spectrum and multigenic burden of exome-level rare variants in a childhood epilepsy cohort. Frontiers in Genetics, 12. https://doi.org/10.3389/fgene.2021.782419

Yao, Y., Yang, Y., He, X., & Wang, X. (2017). Mir-16-1 expression, heat shock protein 70 and inflammatory reactions in astrocytes of mice with epilepsy induced by encephalitis b virus infection. Experimental and Therapeutic Medicine, 14(1), 495–498. https://doi.org/10.3892/etm.2017.4513

Zeibich, R., O'Brien, T., Perucca, P., Kwan, P., & Anderson, A. (2025). Identification of dysregulated transcription factor activity in temporal lobe epilepsy. https://doi.org/10.1101/2025.04.24.25326321

Zhang, H, Liu, H., Chen, W., Liu, G., Zhou, B., & Liang, H. (2019). Gray matter abnormalities in pediatric focal and generalized epilepsy: A voxel-based morphometry study with MRI at 3.0 t. https://doi.org/10.21203/rs.2.14484/v1.

Zhang, M., Eichhorn, S.W., Zingg, B., Yao, Z., Cotter, K., & Zeng, H., et al. (2021). Spatially resolved cell atlas of the mouse primary motor cortex by MERFISH. Nature, 598(7879), 137–43.

Zheng, P., Shultz, S., Hovens, C., Velakoulis, D., Jones, N., & O'Brien, T. (2013). Hyperphosphorylated tau is implicated in acquired epilepsy and neuropsychiatric comorbidities. Molecular Neurobiology, 49(3), 1532–1539. https://doi.org/10.1007/s12035-013-8601-9

Zybura, A., Sahoo, F., Hudmon, A., & Cummins, T. (2022). CAMKII inhibition attenuates distinct gain-of-function effects produced by mutant NAV1.6 channels and reduces neuronal excitability. Cells, 11(13), 2108. https://doi.org/10.3390/cells11132108

6 Novel Approaches to Epilepsy Management and Therapeutic Advances

Iman Imtiyaz Ahmed Juvale, Mohd. Farooq Shaikh, and Alina Arulsamy

6.1 INTRODUCTION

Epilepsy, affecting over 50 million people worldwide, is marked by recurrent, unpredictable seizures with significant neurological and psychosocial risks, including cognitive decline and increased mortality, such as sudden unexpected death in epilepsy (SUDEP).[1,2] Anti-seizure drugs (ASDs) are the mainstay of treatment, yet about one-third of patients experience DRE,[3] highlighting the need for more effective therapies. The heterogeneity of epilepsy, with its diverse etiologies and seizure types, complicates treatment, limiting the efficacy of one-size-fits-all approaches. For DRE, alternative therapies like vagus nerve stimulation (VNS), responsive neurostimulation (RNS), and the ketogenic diet can reduce seizure frequency but do not address the underlying mechanisms.[4] Emerging treatments aim to modify the molecular and cellular dysfunctions that drive epilepsy, offering the potential for disease modification rather than just symptom control. There is an urgent need for novel therapies that target the root causes of epilepsy to improve patient outcomes and quality of life, advancing toward more personalized and curative treatments.

6.2 EMERGING TRENDS IN EPILEPSY THERAPY

6.2.1 Role of Prebiotics and Probiotics

Current epilepsy treatments, while effective, often have limitations, leading researchers to explore more personalized therapies. Recent findings on the gut-brain axis suggest that gut dysbiosis and neuroinflammation play key roles in epilepsy, making the gut microbiota a promising therapeutic target. Short-chain fatty acids (SCFAs) produced by gut bacteria, such as acetate, propionate, and butyrate, have been shown to reduce neuroinflammation and protect neurons.[5] These SCFAs can be increased through prebiotics (dietary fibers) and probiotics (beneficial bacteria), which promote a healthier gut environment.[6] Research shows that SCFAs, particularly butyrate, have anti-inflammatory and neuroprotective effects by modulating immune responses and epigenetic regulation.[7–12] Animal models of epilepsy have demonstrated reduced seizure frequency and severity when SCFA-producing probiotics were introduced.[13,14] Probiotic strains of bacteria such as *Lactobacillus acidophilus*, *Lactobacillus plantarum*, *Lactobacillus casei*, *Lactobacillus helveticus*, *Lactobacillus brevis*, *Bifidobacterium lactis*, *Streptococcus salivarius* subsp., and *Thermophilus* have been reported to have a potential therapeutic effect in patients with epilepsy.[15,16] SCFAs may also work synergistically with the ketogenic diet, enhancing its anticonvulsant effects. Though clinical trials are still in the early stages, prebiotics and probiotics show promise as safe, non-invasive adjunct therapies for epilepsy, offering hope for improved seizure control by targeting gut health.

DOI: 10.1201/9781003500025-6

6.2.2 Fecal Microbiota Transplantation

Another technique that has gained significant attention as a potential therapeutic strategy for neurological disorders, including epilepsy, is FMT. This procedure involves transplanting healthy gut microbiota from a donor into the gastrointestinal tract of a patient with dysbiosis, thereby restoring a balanced microbial environment.[17] Originally rooted in ancient Chinese medicine,[18] FMT has more recently been approved by the U.S. Food and Drug Administration (FDA) for treating recurrent *Clostridium difficile* infections, and ongoing clinical trials are exploring its potential anticonvulsant effects in patients with epilepsy.[19] Epilepsy is increasingly recognized as a condition linked to neuroinflammation and gut dysbiosis, and the gut-brain axis plays a crucial role in these processes. By restoring a healthy gut microbiome, FMT can potentially reduce neuroinflammation, disrupting the pathophysiological cascade that leads to seizure activity. The therapeutic effects of FMT stem from rebalancing microbial metabolites such as SCFAs, which are known to exert anti-inflammatory and neuroprotective effects, as well as modulating the immune system and maintaining the integrity of the blood-brain barrier.[20–23] Several clinical studies have begun to investigate the potential of FMT in epilepsy. In an isolated case study, a patient became seizure-free after undergoing FMT,[24] highlighting its potential efficacy in restoring neurological health. Preclinical research further supports this concept,[25] showing that a healthy gut microbiome can modulate neuronal excitability and reduce pro-inflammatory markers, critical in managing seizure disorders. Although research on FMT for epilepsy is still in its early stages, the treatment's ability to directly alter the gut microbiota offers a promising, personalized approach. This technique could complement existing therapies by targeting the underlying gut dysbiosis contributing to neuroinflammation and seizure activity. Ongoing clinical trials are expected to further elucidate the role of FMT in epilepsy treatment.

6.2.3 Cannabidiol Therapy

CBD has emerged as a novel and promising treatment option for epilepsy, especially after its approval in 2018 for managing severe forms of the disorder, such as Dravet syndrome and Lennox-Gastaut syndrome.[26] Despite the controversy surrounding CBD, largely due to its association with cannabis and concerns over misuse, CBD lacks the psychoactive properties of tetrahydrocannabinol (THC), making it a safer therapeutic candidate.[27] Its anticonvulsant potential is of great interest, even though its precise mechanism of action remains incompletely understood. CBD is thought to exert its therapeutic effects by interacting with various ion channels and neuroreceptors critical for neuronal signaling. Among these, CBD has a high affinity for transient receptor potential cation channel subfamily V member 1 (TRPV1) receptors, which regulate pain and inflammation, as well as ion transport.[28] This interaction may reduce neuronal hyperexcitability, a key factor in the development of seizures. Additionally, CBD is known to inhibit other ion transport channels, including T-type voltage-gated calcium channels (VGCC) and voltage-gated sodium channels (VGSC), both of which play critical roles in neuronal excitability.[29,30] By inhibiting these channels, CBD helps to regulate the influx of calcium and sodium ions, thereby stabilizing neuronal activity and reducing the likelihood of seizures. CBD also acts on the serotonin 2A (5-HT2A) receptor that, when activated, can lead to increased calcium entry into neurons. By inhibiting 5-HT2A activity, CBD may further reduce intracellular calcium levels, helping to dampen excitatory signaling in the brain.[28] This multi-targeted approach suggests that CBD may work through various mechanisms to achieve its anticonvulsant effects, even though the exact pathways remain to be fully delineated. Clinical evidence supports the efficacy of CBD in reducing seizure frequency, particularly in cases of DRE. Notably, studies have shown 43% of patients treated with CBD achieved at least a 50% reduction in seizure frequency, compared to 27% in the placebo group.[31] These findings highlight CBD's significant potential as an adjunctive treatment for patients who do not respond to conventional therapies. Beyond its direct impact on seizures, CBD may offer additional neuroprotective benefits by reducing inflammation and oxidative stress,[32,33] both of which are implicated in the pathophysiology of epilepsy. Its ability to interact with multiple molecular targets without the euphoric effects associated with THC makes CBD an attractive therapeutic option for a broad range of patients, including children with severe epilepsy syndromes. However, despite the promising data, the use of CBD in epilepsy is not without challenges. Variability in individual responses, optimal dosing strategies, and long-term safety concerns remain areas of active investigation.

6.2.4 Transcranial Magnetic Stimulation

TMS initially developed in the UK in 1985 for brain mapping and activity studies,[34] has evolved into a promising therapeutic tool for a range of neurological and psychiatric disorders, including epilepsy. Originally employed to probe cortical excitability in epilepsy and assess the impact of ASDs on brain function, TMS has since garnered attention for its potential as a treatment modality for epilepsy. The technique involves positioning a large electromagnetic coil against the scalp, generating magnetic fields that induce electrical currents in targeted brain regions.[35] These currents modulate neuronal activity, with the capability to either enhance or inhibit neural excitability over extended periods, thus holding the potential to alter the excitability of cortical areas implicated in seizure generation. By modulating neural circuits involved in seizure propagation, TMS could offer a non-invasive means to reduce seizure frequency and improve overall symptom management,[36] making it a compelling area of research for optimizing epilepsy treatment strategies. Multiple randomized controlled trials have demonstrated the efficacy of TMS in reducing seizure frequency in patients with DRE.[37] Additionally, small pilot studies have explored the combination of TMS with EEG (electroencephalogram), suggesting its potential for developing quantitative biomarkers of cortical hyperexcitability,[38–40] a hallmark of epilepsy. TMS is already FDA-approved for presurgical mapping of the eloquent cortex, and preliminary data indicate that navigated TMS could significantly enhance preoperative functional planning by accurately identifying critical brain regions.[41] However, further research is necessary to optimize stimulation protocols, clarify the underlying mechanisms of TMS, evaluate its long-term effects, and determine its precise role and efficacy in epilepsy treatment. These future studies are essential for refining TMS as a therapeutic tool and expanding its application in clinical practice.

6.2.5 Stem Cell Therapy

Personalized medicine, which tailor the treatments to a patient's genetic, environmental, and lifestyle factors, is transforming modern health care by enhancing therapeutic efficacy and reducing side effects. In epilepsy, personalized approaches offer the potential to target specific genetic mutations or inflammatory pathways, especially in drug-resistant cases. Advances in biotechnology, such as stem cell therapy, gene therapy, and immunotherapy, represent cutting-edge treatments that address the cellular and molecular mechanisms underlying epilepsy. These therapies, though still experimental, hold significant promise for revolutionizing care, particularly for patients unresponsive to traditional ASDs. Stem cell therapy offers the potential to repair or replace damaged neurons in the brain, addressing the neuronal loss and dysfunction that contribute to epilepsy.[42] Stem cells, particularly pluripotent stem cells (PSCs) and neural stem cells (NSCs), can differentiate into various types of neural cells, including inhibitory neurons that play a critical role in maintaining the balance of excitatory and inhibitory signaling in the brain.[43] In epilepsy, this balance is often disrupted, leading to hyperexcitability and recurrent seizures. Transplanting stem cells capable of differentiating into inhibitory neurons (e.g., GABAergic interneurons) has shown promise in animal models, where it has reduced seizure frequency and severity.[44] Stem cells may also support the regeneration of damaged brain tissue, promote synaptic repair, and enhance neuroplasticity, creating a more stable neuronal network.[45] Another promising avenue is using mesenchymal stem cells (MSCs), which possess immunomodulatory and anti-inflammatory properties.[46] Since neuroinflammation plays a key role in the pathophysiology of epilepsy, MSCs could help modulate the inflammatory environment, reducing neuronal damage and seizure susceptibility. However, challenges remain, including the precise control of stem cell differentiation, integration into existing neural circuits, and the risk of tumorigenesis.[47] Future research must focus on refining techniques for directing stem cell differentiation and ensuring their long-term safety in human patients.

6.2.6 Gene Therapy

Gene therapy has the potential to address the genetic mutations and abnormalities that underlie certain forms of epilepsy. This therapeutic approach involves the delivery of specific genes into the patient's brain cells to correct or compensate for defective genes responsible for seizure activity.[48] Techniques such as viral vector-mediated gene transfer and CRISPR/Cas9 gene editing are being explored to alter the genetic landscape of neurons and restore normal neuronal function.[49] One promising application of gene therapy in epilepsy involves targeting ion channels that regulate neuronal excitability. Many forms of epilepsy are linked to mutations in genes encoding voltage-gated sodium or potassium channels, leading to hyperexcitability of neurons.[50] Gene therapy could correct these mutations or upregulate the expression of inhibitory ion channels, restoring the proper balance of neuronal signaling and reducing seizure frequency. Another exciting prospect is the use of gene therapy to enhance the production of inhibitory neurotransmitters, such as GABA. By upregulating GABA synthesis or promoting the expression of GABA receptors,[51] gene therapy could directly increase inhibitory signaling in the brain, counteracting the excessive excitatory activity characteristic of epilepsy. While preclinical studies have demonstrated the potential of gene therapy to reduce seizures in animal models,[52–54] several challenges remain. These include ensuring precise targeting of the affected brain regions, minimizing off-target effects, and overcoming immune responses to viral vectors.[55] Ongoing research is needed to refine gene delivery methods and explore the long-term safety and efficacy of these treatments in human trials.

6.2.7 Immunotherapy

Immunotherapy represents another promising frontier for epilepsy treatment, particularly in cases where neuroinflammation plays a significant role in disease progression. Inflammatory processes, driven by both innate and adaptive immune responses, have been implicated in the development of epilepsy, especially in forms of the disorder associated with autoimmune conditions or infections.[56] Immunotherapy aims to modulate the immune response to reduce inflammation and prevent neuronal damage. One approach involves using monoclonal antibodies (mAbs) to target specific inflammatory mediators, such as cytokines or immune cells, contributing to seizure activity.[57] For example, mAbs directed against interleukin-1β (IL-1β) or tumor necrosis factor-alpha (TNF-α) have shown promise in reducing neuroinflammation and seizure susceptibility.[58] Another potential immunotherapeutic strategy involves using vaccines to induce tolerance to specific autoantigens implicated in autoimmune epilepsy.[59] By dampening the immune response to these antigens, vaccines could reduce the autoimmune attack on neurons, thereby reducing seizure frequency and severity. Despite its potential, immunotherapy for epilepsy is still in the early stages of development. Future research will need to focus on identifying the specific immune pathways involved in different types of epilepsy and determining how best to modulate these pathways to achieve therapeutic effects. Additionally, long-term studies will be required to assess the safety and efficacy of immunotherapies in human patients (Figure 6.1).

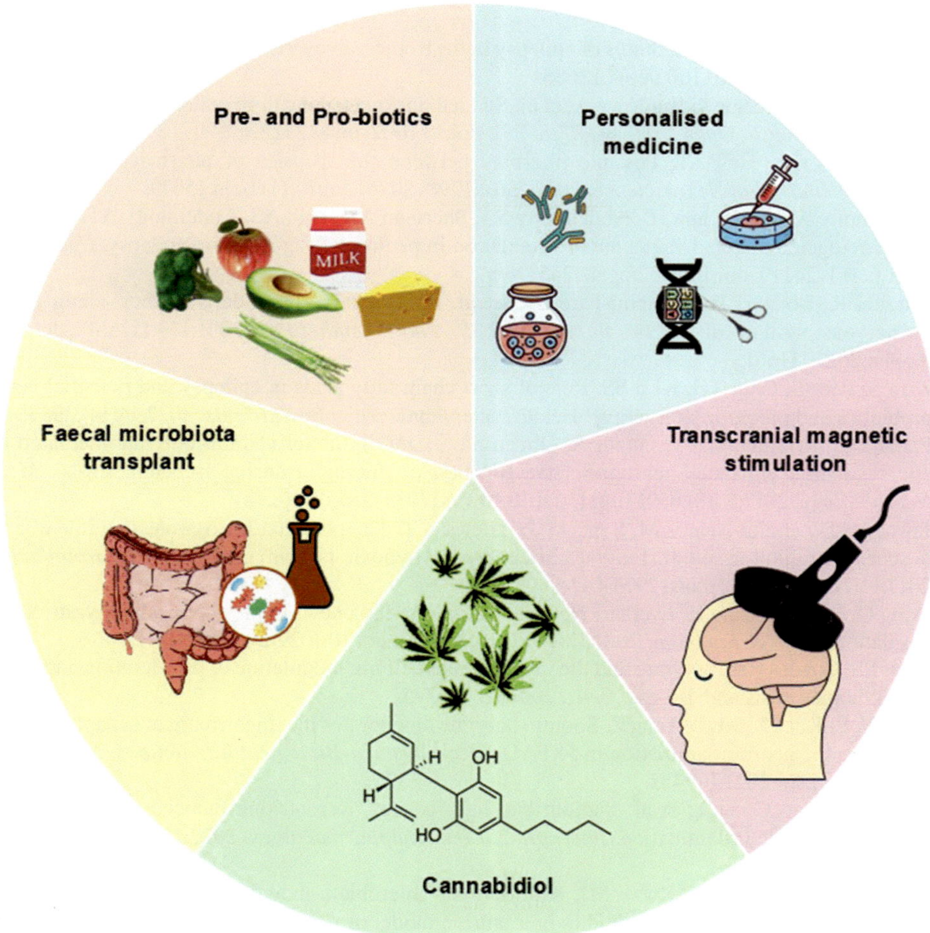

FIGURE 6.1 Summary of the current novel approaches for epilepsy management.

6.3 CONCLUSION

Gone are the days when seizures were thought to be the work of mischievous spirits, and the remedy involved ritualistic exorcisms. Fast forward to today, and we're witnessing a thrilling era of innovation that brings hope and tangible progress for those living with epilepsy. Modern medicine now boasts a diverse arsenal of therapies, from traditional antiepileptic drugs to advanced treatments like stem cell therapy and gene editing. Adding to the excitement are new developments, such as AI-driven diagnostic tools,[60] which are revolutionizing our understanding of epilepsy by analyzing vast amounts of data to predict seizures and tailor treatments with unprecedented precision. And let's not forget the burgeoning field of seizure response therapy dogs[61]—yes, these amazing canines are being trained to detect seizures and provide life-saving assistance, proving that man's best friend is indeed a superhero in disguise. On top of this, smartphone apps are transforming how patients manage their conditions.[62] These digital allies can track seizure activity, monitor medication adherence, and even connect users with support communities, all from the palm of their hands. As we continue to innovate and integrate these advancements, the horizon for epilepsy treatment looks brighter than ever.

REFERENCES

1. Liu J, Zhang P, Zou Q, et al. Status of epilepsy in the tropics: an overlooked perspective. *Epilepsia Open*. 2023;8(1):32–45. doi:10.1002/epi4.12686.
2. Giussani G, Falcicchio G, La Neve A, et al. Sudden unexpected death in epilepsy: a critical view of the literature. *Epilepsia Open*. 2023;8(3):728–757. doi:10.1002/epi4.12722.
3. Juvale IIA, Che Has AT. Possible interplay between the theories of pharmacoresistant epilepsy. *European Journal of Neuroscience*. 2021;53(6):1998–2026. doi:10.1111/ejn.15079.
4. Abdelmoity AT, Le Pichon JB, Abdelmoity SA, Sherman AK, Hall AS, Abdelmoity AT. Combined use of the ketogenic diet and vagus nerve stimulation in pediatric drug-resistant epilepsy. *Epilepsia Open*. 2021;6(1):112–119. doi:10.1002/epi4.12453.
5. Mirzaei R, Bouzari B, Hosseini-Fard SR, et al. Role of microbiota-derived short-chain fatty acids in nervous system disorders. *Biomedicine & Pharmacotherapy*. 2021;139:111661. doi:10.1016/j.biopha.2021.111661.
6. Kim S, Park S, Choi TG, Kim SS. Role of short chain fatty acids in epilepsy and potential benefits of probiotics and prebiotics: targeting "Health" of epileptic patients. *Nutrients*. 2022;14(14):2982.
7. Patnala R, Arumugam TV, Gupta N, Dheen ST. HDAC inhibitor sodium butyrate-mediated epigenetic regulation enhances neuroprotective function of microglia during ischemic stroke. *Molecular Neurobiology*. 2017;54(8):6391–6411. doi:10.1007/s12035-016-0149-z.
8. Jaworska J, Ziemka-Nalecz M, Sypecka J, Zalewska T. The potential neuroprotective role of a histone deacetylase inhibitor, sodium butyrate, after neonatal hypoxia-ischemia. *Journal of Neuroinflammation*. 2017;14(1):34. doi:10.1186/s12974-017-0807-8.
9. Dang G, Wu W, Zhang H, Everaert N. A new paradigm for a new simple chemical: butyrate & immune regulation. *Food & Function*. 2021;12(24):12181–12193. doi:10.1039/D1FO02116H.
10. Salvi PS, Cowles RA. Butyrate and the intestinal epithelium: modulation of proliferation and inflammation in homeostasis and disease. *Cells*. 2021;10(7):1775.
11. Jiang Y, Li K, Li X, Xu L, Yang Z. Sodium butyrate ameliorates the impairment of synaptic plasticity by inhibiting the neuroinflammation in 5XFAD mice. *Chemico-Biological Interactions*. 2021;341:109452. doi:10.1016/j.cbi.2021.109452.
12. Wang X, Duan C, Li Y, et al. Sodium butyrate reduces overnutrition-induced microglial activation and hypothalamic inflammation. *International Immunopharmacology*. 2022;111:109083. doi:10.1016/j.intimp.2022.109083.
13. Eor JY, Tan PL, Son YJ, Kwak MJ, Kim SH. Gut microbiota modulation by both Lactobacillus fermentum MSK 408 and ketogenic diet in a murine model of pentylenetetrazole-induced acute seizure. *Epilepsy Research*. 2021;169:106506. doi:10.1016/j.eplepsyres.2020.106506.
14. Wlaź P, Wiater A, Majewska M, et al. Effect of dietary supplementation with *Lactobacillus helveticus* R0052 on seizure thresholds and antiseizure potency of sodium valproate in mice. *Psychopharmacology*. 2024;241(2):327–340. doi:10.1007/s00213-023-06489-2.
15. Gómez-Eguílaz M, Ramón-Trapero JL, Pérez-Martínez L, Blanco JR. The beneficial effect of probiotics as a supplementary treatment in drug-resistant epilepsy: a pilot study. *Beneficial Microbes*. 2018;9(6):875–881. doi:10.3920/bm2018.0018.
16. Yeom JS, Park JS, Kim YS, et al. Neonatal seizures and white matter injury: role of rotavirus infection and probiotics. *Brain and Development*. 2019;41(1):19–28. doi:10.1016/j.braindev.2018.07.001.
17. Onifade IA, Sunbare-Funto OJ, Mbah CE, et al. Faecal microbial transplant. *Advances in Biomarker Sciences and Technology*. 2024;6:20–34. doi:10.1016/j.abst.2024.02.001.
18. Shi YC, Yang YS. Fecal microbiota transplantation: current status and challenges in China. *JGH Open*. 2018;2(4):114–116. doi:10.1002/jgh3.12071.
19. Kempler C. *FDA approves first orally administered fecal microbiota product for the prevention of recurrence of Clostridioides difficile infection* [updated 04/26/2023; accessed 10/09/2024, 2024]. https://www.fda.gov/news-events/press-announcements/fda-approves-first-orally-administered-fecal-microbiota-product-prevention-recurrence-clostridioides
20. Jing Y, Yu Y, Bai F, et al. Effect of fecal microbiota transplantation on neurological restoration in a spinal cord injury mouse model: involvement of brain-gut axis. *Microbiome*. 2021;9(1):59. doi:10.1186/s40168-021-01007-y.
21. Su S-H, Chen M, Wu Y-F, et al. Fecal microbiota transplantation and short-chain fatty acids protected against cognitive dysfunction in a rat model of chronic cerebral hypoperfusion. *CNS Neuroscience & Therapeutics*. 2023;29(S1):98–114. doi:10.1111/cns.14089.

22. Mathias K, Machado RS, Stork S, et al. Short-chain fatty acid on blood-brain barrier and glial function in ischemic stroke. *Life Sciences*. 2024;354:122979. doi:10.1016/j.lfs.2024.122979.
23. Hediyal TA, Vichitra C, Anand N, et al. Protective effects of fecal microbiota transplantation against ischemic stroke and other neurological disorders: an update. *Frontiers in Immunology*. 2024;15. doi:10.3389/fimmu.2024.1324018.
24. He Z, Cui BT, Zhang T, et al. Fecal microbiota transplantation cured epilepsy in a case with Crohn's disease: the first report. *World J Gastroenterol*. 2017;23(19):3565–3568. doi:10.3748/wjg.v23.i19.3565.
25. Olson CA, Vuong HE, Yano JM, Liang QY, Nusbaum DJ, Hsiao EY. The gut microbiota mediates the anti-seizure effects of the ketogenic diet. *Cell*. 2018;173(7):1728–1741.e13. doi:10.1016/j.cell.2018.04.027.
26. Lattanzi S, Trinka E, Striano P, et al. Highly purified cannabidiol for epilepsy treatment: a systematic review of epileptic conditions beyond Dravet Syndrome and Lennox–Gastaut Syndrome. *CNS Drugs*. 2021;35(3):265–281. doi:10.1007/s40263-021-00807-y.
27. Abu-Sawwa R, Scutt B, Park Y. Emerging use of Epidiolex (Cannabidiol) in epilepsy. *The Journal of Pediatric Pharmacology and Therapeutics*. 2020;25(6):485–499. doi:10.5863/1551-6776-25.6.485.
28. Gaston TE, Friedman D. Pharmacology of cannabinoids in the treatment of epilepsy. *Epilepsy & Behavior*. 2017;70:313–318. doi:10.1016/j.yebeh.2016.11.016.
29. Ross HR, Napier I, Connor M. Inhibition of recombinant human T-type calcium channels by Delta9-tetrahydrocannabinol and cannabidiol. *Journal of Biological Chemistry*. 2008;283(23):16124–16134. doi:10.1074/jbc.M707104200.
30. Patel RR, Barbosa C, Brustovetsky T, Brustovetsky N, Cummins TR. Aberrant epilepsy-associated mutant Nav1.6 sodium channel activity can be targeted with cannabidiol. *Brain*. 2016;139(8):2164–2181. doi:10.1093/brain/aww129.
31. Devinsky O, Cross JH, Laux L, et al. Trial of cannabidiol for drug-resistant seizures in the Dravet syndrome. *New England Journal of Medicine*. 2017;376(21):2011–2020. doi:10.1056/NEJMoa1611618.
32. Cocetta V, Governa P, Borgonetti V, et al. Cannabidiol isolated from *Cannabis sativa* L. protects intestinal barrier from in vitro inflammation and oxidative stress. *Frontiers in Pharmacology*. 2021;12. doi:10.3389/fphar.2021.641210.
33. Kopustinskiene DM, Masteikova R, Lazauskas R, Bernatoniene J. *Cannabis sativa* L. bioactive compounds and their protective role in oxidative stress and inflammation. *Antioxidants*. 2022;11(4):660.
34. Hamed SA. Cortical excitability in epilepsy and the impact of antiepileptic drugs: transcranial magnetic stimulation applications. *Expert Review of Neurotherapeutics*. 2020;20(7):707–723. doi:10.1080/14737175.2020.1780122.
35. Walton D, Spencer DC, Nevitt SJ, Michael BD. Transcranial magnetic stimulation for the treatment of epilepsy. *Cochrane Database of Systematic Reviews*. 2021;4(4):CD011025. doi:10.1002/14651858.CD011025.pub3
36. VanHaerents S, Chang BS, Rotenberg A, Pascual-Leone A, Shafi MM. Noninvasive brain stimulation in epilepsy. *Journal of Clinical Neurophysiology*. 2020;37(2):118–130. doi:10.1097/wnp.0000000000000573.
37. VanHaerents S, Chang BS, Rotenberg A, Pascual-Leone A, Shafi MM. Noninvasive brain stimulation in epilepsy. *Journal of Clinical Neurophysiology*. 2020;37(2):118–130.
38. Passera B, Buss SS, Ross JM, Hagan B, Press DZ, Shafi MM. TMS-EEG as a measure of cortical hyperexcitability in motor and parietal cortex in Alzheimer's disease: a pilot study. *Brain Stimulation: Basic, Translational, and Clinical Research in Neuromodulation*. 2021;14(6):1658–1659. doi:10.1016/j.brs.2021.10.223.
39. Casula EP, Borghi I, Maiella M, et al. Regional precuneus cortical hyperexcitability in Alzheimer's disease patients. *Annals of Neurology*. 2023;93(2):371–383. doi:10.1002/ana.26514.
40. Buss SS, Ross J, Hagan B, Press DZ, Shafi M. TMS-EEG as a measure of cortical hyperexcitability in motor and parietal cortex in Alzheimer's disease: a pilot study. *Alzheimer's & Dementia*. 2021;17(S5):e055780. doi:10.1002/alz.055780.
41. Schramm S, Mehta A, Auguste KI, Tarapore PE. Navigated transcranial magnetic stimulation mapping of the motor cortex for preoperative diagnostics in pediatric epilepsy. *Journal of Neurosurgery: Pediatrics*. 2021;28(3):287–294. doi:10.3171/2021.2.Peds20901.
42. Alayli A, Lockard G, Gordon J, et al. Stem cells: recent developments redefining epilepsy therapy. *Cell Transplant*. 2023;32:9636897231158967. doi:10.1177/09636897231158967.
43. Gupta J, Naegele JR. Forming new connections: advances in human stem-cell-derived interneuron therapy for treating epilepsy. *Neuron*. 2023;111(6):758–760. doi:10.1016/j.neuron.2023.02.034.

44. Waloschková E, Gonzalez-Ramos A, Mikroulis A, et al. Human stem cell-derived GABAergic interneurons establish efferent synapses onto host neurons in rat epileptic hippocampus and inhibit spontaneous recurrent seizures. *International Journal of Molecular Sciences.* 2021;22(24):13243.
45. Chang J, Li Y, Shan X, et al. Neural stem cells promote neuroplasticity: a promising therapeutic strategy for the treatment of Alzheimer's disease. *Neural Regeneration Research.* 2024;19(3):619–628. doi:10.4103/1673-5374.380874.
46. Li P, Ou Q, Shi S, Shao C. Immunomodulatory properties of mesenchymal stem cells/dental stem cells and their therapeutic applications. *Cellular & Molecular Immunology.* 2023;20(6):558–569. doi:10.1038/s41423-023-00998-y.
47. Chen X, Jiang S, Wang R, Bao X, Li Y. Neural stem cells in the treatment of Alzheimer's disease: current status, challenges, and future prospects. *Journal of Alzheimer's Disease.* 2023;94(s1):S173–S186. doi:10.3233/jad-220721.
48. Ingusci S, Verlengia G, Soukupova M, Zucchini S, Simonato M. Gene therapy tools for brain diseases. *Frontiers in Pharmacology.* 2019;10:724. doi:10.3389/fphar.2019.00724.
49. Sandoval A Jr., Elahi H, Ploski JE. Genetically engineering the nervous system with CRISPR-Cas. *eNeuro.* 2020;7(2) doi:10.1523/eneuro.0419-19.2020.
50. Ademuwagun IA, Rotimi SO, Syrbe S, Ajamma YU, Adebiyi E. Voltage gated sodium channel genes in epilepsy: mutations, functional studies, and treatment dimensions. *Front Neurol.* 2021;12:600050. doi:10.3389/fneur.2021.600050.
51. Donegan JJ, Boley AM, Yamaguchi J, Toney GM, Lodge DJ. Modulation of extrasynaptic GABAA alpha 5 receptors in the ventral hippocampus normalizes physiological and behavioral deficits in a circuit specific manner. *Nature Communications.* 2019;10(1):2819. doi:10.1038/s41467-019-10800-1.
52. Wykes RC, Heeroma JH, Mantoan L, et al. Optogenetic and potassium channel gene therapy in a rodent model of focal neocortical epilepsy. *Science Translational Medicine.* 2012;4(161):161ra152–161ra152. doi:10.1126/scitranslmed.3004190.
53. Dogbevia GK, Töllner K, Körbelin J, et al. Gene therapy decreases seizures in a model of *Incontinentia pigmenti. Annals of Neurology.* 2017;82(1):93–104. doi:10.1002/ana.24981.
54. Powell KL, Fitzgerald X, Shallue C, et al. Gene therapy mediated seizure suppression in genetic generalised epilepsy: neuropeptide Y overexpression in a rat model. *Neurobiology of Disease.* 2018;113:23–32. doi:10.1016/j.nbd.2018.01.016.
55. Rittiner JE, Moncalvo M, Chiba-Falek O, Kantor B. Gene-editing technologies paired with viral vectors for translational research into neurodegenerative diseases. *Frontiers in Molecular Neuroscience.* 2020;13:148. doi:10.3389/fnmol.2020.00148.
56. Mukhtar I. Inflammatory and immune mechanisms underlying epileptogenesis and epilepsy: from pathogenesis to treatment target. *Seizure.* 2020;82:65–79. doi:10.1016/j.seizure.2020.09.015.
57. Costagliola G, Depietri G, Michev A, et al. Targeting inflammatory mediators in epilepsy: a systematic review of its molecular basis and clinical applications. *Frontiers in Neurology.* 2022:13. doi:10.3389/fneur.2022.741244.
58. Jang Y, Ahn SH, Park K-I, et al. Prognosis prediction and immunotherapy optimisation for cryptogenic new-onset refractory status epilepticus. *Journal of Neurology, Neurosurgery & Psychiatry.* 2024:jnnp-2024-334285. doi:10.1136/jnnp-2024-334285.
59. Jang Y, Kim DW, Yang KI, et al. Clinical approach to autoimmune epilepsy. *Journal of Clinical Neurology.* 2020;16(4):519–529. doi:10.3988/jcn.2020.16.4.519.
60. Handa P, Lavanya, Goel N, Garg N. Software advancements in automatic epilepsy diagnosis and seizure detection: 10-year review. *Artificial Intelligence Review.* 2024;57(7):181. doi:10.1007/s10462-024-10799-y.
61. Catala A, Cousillas H, Hausberger M, Grandgeorge M. Dog alerting and/or responding to epileptic seizures: a scoping review. *PLoS One.* 2018;13(12):e0208280. doi:10.1371/journal.pone.0208280.
62. Mohammadzadeh N, Khenarinezhad S, Ghazanfarisavadkoohi E, Safari MS, Pahlevanynejad S. Evaluation of M-health applications use in epilepsy: a systematic review. *Iranian Journal of Public Health.* 2021;50(3):459–469. doi:10.18502/ijph.v50i3.5586.

7 Ischemic Stroke
Insights into Brain Ischemia

Arshi Waseem and Syed Shadab Raza

7.1 INTRODUCTION

A stroke is an acute neurological disorder resulting from the sudden interruption of cerebral blood flow, leading to a rapid and critical deprivation of oxygen and essential nutrients in affected brain regions. This ischemic insult initiates cellular energy failure and triggers a cascade of detrimental processes, including excitotoxicity, oxidative stress, and neuroinflammation, ultimately culminating in neuronal death and neurological impairment (Raza, 2024). Ischemic stroke, caused by the obstruction of a cerebral artery, accounts for approximately 85% of all stroke cases and represents a major cause of long-term disability and mortality globally (Waseem et al., 2023). The World Health Organization (WHO) estimates that over 12 million people suffer a stroke each year, with nearly 6.5 million being ischemic in nature. Stroke is the second leading cause of death and the third leading cause of disability-adjusted life years (DALYs) worldwide (Salaudeen et al., 2024; Feigin et al., 2025). The burden is especially pronounced in low- and middle-income countries, where limited access to healthcare infrastructure and preventive strategies exacerbates outcomes. Beyond its medical implications, ischemic stroke imposes substantial socioeconomic and emotional strain on patients, caregivers, and healthcare systems alike.

Ischemic strokes can be classified based on their underlying etiology. The most commonly used classification system is the Trial of Org 10172 in Acute Stroke Treatment (TOAST) criteria, which categorizes ischemic strokes into the following subtypes: Large artery atherosclerosis—caused by thrombotic occlusion or emboli originating from atherosclerotic plaques in major arteries. Cardioembolic stroke—due to emboli arising from cardiac sources, often associated with atrial fibrillation, valvular heart disease, or myocardial infarction. Small vessel occlusion (lacunar stroke)—resulting from occlusion of small penetrating arteries, typically linked with chronic hypertension or diabetes. Stroke of other determined etiology—including uncommon causes such as arterial dissection, vasculitis, or prothrombotic states (Barthels and Das, 2020). Stroke of undetermined etiology—when no clear cause is identified after standard evaluation. Understanding the classification of ischemic stroke is crucial for tailoring appropriate diagnostic, preventive, and therapeutic strategies in clinical practice (Majumder, 2024).

7.2 PATHOPHYSIOLOGY AND ETIOLOGY OF ISCHEMIC STROKE

The pathophysiology of any disease condition encompasses its etiology, pathogenesis, and the associated molecular, biochemical, and structural alterations. In the case of ischemic stroke, the pathological process is initiated almost immediately—within minutes—following the occlusion of cerebral blood flow. This abrupt interruption leads to a critical reduction in oxygen and glucose delivery, triggering a cascade of cellular and molecular events that result in rapid neuronal injury and tissue damage. Tissues as a result of thrombus formation on fatty deposits known as atherosclerotic plaque or blood clots created by atrial fibrillation obstructing the arteries supplying the brain (Salaudeen et al., 2024). The ischemic core is frequently thought to be the afflicted area of the brain. In this case, the majority of the cells die irreversibly before the neuroprotective agent's effect or effects are known. The ischemic penumbra, an area of salvageable cells that

DOI: 10.1201/9781003500025-7

surrounds the ischemic core, is frequently the focus of therapeutic interventions. Hemiplegia, paraplegia, dysarthria, and paresis are among the phenotypic manifestations that arise from the interaction of intricate molecular and cellular pathways. Depending on where the area of the brain receives blood flow from the blocked arteries, other symptoms could appear (Salaudeen et al., 2024). Ischemic stroke is typified by a variety of alterations in both the affected ischemic core and the surrounding penumbra, much like many other neurodegenerative diseases. The five general terms neuroinflammation, excitotoxicity, oxidative stress, apoptosis, and autophagy are frequently used to describe these microscopic and macro-alterations (Figure 7.1). These separate yet mutually reinforcing sequences of pathogenic events interact intricately to cause cell death in ischemic stroke (Sarkar et al., 2019).

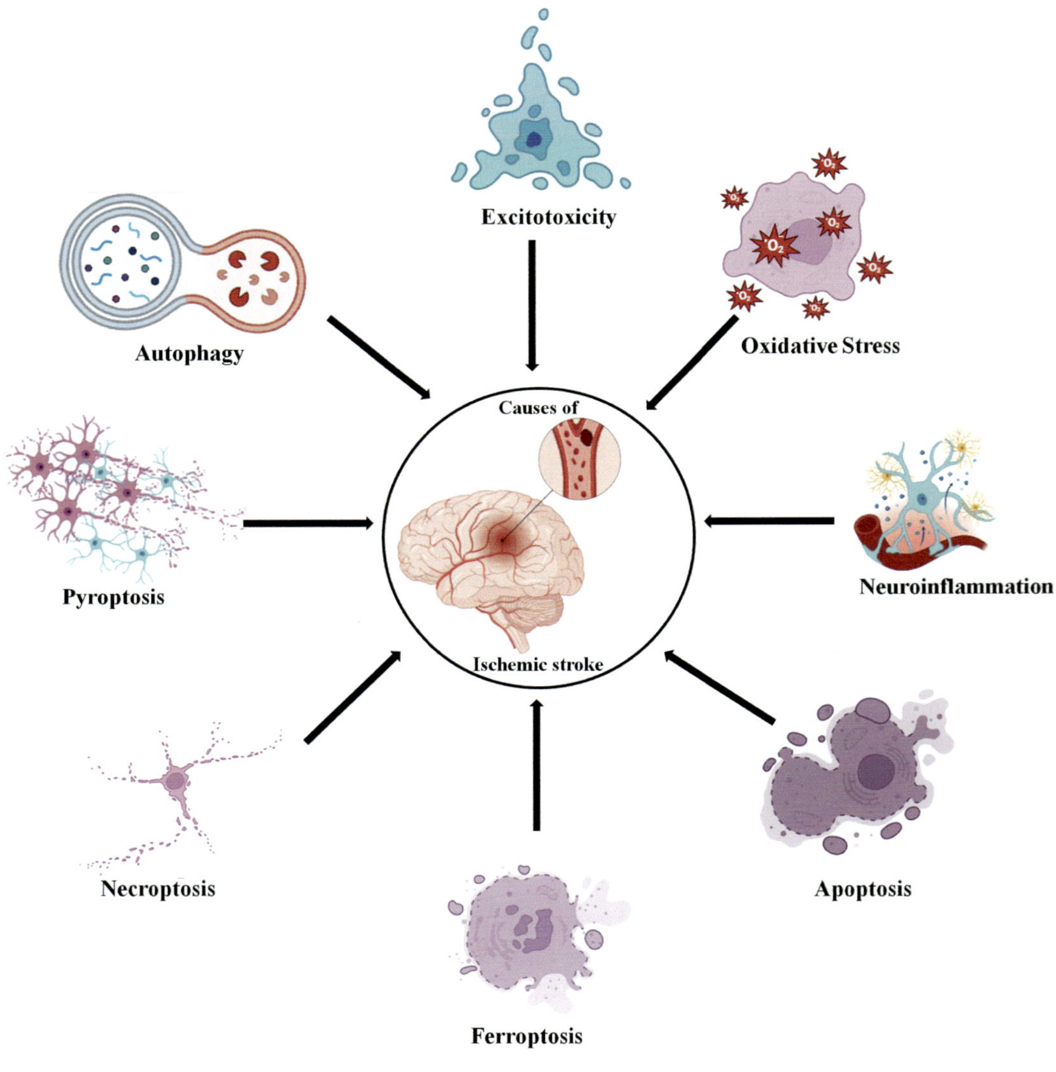

FIGURE 7.1 Pathophysiological mechanisms underlying ischemic stroke. This figure shows the main cellular and molecular reasons of ischemic stroke. Important processes include excitotoxicity, oxidative stress, and inflammation in the brain. These can lead to different types of cell death, such as apoptosis, ferroptosis, necroptosis, pyroptosis, and problems with autophagy. Together, these processes damage brain cells and affect brain function.

7.2.1 Excitotoxicity

The brain depends on efficient blood circulation to operate properly, and a steady supply of blood is essential to its survival (Fan et al., 2022). The damaged brain region receives less blood when a major cerebral artery is stopped. By interfering with adenosine triphosphate (ATP) synthesis, the reduced circulation leads to hypoxia and ischemia, which disrupt energy. This leads to disruption of ionic gradients of ion channels on the plasma and organelle membranes of neurones, including calcium ATPase, sodium/calcium exchange, and sodium/potassium ATPase (van Putten et al., 2021). This results in an overabundance of calcium entering the neurones, which in turn triggers the release of excess glutamate and a decrease in its reuptake by activating calcium ion-dependent enzymes (Rose et al., 2020; Ludhiadch et al., 2022). Excitotoxicity is the result of a sequence of events that include excessive stimulation of N-methyl-D-aspartate receptors (NMDAR) in postsynaptic neuron membranes, which generates reactive oxygen species (ROS) and oxidative stress, which disrupts mitochondrial function and causes neuronal death (Feske et al., 2021). The impairment of neural plasticity caused by excessive NMDA receptor activation also affects learning, memory, and aging, which results in the cognitive decline linked to stroke (Shen et al., 2022).

7.2.2 Oxidative Stress

A disturbance in blood flow to the brain results in oxidative stress damage and poor energy metabolism. Oxidative stress-induced damage has been linked to recanalization and reperfusion after blood flow restriction. The oxidant–antioxidant equilibrium is upset by oxidative stress, a key mechanism in ischemic stroke, especially in brain cells that are high in polyunsaturated fatty acids. Worsened oxidative damage is caused by a number of factors, including inadequate antioxidants, excessive pro-oxidants (such iron), and accelerated oxidative metabolism (Pawluk et al., 2022). The brain is severely damaged by the excess ROS and hydroxyl radicals caused by this imbalance in oxidants and antioxidants (Lee et al., 2025). Because of glucose and oxygen deprivation, cellular ROS production rises during ischemic stroke, aggravating oxidative stress and brain injury (Cheng et al., 2018). Xanthine oxidase (XO) and NADPH oxidase (NOX) are the main enzymes responsible for the generation of superoxide anion during ischemia. ROS formation results from the buildup of hypoxanthine and xanthine, which are substrates for XO, brought on by ATP depletion during ischemia (Yu et al., 2023). Following an ischemic stroke, there has been evidence of elevated XO expression in the infarcted area. Following a stroke, NOX, another important source of ROS, is increased. NOX2 has been found to be the main source of superoxide generation that is triggered by the NMDAR (Vermot et al., 2021).

7.2.3 Neuroinflammation

Another important factor in ischemic stroke is neuroinflammation, which involves several immune cells, including innate and adaptive immune cells. Following an ischemic stroke, the brain is insulted by necrosis and apoptosis, which triggers an inflammatory response regulated by the release of cytokines, chemokines, and ROS. Neuronal death results from this process, which arises in the microcirculation and involves a variety of cytotypes, including innate immune cells (microglia) and adaptive immune cells (lymphocytes) (Shaheryar et al., 2021). The scene, time, and progression of the neurological insult all influence the neuroinflammation process. In neuroinflammation during the acute phase of stroke onset, microglia have two roles. By focusing on microglia, microRNAs like miR-203 have been shown to lessen cerebral ischemia–reperfusion injury (Lian et al., 2021). Additionally, cerebral ischemia has been linked to the polarization of microglia, namely M1 polarization (Xue et al., 2021). The severe neuroinflammation that occurs during the acute phase of stroke is associated with poor outcomes, neuronal damage, and collapse of the blood–brain barrier (BBB) (Candelario-Jalil et al., 2022). The ultimate cause of ischemic stroke-induced morbidity and mortality is neuronal death, and the degree to which it may be avoided determines how well ischemic stroke can be managed.

7.2.4 Apoptosis

In order to create apoptotic bodies, apoptosis entails a sequence of internal and/or external processes that cause neurones to shrink and their cytoplasm to compress, ultimately shattering their nuclear membrane. In the intrinsic pathway, the usual glycolytic oxidative phosphorylation pathway that produces ATP is disrupted by a diminished availability of nutrients and oxygen to the cell. Because of this, the anaerobic pathway is more prevalent and the ATP generated is not enough to sustain cellular functions. Calcium ions build up in the cell as a result of this ionic imbalance (Na^+/Ca^{2+} influx and K^+ outflow), which leads to an excessive release of excitatory amino acid neurotransmitters, particularly glutamate, into the extracellular environment. Following this process, a series of cytotoxic events take place in the nucleus and cytoplasm, such as calpain activation (calpain-mediated), the production of ROS-mediated from mitochondrial metabolism, which damages the cellular membrane, and DNA breakage (DNA damage-mediated events) (Tuo et al., 2022). Meanwhile, the extrinsic pathway, which frequently happens either independently or in conjunction with the intrinsic pathway, involves the activity of the inflammatory signaling factors released by astrocytes, microglia, and oligodendrocytes as a result of cerebrovascular damage. Proinflammatory cytokines and receptors such as tumor necrosis factor (TNF)-α/β, chemokines, interleukin (IL)-1β, TNF-related apoptosis-inducing ligand receptor (TRAIL-R), and Fas ligand (FasL) are examples of these inflammatory signaling factors (Jiang et al., 2018). By activating the downstream effector caspase-3 or BID, which promotes apoptosis through the mitochondrial-dependent route, these receptors at the neuronal cell membrane initiate an apoptotic process (Xu et al., 2017).

Following an ischemic stroke, cell death can also happen through ferroptosis, phagoptosis, parthanatos, pyroptosis, and necroptosis, in addition to apoptosis. Developing focused therapeutic strategies requires an understanding of the complex interactions between these many cell death processes in the setting of ischemic stroke. Integrating knowledge from these processes could lead to better ways to reduce neuronal damage and enhance outcomes for people who have had ischemic stroke.

7.2.4.1 Ferroptosis

Ferroptosis is a newly identified type of cell death that has been linked to the etiology of ischemic stroke. It is distinguished by the buildup of iron-dependent ROS and lipid peroxides, which results in oxidative damage and cell death (Wu et al., 2018). Ferroptosis has been demonstrated to play a role in tissue damage and neuronal death in ischemic stroke (Yan et al., 2021). For instance, following a stroke, the ischemic area experiences a reduction in the amount of soluble tau protein, which facilitates iron transfer, resulting in iron buildup and neuronal death (Chen et al., 2019). In animal models of stroke, it has been demonstrated that utilizing certain inhibitors to prevent ferroptosis protects against neuronal injury (Jin et al., 2021). Ferroptosis is largely caused by the control of lipid peroxidation and iron metabolism. Lipid peroxidation and ferroptotic cell death can be induced by excessive iron buildup and compromised antioxidant defence systems, such as decreased glutathione peroxidase 4 (GPx4) activity. Thus, focusing on the mechanisms involved in lipid peroxidation and iron metabolism may be viable treatment approaches for ischemic stroke (Ursini et al., 2020).

7.2.4.2 Necroptosis

Necroptosis is a controlled type of necrosis that happens in reaction to a number of factors, including as inflammation and ischemia. The phosphorylation and activation of mixed lineage kinase domain-like protein (MLKL) occurs as a result of the activation of receptor-interacting protein kinase 1 (RIPK1) and RIPK3. After that, MLKLs go to the plasma membrane, where they compromise membrane integrity and cause cell death (Yu et al., 2021). It has been demonstrated that necrotic death plays a role in neuronal death in ischemic stroke (Hribljan et al., 2019; Wang et al., 2024).

7.2.4.3 Pyroptosis

Neuronal death during ischemic stroke is linked to this type of planned cell death. When inflammasomes develop in response to cerebral ischemia, caspase-1 is activated, which in turn causes pyroptosis (Ye et al., 2022). The sensor proteins NLRP1, NLRP3, and NLRP4 are part of multiprotein complexes called inflammasomes, which are involved in the processing of pro-inflammatory cytokines (Moltrasio et al., 2022). When inflammasomes are activated during cerebral ischemia, caspase-1 is triggered. This caspase-1 cleaves pro-IL-1β and IL-17 to produce IL-1β and IL-17, respectively. These are important inflammatory cytokines that cause neuronal death during pyroptosis, along with other pro-inflammatory factors (Piancone et al., 2021).

7.2.4.4 Parthanatos

One potential cause of neuronal death in ischemic stroke could be this type of controlled cell death (Koehler et al., 2021). It depends on the enzyme poly ADP-ribose polymerase 1 (PARP1) and is triggered by chromatinolysis and DNA damage brought on by oxidative stress. Parthanatos, in contrast to apoptosis, is characterized by rupture of the plasma membrane and does not produce apoptotic bodies or little DNA fragments (Prokhorova et al., 2020). A nuclear, chromatin-associated protein called PARP1 is essential for parthanatos because it uses nicotinamide adenine dinucleotide (NAD^+) and ATP in the poly (ADP-ribosyl)ation process to identify and repair DNA breaks (Kumar et al., 2022).

7.2.4.5 Phagoptosis

Microglia recognize, engulf, and degrade neurones as part of the phagoptosis pathway in ischemic stroke. Microglia are able to identify and consume neurones that release "eat me" signals because they are always watching the surface of neurones. Phosphatidylserine (PS) on the cell surface is one of the common "eat me" cues. Oxidative stress, elevated calcium, or ATP depletion can all cause PS expression (Merelli et al., 2021). Opsonins, such as growth arrest-specific factor 6 and Mer receptor tyrosine kinase (MERTK) receptors or milk fat globule EGF-like factor 8 (MFG-E8) and vitronectin receptors, recognize neurones that express PS (Fricker et al., 2012; Neher et al., 2013).

7.2.5 Autophagy

An important factor in ischemic stroke is autophagy, a cellular mechanism that breaks down and recycles damaged or superfluous cellular components by packing them into autophagosomes and assembling them in lysosomes (Kaur et al., 2015). In IS, oxygen and food deprivation set off a series of cellular processes, including autophagy (Adhami et al., 2006). The major inhibitor of autophagy, the mammalian target of rapamycin complex 1 (mTORC1), is not activated during cerebral ischemia due to the low supply of insulin and amino acids (Hwang et al., 2017). This stimulates AMP-activated protein kinase (AMPK), which in turn promotes autophagy, in conjunction with the elevated ratio of adenosine monophosphate (AMP) to ATP (Herzig et al., 2018). Autophagy is also triggered in response to cerebral ischemia by endoplasmic reticulum (ER) stress, mitochondrial malfunction, and the buildup of ROS (Feng et al., 2017; Yin et al., 2017). Autophagy is also triggered in ischemic stroke by the hypoxia-inducible factor 1α (HIF-1α) signaling pathway. Hypoxia-induced activation of HIF-1α can trigger the transcription of autophagy-related genes. By promoting the transcription of genes that improve oxygen supply to hypoxic regions, it reacts to systemic oxygen levels (Kim et al., 2009; Jiang et al., 2020). In ischemic stroke, autophagy has been demonstrated to have both positive and negative consequences.

7.3 CURRENT THERAPIES FOR ISCHEMIC STROKE AND THEIR TARGETS

Current therapeutic interventions for ischemic stroke are primarily directed toward the restoration of cerebral perfusion, prevention of secondary thrombotic events, and facilitation of neuroprotection and tissue repair. These strategies target multiple phases of the ischemic cascade, including thrombolysis, inhibition of platelet activation, attenuation of neuroinflammation, and promotion of neuroregenerative mechanisms. Conventional pharmacological treatments—such as thrombolytic agents, antithrombotics, and antiplatelet drugs—constitute the cornerstone of acute management. In parallel, advanced therapeutic approaches including stem cell-based interventions, nanotechnology-enabled drug delivery systems, and gene modulation techniques are being explored to enhance long-term recovery and functional neurorestoration. This section provides a comprehensive overview of the principal therapeutic modalities for ischemic stroke and their respective molecular and cellular targets (Figure 7.2).

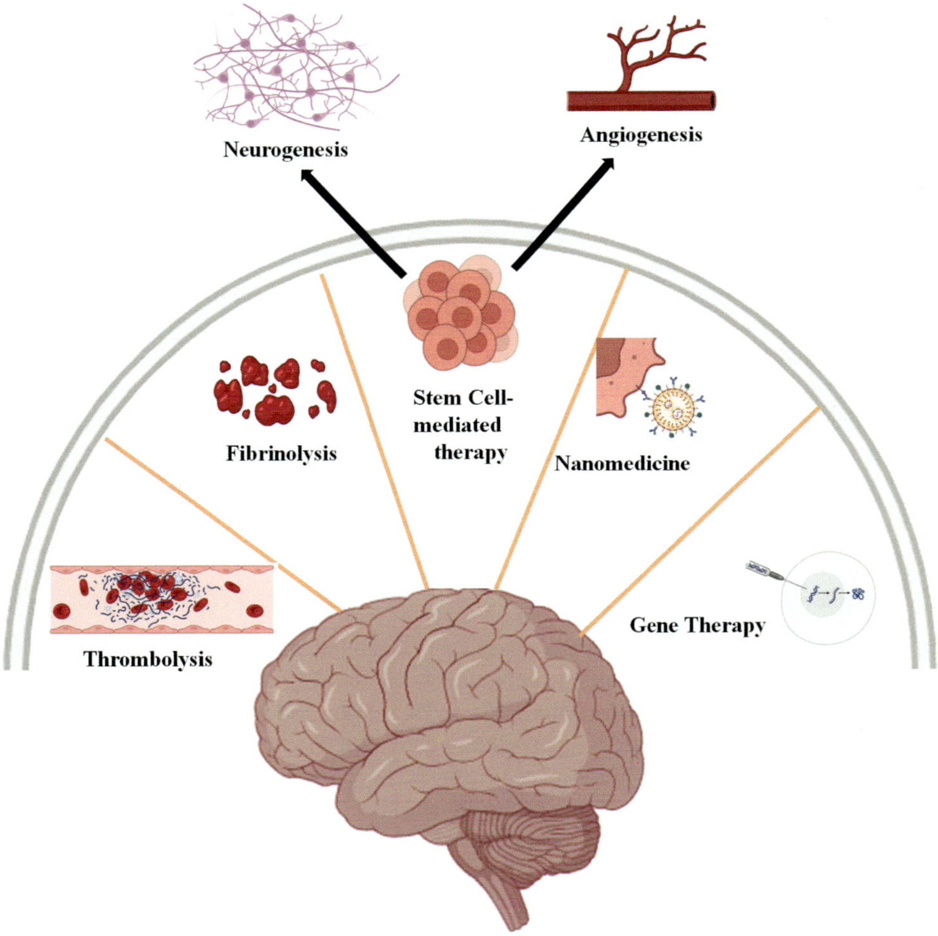

FIGURE 7.2 Central Therapeutic strategies for ischemic stroke. This figure shows the key treatments used to manage ischemic stroke. These include thrombolytics and fibrinolytic drugs that help reopen blocked blood vessels. Newer treatments, like stem cell therapy, support the growth of new brain cells and blood vessels. Other advanced approaches, such as nanomedicine and Gene therapy, aim to deliver protective medicines directly to the brain and help repair the damage caused by stroke.

7.3.1 Thrombolytic Agents and Thrombolytics

Intravenous thrombolysis is the main treatment for acute ischemic stroke, and endovascular thrombolysis is occasionally used to improve channel recanalization (Tsivgoulis et al., 2023). Initially created to treat coronary thrombolysis, the intravenous thrombolytic (IVT) treatment paradigm was later discovered to be beneficial for stroke patients. Numerous factors, such as the age of the clot, the thrombolytic agent's specificity for fibrin, and the existence and half-life of neutralizOing antibodies, affect how effective thrombolytic medications are (Barreto et al., 2011). Alteplase, a second-generation thrombolytic, is used to dissolve blood clots using intravenous thrombolysis. Prourokinase (Song et al., 2022), tenecteplase (Menon et al., 2022), and staphylokinase (Gusev et al., 2021) are other thrombolytics that are now in clinical trials and have demonstrated similar safety and effectiveness to alteplase. These fibrinolytics are given intravenously and are a possible future substitute for alteplase, with the exception of prourokinase, an intra-arterial thrombolytic. In addition to intravenous thrombolysis, endovascular thrombolysis is sometimes used to unclog big veins (Zi et al., 2021). These thrombolytics are usually plasminogen activators that work to encourage the production of fibrinolysin. In order to break down fibrin and fibrinogen in the thrombi obstructing the damaged cerebral channel or vessels, they convert plasminogen to soluble plasmin, a proteolytic enzyme (Collen et al., 2005).

7.3.2 Antithrombotic Agents

To increase the effectiveness of alteplase and any other thrombolytics delivered, ischemic stroke patients are also given other medications known as anti-thrombotic medicines to stop new clots from forming. By inhibiting glycoprotein IIb/IIIa, these medications prevent clot formation and guarantee full vessel recanalization. Tirofiban, eptifibatide, glenzocimab, and argatroban are antithrombotic medications that are utilized as supplements to alteplase-assisted intravenous thrombolysis at various stages of clinical trials (Barreto et al., 2017; Voors-Pette et al., 2019).

7.3.3 Treatment with Antiplatelets

This treatment is used to treat acute ischemic stroke and reduce the risk of stroke. It is also essential for managing transient ischemic attacks (TIA) and non-cardioembolic IS. The most commonly prescribed medications to stroke patients in the initial days following an attack are antiplatelet medicines such as aspirin, ticagrelor, and clopidogrel (Stringberg et al., 2019). Because they suppress platelet aggregation and blood clot formation, they are essential in the treatment of ischemic stroke. These medications mainly work by disrupting the coagulation cascade and platelet activation. Arachidonic acid is transformed into prostaglandins by cyclooxygenase (COX), which contributes to platelet activation and aggregation. Aspirin irreversibly inhibits platelet COX, hence lowering the formation of thromboxane A2, a potent platelet aggregator. A thienopyridine derivative called clopidogrel prevents adenosine diphosphate (ADP)-induced platelet aggregation by permanently blocking the P2Y12 ADP receptor on the platelet surface. Similar to clopidogrel, ticagrelor has a reversible mode of action.

Combining ticagrelor, prasugrel, or clopidogrel with aspirin is known as dual antiplatelet therapy, and it has gained popularity. Numerous studies have examined the safety and effectiveness of this dual therapy. Clopidogrel and aspirin combination therapy appears to be most effective when started within 24 h after a stroke and sustained for 4–12 weeks, according to trial results (Mehta et al., 2001). According to several antiplatelet therapy studies, clopidogrel, either by itself or in combination with aspirin, is more successful than aspirin alone at preventing blood clots during acute coronary syndrome and following percutaneous coronary intervention (Peters et al., 2003). In contrast, the MATCH study found that using clopidogrel and aspirin together did not lower the risk of a subsequent stroke when compared to taking clopidogrel by itself. Rather, the combination raised the risk of potentially fatal bleeding problems. Therefore, following an initial incident or TIA, the trial did not advise using clopidogrel and aspirin combined to prevent recurrent ischemic stroke (Diener et al., 2004). These results were in line with those of the CHARISMA trial, which examined the safety and effectiveness of low-dose aspirin (75–162 mg/day) plus clopidogrel (75 mg/day) versus low-dose aspirin alone in patients with recorded vascular disease, such as MI, stroke, TIA, and symptomatic peripheral vascular disease, and a high risk of atherothrombotic cerebrovascular events (Bhatt et al., 2004).

7.3.4 Fibrinogen-Lowering Therapies

Research findings have repeatedly shown that high fibrinogen levels in stroke patients are associated with a poor prognosis and adverse clinical outcomes (Woodward et al., 2005; Chen et al., 2018). Fibrinogen-depleting medicines increase blood flow and decrease blood thickness by lowering blood plasma levels of fibrinogen. Additionally, they clear the artery of blood clots and restore blood flow to the brain's afflicted areas. While some defibrinogenation treatment randomized clinical trials found that fibrinogen-depleting medications helped stroke patients, other studies did not find that they improved clinical outcomes after stroke (Hao et al., 2012). Additionally, bleeding following defibrinogenating drug treatment was noted in a few investigations. The potential of ancrod, a defibrinogenating substance made from snake venom, to cure ischemic stroke within three hours of onset has been investigated (Sherman et al., 2000). According to the results of the European Stroke Treatment with Ancrod Trial (ESTAT), ancrod administered under controlled conditions at 70 mg/dL fibrinogen was safe, effective, and reduced the prevalence of Intracerebral Hemorrhage (ICH) compared to lower fibrinogen levels (Levy et al., 2009).

7.3.5 Emerging Paradigms in Cellular Therapies for Ischemic Stroke

Because of their profound impact on enhancing patient recovery, cellular treatments are significant therapeutic options to be taken into consideration in the therapy of ischemic stroke. Stem cells can lessen neuroinflammation, repair damaged brain tissue, and replace lost neurones. The four primary cells that have been tested in stem cell therapy for the treatment of ischemic stroke are mesenchymal stem cells (MSCs), hematopoietic stem cells (blood stem cells), neural stem cells (NSCs), and epithelial stem cells. Stem cell therapy is widely recognized for its therapeutic potential due to the intrinsic properties of self-renewal and multilineage differentiation. These capabilities enable stem cells to contribute to tissue repair and regeneration, making them a promising strategy for restoring function in damaged or diseased tissues (Salaudeen et al., 2024). Stem cell therapy provide stroke patients with efficacious, safe, and promising therapy options. The potential for tissue regeneration, maintenance, migration, and proliferation, brain circuit rewiring, and physical and behavioral rejuvenation has been evaluated in studies on embryonic stem cells (ESCs), MSCs, and induced pluripotent stem cells (iPSCs) (Borlongan et al., 1997). Multilineage differentiating stress-enduring (Muse) cells are a novel subset of MSCs that have recently been discovered in connective tissues. These cells have been studied as a stroke treatment and show a high ability for regeneration. Muse cells have been shown to engraft into injured host tissue and develop to offer functional recovery in the host after intravenous transplantation in a mouse model (Park et al., 2020). By promoting neurogenesis, restoring neuronal networks, and enhancing axonal growth and synaptogenesis, well-designed experimental and clinical studies on stem cell therapies are expected to initiate a paradigm shift in the treatment of stroke (Aizman et al., 2009). An alternative treatment to neuroprotection is neural healing. It is not time-bound because it is used to repair tissue after damage has already occurred, however it works best when given 24 h following a stroke. In an effort to promote neurogenesis and start the neural repair process, numerous animal models have been employed (Chopp et al., 2009). By integrating cells into the wound or blocking neuronal growth inhibitors with neurotrophic factors, stem cell treatment is used in neural repair to start the repair processes. To improve synaptic connection, these cells can be directed to any damaged area. NSCs therapy has shown promise in stroke patients' clinical studies.

7.3.5.1 Embryonic Stem Cells (ESCs)

ESCs are pluripotent stem cells derived from early-stage blastocysts. Due to their ability to differentiate into any cell type, including neurons and glia, they are considered a promising source for neural regeneration. Studies have shown that ESC-derived neurons can integrate with host neural circuits and aid in functional recovery. However, major challenges such as ethical concerns regarding embryo use, risk of teratoma formation, immune rejection, and difficulties in long-term preservation limit their clinical application (Liu et al., 2020).

7.3.5.2 Hematopoietic Stem Cells (HSCs)

HSCs are isolated from bone marrow, peripheral blood, or umbilical cord blood. They are known for their ability to support vascular repair and neurogenesis. Transplantation of $CD34^+$ HSCs has been shown to enhance neovascularization and reduce infarct volume in preclinical models. HSCs can be used in both autologous and allogeneic settings without major ethical issues. However, variability in cell quality and quantity, particularly when sourced from cord blood, remains a limitation (Penny et al., 2024).

7.3.5.3 Neural Stem Cells (NSCs)

NSCs are multipotent cells that can give rise to neurons, astrocytes, and oligodendrocytes. When transplanted, they contribute to tissue repair, modulate inflammation, and promote recovery of neurological function. Most exogenous NSCs are derived from embryonic or fetal tissue, raising ethical concerns. Additionally, their survival rate post-transplantation is often low, which can hinder therapeutic success. Despite these issues, NSC transplantation has shown significant benefits in experimental stroke models with minimal risk of tumorigenesis (Kaminska et al., 2022).

7.3.5.4 Mesenchymal Stem Cells (MSCs)

MSCs can be obtained from various tissues including bone marrow, adipose tissue, umbilical cord blood, and peripheral blood. They are widely studied due to their immunomodulatory properties and ability to secrete growth factors and cytokines that support neuroprotection, angiogenesis, and neurogenesis. MSCs are generally safe, with low tumorigenic potential and minimal ethical concerns, particularly in autologous applications. However, their expansion *in vitro* to clinically relevant numbers can take time, limiting their immediate use in acute stroke settings (Krawczenko and Klimczak, 2022).

7.3.5.5 Others (Induced Pluripotent Stem Cells and NT2N Cells)

Induced pluripotent stem cells (iPSCs) are generated by reprogramming adult somatic cells into a pluripotent state. iPSCs can differentiate into various neural cell types and improve neurological outcomes in animal stroke models. They bypass ethical issues and immune rejection when autologous, but concerns remain regarding reprogramming efficiency and tumorigenicity. NT2N cells, derived from human teratocarcinoma cells, also show potential in ischemic stroke therapy due to their ability to differentiate into functional neurons. However, their clinical use is still under evaluation (Al Abbar et al., 2020).

7.3.6 Stem Cell-Mediated Neuroprotection in Ischemic Stroke

Numerous studies have been conducted on the neuroprotective effects of stem cells in ischemic stroke (Xu et al., 2017; Liu et al., 2021; Sun et al., 2024). The mechanisms underlying these effects include antiapoptosis, anti-inflammation, angiogenesis and neurogenesis promotion, the formation of new neural cells and neuronal circuitry, antioxidation, and BBB protection. These molecular pathways will be specifically examined in the following.

7.3.6.1 New Neuronal Circuitry Formation and Damage Tissue Replacement

Replacing damaged tissues with newly developed neural cells derived from stem cells is the most direct method of regaining neurological functioning. *In vivo*, brain-derived neurotrophic factor (BDNF) and epithelial growth factor could cause the MSCs to develop into neural cells. Furthermore, it was shown by Ishibashi and his associates that totipotent stem cells may migrate to ischemic regions, develop into adult neurones, and create new neural circuits, all of which could enhance neurological capabilities (Ishibashi et al., 2004). Some researchers questioned the effectiveness of this impact, though. They believed that the stem cells that underwent differentiation into new neural cells were primarily found *in vivo* in particular cultures. Since just a limited number of cells were monitored around the ischemic areas, the brain cells that were transplanted may have neuroprotective benefits through different methods (Chen et al., 2002; Raedt et al., 2007). Thus, additional research on the functions of stem cells in repairing damaged tissues and developing new neural circuits ought to be initiated.

7.3.6.2 Angiogenesis Promotion

When treating ischemic stroke, attention should be paid to the supporting tissues, such as blood vessels, in addition to the regeneration of brain cells. Following a stroke, angiogenesis was increased. Vascular volume increased from 3% before the stroke to 6% 90 days after the stroke, according to Zhang and his colleagues' study of the structural alterations following a stroke (Zhang et al., 2014). Following the transplantation of stem cells around infarcted areas, angiogenesis has also been noted. Increases in vascular endothelial growth factor (VEGF) or other endogenous factors, such as fibroblast growth factor (FGF) and BDNF, were the primary mechanisms of action. Furthermore, via controlling the production of Notch 1, angiopoietin-1, angiopoietin-2, and other proteins, stem cells may also play a part in angiogenesis (Hegde et al., 2024).

7.3.6.3 Neurogenesis Promotion

Following an ischemic attack, endogenous neurogenesis—the process of self-healing—increases. Patients with cerebral ischemia require neurogenesis in order to recover neurologically (Arvidsson et al, 2002). By expressing the BDNF, Jeong and his colleagues demonstrated that MSCs not only decreased the number of cells that underwent apoptosis but also helped to increase endogenous neurogenesis (Jeong et al., 2014). Furthermore, Zhao et al. discovered that electroacupuncture (EA) therapy might stimulate endogenous neurogenesis and NSCs in rats' dentate gyrus (Zhao et al., 2015). Additionally, by altering cerebral plasticity through the neurotrophic impact and creating new connections with the host brain, MSCs have been shown to improve neurological deficiency in rats (Jiang et al., 2023).

7.3.6.4 Anti-Inflammatory Effects of Stem Cells

An intricate immunological reaction to an injury is inflammation. In a healthy state, inflammation may aid in scavenging dead tissues or cells and starting the healing process (Turrin et al., 2006). Excessive immune response activation, however, can be damaging to organisms and result in injury (Brown et al., 2010). By increasing the expression of anti-inflammatory cytokines and decreasing that of proinflammatory cytokines, stem cells have demonstrated their dual functions in controlling inflammation. In the early stages of cerebral ischemia assault, stem cells have been shown in numerous studies to decrease the expression of TNF-α, IL-1β, and IL-6. Additionally, some researchers proposed that stem cell infusion should occur early because the inflammatory responses would eventually subside. Furthermore, Zhu et al. showed that human umbilical cord blood mesenchymal stem cells (hUCB-MSCs) may simultaneously downregulate TNF-α, IL-1β, and IL-6 and upregulate IL-10 in peri-ischemic brain regions (Zhu et al., 2014). Furthermore, allotransplantation may inhibit the receptors' immunological response. By inhibiting T cells, Vendrame and his associates demonstrated that hUCBCs reduced inflammation in the brain following a stroke (Vendrame et al., 2005).

7.3.6.5 Antiapoptotic Effects of Stem Cells

One form of cell death that is distinguished by energy dependence and programmed cell death is apoptosis (Elmore et al., 2007). Kerr et al. were the first to describe "apoptosis" (Kerr et al., 1972). Normal physiological metabolism, growth, and development depend on apoptosis. It also maintains hemostasis by scavenging aging or damaged cells, shapes organs, and regulates the immune system by eliminating excess and faulty cells (Rathmell et al., 2002). Uncontrolled apoptosis, on the other hand, can lead to a number of pathogenic processes in a variety of disorders, including stroke, Alzheimer's disease, and cancer (Reed et al., 2002). Numerous experimental models have confirmed the impact of stem cell antiapoptosis on ischemic stroke. Modifying the expression of certain trophins, such as FGF, BDNF, VEGF, glial cell line-derived factor (GDNF), and nerve growth factor (NGF), may mediate this impact. By upregulating the expression of Bcl-2, Zhang and his colleagues demonstrated that NSCs transplantation could considerably lower the amount of apoptotic cells in the penumbra at 7 days (Zhang et al., 2009). Additionally, Zhu and his associates confirmed that hUCB-MSCs has an antiapoptotic impact on ischemic stroke in the vicinity of the ischemic zone (Zhu et al., 2014). In animal models, the neurological recovery was linked to the antiapoptotic impact (Lladó et al., 2004; Kurozumi et al., 2005).

7.3.7 Nanomedicine and Targeted Drug Delivery in Stroke Therapy

Nanomedicine is an emerging field that utilizes nanotechnology-based tools and platforms for diagnosis, monitoring, and treatment. In ischemic stroke, nanomedicine offers the unique potential to improve drug delivery across the BBB, which is typically compromised after ischemia, yet still poses a challenge for many systemically delivered therapeutics (Wang et al., 2024). Various nanocarriers such as liposomes, polymeric nanoparticles, dendrimers, micelles, gold nanoparticles, and solid lipid nanoparticles have been engineered to deliver drugs precisely to the ischemic area. These nanocarriers can encapsulate neuroprotective agents, antioxidants, anti-inflammatory drugs, and growth factors, shielding them from premature degradation and improving their bioavailability at the target site (Correa-Paz et al., 2021). One of the key advantages of nanocarrier systems in stroke is targeted delivery. Functionalization of nanoparticles with ligands or antibodies that recognize ischemia-induced molecular markers (e.g., ICAM-1, VCAM-1, or integrins) allows for selective accumulation in the damaged brain tissue (Patra et al., 2018). Additionally, many nanocarriers can be designed to release their payload in response to specific stimuli such as pH changes, oxidative stress, or enzymatic activity, which are characteristic of ischemic brain environments (Zhang et al., 2024). For instance, delivery of tissue plasminogen activator (tPA) using nanoparticles has been explored to enhance thrombolytic efficacy while reducing hemorrhagic risk. Similarly, nanoformulations of curcumin, resveratrol, and quercetin have demonstrated antioxidant and anti-inflammatory effects in preclinical models of stroke. Moreover, exosomes, naturally derived nanovesicles, are gaining attention due to their ability to cross the BBB and deliver functional RNAs or proteins, making them a promising tool for regenerative therapy (Nguyen et al., 2021). Despite these advancements, challenges remain in translating nanomedicine from bench to bedside. These include the need for standardized production protocols, safety and toxicity assessments, and understanding long-term effects on the brain and body (Fakhri et al., 2022).

7.3.8 Gene Therapy and Molecular Modulators for the Treatment of Ischemic Stroke

Gene therapy involves the introduction, suppression, or modification of specific genes to alter cellular function and enhance therapeutic outcomes. In the context of ischemic stroke, gene therapy holds potential for neuroprotection, angiogenesis, anti-inflammatory action, and functional recovery. Viral vectors such as adenovirus, adeno-associated virus (AAV), and lentivirus are commonly used to deliver therapeutic genes into the brain. These vectors have been employed to overexpress genes encoding BDNF, VEGF, GDNF, and insulin-like growth factor-1 (IGF-1)—all of which are known to support neuronal survival, neurogenesis, and synaptic plasticity (Wang et al., 2023). One promising direction is the upregulation of antioxidant genes (such as SOD2 or catalase) to counteract oxidative stress-induced damage in ischemic regions. Another strategy involves delivering anti-apoptotic genes (e.g., Bcl-2) or suppressing pro-apoptotic factors (e.g., caspase-3) to limit cell death (Briones-Valdivieso et al., 2024). Modulation of inflammatory pathways through genetic regulation of NLRP3, NF-κB, or TLR4 has also shown efficacy in reducing neuroinflammation and preserving tissue integrity. In addition to gene delivery, molecular modulators—including small molecules, peptides, and biologics—are being developed to selectively regulate key signaling pathways disrupted by ischemic stroke. For example: NLRP3 inflammasome inhibitors like MCC950 reduce microglial activation and pyroptosis (Coll et al., 2022). Histone deacetylase (HDAC) inhibitors promote neuroplasticity and repair. Modulators of autophagy such as rapamycin or metformin can enhance cellular clearance and mitochondrial quality control (Zhang et al., 2024). Mitochondrial-targeted antioxidants (e.g., MitoQ) and NOX inhibitors (e.g., apocynin) aim to reduce ROS-induced injury (Apostolova et al., 2015). Gene silencing technologies, such as siRNA, antisense oligonucleotides (ASOs), and CRISPR/Cas9-based editing, offer precise ways to downregulate harmful genes post-stroke. For instance, silencing NOX2, HMGB1, or TNF-α has demonstrated reduced infarct size and improved behavioral outcomes in animal models (Yingze et al., 2022; Zhu et al., 2022; Gao et al., 2023). Although gene and molecular therapies have shown great promise in preclinical settings, their clinical application is still in early stages. Key hurdles include delivery across the BBB, vector immunogenicity, long-term gene expression control, and safety concerns such as insertional mutagenesis (Blevins et al., 2022).

7.4 CONCLUSION

Ischemic stroke represents a complex neurovascular disorder characterized by a cascade of pathological events that contribute to neuronal injury and functional impairment. Understanding the intricate mechanisms involved—from excitotoxicity and oxidative stress to neuroinflammation and multiple forms of programmed cell death—offers crucial insights for developing targeted interventions. While current treatments such as thrombolytics and antiplatelets are time-sensitive and limited by efficacy and safety concerns, newer strategies including stem cell therapy, nanomedicine-based drug delivery, and gene modulation hold considerable promise for enhancing neuroprotection and tissue repair. Stem cells, particularly MSCs and NSCs, have demonstrated encouraging neuroregenerative and anti-inflammatory effects in preclinical and early clinical studies. Likewise, nanocarriers and gene-editing technologies are being optimized to overcome the BBB and deliver therapeutic agents with precision. Continued research integrating these novel approaches with conventional care paradigms may pave the way for more effective and personalized stroke therapies, ultimately improving prognosis and quality of life for stroke survivors. However, translating these advancements from bench to bedside requires addressing critical challenges, including long-term safety, delivery efficiency, immunogenicity, and cost-effectiveness. Collaborative efforts between neuroscientists, clinicians, and bioengineers will be essential to refine these technologies and develop standardized clinical protocols. In parallel, the identification of reliable biomarkers for early diagnosis and patient stratification could further enhance the therapeutic impact of targeted interventions. As our understanding of stroke pathophysiology deepens and innovative therapeutic platforms evolve, the future of ischemic stroke management is poised to shift from damage control to true neural repair and functional restoration.

ABBREVIATIONS

AAV	Adeno-Associated Virus
AMP	Adenosine Monophosphate
AMPK	AMP-Activated Protein Kinase
ASOs	Antisense Oligonucleotides
ATP	Adenosine Triphosphate
BBB	Blood–Brain Barrier
BDNF	Brain-Derived Neurotrophic Factor
COX	Cyclooxygenase
DALYs	Disability-Adjusted Life Years
ER	Endoplasmic Reticulum
ESCs	Embryonic Stem Cells
ESTAT	European Stroke Treatment with Ancrod Trial
FasL	Fas Ligand
FGF	Fibroblast Growth Factor
GDNF	Glial Cell Line-Derived Factor
GPx4	Glutathione Peroxidase 4
HDAC	Histone Deacetylase
HIF-1α	Hypoxia-Inducible Factor 1α
HSCs	Hematopoietic Stem Cells
hUCB-MSCs	Human Umbilical Cord-Blood Mesenchymal Stem Cells
IL	Interleukin
iPSCs	Induced Pluripotent Stem Cells
IVT	Intravenous Thrombolytic
MERTK	Mer Receptor Tyrosine Kinase
MFG-E8	Milk Fat Globule EGF-Like Factor 8
MLKL	Mixed Lineage Kinase Domain-Like Protein
MSCs	Mesenchymal Stem Cells
mTORC1	Mammalian Target of Rapamycin Complex 1
NAD	Nicotinamide Adenine Dinucleotide
NGF	Nerve Growth Factor
NMDAR	N-methyl-D-Aspartate Receptors
NOX	NADPH Oxidase
NSCs	Neural Stem Cells
PARP1	Poly ADP-Ribose Polymerase 1
PS	Phosphatidylserine
RIPK1	Receptor-Interacting Protein Kinase 1
ROS	Reactive Oxygen Species
TIA	Transient Ischemic Attacks
TNF	Tumor Necrosis Factor
TOAST	Trial of Org 10172 in Acute Stroke Treatment
tPA	Tissue Plasminogen Activator
TRAIL-R	TNF-Related Apoptosis-Inducing Ligand Receptor
VEGF	Vascular Endothelial Growth Factor
WHO	World Health Organization
XO	Xanthine Oxidase

REFERENCES

Adhami F, Liao G, Morozov YM, et al. Cerebral Ischemia-Hypoxia Induces Intravascular Coagulation and Autophagy. *Am J Pathol*. 2006;169(2):566–583. doi:10.2353/ajpath.2006.051066.

Aizman I, Tate CC, McGrogan M, Case CC. Extracellular Matrix Produced by Bone Marrow Stromal Cells and by Their Derivative, SB623 Cells, Supports Neural Cell Growth. *J Neurosci Res*. 2009;87(14):3198–3206. doi:10.1002/jnr.22146.

Al Abbar A, Ngai SC, Nograles N, Alhaji SY, Abdullah S. Induced Pluripotent Stem Cells: Reprogramming Platforms and Applications in Cell Replacement Therapy. *Biores Open Access*. 2020;9(1):121–136. doi:10.1089/biores.2019.0046.

Apostolova N, Victor VM. Molecular Strategies for Targeting Antioxidants to Mitochondria: Therapeutic Implications. *Antioxid Redox Signal*. 2015;22(8):686–729. doi:10.1089/ars.2014.5952.

Arvidsson A, Collin T, Kirik D, Kokaia Z, Lindvall O. Neuronal Replacement from Endogenous Precursors in the Adult Brain after Stroke. *Nat Med*. 2002;8(9):963–970. doi:10.1038/nm747.

Barreto AD. Intravenous Thrombolytics for Ischemic Stroke. *Neurotherapeutics*. 2011;8(3):388–399. doi:10.1007/s13311-011-0049-x.

Barreto AD, Ford GA, Shen L, et al. Randomized, Multicenter Trial of ARTSS-2 (Argatroban with Recombinant Tissue Plasminogen Activator for Acute Stroke). *Stroke*. 2017;48(6):1608–1616. doi:10.1161/STROKEAHA.117.016720.

Barthels D, Das H. Current Advances in Ischemic Stroke Research and Therapies. *Biochim Biophys Acta Mol Basis Dis*. 2020;1866(4):165260. doi:10.1016/j.bbadis.2018.09.012.

Bhatt DL, Topol EJ. Clopidogrel for High Atherothrombotic Risk and Ischemic Stabilization, Management, and Avoidance Executive Committee. Clopidogrel Added to Aspirin Versus Aspirin Alone in Secondary Prevention and High-risk Primary Prevention: Rationale and Design of the Clopidogrel for High Atherothrombotic Risk and Ischemic Stabilization, Management, and Avoidance (CHARISMA) Trial. *Am Heart J*. 2004;148(2):263–268. doi:10.1016/j.ahj.2004.03.028.

Blevins HM, Xu Y, Biby S, Zhang S. The NLRP3 Inflammasome Pathway: A Review of Mechanisms and Inhibitors for the Treatment of Inflammatory Diseases. *Front Aging Neurosci*. 2022;14:879021. doi:10.3389/fnagi.2022.879021.

Borlongan CV, Koutouzis TK, Jorden JR, et al. Neural Transplantation as an Experimental Treatment Modality for Cerebral Ischemia. *Neurosci Biobehav Rev*. 1997;21(1):79–90. doi:10.1016/0149-7634(95)00063-1.

Briones-Valdivieso C, Briones F, Orellana-Urzúa S, Chichiarelli S, Saso L, Rodrigo R. Novel Multi-Antioxidant Approach for Ischemic Stroke Therapy Targeting the Role of Oxidative Stress. *Biomedicines*. 2024;12(3):501. doi:10.3390/biomedicines12030501.

Brown GC, Neher JJ. Inflammatory Neurodegeneration and Mechanisms of Microglial Killing of Neurons. *Mol Neurobiol*. 2010;41(2–3):242–247. doi:10.1007/s12035-010-8105-9.

Candelario-Jalil E, Dijkhuizen RM, Magnus T. Neuroinflammation, Stroke, Blood-Brain Barrier Dysfunction, and Imaging Modalities. *Stroke*. 2022;53(5):1473–1486. doi:10.1161/STROKEAHA.122.036946.

Chen J, Sun D, Liu M, Zhang S, Ren C. Defibrinogen Therapy for Acute Ischemic Stroke: 1332 Consecutive Cases. *Sci Rep*. 2018;8(1):9489. doi:10.1038/s41598-018-27856-6.

Chen X, Jiang H. Tau as a Potential Therapeutic Target for Ischemic Stroke. *Aging (Albany NY)*. 2019;11(24):12827–12843. doi:10.18632/aging.102547.

Chen X, Li Y, Wang L, et al. Ischemic Rat Brain Extracts Induce Human Marrow Stromal Cell Growth Factor Production. *Neuropathology*. 2002;22(4):275–279. doi:10.1046/j.1440-1789.2002.00450.x.

Cheng Z, Wang L, Qu M, li, et al. Mesenchymal Stem Cells Attenuate Blood-Brain Barrier Leakage after Cerebral Ischemia in Mice. *J Neuroinflammation*. 2018;15(1):135. doi:10.1186/s12974-018-1153-1.

Chopp M, Li Y, Zhang ZG. Mechanisms Underlying Improved Recovery of Neurological Function after Stroke in the Rodent after Treatment with Neurorestorative Cell-Based Therapies. *Stroke*. 2009;40(3 Suppl):S143–S145. doi:10.1161/STROKEAHA.108.533141.

Coll RC, Schroder K, Pelegrín P. NLRP3 and Pyroptosis Blockers for Treating Inflammatory Diseases. *Trends Pharmacol Sci*. 2022;43(8):653–668. doi:10.1016/j.tips.2022.04.003.

Collen D, Lijnen HR. Thrombolytic Agents. *Thromb Haemost*. 2005;93(4):627–630. doi:10.1160/TH04-11-0724.

Correa-Paz C, da Silva-Candal A, Polo E, et al. New Approaches in Nanomedicine for Ischemic Stroke. *Pharmaceutics*. 2021;13(5):757. doi:10.3390/pharmaceutics13050757.

Diener HC, Bogousslavsky J, Brass LM, et al. Aspirin and Clopidogrel Compared with Clopidogrel Alone after Recent Ischaemic Stroke or Transient Ischaemic Attack in High-risk Patients (MATCH): Randomised, Double-blind, Placebo-controlled Trial. *Lancet*. 2004;364(9431):331–337. doi:10.1016/S0140-6736(04)16721-4.

Elmore S. Apoptosis: A Review of Programmed Cell Death. *Toxicol Pathol*. 2007;35(4):495–516. doi:10.1080/01926230701320337.

Fakhri S, Abdian S, Zarneshan SN, Moradi SZ, Farzaei MH, Abdollahi M. Nanoparticles in Combating Neuronal Dysregulated Signaling Pathways: Recent Approaches to the Nanoformulations of Phytochemicals and Synthetic Drugs against Neurodegenerative Diseases. *Int J Nanomedicine*. 2022;17:299–331. doi:10.2147/IJN.S347187.

Fan JL, Brassard P, Rickards CA, et al. Integrative Cerebral Blood Flow Regulation in Ischemic Stroke. *J Cereb Blood Flow Metab*. 2022;42(3):387–403. doi:10.1177/0271678X211032029.

Feigin VL, Brainin M, Norrving B, et al. World Stroke Organization: Global Stroke Fact Sheet 2025. *Int J Stroke*. 2025;20(2):132–144. doi:10.1177/17474930241308142.

Feng D, Wang B, Wang L, et al. Pre-ischemia Melatonin Treatment Alleviated Acute Neuronal Injury after Ischemic Stroke by Inhibiting Endoplasmic Reticulum Stress-Dependent Autophagy via PERK and IRE1 Signalings. *J Pineal Res*. 2017;62(3):10.1111/jpi.12395. doi:10.1111/jpi.12395.

Feske SK. Ischemic Stroke. *Am J Med*. 2021;134(12):1457–1464. doi:10.1016/j.amjmed.2021.07.027.

Fricker M, Neher JJ, Zhao JW, Théry C, Tolkovsky AM, Brown GC. MFG-E8 Mediates Primary Phagocytosis of Viable Neurons during Neuroinflammation. *J Neurosci*. 2012;32(8):2657–2666. doi:10.1523/JNEUROSCI.4837-11.2012.

Gao B, Wang S, Li J, et al. HMGB1, Angel or Devil, in Ischemic Stroke. *Brain Behav*. 2023;13(5):e2987. doi:10.1002/brb3.2987.

Gusev EI, Martynov MY, Nikonov AA, et al. Non-Immunogenic Recombinant Staphylokinase versus Alteplase for Patients with Acute Ischaemic Stroke 4·5 h after Symptom Onset in Russia (FRIDA): A Randomised, Open Label, Multicentre, Parallel-Group, Non-Inferiority Trial. *Lancet Neurol*. 2021;20(9):721–728. doi:10.1016/S1474-4422(21)00210-6.

Hao Z, Liu M, Counsell C, Wardlaw JM, Lin S, Zhao X. Fibrinogen Depleting Agents for Acute Ischaemic Stroke. *Cochrane Database Syst Rev*. 2012;2012(3):CD000091. doi:10.1002/14651858.CD000091.pub2.

Hegde M, Singh AK, Kannan S, Kolkundkar U, Seetharam RN. Therapeutic Applications of Engineered Mesenchymal Stromal Cells for Enhanced Angiogenesis in Cardiac and Cerebral Ischemia. *Stem Cell Rev Rep*. 2024;20(8):2138–2154. doi:10.1007/s12015-024-10787-3.

Herzig S, Shaw RJ. AMPK: Guardian of Metabolism and Mitochondrial Homeostasis. *Nat Rev Mol Cell Biol*. 2018;19(2):121–135. doi:10.1038/nrm.2017.95.

Hribljan V, Lisjak D, Petrović DJ, Mitrečić D. Necroptosis Is One of the Modalities of Cell Death Accompanying Ischemic Brain Stroke: From Pathogenesis to Therapeutic Possibilities. *Croat Med J*. 2019;60(2): 121–126. doi:10.3325/cmj.2019.60.121.

Hwang JY, Gertner M, Pontarelli F, et al. Global Ischemia Induces Lysosomal-Mediated Degradation of mTOR and Activation of Autophagy in Hippocampal Neurons Destined to Die. *Cell Death Differ*. 2017;24(2):317–329. doi:10.1038/cdd.2016.140.

Ishibashi S, Sakaguchi M, Kuroiwa T, et al. Human Neural Stem/Progenitor Cells, Expanded in Long-Term Neurosphere Culture, Promote Functional Recovery after Focal Ischemia in Mongolian Gerbils. *J Neurosci Res*. 2004;78(2):215–223. doi:10.1002/jnr.20246.

Jeong CH, Kim SM, Lim JY, Ryu CH, Jun JA, Jeun SS. Mesenchymal Stem Cells Expressing Brain-Derived Neurotrophic Factor Enhance Endogenous Neurogenesis in an Ischemic Stroke Model. *Biomed Res Int*. 2014;2014:129145. doi:10.1155/2014/129145.

Jiang Q, Geng X, Warren J, et al. Hypoxia Inducible Factor-1α (HIF-1α) Mediates NLRP3 Inflammasome-Dependent-Pyroptotic and Apoptotic Cell Death Following Ischemic Stroke. *Neuroscience*. 2020;448:126–139. doi:10.1016/j.neuroscience.2020.09.036.

Jiang X, Andjelkovic AV, Zhu L, et al. Blood-Brain Barrier Dysfunction and Recovery after Ischemic Stroke. *Prog Neurobiol*. 2018;163-164:144–171. doi:10.1016/j.pneurobio.2017.10.001.

Jiang X, Zhou L, Sun Z, et al. MSCs Overexpressing GDNF Restores Brain Structure and Neurological Function in Rats with Intracerebral Hemorrhage. *Mol Biomed*. 2023;4(1):43. doi:10.1186/s43556-023-00159-7.

Jin Y, Zhuang Y, Liu M, Che J, Dong X. Inhibiting Ferroptosis: A Novel Approach for Stroke Therapeutics. *Drug Discov Today*. 2021;26(4):916–930. doi:10.1016/j.drudis.2020.12.020.

Kaminska A, Radoszkiewicz K, Rybkowska P, Wedzinska A, Sarnowska A. Interaction of Neural Stem Cells (NSCs) and Mesenchymal Stem Cells (MSCs) as a Promising Approach in Brain Study and Nerve Regeneration. *Cells*. 2022;11(9):1464. doi:10.3390/cells11091464.

Kaur J, Debnath J. Autophagy At the Crossroads of Catabolism and Anabolism. *Nat Rev Mol Cell Biol*. 2015;16(8):461–472. doi:10.1038/nrm4024.

Kerr JF, Wyllie AH, Currie AR. Apoptosis: A Basic Biological Phenomenon with Wide-Ranging Implications in Tissue Kinetics. *Br J Cancer*. 1972;26(4):239–257. doi:10.1038/bjc.1972.33.

Kim HA, Mahato RI, Lee M. Hypoxia-Specific Gene Expression for Ischemic Disease Gene Therapy. *Adv Drug Deliv Rev*. 2009;61(7–8):614–622. doi:10.1016/j.addr.2009.04.009.

Koehler RC, Dawson VL, Dawson TM. Targeting Parthanatos in Ischemic Stroke. *Front Neurol.* 2021;12:662034. doi:10.3389/fneur.2021.662034.

Krawczenko A, Klimczak A. Adipose Tissue-Derived Mesenchymal Stem/Stromal Cells and Their Contribution to Angiogenic Processes in Tissue Regeneration. *Int J Mol Sci.* 2022;23(5):2425. doi:10.3390/ijms23052425.

Kumar V, Kumar A, Mir KUI, Yadav V, Chauhan SS. Pleiotropic Role of PARP1: An Overview. *3 Biotech.* 2022;12(1):3. doi:10.1007/s13205-021-03038-6.

Kurozumi K, Nakamura K, Tamiya T, et al. Mesenchymal Stem Cells That Produce Neurotrophic Factors Reduce Ischemic Damage in the Rat Middle Cerebral Artery Occlusion Model. *Mol Ther.* 2005;11(1):96–104. doi:10.1016/j.ymthe.2004.09.020.

Lee KH, Kim UJ, Lee BH, Cha M. Safeguarding the Brain from Oxidative Damage. *Free Radic Biol Med.* 2025;226:143–157. doi:10.1016/j.freeradbiomed.2024.11.019.

Levy DE, Trammel J, Wasiewski WW; Ancrod Stroke Program (ASP) Study Team. Ancrod for Acute Ischemic Stroke: A New Dosing Regimen Derived from Analysis of Prior Ancrod Stroke Studies. *J Stroke Cerebrovasc Dis.* 2009;18(1):23–27. doi:10.1016/j.jstrokecerebrovasdis.2008.07.009.

Lian L, Zhang Y, Liu L, et al. Neuroinflammation in Ischemic Stroke: Focus on MicroRNA-Mediated Polarization of Microglia. *Front Mol Neurosci.* 2021;13:612439. doi:10.3389/fnmol.2020.612439.

Liu G, David BT, Trawczynski M, Fessler RG. Advances in Pluripotent Stem Cells: History, Mechanisms, Technologies, and Applications. *Stem Cell Rev Rep.* 2020;16(1):3–32. doi:10.1007/s12015-019-09935-x.

Liu H, Reiter S, Zhou X, et al. Insight into the Mechanisms and the Challenges on Stem Cell-Based Therapies for Cerebral Ischemic Stroke. *Front Cell Neurosci.* 2021;15:637210. doi:10.3389/fncel.2021.637210.

Lladó J, Haenggeli C, Maragakis NJ, Snyder EY, Rothstein JD. Neural Stem Cells Protect against Glutamate-Induced Excitotoxicity and Promote Survival of Injured Motor Neurons through the Secretion of Neurotrophic Factors. *Mol Cell Neurosci.* 2004;27(3):322–331. doi:10.1016/j.mcn.2004.07.010.

Ludhiadch A, Sharma R, Muriki A, Munshi A. Role of Calcium Homeostasis in Ischemic Stroke: A Review. *CNS Neurol Disord Drug Targets.* 2022;21(1):52–61. doi:10.2174/1871527320666210212141232.

Majumder D. Ischemic Stroke: Pathophysiology and Evolving Treatment Approaches. *Neurosci Insights.* 2024;19:26331055241292600. doi:10.1177/26331055241292600.

Mehta SR, Yusuf S, Peters RJ, et al. Effects of Pretreatment with Clopidogrel and Aspirin Followed by Long-Term Therapy in Patients Undergoing Percutaneous Coronary Intervention: The PCI-CURE Study. *Lancet.* 2001;358(9281):527–533. doi:10.1016/s0140-6736(01)05701-4.

Menon BK, Buck BH, Singh N, et al. Intravenous Tenecteplase Compared with Alteplase for Acute Ischaemic Stroke in Canada (AcT): A Pragmatic, Multicentre, Open-Label, Registry-Linked, Randomised, Controlled, Non-Inferiority Trial. *Lancet.* 2022;400(10347):161–169. doi:10.1016/S0140-6736(22)01054-6.

Merelli A, Repetto M, Lazarowski A, Auzmendi J. Hypoxia, Oxidative Stress, and Inflammation: Three Faces of Neurodegenerative Diseases. *J Alzheimers Dis.* 2021;82(s1):S109–S126. doi:10.3233/JAD-201074.

Moltrasio C, Romagnuolo M, Marzano AV. NLRP3 Inflammasome and NLRP3-Related Autoinflammatory Diseases: From Cryopyrin Function to Targeted Therapies. *Front Immunol.* 2022;13:1007705. doi:10.3389/fimmu.2022.1007705.

Neher JJ, Emmrich JV, Fricker M, Mander PK, Théry C, Brown GC. Phagocytosis Executes Delayed Neuronal Death after Focal Brain Ischemia. *Proc Natl Acad Sci U S A.* 2013;110(43):E4098–E4107. doi:10.1073/pnas.1308679110.

Nguyen TT, Dung Nguyen TT, Vo TK, et al. Nanotechnology-Based Drug Delivery for Central Nervous System Disorders. *Biomed Pharmacother.* 2021;143:112117. doi:10.1016/j.biopha.2021.112117.

Park YJ, Niizuma K, Mokin M, Dezawa M, Borlongan CV. Cell-Based Therapy for Stroke: Musing with Muse Cells. *Stroke.* 2020;51(9):2854–2862. doi:10.1161/STROKEAHA.120.030618.

Patra JK, Das G, Fraceto LF, et al. Nano Based Drug Delivery Systems: Recent Developments and Future Prospects. *J Nanobiotechnology.* 2018;16(1):71. doi:10.1186/s12951-018-0392-8.

Pawluk H, Kołodziejska R, Grześk G, et al. Increased Oxidative Stress Markers in Acute Ischemic Stroke Patients Treated with Thrombolytics. *Int J Mol Sci.* 2022;23(24):15625. doi:10.3390/ijms232415625.

Penny TR, Jenkin G, Miller SL, McDonald CA. Umbilical Cord Blood Derived Cell Expansion: A Potential Neuroprotective Therapy. *Stem Cell Res Ther.* 2024;15(1):234. doi:10.1186/s13287-024-03830-0.

Peters RJ, Mehta SR, Fox KA, et al. Effects of Aspirin Dose When Used Alone or in Combination with Clopidogrel in Patients with Acute Coronary Syndromes: Observations from the Clopidogrel in Unstable Angina to Prevent Recurrent Events (CURE) Study. *Circulation.* 2003;108(14):1682–1687. doi:10.1161/01.CIR.0000091201.39590.CB.

Piancone F, La Rosa F, Marventano I, Saresella M, Clerici M. The Role of the Inflammasome in Neurodegenerative Diseases. *Molecules.* 2021;26(4):953. doi:10.3390/molecules26040953.

Prokhorova EA, Egorshina AY, Zhivotovsky B, Kopeina GS. The DNA-Damage Response and Nuclear Events as Regulators of Nonapoptotic Forms of Cell Death. *Oncogene.* 2020;39(1):1–16. doi:10.1038/s41388-019-0980-6.

Raedt R, Pinxteren J, Van Dycke A, et al. Differentiation Assays of Bone Marrow-Derived Multipotent Adult Progenitor Cell (MAPC)-Like Cells towards Neural Cells Cannot Depend on Morphology and a Limited Set of Neural Markers. *Exp Neurol.* 2007;203(2):542–554. doi:10.1016/j.expneurol.2006.09.016.

Rathmell JC, Thompson CB. Pathways of Apoptosis in Lymphocyte Development, Homeostasis, and Disease. *Cell.* 2002;109(Suppl):S97–S107. doi:10.1016/s0092-8674(02)00704-3.

Raza SS. Rat Model of Middle Cerebral Artery Occlusion. *Methods Mol Biol.* 2024;2761:623–633. doi:10.1007/978-1-0716-3662-6_41.

Reed JC. Apoptosis-Based Therapies. *Nat Rev Drug Discov.* 2002;1(2):111–121. doi:10.1038/nrd726.

Rose CR, Ziemens D, Verkhratsky A. On the Special Role of NCX in Astrocytes: Translating Na^+-Transients into Intracellular Ca^{2+} Signals. *Cell Calcium.* 2020;86:102154. doi:10.1016/j.ceca.2019.102154.

Salaudeen MA, Allan S, Pinteaux E. Hypoxia and Interleukin-1-Primed Mesenchymal Stem/Stromal Cells as Novel Therapy for Stroke. *Hum Cell.* 2024;37(1):154–166. doi:10.1007/s13577-023-00997-1.

Salaudeen MA, Bello N, Danraka RN, Ammani ML. Understanding the Pathophysiology of Ischemic Stroke: The Basis of Current Therapies and Opportunity for New Ones. *Biomolecules.* 2024;14(3):305. doi:10.3390/biom14030305.

Sarkar S, Chakraborty D, Bhowmik A, Ghosh MK. Cerebral Ischemic Stroke: Cellular Fate and Therapeutic Opportunities. *Front Biosci (Landmark Ed).* 2019;24(3):435–450. doi:10.2741/4727.

Shaheryar ZA, Khan MA, Adnan CS, Zaidi AA, Hänggi D, Muhammad S. Neuroinflammatory Triangle Presenting Novel Pharmacological Targets for Ischemic Brain Injury. *Front Immunol.* 2021;12:748663. doi:10.3389/fimmu.2021.748663.

Shen Z, Xiang M, Chen C, et al. Glutamate Excitotoxicity: Potential Therapeutic Target for Ischemic Stroke. *Biomed Pharmacother.* 2022;151:113125. doi:10.1016/j.biopha.2022.113125.

Sherman DG, Atkinson RP, Chippendale T, et al. Intravenous Ancrod for Treatment of Acute Ischemic Stroke: The STAT Study: A Randomized Controlled Trial. Stroke Treatment with Ancrod Trial. *JAMA.* 2000;283(18):2395–2403. doi:10.1001/jama.283.18.2395.

Song H, Wang Y, Ma Q, et al. Efficacy and Safety of Recombinant Human Prourokinase in Acute Ischemic Stroke: A Phase IIa Randomized Clinical Trial. *Transl Stroke Res.* 2022;13(6):995–1004. doi:10.1007/s12975-022-01012-9.

Stringberg A, Camden R, Qualls K, Naqvi SH. Update on Dual Antiplatelet Therapy for Secondary Stroke Prevention. *Mo Med.* 2019;116(4):303–307.

Sun Y, Jiang X, Gao J. Stem Cell-Based Ischemic Stroke Therapy: Novel Modifications and Clinical Challenges. *Asian J Pharm Sci.* 2024;19(1):100867. doi:10.1016/j.ajps.2023.100867.

Tsivgoulis G, Katsanos AH, Sandset EC, et al. Thrombolysis for Acute Ischaemic Stroke: Current Status and Future Perspectives. *Lancet Neurol.* 2023;22(5):418–429. doi:10.1016/S1474-4422(22)00519-1.

Tuo QZ, Zhang ST, Lei P. Mechanisms of Neuronal Cell Death in Ischemic Stroke and Their Therapeutic Implications. *Med Res Rev.* 2022;42(1):259–305. doi:10.1002/med.21817.

Turrin NP, Rivest S. Molecular and Cellular Immune Mediators of Neuroprotection. *Mol. Neurobiol.* 2006;34(3):221–242. doi:10.1385/MN:34:3:221.

Ursini F, Maiorino M. Lipid Peroxidation and Ferroptosis: The Role of GSH and GPx4. *Free Radic Biol Med.* 2020;152:175–185. doi:10.1016/j.freeradbiomed.2020.02.027.

van Putten MJAM, Fahlke C, Kafitz KW, Hofmeijer J, Rose CR. Dysregulation of Astrocyte Ion Homeostasis and Its Relevance for Stroke-Induced Brain Damage. *Int J Mol Sci.* 2021;22(11):5679. doi:10.3390/ijms22115679.

Vendrame M, Gemma C, de Mesquita D et al. Anti-Inflammatory Effects of Human Cord Blood Cells in a Rat Model of Stroke. *Stem Cells Dev.* 2005;14(5):595–604. doi:10.1089/scd.2005.14.595.

Vermot A, Petit-Härtlein I, Smith SME, Fieschi F. NADPH Oxidases (NOX): An Overview from Discovery, Molecular Mechanisms to Physiology and Pathology. *Antioxidants (Basel).* 2021;10(6):890. doi:10.3390/antiox10060890.

Voors-Pette C, Lebozec K, Dogterom P, et al. Safety and Tolerability, Pharmacokinetics, and Pharmacodynamics of ACT017, an Antiplatelet GPVI (Glycoprotein VI) Fab. *Arterioscler Thromb Vasc Biol.* 2019;39(5):956–964. doi:10.1161/ATVBAHA.118.312314.

Wang J, Zhu M, Sun J et al. Gene Therapy of Adeno-Associated Virus (AAV) Vectors in Preclinical Models of Ischemic Stroke. *CNS Neurosci Ther.* 2023;29(12):3725–3740. doi:10.1111/cns.14392.

Wang Q, Yang F, Duo K, et al. The Role of Necroptosis in Cerebral Ischemic Stroke. *Mol Neurobiol.* 2024;61(7):3882–3898. doi:10.1007/s12035-023-03728-7.

Wang Y, Che H, Qu L, et al. The Role of Nanomaterials in Revolutionizing Ischemic Stroke Treatment: Current Trends and Future Prospects. *iScience.* 2024;27(12):111373. doi:10.1016/j.isci.2024.111373.

Waseem A, Khan AQ, Khan MA, et al. Unveiling the Therapeutic Potential of Non-Coding RNAs in Stroke-Induced Tissue Regeneration. *Stem Cells.* 2023;41(11):987–1005. doi:10.1093/stmcls/sxad062.

Woodward M, Lowe GD, Campbell DJ, et al. Associations of Inflammatory and Hemostatic Variables with the Risk of Recurrent Stroke. *Stroke.* 2005;36(10):2143–2147. doi:10.1161/01.STR.0000181754.38408.4c.

Wu JR, Tuo QZ, Lei P. Ferroptosis, a Recent Defined Form of Critical Cell Death in Neurological Disorders. *J Mol Neurosci.* 2018;66(2):197–206. doi:10.1007/s12031-018-1155-6.

Xu W, Jin W, Zhang X, Chen J, Ren C. Remote Limb Preconditioning Generates a Neuroprotective Effect by Modulating the Extrinsic Apoptotic Pathway and TRAIL-Receptors Expression. *Cell Mol Neurobiol.* 2017;37(1):169–182. doi:10.1007/s10571-016-0360-5.

Xu W, Zheng J, Gao L, Li T, Zhang J, Shao A. Neuroprotective Effects of Stem Cells in Ischemic Stroke. *Stem Cells Int.* 2017;2017:4653936. doi:10.1155/2017/4653936.

Xue Y, Nie D, Wang LJ, et al. Microglial Polarization: Novel Therapeutic Strategy against Ischemic Stroke. *Aging Dis.* 2021;12(2):466–479. doi:10.14336/AD.2020.0701.

Yan HF, Zou T, Tuo QZ, et al. Ferroptosis: Mechanisms and Links with Diseases. *Signal Transduct Target Ther.* 2021;6(1):49. doi:10.1038/s41392-020-00428-9.

Ye X, Song G, Huang S, et al. Caspase-1: A Promising Target for Preserving Blood-Brain Barrier Integrity in Acute Stroke. *Front Mol Neurosci.* 2022;15:856372. doi:10.3389/fnmol.2022.856372.

Yin Y, Sun G, Li E, Kiselyov K, Sun D. ER Stress and Impaired Autophagy Flux in Neuronal Degeneration and Brain Injury. *Ageing Res Rev.* 2017;34:3–14. doi:10.1016/j.arr.2016.08.008.

Yingze Y, Zhihong J, Tong J, et al. NOX2-Mediated Reactive Oxygen Species are Double-Edged Swords in Focal Cerebral Ischemia in Mice. *J Neuroinflammation.* 2022;19(1):184. doi:10.1186/s12974-022-02551-6.

Yu H, Chen X, Guo X, et al. The Clinical Value of Serum Xanthine Oxidase Levels in Patients with Acute Ischemic Stroke. *Redox Biol.* 2023;60:102623. doi:10.1016/j.redox.2023.102623.

Yu Z, Jiang N, Su W, Zhuo Y. Necroptosis: A Novel Pathway in Neuroinflammation. *Front Pharmacol.* 2021;12:701564. doi:10.3389/fphar.2021.701564.

Zhang J, Chen Z, Chen Q. Advanced Nano-Drug Delivery Systems in the Treatment of Ischemic Stroke. *Molecules.* 2024;29(8):1848. doi:10.3390/molecules29081848.

Zhang LY, Zhang SY, Wen R, Zhang TN, Yang N. Role of Histone Deacetylases and Their Inhibitors in Neurological Diseases. *Pharmacol Res.* 2024;208:107410. doi:10.1016/j.phrs.2024.107410.

Zhang P, Li J, Liu Y, et al. Human Neural Stem Cell Transplantation Attenuates Apoptosis and Improves Neurological Functions after Cerebral Ischemia in Rats. *Acta Anaesthesiol Scand.* 2009;53(9):1184–1191. doi:10.1111/j.1399-6576.2009.02024.x.

Zhang RL, Chopp M, Roberts C, et al. Stroke Increases Neural Stem Cells and Angiogenesis in the Neurogenic Niche of the Adult Mouse. *PLoS One.* 2014;9(12):e113972. doi:10.1371/journal.pone.0113972.

Zhao J, Sui M, Lü X, Jin D, Zhuang Z, Yan T. Electroacupuncture Promotes Neural Stem Cell Proliferation and Neurogenesis in the Dentate Gyrus of Rats Following Stroke via Upregulation of Notch1 Expression. *Mol Med Rep.* 2015;12(5):6911–6917. doi:10.3892/mmr.2015.4279.

Zhu H, Hu S, Li Y, et al. Interleukins and Ischemic Stroke. *Front Immunol.* 2022;13:828447. doi:10.3389/fimmu.2022.828447.

Zhu Y, Guan YM, Huang HL, Wang QS. Human Umbilical Cord Blood Mesenchymal Stem Cell Transplantation Suppresses Inflammatory Responses and Neuronal Apoptosis during Early Stage of Focal Cerebral Ischemia in Rabbits. *Acta Pharmacol Sin.* 2014;35(5):585–591. doi:10.1038/aps.2014.9.

Zi W, Qiu Z, Li F, et al. Effect of Endovascular Treatment Alone vs Intravenous Alteplase Plus Endovascular Treatment on Functional Independence in Patients with Acute Ischemic Stroke: The DEVT Randomized Clinical Trial. *JAMA.* 2021;325(3):234–243. doi:10.1001/jama.2020.23523.

8 Hemorrhagic Stroke
Unraveling the Intricacies of Intracerebral Bleeding

Ushmita Mukherjee, Nikita Malik, Bijoyani Ghosh, Aishika Datta, and Pallab Bhattacharya

8.1 INTRODUCTION

With an incidence rate of approximately 13%, hemorrhagic stroke is denoted as the most disproportionately lethal stroke subtype[1,2]. Hemorrhagic stroke is predominantly associated with bleeding into the brain due to various underlying factors[3]. Bleeding into the various brain structures, result in primary injury followed by the mechanical compression of the surrounding tissue architecture and enhanced intracranial pressure (ICP)[3]. Further, the thrombin activated from the resultant bleeding causes the activation of inflammatory cascade and increases oxidative stress, which leads to secondary brain injury followed by herniation and death[4,5]. Hemorrhagic stroke presents itself with symptoms like motor impairment, dysphagia, sudden loss of consciousness, bleeding, headache, nausea, etc., which are common to several other pathologies (rupture of aneurysms, cerebral amyloid angiopathy (CAA), arteriovenous malformation (AVM), etc.)[5–7]. Thus, making it tough to get a proper diagnosis and effective treatment.

The secondary injury that results from hemorrhagic stroke, leads to overactivation of inflammatory response, coagulation cascade as well as the activation or inhibition of the associated systems namely, renin angiotensin aldosterone system (RAAS), dopaminergic system, glutaminergic system, etc.[5,8,9] The dysregulation of these systems, result in osmotic imbalance, cognitive impairment, excitotoxicity, inflammation, enhanced oxidative stress, etc.[4,8,9] This further leads to an increased brain damage, enhanced ICP, herniation and diminishes the chances of survival post-hemorrhagic stroke[9,10].

The conventional therapies mostly work by controlling the blood pressure, reducing the ICP and surgical evacuation of the hematoma[9–11]. Most of which, have not proven to be very effective in improving functional recovery and reducing mortality, in the long run. Thus, further investigation of the underlying causes and the targets (receptors, hormones, etc.) associated with hemorrhagic stroke are needed, in order to create an effective therapeutic strategy to manage and treat hemorrhagic stroke.

8.2 CLASSIFICATION OF HEMORRHAGIC STROKE

Hemorrhagic stroke are classified into following[12]:

- *Intracerebral hemorrhage (ICH)*

  ICH occurs due to rupture of blood vessels within brain parenchyma. Most common risk factors of ICH are CAA, excessive alcohol intake, hypocholesterolemia, hypertension, and tobacco use[13]. ICH can occur due to pre-existing lesion, such as tumor or vascular malformation, which is known as secondary ICH[14]. The larger number of ICH cases are due to CAA and hypertensive vasculopathy[15].

DOI: 10.1201/9781003500025-8

- *Subarachnoid hemorrhage (SAH)*

  Bleeding into the subarachnoid space between the arachnoid membrane and the pia mater, a delicate lining directly linked to the surface of the brain, is referred to as SAH[16]. Risk factors of SAH includes hypertension, alcohol abuse, smoking history, cocaine and caffeine consumption. Symptoms of SAH includes severe headache, loss of consciousness, vomiting, seizures and meningismus[17].
- *Intraventricular hemorrhage (IVH)*

  IVH is constricted to ventricles and is more frequently linked with ICH in basal ganglia and thalamus[15]. IVH initiates when hemorrhage occurs and the underlying ependyma breaks, the germinal matrix hemorrhage converts to IVH as blood fills in the cerebral ventricle[18].

8.2.1 Epidemiology

The average occurrence of hemorrhagic stroke is 15–40 per 100,000 individuals of the population. Hemorrhagic stroke represents approximately 15% of total stroke in which ICH represent 10–15% compared to SAH represent 5%[12]. The incidence rate of ICH per 100,000 person-years was found to be 19.6 in Hispanics, 51.8 in Asians, 24.2 in Whites and 22.9 in Blacks. The significant proportion of ICH death and incident cases happened in middle- and low-income countries includes Central Asia, Southeast Asia and Sub-Saharan Africa[19]. Incidence rate of ICH in men is higher compared to women in the studies of Japan, Italy, China, Spain, Changsha, Greece, Portugal, and Croatia[20]. Incidence of SAH is found to be age related with higher incidence occurs among individuals between 40 and 60 years age peak at aged 55 year[20]. It also appears that cases of SAH to be gender based in female having an incidence rate of 1.6 times higher compared to men. Most studies reported that mortality rate of SAH differs from 25% to 35% in countries of high income, however low income countries show mortality rates of 48% approximately[21].

8.2.2 Pathophysiology of Hemorrhagic Stroke

Hemorrhagic stroke is associated with bleeding into the brain following the rupture of a blood vessel[22]. After a hemorrhagic stroke, neurotoxic blood-derived substances such as fibrin, thrombin, and erythrocytes initiate a cascade that leads to additional disruption of the blood–brain barrier (BBB) through neuroinflammation, cell excitotoxicity, oxidative stress and cell edema. Thrombin binding to protease activated receptors 1 (PAR1) cause BBB disruption during hemorrhagic stroke by activating microglia and phosphorylating SRC kinases[23] and also giving rise to vasogenic edema that may lead to cell membrane rupture resulting in onset of secondary injury[24]. Fibrin derived from fibrinogen reported to play important role in microglial activation and neuroinflammation[23]. Further in cerebral hemorrhage, the hemoglobin burden is digested by heme oxygenase releasing iron, which is a neurotoxin. This Iron enhances edema, inflammation, BBB disruption, further effecting the interaction of veins and endothelial-pericytes, leading to axonal and glial injury[22]. Through the overexpression of lipocalin 2 and transferrin, thrombin promotes cellular uptake of iron[4]. In the central region of hemorrhagic and perihematomal stroke lesions, microglia are activated. The activated microglia produce numerous of inflammatory mediators that further initiates cascade of various inflammatory signal pathways such as NLRP3 pathway, after injury increasing TREM2 endogenous expression reaching a peak at 24 h following stroke[25].

Traditionally injury from ICH is further divided into primary injury and secondary injury. Primary injury is initial damage relevant to mass effect from initial hematoma. The hematoma expansion and hydrocephalus occurs instantly within first few days after hemorrhage[11]. Secondary injury arises in the existence of intraparenchymal blood and relies on initial hematoma volume, ventricular volume and age. Various pathological pathways are activated during secondary injury includes blood cytotoxicity, oxidative stress, spreading depression, hypermetabolism and inflammation[26]. During ICH mass effect cause physical disruption of the brain, and also cause shift in midline of the brain which cause compaction of crucial structure in the brain and herniation. In the context of secondary injury, perihematomal edema (PHE) keeps growing over 7–14 days, causing the release of thrombin, iron-induced damage, and inflammation.[11] NLRP3 inflammasome in ICH enhanced inflammatory response by facilitating processing of IL-1β and caspase-1 activation.[27] Cytokines after ICH are found to promotes neurotoxic immune responses, such as tumor necrosis factor (TNF-α), IL-6, HMGB1, colony-stimulating factor-1 (CSF-1), etc. Neutrophils, microglia, and monocytes-derived macrophages produced TNF with in first day of ICH (Figure 8.1)[10].

Further numerous pathophysiologic alterations, such as hydrocephalus, aneurysmal rebleeding, and delayed neurologic deterioration with or without cerebral vasospasm, occur in aneurysmal SAH. During SAH BBB is disrupted in the form of pericytes and endothelial dysfunction. After SAH-mediated neuroinflammation, cytokines are generated locally, breaking down cell adhesion molecules (i.e., intercellular adhesion molecule–vascular cell adhesion molecule [ICAM-VCAM]), cause leaky BBB, decreased tight junction protein level, i.e., ZO-1, junctional adhesion molecule (JAM), and occludin. In SAH breakdown of glycocalyx occurs by enhancing pro-inflammatory cytokines (e.g., IL-6, TNF-α, and IL-1β). Endothelial dysfunction in SAH characterized by uncontrolled coagulation, loss of NO production and increased permeability due to impaired glycocalyx[28]. Procoagulant activity has leads to delayed cerebral Ischemia with enhanced levels of platelet-activating factors observed on 4th day post-SAH as well as increased Willebrand factor seen within 72 h after ictus and also loss of collagen type IV followed by platelet aggregation in parenchymal vessels seen within 10 min after experimental SAH[29]. Vascular endothelial growth factor (VEGF)

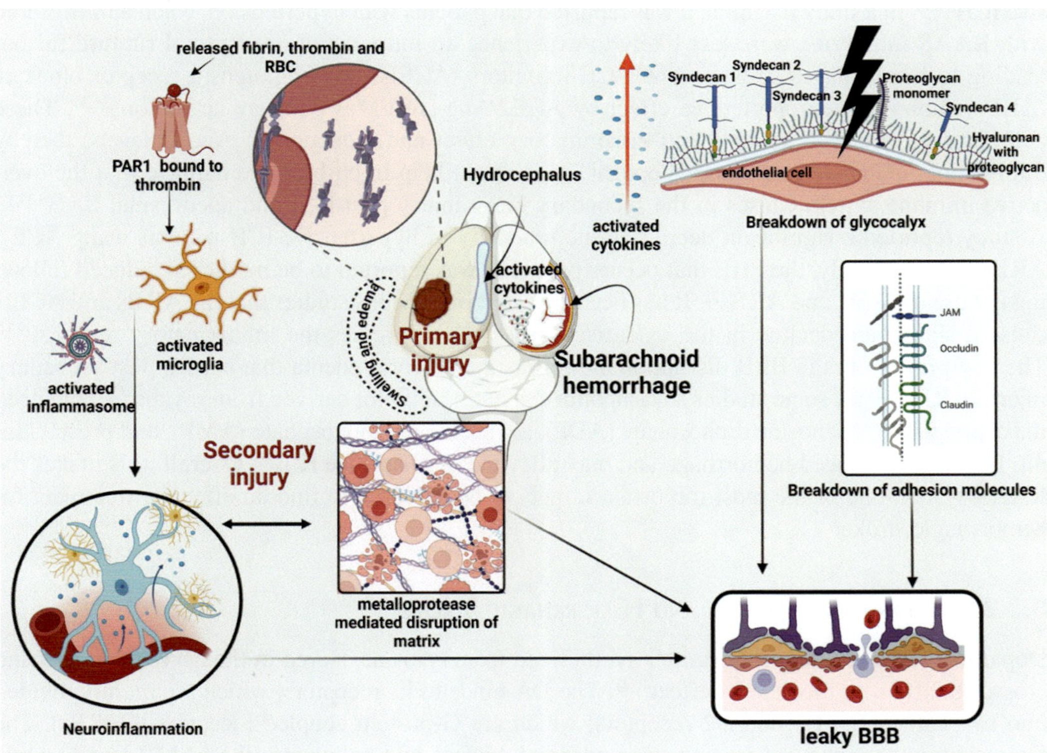

FIGURE 8.1 Pathophysiology of hemorrhagic stroke-primary injury resulting from the rupture of blood vessels leads to hydrocephalus. The thrombin released with the blood activates the inflammatory cascade, thereby triggering the secondary injury. This leads to the disruption of the endothelial glycocalyx and the blood-brain barrier, further worsening the outcome post-stroke.

expression also upregulated in blood vessel during early stage of experimental SAH results in increased intracellular adhesion molecule 1 expression and leukocyte stasis on blood vessel wall. In addition elevated VCAM-1, L-selectin, E-selectin levels and it was reported that leukocytes concentration was increased in CSF in SAH[30].

8.3 INVOLVEMENT OF DIFFERENT SYSTEMS IN HEMORRHAGIC STROKE

8.3.1 Raas System in Hemorrhagic Stroke

The RAAS is primarily associated with hypertension, the most crucial risk factor in hemorrhagic stroke[9,31]. The RAAS is also involved in the modulation of baroreceptor reflex, vasoconstriction, sodium uptake, osmotic balance as well as cognitive functions[31]. Angiotensin II (Ang II), aldosterone, angiotensin converting enzyme (ACE), the AT1, AT2, as well as the Mas receptors are the mediators responsible for the effect of the RAAS system on the adrenal cortex as well as the arterioles[9,31]. An imbalance in the levels of the mediators of RAAS is associated with various pathologies[9,31]. The ACE regulates the proteolytic conversion of Ang II from its precursor angiotensin I (Ang I), which further regulates blood pressure[9,31,32]. The involvement of AT1 receptors in vasoconstriction, upregulation of inflammatory response, oxidative stress, fibrosis and hypertension, has been reported in several studies[32,33].

The inhibition of the overexpressed RAAS mediators, has been reported to show improvements in pathologies like hyperaldosteronism, hypertension, inflammatory response as well as oxidative stress[33–35]. These pathologies increase the risk of mortality associated with stroke[9,35]. Antihypertensives, like RAAS inhibitors have been reported to be beneficial in treating hypertensive ICH[35,36]. In a study in China, it was reported that patients with hypertension, when administered with RAAS inhibitors, were less likely to experience an intracranial aneurysmal rupture further leading to hemorrhagic stroke[36]. The ACE inhibitors (ACEIs) and angiotensin receptor blockers (ARBs) shows its neuroprotective effect by ACE2/Ang-(1-7)/Mas pathway activation[9,36,37]. These neuroprotective effects include anti-inflammatory effect and reduction in Ang II levels, thereby diminishing oxidative stress and apoptosis. This might help in curbing the damage that the overactive immune system causes in the secondary brain injury post-ICH and aneurysmal SAH[9,37,38]. A study reported a significant decline in the mortality of hypertensive ICH patients using ACEI/ARBs[37]. Additionally, the PHE that occurs post-ICH was reported to be markedly reduced following the use of ACEI and ARBs[35]. It has been reported in previous studies that the ARBs and ACEIs cause a significant decline in the oxidative stress and suppresses the inflammatory response[9,37]. Thereby preventing the BBB disruption as well as the cerebral edema that ensues post-secondary injury in ICH[37,38]. In some studies soyasaponin 1, a renin inhibitor derived from soyabean, enhanced the expression of adenosine diphosphate (ADP), adenosine monophosphate (AMP), and prostaglandin E_2 (PGE_2), reduced hemorrhage and may alleviate hypertensive ICH[39]. Overall, this makes the RAAS system one of the most important targets to be explored to find an effective treatment for hemorrhagic stroke.

8.3.2 Dopaminergic System and Hemorrhagic Stroke

Dopamine (DA) is a neurotransmitter synthesized from tyrosine, stored in the nerve terminals and crosses the BBB to mediate its effect[40,41]. The DA binds to its receptors, which are mainly divided into two categories: D1 and D2 receptors, which are G-protein coupled receptors (GPCRs). The D1 receptors are involved in activating adenylyl cyclase and enhancing the cAMP levels, while the D2 receptors downregulate adenylyl cyclase and reduce cAMP levels, leading to an inhibitory effect[40–42].

The cognitive impairment, depression and dementia that ensues post-hemorrhagic stroke is where the dopaminergic system comes into play[40,42]. The dopaminergic system is responsible for regulation of movement, reward and punishment behavior[42,43]. DA has also been reportedly involved in the regulation of memory, cognition, motivation and mood[44–46]. It was observed/reported by Wang et al. that D1 and D4 receptors are involved in improving cognition, in contrast to D2 receptors that are involved in motor functions[46,47]. Thus, the disruption in DA receptors as well as the imbalance in DA levels often lead to mood and movement disorder as well as cognitive impairment, as is seen mostly post-stroke[44,45]. Studies have reported that DA receptor regulation might prove to be an essential target to alleviate post-stroke motor impairment[48].

DA regulates immune response by modulating the expression of macrophages, T cells, B cells, dendritic cells, etc., in various pathologies[41,49]. Diminishing/Depleting levels of DA and loss of Dopaminergic neurons lead to an exacerbated oxidative stress and inflammatory response by over activating the RAAS[41,42,50]. It has been noted that D1 receptor agonists lower blood pressure and diminish sodium and water retention, whereas D2 receptor agonists are known to reduce afterload and promote vasodilation[47,50]. Both these facts hint to the their anti-hypertensive affect, which could reduce the risk of hemorrhagic stroke[46,48–51]. In a cohort study, a significant increase was observed in the incidence rate of new-onset dementia, in patients with lobar ICH[52,53]. Additionally, the impairment rates of global cognitive tests ranges between 26% and 43% post-SAH[54,55]. It has also been reported that damage to midbrain post-ICH, results in loss of DA neurons and mood disorders at day 14 following stroke[46,51,53,55]. With DA being one of the major neuropeptides associated with swallowing, DA agonists might help in motor recovery to help resolve neurogenic dysphagia that ensues post-hemorrhagic stroke[47,56]. The dementia that occurs post-stroke, when treated with antipsychotics in the long run show side effects like impotence, gynecomastia, fatigue, stroke, etc.[48,49,57] This duality of DA receptor agonists highlights the need to explore therapeutic effectiveness the DA receptor agonists might have in treating the cognitive impairment and movement disorder that ensues post-stroke.

8.3.3 Prolactin and Hemorrhagic Stroke

Hemorrhagic stroke has a high incidence rate of approximately 41% out of all pregnancy related strokes[58]. With prolactin being one of the major neuropeptides involved in pregnancy, it becomes an important target to explore[58]. Prolactin is involved majorly in the development of mammary glands and lactogenesis[58,59]. Prolactin inhibits DA synthesis and release and visa-versa[58–60]. Oxytocin, vasopressin as well as DA antagonists have been known to cause hyperprolactinemia[60]. An imbalance in prolactin levels leads to various pathologies[58]. The hyperprolactinemia that ensues post-imbalance, further leads to loss of DA inhibition and destruction of the hypothalamus/hypothalamic-hypophyseal tract[60,62].

Various studies have reported that a hypothalamic-pituitary insufficiency ensues post aneurysmal SAH[62]. Prolactin has been reported to have a relationship with the induction of pregnancy related hypertensive pathologies, a major risk factors of hemorrhagic stroke[59]. Prolactin shows pro-inflammatory effect by modulating the expression of microglia, macrophages and proinflammatory cytokines[61,63,64]. This can be attributed to the fact that prolactin acts through its receptors, which belong to a class-1 cytokine receptor superfamily. Prolactin is also known to enhance IFN-γ proliferation in NK cells as well as oxidative stress[61,64,65]. In combination with cytokines, prolactin leads to microglial activation by enhanced TNF-α and glial fibrillary acidic protein (GFAP) expression in astrocytes[60–62]. However, contradictory results have been reported in several cases of brain injury, where prolactin has shown a neuroprotective and anti-inflammatory effect[66]. In a study it was reported that, the administration of prolactin in ovariectomized rats decreased microglial activation[63,67]. Prolactin has also known to reduce oxidative stress, by enhancing glutathione levels[66]. Prolactin enhances regeneration, remyelination, neurogenesis and ameliorate excitotoxicity, post-brain injury[66,68,69].

Various studies have reported that enhanced levels of prolactin levels 12 h post-SAH[62,68]. Prolactin has been reported to increase approximately three times in the cerebrospinal fluid (CSF) of affected brain-damaged patients[62]. The intracerebroventricular (ICV) administration of prolactin post-brain injury, has been reported to reduce nerve damage, cognitive impairment, improve GFAP and apolipoprotein E (APOE) expression and proved to be beneficial in movement disorders post-ICH[68]. These dual traits of prolactin make it a beneficial target to explore in treating hemorrhagic stroke.

8.3.4 Glutametergic System and Hemorrhagic Stroke

Glutamate is one of the most prominent excitatory neurotransmitters that are responsible for inducing excitotoxicity post-stroke, by overactivation of glutamate receptors (NMDA, α-amino-3-hydroxy-5-methyl-4-isoxazolepropionic acid [AMPA], and kainic acid)[70]. Glutamate is known to regulate mood, long term potentiation, memory, excitatory synaptic transmission and cognition in normal physiological conditions[70–72]. Pathologies like hemorrhagic stroke upset the natural balance, leading to excitotoxicity and further neuronal death[70,71]. while the inhibition of the overactivated N-methyl-1D aspartic acid (NMDA) glutamate (GluN) may show a neuroprotective effect, it might also hinder the clearing of the excess glutamate[71,73–75]. It has been reported that the PAR1 receptor activation may be involved in the cognitive impairment post-ICH[74].

Various studies have reported that the glutamate levels in the CSF and blood were enhanced post-brain injury that follows post-stroke[70,74]. In a clinical study, the blood glutamate levels of stroke patients were enhanced and kept increasing for a period of 7 days following stroke[74–76]. However, the blood glutamate levels in hemorrhagic stroke (ICH and SAH) patients were observed to be significantly lower than that of acute ischemia stroke[75]. Further, a study has proposed a connection between glutamate and the gut-brain axis[75]. This might explain the post-stroke depression that occurs following the imbalance in glutamate levels following secondary brain injury post-stroke[72,75].

It has also been reported in some studies that there is a relationship between Aquaporin-4 and glutamate transporter-1 (GLT-1), involved in water and glutamate buffering pathway in astrocytes[77,78]. In ICH, the polarity and expression of astrocytes are modulated post-secondary brain injury[79]. The excess blood glutamate crosses the BBB post-secondary brain injury, thereby leading to an enhanced glutamate levels in the brain parenchyma[71]. The damaged BBB and the loss of neurons and astrocytes, hinder the clearance of the excessive glutamate by GLT-1 and excitatory amino acid transporters (EAAT) present on astrocytes and further worsening the glutamate accumulation[73,78]. Thus, exacerbating the glutamate-excitotoxicity and the secondary brain injury that occurs post-stroke[70,73,78]. Administration of blood glutamate scavengers (BGS) like aspartate aminotransferase, pyruvate and oxaloacetate, etc., are reported to decrease glutamate levels in the blood as well as accelerating glutamate clearance from the brain parenchyma[73]. These BGS's efficacy has been reported to improve stroke outcome by restoring the blood- brain glutamate homeostasis[73,78]. This decline in glutamate levels prevents the induction of secondary brain injury associated with the glutamate excitotoxicity that follows post-stroke[78]. Various studies have reported that Ifenprodil and perampanel, glutamate receptor blockers, may have improve the neurotoxicity that occurs following SAH[73,78]. Thus, the modulation of glutamate levels prove to be a lucrative target to explore for stroke therapy.

8.4 DIFFERENT RECEPTORS INVOLVED IN HEMORRHAGIC STROKE

8.4.1 Matrix Metalloproteinases

Matrix metalloproteinases (MMPs) belong to a family of transmembrane ubiquitous zinc dependent endopeptidases that are involved in degradation of the extracellular matrix[80,81]. There are reported to be about 20 or more categories of MMPs that have been identified[80–82]. MMPs are not released or activated under physiological conditions and are associated with cell differentiation, proliferation, angiogenesis, activation or inhibition of cytokines and apoptosis[81–84]. MMPs can be activated by stimuli, such as inflammatory cytokines, VEGF, free radicles, stress, etc., which occurs mostly post-brain injury[83,85].

In case of ICH, MMPs have been reported to have a biphasic nature depending upon the phase of hematoma development and expansion[82]. Certain experimental studies have reported an upregulated expression of MMPs post-hemorrhagic stroke[82]. A study by Power et al. has reported an enhanced expression of MMPs (MMP-2, MMP-3, MMP-7, MMP-9, MMP-12, etc.), 24 h post-collagenase induced ICH rodent model[82,86]. In another study by Wells and colleagues, an upregulated expression of MMP3 and MMP 12 in collagenase- induced ICH rodent model was reported[86,87]. In the autologous blood rodent model of ICH, an enhanced expression of MMP2 and MMP9 was reported at 6 and 24 h post-injury[81,87]. Additionally, in a study by Wu et al., it was reported that MMP9 levels were enhanced from 2 h to 5 days, with a peak at day 2, post-autologous blood injection induced ICH in rodents[82]. It was also observed that along with the MMP9 expression, there is an elevation in the brain water content as well, at day 2 post-ICH[82,87]. The biphasic nature of the MMP leads to a peak in its expression at day 1–3 as well as day 7 post-ICH, followed by a further decline to normal levels in the long run[87]. These peaks of MMP expression result in BBB disruption, hyper-permeability, PHE, hematoma expansion, exacerbation of inflammatory response and oxidative stress, which further exacerbates the secondary brain injury and lead to neuronal loss post-ICH[82,81,82,86]. There have been reports of MMP9 and thrombin interactions post-ICH, under the influence of MMP3, leading to neurotoxicity[86].

There have been reports of improvement in brain damage in MMP3 and MMP9 knockout mice, which has led to the observation that inhibition of multiple MMPs prove to be beneficial in ameliorating the secondary brain injury post-ICH[81,82,86]. On the contrary, there have been reports of exacerbated PHE, hemorrhage and neuronal death following the inhibition of MMPs[81,87]. This might be due to the disruption of the extracellular matrix homeostasis that ensues following the long-term inhibition of MMPs[87]. Additionally, MMPs have also been observed to have neuro-reparative effect by modulating neurogenesis, gliogenesis, remyelination and angiogenesis[84,85,88]. This might have a neuroprotective effect by leading to neurovascular remodeling at the injury site post-ICH[82]. There have been reports of decline in the levels of neurovascular remodeling biomarkers post-inhibition of MMPs, thus exacerbating brain injury post-stroke[81,83]. These contradictory effects of MMPs suggest that modulating MMP levels at different time points post-ICH may prove to be a therapeutic approach to improve stroke outcome and thus needs to be explored.

8.4.2 Aquaporins

Aquaporins (AQPs) are one of the key players associated with the development as well as the resolution of cerebral edema that follows post-stroke[89,90]. They are found mostly on the astrocytes, in the BBB and CNS[89]. These water channels work on the principle of creating an osmotic gradient and thus, modulating the levels of AQPs may enhance or reduce water permeability and regulate cerebral water homeostasis[89,90]. This feature of AQPs has made them an important target in controlling the peri-hematomal edema that follows in the secondary brain injury, post-stroke[89,91,92].

AQP1 channels have been reported to be upregulated in astrocytes and endothelial cells following SAH and ICH[90,91]. The aquaglyceroporin, AQP9 has been observed to regulate energy metabolism and astrogliosis post-brain injury, however its direct role in cerebral edema has not been discovered yet[90]. AQP4 is one of the water channels predominantly involved in the cerebral edema development, which was reported in a study involving water intoxication and brain ischemia models[93,94]. The regulation of AQ4 expression modulated the peri-hematomal edema and astrocytic swelling that follows post-stroke, however it's exact role is still unclear[94,95]. It has been reported that inhibition of AQP4 has been associated with a reduced lesion volume, edema and improvement in cognitive decline, in the long-term post-stroke[96,97].

AQP4 present on the astrocytic endfeet is colocalized with many other receptors, namely, inward rectifying potassium channels (KIR4.1), connexin 43 (Cx43), SUR1-TRPM4, etc.[93] This has given an indication of the reason behind the varying functions of AQP4 in the development of cerebral edema and neuroinflammation in pathologies like stroke[89,95,98]. Various preclinical studies have shown a beneficial effect on brain edema following brain injury induced by stroke, traumatic brain injury, etc.[89,91,98] Various studies have also reported an upregulation of AQP4 levels following the exacerbated inflammatory response post-secondary brain injury[90,92].Inhibition or knock out of AQP4 led to a modulation in astrocytic morphology and hindered activation of reactive astrocytes as well as microglia[94,97]. It has also been reported that proinflammatory cytokines, such as IL-1β, upregulate AQP4 expression[92,98]. Thereby regulating neuroinflammation and resolution of cerebral edema[89,92]. Use of inhibitors like bumetanide, acetazolamide that block AQPs (AQP1 and 4) led to edema resolution and improved stroke outcome[92,95,98]. On the contrary, enhanced AQP4 expression has also been observed to have an effect of edema resolution in vivo[89,95,98]. Thus, regulating the expression of AQPs could prove to be an effective therapeutic approach to treat the resultant edema post-stroke.

8.5 COAGULATION CASCADE AND ITS INVOLVEMENT IN HEMORRHAGIC STROKE

The coagulation cascade is involved in maintaining thrombo-hemorrhagic equilibrium[99,100]. The interactions within the fibrinolytic system, platelets, adhesive proteins and walls of blood vessels maintain the hemostasis, leading to the formation of a platelet plug[99,100]. This platelet plug acts like a surface for the coagulation factors to attach, thereby stabilizing the fibrin-platelet aggregates and further clot retraction[99,101].

Thrombin is one of the predominant enzymes to be released post-activation of the coagulation cascade and is involved in hemostasis[99,101,102]. It is a protease that binds to PARs, which is involved in cleavage of fibrinogen to form fibrin which further forms the insoluble plug[99,102,103]. PAR1 mediates astrocytic and microglial proliferation as well as stimulates the release of COX (cyclooxygenase), iNOS (inducible nitric oxide synthase) and other inflammatory mediators (IL-1β, TNF-α, IL-6, IL-12, etc.)[101,103]. Along with neuroinflammation, PARs also regulate neurodegeneration as well as neuronal death[103]. PAR-1 has also been observed to be upregulated post the autologous whole blood induced ICH, knock down of which leads to a reduction in microglial polarization and thereby decrease neuroinflammatory response[104]. The downregulation of PAR-1, further leads to a decline in glutamate excitotoxicity and the seizures that follow post-stroke[102,104].

Thrombin has also been reported to have functions in downregulating disintegrin as well as thrombospondin type 1 metalloproteases, thereby enhancing platelet adhesion[101–103]. The thrombin and PARs (PAR 1-4) have been reported to modulate brain injury in pathologies like hemorrhagic stroke[99,102,105,106]. It has been reported that, through PARs, thrombin exerts effects like BBB disruption, cerebral edema, neuroinflammation (activation of 1L-1α) as well as neuronal loss, which exacerbates brain injury and enhance motor impairment, following hemorrhagic stroke[105–107]. The neuroinflammatory effect of thrombin is mediated via the activation of the PAR1 receptor, further leading to microglial activation and enhances oxidative stress[107]. However, it has also been reported that thrombin leads to formation of the hemostatic plug stops the bleeding, thereby preventing the expansion of hematoma, enhanced and exacerbation of brain injury and death, post-ICH[99,102]. It has been reported that the levels of thrombin in the perihematomal region post-ICH, are regulated by the levels of hematoma induced-thrombin production from prothrombin and anti-thrombin as well as protease nexin I produced in the astrocytes[99,104]. Thrombin has been reported to have a neuroprotective effect at lower doses which binds to PAR1 and 3, while higher doses of thrombin are associated with exacerbation of brain injury and edema post-stroke[107]. It has also been reported that thrombin is associated with brain development, synaptogenesis as well as synaptic plasticity (Figure 8.2)[99,102].

Another key mediator of the coagulation cascade is fibrinogen, which is reported to have a modulatory effect on the inflammatory response post-hemorrhagic stroke[101,103]. The fibrin formed from fibrinogen under the influence of thrombin, has a role in microglial activation by modulating the

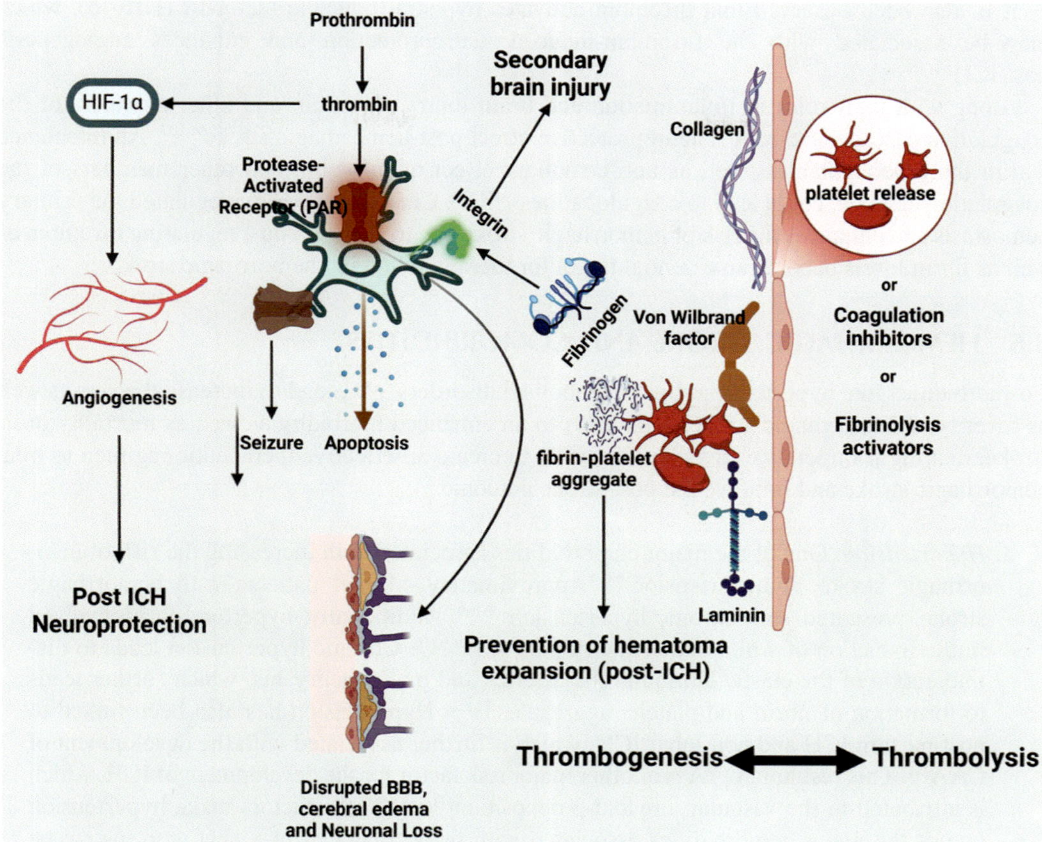

FIGURE 8.2 The effect of the coagulation cascade on intracerebral hemorrhage-following the hematoma formation, the activated thrombin acts on protease activate receptor-1 (PAR-1) triggering the inflammatory response, which further leads to BBB disruption mediated secondary injury. On the contrary, post-ICH neuroprotection is brought upon by the activation hypoxia inducible factor-1 (HIF-1a) and the fibrin-platelet aggregation, thus, preventing hematoma expansion and improving post-stroke prognosis.

expression of CD11b/CD18 integrin receptor, thereby exacerbating secondary brain injury and worsening the stroke outcome[99,101,103]. Various clinical trials have investigated the effect of administration of thrombolytic or fibrinolytic agents post-ICH, which has not shown any distinct improvement in motor coordination[102,105,106]. It has also been observed that oral anti-coagulants like dabigatran and vorapaxar, inhibit PAR-1 and thereby diminishes thrombin-induced BBB hyperpermeability[105,107]. Further, it has also been reported that, administration of argatroban (thrombin inhibitor) at later time points post-ICH, was found to reduce PHE[107]. Additionally, it was observed that administration of thrombin induced neurobehavioral impairment in autologous blood model of ICH[104,105]. On the contrary, it has also been observed that co-administration of thrombin inhibitor (hirudin) with whole blood, has shown a decline in neurobehavioral defects[102,104]. It has been observed that entry of prothrombin into the ventricles following IVH, further activating the coagulation cascade further leading to hematoma expansion, impairing CSF circulation and further leading to hydrocephalous, worsening the stroke outcome[104,106,107]. The thrombin levels are also elevated in the CSF of SAH patients and is associated with PHE and neurological damage[107]. It has been reported in a study that administration of heparin at low doses intravenously, the inflammatory response was significantly reduced in SAH rodent model[107].

It is also been observed that thrombin activates hypoxia inducible factor-1α (HIF-1α), which may be associated with the thrombin-induced neuroprotection and enhances angiogenesis post-ICH[104,104,107].

Along with their roles in inflammation and brain injury, thrombin and other members of the coagulation cascade also have a neuroprotective effect post-hemorrhagic stroke[99,103]. An imbalance within the procoagulant as well as anti-coagulant effect of thrombin and other members of the coagulation cascade, could lead to coagulopathies[99,101,102]. Coagulopathies dysregulated the primary hemostasis and enhances the risk of hemorrhagic stroke occurrence[107]. Thus, regulating thrombin as well as fibrin levels become an essential target for therapy following hemorrhagic stroke.

8.6 HEMORRHAGIC STROKE AND COMORBIDITIES

Co-morbidities like hypertension, lipid metabolism disorders, etc., tend to increase the risk as well as severity of hemorrhagic stroke. This leads to an enhanced morbidity as well as mortality post-stroke, making it imperative to explore in order to create an effective therapeutic regimen to treat hemorrhagic stroke and improve the post-stroke outcome.

a. *Hypertension*: One of the major comorbidities associated with increasing the risk of hemorrhagic stroke, is hypertension[108]. Approximately 44% of patients with hemorrhagic stroke, presented with chronic hypertension[108,109]. Additionally, hypertension is involved in the induction of white matter injury post-ICH[109,110]. Chronic hypertension leads to disintegration of the elastic lamina in the arteries and microaneurysms, which further leads to formation of fibrin and platelet aggregates[111,112]. Hypertension has also been linked to post-partum ICH and non-lobar ICH, which is further associated with the development of CAA[111]. This resultant CAA is another major risk factor for the development of ICH, which is attributed to the vascular amyloid-β deposition[111–113]. These factors make hypertension one of the major comorbidities associated with hemorrhagic stroke, that worsens stroke outcome and enhances mortality following stroke.
b. *Lipid metabolism disorders*: Mutations in the APOE gene enhances the risk of lobar ICH development[114,115]. Studies have reported the increment in the ICH cases, hematoma volume and mortality with an overexpressed APOE ε2 in Asian population[116]. Additionally, the overexpressed APOE alleles have reported a greater risk of developing lobar ICH, post-warfarin administration[114–117]. An allele of APOE H has been reported to be enhanced in ICH patients with chronic hypertension as comorbidity[116,118]. Regulation of endoplasmic reticulum lipid raft-associated 1 (ERLIN1) and low-density lipoprotein receptor (LDLR) is essential for cholesterol metabolism and have been reported to diminish the incidence of ICH as well as the associated mortality[113,118]. Hypercholesteremia, which shows an enhanced concentration of high-density lipoprotein (HDL), is primarily associated with an enhanced risk of ICH development[113,118]. Inhibition of cholesteryl ester transfer protein (CETP) gene expression, which mediates plasma HDL levels, might prove to be an effective strategy to treat atherosclerosis and prevent the development of hemorrhagic stroke[116,117]. However, further studies are required to explore disorders of lipid metabolism as potential targets to reduce the incidence and mortality that follows post-stroke.
c. *Inflammation*: Neuroinflammation is another underlying cause for enhanced brain damage and mortality following the hemorrhagic stroke pathology, further worsening the stroke outcome[119]. Overactivation of inflammatory response with an excess of circulating proinflammatory cytokines (TNF-α, IL-1β, etc.) that further results in secondary brain damage and mortality post-stroke[119–121]. Studies have reported an overactivation of NLRP3 inflammasome and NLRP3 pathway to exacerbate brain damage post-ICH[108,121–123]. Studies have reported a decline in NLRP3 expression on treatment with P2X7R siRNA and P2X7R inhibitor, leading to a decline in NOX2, Inos, etc., and further improve outcome post-SAH

and ICH[108,121–123]. The inhibition of NMDAR1 by, MK801, resulted in the downregulation of NLRP3 inflammasome as well as reduced IL-1β activation and secretion post-ICH[121,123,124]. Further, downregulation of nuclear factor E2-related factor 2 (Nrf2) and modulation of NF-κB leads to upregulation of NLRP3 inflammasome expression, enhancement of oxidative stress and mortality post-ICH[108,113,121]. The association of neuroinflammation-induced aneurysm rupture post-SAH, has been reported in recent studies[119,122–124]. Additionally, inhibition of RIP1, a receptor interacting protein, by Necrostatin-1 resulted in downregulation of the NLRP3 inflammasome and inhibits neuroinflammation post-SAH[123]. These factors together highlight the importance of modulation of NLRP3 inflammasome expression, as a therapeutic target to reduce inflammation induced secondary brain damage and improve outcome post-HS.

d. *Obesity*: Obesity is another risk factor, which impacts the risk of development and progression of stroke indirectly[125,126]. It has been reported that obesity has been associated with excessive cerebral bleeding, resulting in deep ICH[126,127]. Hypertension is one such linker that associates obesity with a risk of developing deep ICH, indicating the involvement of vascular damage[113,125,127,128]. Various studies have evaluated the role of BMI in the stroke pathology, which has shown the effect of obesity on hypertension vasculopathy-independent ICH (associated with CAA)[125–127]. There also might be an effect of obesity on the genetic factors that may contribute to the risk of developing a cerebral bleed and further progression of the hematoma in the stroke pathology[128]. On the contrary, it has been observed in a prospective study that women with a high BMI showed a decline in the risk of hemorrhagic stroke incidence, with an equal decline in both ICH and SAH cases[113,125,127,128]. This portrays the protective effect obesity might have to reduce the risk of hemorrhagic stroke[125,128]. However, some studies have demonstrated a decreased protective effect associated with obesity as well as an enhanced mortality post-stroke[113]. This hints at weight reduction being an effective preventive and management strategy to combat comorbidities to reduce the risk of developing cerebral hemorrhage[113,126,127].

e. *Others*: Small vessel diseases like CAA and AVM contribute to approximately one-third of the incidence of hemorrhagic stroke cases[108,129]. CAA is one of the primary causes of development of lobar ICH in geriatric patients[108,111,130–132]. The occurrence of non-traumatic spontaneous ICH is associated with CAA as a comorbidity, resulting in fatal lobar hemorrhage, the risk of which increases with age[111,131,133]. It has been reported that amyloid plaques deposition in amyloid precursor protein (APPA) transgenic mouse model results in an increased risk of CAA-associated spontaneous hemorrhage[111,132,133]. On the other hand, AVM is associated with structural abnormality of the vasculature that transmits at a higher rate per unit volume of blood as compared to the neighboring vessels, resulting in ICH[129,134,135]. A population-based study investigated AVM associated hemorrhage (including ICH, SAH, and IVH), revealed ICH cases associated with AVM as a comorbidity was involved in cognitive impairment and mortality[134,136,137]. In AVM, ruptured aneurysms result in enhanced arterial pressure and vasospasm further lead to worsening of stroke outcome, post-ICH and SAH[136–138]. This has given way to explore AVM as a therapeutic target to prevent and treat ICH cases (new as well as recurrent).

8.7 DIAGNOSIS AND EVALUATION

With the of symptoms of hemorrhagic stroke (bleeding, abnormal speech, movement impairment, seizures, headache, syncope, etc.) being common with most pathologies, it is often misdiagnosed[139]. The misdiagnosis includes hypertensive crisis, hemorrhagic neoplasms, hemorrhagic infarct, cerebral venous thrombosis, AVMs, etc.[139,140] This misdiagnosis can be prevented by the use of computed tomography (CT) scan and magnetic resonance imaging (MRI)[108,140–144]. A useful feat of MRI is its ability to distinguish between hemorrhagic transformation post-Ischemic stroke and primary hemorrhage[145,146]. Additionally, MRI may also be used to detect the root causes of secondary hemorrhages like tumors, AVMS, cerebral vein thrombosis, etc.[141,145–148] Further, investigations of blood parameter like bleeding and clotting time, peripheral smear, activated partial thromboplastin time, platelet count, etc., are essential to detect any coagulopathies that might lead to hemorrhage[146,147,149]. Additionally, renal and hepatic function test may be required to rule out any underlying disorders that may result in hemorrhage[146,147]. Evaluation of immunoglobulins, histone antibodies, rheumatoid factor, anti-endothelial antibody, etc., may be essential to exclude vasculitis as an underlying pathology responsible for hemorrhage[139,142,145,147]. The early deterioration as well as locked in syndrome post-ICH is associated with enhanced brain damage and mortality[139,142]. The prognosis and mortality post-ICH, is predicted on the basis of ICH score[144,146,149]. Thus, early detection and correct diagnosis and evaluation prove to be critical in reducing mortality post-stroke.

8.8 THERAPEUTIC APPROACHES FOR HEMORRHAGIC STROKE TREATMENT

A range of therapies are available to treat and manage hemorrhagic stroke, however none of them have proved to be an effective therapy in the long run[108,150]. This has resulted in various trails to explore therapeutic options following hemorrhagic stroke, namely, antihypertensive treatment in acute cerebral hemorrhage (ATACH), intensive blood pressure reduction in acute cerebral hemorrhage trial (INTERACT), surgical trial in intracerebral hemorrhage (STICH), etc.[151,152] The ATACH study has reported no significant connection between BP control and hematoma expansion, while the INTERACT study reported that early BP control helps in suppresses hematoma growth over the duration of 72 h[150,151,153]. Blood pressure control and monitoring is one of the most effective treatment strategies for hemorrhagic stroke to prevent neuronal damage and death[150,154]. Medications like beta blockers, calcium channel blockers are used to bring the BP down to 150/90 mmHg gradually and prevent brain damage[150,153,154]. Further, increased ICP was monitored and treated using osmotherapy (20% mannitol, hypertonic saline) and elevating the head of patient's bed at an angle of 30°, initially. Further, in the FAST study it was observed that recombinant FactorV11a (rFVIIa) diminished hematoma growth with no reduction in mortality post-stroke[154]. In order to treat the post-stroke clinical seizures resulting from a lobar hematoma and it's expansion, anti-epileptic medications are administered to the patients along with EEG monitoring[155,156].

Surgical interventions like craniotomy, decompressive craniectomy, hematoma evacuation, catheter aspiration, stereotactic aspiration, etc., are utilized to reduce the ICP post-hemorrhagic str oke[152,157,158]. The STICH study reported no improvement in survival post-surgery in supratentorial ICH as compared to conventional treatments[157,158]. MISTIE is another trail, in which removal of the hematoma resulting in decreased PHE[155,158]. In a randomized controlled trail, decompressive craniectomy is used to evacuate the hypertensive hemispheric ICH, in patients with GCS scores of about 8 as well as a hematoma volume >60 ml, improving neurological function[155,159,160]. However, the surgical procedures are invasive and are not mostly suited for situations like IVH, which results in a high mortality rate post-treatment[155,159].

The exacerbated inflammatory response and oxidative stress that follows the secondary brain injury, is the main reason for the decreased survival rate associated with hemorrhagic stroke[108,150,156]. Thus, drugs like celecoxib, misoprostol, flavonoid, etc., are being used to reduce the damage that results from exacerbation of inflammation and oxidative stress[108,150]. Rosuvastatin, an HMG CoA reductase inhibitor helps to manage hypercholesteremia (risk factor for hemorrhagic stroke) and thus, resulted in better outcome post-stroke[156]. Nimodipine, a calcium channel blocker, helps to manage hypertension and has shown a neuroprotective effect in SAH[161]. However, even with the existing approaches, the treatment is not effective to manage and treat hemorrhagic stroke[153]. This gives rise to a search for better and safer therapeutic alternatives to treat hemorrhagic stroke and increase the survival rate (Table 8.1).

TABLE 8.1
Established and Novel Therapies for Hemorrhagic Stroke

Therapy	Mechanism	Results	References
Memantine	Low-affinity channel-blocking uncompetitive NMDA receptor antagonist	In a collagenase induced hemorrhage model, the drug improved behavioral outcome. Also when administered at a dose of 20 mg/kg/day, it decreased the hematoma volume, apoptotic cells numbers, activation of caspase-3 and MMP9 levels and other inflammatory markers while increasing expression of Bcl-2.	1. Lapchak, Paul A, Araujo, Dalia M. (2007). Advances in hemorrhagic stroke therapy: conventional and novel approaches. *Exp. Opin. Emerg. Drugs 12*(3), 389–406. doi:10.1517/14728214.12.3.389 2. Lee ST, Chu K, Jung KH, et al. (2006). Memantine reduces hematoma expansion in experimental intracerebral hemorrhage, resulting in functional improvement. *J. Cereb. Blood Flow Metab. 26*(4), 536–544. 3. Sinn DI, Lee ST, Chu K, et al. (2007). Combined neuroprotective effects of celecoxib and memantine in experimental intracerebral hemorrhage. *Neurosci. Lett.* 411(3), 238–242.
Minocycline	It is a wide-spectrum antimicrobial agent that binds to the 30S ribosomal subunit of bacteria and hinders its protein synthesis.	In the clinical trial, use of minocycline in intracerebral hemorrhage, the drug was reported to reduce the MMP-9 levels although no functional improvement was observed. Whereas in an another clinical trial, a pilot study of minocycline in intracerebral hemorrhage patients (MACH), there drug showed no alterations in MMP-9 and interleukin 6 levels among the groups.	1. Zille M, Farr TD, Keep RF, Römer C, Xi G, Boltze, J. (2022). Novel targets, treatments, and advanced models for intracerebral haemorrhage. *EBioMedicine*, 76.
MMP inhibitors (BB-2284, BB-94) and TACE inhibitors (Ro32-7315 and DPH-067517)	MMPs and TNF-α-converting enzyme (TACE) cleaves pro-TNF-α to a biologically active form TNF-α, which conversely induces MMP-9 in the CNS, leading to membrane damage. Thus, these specific inhibitors block the production of TNF-α.	In a collagenase induced hemorrhage rat model, BB-94 (non-selective MMP inhibitor) was reported to reduce hematoma volume. In a large clot embolism model of rabbit and a rodent stroke mode, BB-94 hindered the t-PA-induced hemorrhagic transformation. This result was correlative to the significant reduction of TNF- α levels in brain by BB-94.	1. Lapchak, PA, Araujo, DM. (2007). Advances in hemorrhagic stroke therapy: conventional and novel approaches. *Exp. Opin. Emerg. Drugs*, 12(3), 389–406. doi:10.1517/14728214.12.3.389 2. Lapchak, PA. (2002). Hemorrhagic transformation following ischemic stroke: Significance, causes, and relationship to therapy and treatment. *Curr Neurol Neurosci Rep.*, 2(1), 38–43. doi:10.1007/s11910-002-0051-0 3. Gearing AJ, Beckett P, Christodoulou M et al. (1994). Processing of TNF-α precursor by metalloproteinases. Nature 370(6490), 555–557. 4. Corbel M, Lanchou J, Germain N, Malledant Y, Boichot E, Lagente V. (2001). Modulation of airway remodeling-associated mediators by the antifibrotic compound, pirfenidone, and the matrix metalloproteinase inhibitor, batimastat, during acute lung injury in mice. *Eur. J. Pharmacol.* 426(1–2): 113–121. 5. Mayer SA, Rincon F. (2005). Treatment of intracerebral haemorrhage. *Lancet Neurol.* 4(10): 662–672. 6. Sumii T, Lo EH. (2002). Involvement of matrix metalloproteinase in thrombolysis-associated hemorrhagic transformation after embolic focal ischemia in rats. *Stroke* 33(3): 831–836.

(Continued)

TABLE 8.1 (*Continued*)
Established and Novel Therapies for Hemorrhagic Stroke

Therapy	Mechanism	Results	References
NXY-059 (Disodium-[(tert-butylimino) methyl] benzene1,3-disulfonate N-oxide)	A broad spectrum nitrone-based free radical scavenger by Centaur Pharmaceuticals, which is an analogs of alpha-phenyl-N-t-butylnitrone.	In a large clot embolism model of rabbit NXY-059 reduced t-PA induced hemorrhages. In a collagenase induced hemorrhage rat model, the drug improved behavioral outcome although with no significant neuroprotection. The CHANT (Cerebral Hemorrhagic and NXY-059 Treatment) clinical trial reported no significant difference among treated and untreated patients, although the study was powered for safety, not efficacy. Other clinical trials with NXY-059 have been aborted by Renovis and AstraZeneca, since it failed in treatment of acute ischemic stroke.	1. Lapchak, PA, Araujo, DM. (2007). Advances in hemorrhagic stroke therapy: conventional and novel approaches. *Exp. Opin. Emerg. Drugs*, 12(3), 389–406. doi:10.1517/14728214.12.3.389 2. Lapchak PA, Araujo DM, Song D, Wei J, Zivin JA. (2002). Neuroprotective effects of the spin trap agent disodium-[(tertbutylimino)methyl]benzene-1,3-disulfonate N-oxide (generic NXY-059) in a rabbit small clot embolic stroke model: combination studies with the thrombolytic tissue plasminogen activator. *Stroke* 33(5): 1411–1415. 3. Peeling J, Del Bigio MR, Corbett D, Green AR, Jackson DM. (2001). Efficacy of disodium 4-[(tert-butylimino)methyl]benzene-1,3-disulfonate N-oxide (NXY-059), a free radical trapping agent, in a rat model of hemorrhagic stroke. *Neuropharmacology* 40(3): 433–439. 4. Nilsson D, Wemer J, Cheng YF et al. (2007). NXY-059 does not affect bleeding time in healthy volunteers: a randomized, double-blind, placebo-controlled, 3-period crossover phase I study. *J. Clin. Pharmacol.* 47(2): 264–272.
Edaravone (3-methyl-1-phenyl-2-pyrazolin-5-one)	It is a potent lipid-soluble hyrdoxyl and peroxyl radical scavenger that can cross the blood–brain barrier to inhibit lipid peroxidation and prevent endovascular cellular damage.	It is reported to reduce incidences of ICH when administered prior to t-PA. It was reported to alleviated neurological deficits while reducing hematoma volume in patients, although the long-term functional outcome of this treatment is yet to be explored. The use of edaravone has been associated with high risk of liver failure, thus it is approached with caution of stroke treatment.	1. Lapchak, PA, Araujo, DM. (2007). Advances in hemorrhagic stroke therapy: Conventional and novel approaches. *Exp. Opin. Emerg. Drugs*, 12(3), 389–406. doi:10.1517/14728214.12.3.389 2. Edaravone Acute Infarction Study Group. (2003). Effect of a novel free radical scavenger, edaravone (MCI-186), on acute brain infarction. Randomized, placebo-controlled, double-blind study at multicenters. *Cerebrovasc. Dis.* 15(3): 222–229. 3. Yoshida H, Yanai H, Namiki Y, Fukatsu-Sasaki K, Furutani N, Tada N. (2006). Neuroprotective effects of edaravone: A novel free radical scavenger in cerebrovascular injury. *CNS Drug Rev.*, 12(1): 9–20. 4. Qin M, Feng L, Yang C, Wei D, Li T, Jiang P, Guan J, Zhang X, Shi X, Liang N, Lai X. (2022). Edaravone use in acute intracerebral hemorrhage: A systematic review and meta-analysis of randomized controlled trials. *Front. Pharmacol.*, 13: 935198.

Statins	3-Hydroxy-3-methylglutaryl coenzyme A (HMG-CoA reductase) inhibitors	Atorvastatin improved neurological functional outcome, reducing neuronal loss leading to regional cellular plasticity when administered to ICH induced rats. Although published data by Stroke Prevention by Aggressive Reduction in Cholesterol Levels (SPARCL) study reported an increased incidences of hemorrhagic strokes in patients on statin treatment such as rosuvastatin. Further clinical trials such as statin for neuroprotection in spontaneous intracerebral hemorrhage (STATIC) and statins in intracerbral hemorrhage (SATURN) are ongoing.	1. Seyfried D, Han Y, Lu D, Chen J, Bydon A, Chopp M. (2004). Improvement in neurological outcome after administration of atorvastatin following experimental intracerebral hemorrhage in rats. *J. Neurosurg*. 101(1): 104–107. 2. Jung KH, Chu K, Jeong SW et al. (2004). HMG-CoA reductase inhibitor, atorvastatin, promotes sensorimotor recovery, suppressing acute inflammatory reaction after experimental intracerebral hemorrhage. *Stroke* 35(7): 1744–1749. 3. Vergouwen MD, De Haan RJ, Vermeulen M, Roos YB. (2008). Statin treatment and the occurrence of hemorrhagic stroke in patients with a history of cerebrovascular disease. *Stroke*, 39(2), 497–502.
PP1	Src protein tyrosine kinase inhibitor	In a SD rat model of ICH by injection of autologous blood (50 μL) into striatum, Src antagonist PP1 was reported to decrease neuronal death and glucose metabolic activation. This study also indicated how early activation of SRC kinase leads to phosphorylation of NMDA receptors thus affecting the ICH injury.	1. Ardizzone TD, Zhan X, Ander BP, Sharp FR. (2007). SRC kinase inhibition improves acute outcomes after experimental intracerebral hemorrhage. *Stroke*, 38(5), 1621–1625.
Proteosome inhibitors (bortezomib, PS-519 and UK-279,276)	It degrades the inhibitor of kappaβ (I-κ β) leading to inhibition of NF-κ β, which is involved in expressing inflammatory molecules such as IL-1β, IL-6, TNF-α, E-selectin, and intercellular adhesion molecule-1.	Bortezomib was reported to subside the hematoma growth and volume along with the cerebral edema.	1. Zhang L, Zhang ZG, Liu X, Hozeska A, Stagliano N, Riordan W, Lu, M, Chopp M. (2006). Treatment of embolic stroke in rats with bortezomib and recombinant human tissue plasminogen activator. *Thromb Haemost*, 95(01), 166–173.

(*Continued*)

TABLE 8.1 (*Continued*)
Established and Novel Therapies for Hemorrhagic Stroke

Therapy	Mechanism	Results	References
Poloxamer 188 or P-188 (membrane-active triblock polymer with a formula of poly(ethylene oxide)-poly(propylene oxide)-poly(ethylene oxide))	It is a synthetic surfactant which decreases the blood viscosity	It was administered intracisternally to ICH induced rats, and was reported to reduce the lesion volumes with multiple doses.	1. Watanabe M, Okada T. (2003). Lysophosphatidylcholine-induced myocardial damage is inhibited by pretreatment with poloxamer 188 in isolated rat heart. *Mol. Cell Biochem.* 248(1–2), 209–215.
MDL 101,002	It is a conformationally constrained cyclic analog of PBN (free radical–spin trap compounds)	It has been reported to reduce the hemorrhagic infarction in rabbit thromboembolic model, also in rat collagenase and embolism models.	1. Lapchak, PA. (2002). Hemorrhagic transformation following ischemic stroke: Significance, causes, and relationship to therapy and treatment. *Curr. Neurol. Neurosci. Rep.*, 2(1), 38–43. doi:10.1007/s11910-002-0051-0 2. Asahi M, Asahi K, Wang X, Lo EH. (2000). Reduction of tissue plasminogen activator-induced hemorrhage and brain injury by free radical spin trapping after embolic focal cerebral ischemia in rats. *J. Cereb. Blood Flow Metab.*, 20:452–457. 3. Peeling J, Yan HJ, Chen SG, Campbell M, Del Bigio MR. (1998). Protective effects of free radical inhibitors in intracerebral hemorrhage in rat. *Brain Res.*, 795:63–70. 4. Hu BR, Liu C, Zivin JA. (1999). Reduction of intracerebral hemorrhaging in a rabbit embolic stroke model. *Neurology*, 53:2140–2145.

Cell-based therapy	To assist in regeneration neuronal tissue	In a rodent model of ICH by injection of autologous blood into striatum, intravenous administration of human bone marrow stromal cells improved neurological functions. Further studies also reported that human neural stem cells (NSCs) when administered through i.v. can migrate specifically to the damaged cerebral tissue and lead to behavioral recovery. In a collagenase-induced ICH model, NSCs engineered to overexpress VEGF were transplanted leading to effective neuronal regeneration. A study with immortalized human neural stem cell line derived from the human fetal telencephalon reported similar motor function recovery in ICH models.	1. Seyfried D, Ding J, Han Y, Li Y, Chen J, Chopp M. (2006). Effects of intravenous administration of human bone marrow stromal cells after intracerebral hemorrhage in rats. *J. Neurosurg.* 104(2):313–318. 2. Lee HJ, Kim KS, Park IH, Kim SU. (2007). Human neural stem cells over-expressing VEGF provide neuroprotection, angiogenesis and functional recovery in mouse stroke model. *PLoS One* 2:E156. 3. Lee HJ, Kim KS, Kim EJ et al. (2007). Brain transplantation of immortalized human neural stem cells promotes functional recovery in mouse intracerebral hemorrhage stroke model. *Stem Cells*.
Desmopressin	It is a synthetic form of vasopressin (AVP) and acts as selective agonist of V2 receptors.	In a clinical trial, Improving Platelet Activity for Cerebral Hemorrhage Treatment – DDAVP Proof of Concept (IMPACT), this drug was reported to enhance platelet activity, with significant changes in the hematoma growth. Another trial, Desmopressin for Reversal of Antiplatelet Drugs in Stroke Due to Hemorrhage (DASH) is still ongoing at Phase 2.	1. Zille M, Farr TD, Keep RF, Römer C, Xi, G, Boltze J. (2022). Novel targets, treatments, and advanced models for intracerebral haemorrhage. *EBioMedicine*, 76.
Clevidipine	It is a dihydropyridine L-type calcium channel blocker (antihypertensive)	The trial, Clevidipine in the Treatment of Patients with Acute Hypertension and Intracerebral Hemorrhage (ACCELERATE), reported that the monotherapy of the drug lead to rapid blood pressure reduction in the ICH patients along with small control in hematoma volume. Further clinical comparison of clevidipine with nicardipine (another calcium channel blocker) have reported similar efficacy in reducing the systolic blood pressure by both, whereas nicardipine has been reported to have decrease in rebound hypertension.	1. Graffagnino C, Bergese S, Love J, Schneider D, Lazaridis C, LaPointe M, Lee K, Lynch G, Hu MY, Williams GC. (2013). Clevidipine rapidly and safely reduces blood pressure in acute intracerebral hemorrhage: The ACCELERATE trial. *Cerebrovas. Dis.*, 36(3), 173–180. 2. Saldana S, Breslin J, Hanify J, Heierman T, Larizadeh K, Sanchez M, Phipps W. (2022). Comparison of clevidipine and nicardipine for acute blood pressure reduction in hemorrhagic stroke. *Neurocrit. Care*, 36(3), 983–992.

(*Continued*)

TABLE 8.1 (*Continued*)
Established and Novel Therapies for Hemorrhagic Stroke

Therapy	Mechanism	Results	References
Tranexamic acid (TXA)	It is a synthetic derivative of lysine that binds to the plasminogen, leading inhibition of plasmin formation, which is responsible for stabilization of the fibrin matrix.	In a phase 3 trial of Tranexamic acid for hyperacute primary IntraCerebral Hemorrhage (TICH-2), it was reported to reduce the hematoma expansion in patients along with functional recovery. Although in other trials such as Treatment of Intracerebral Hemorrhage in Patients on Non-vitamin K Antagonist (TICH-NOAC) and The Spot Sign and Tranexamic Acid On Preventing ICH Growth – AUStralasia Trial (STOP-AUST), so evidence of prevention of hematoma expansion was reported. Reports from Tranexamic Acid for Spontaneous Acute Cerebral Hemorrhage Trial (TRANSACT) at phase 3 and stopping hemorrhage with tranexamic acid for hyperacute onset presentation, including mobile stroke units (STOP-MSU) phase 2 are yet to be received.	1. Polymeris AA, Karwacki GM, Siepen BM, Schaedelin S, Tsakiris DA, Stippich C, Guzman R, Nickel CH, Sprigg N, Kägi G, Vehoff J. (2023). Tranexamic acid for intracerebral hemorrhage in patients on non-vitamin K antagonist oral anticoagulants (TICH-NOAC): A multicenter, randomized, placebo-controlled, phase 2 trial. *Stroke*, 54(9), 2223–2234. 2. Zille M, Farr TD, Keep RF, Römer C, Xi G, Boltze J. (2022). Novel targets, treatments, and advanced models for intracerebral haemorrhage. *EBioMed.* 76.
Erythropoietin	An erythropoietic protein	In a collagenase-induced hemorrhage rodent model, recombinant human EPO was reported to improve their functional recovery, reduce hematoma volume, apoptosis, and inflammation while increasing pSTAT3 and pERK levels. Further clinical studies on using EPO for SAH treatment yielded inconclusive results due to restricted population size, along with reports of tumor formation.	1. Lee ST, Chu K, Sinn DI, et al. (2006). Erythropoietin reduces perihematomal inflammation and cell death with eNOS and STAT3 activations in experimental intracerebral hemorrhage. *J. Neurochem.* 96(6): 1728–1739. 2. Souvenir R, Doycheva D, Zhang JH, Tang J. (2015). Erythropoietin in stroke therapy: Friend or foe. *Curr Med Chem.* 22(10):1205–13. doi: 10.2174/0929867322666150114152134

Recombinant activated factor VII (rFVIIa)	It is a recombinant form of activated Factor VII, acting as a rapid procoagulant.	In a clinical trial, it was reported that treating the ICH patient within four hours of its onset with rFVIIa resulted in a reduction in the hematoma growth along with improved functional outcome following 90days, leading to lower mortality (although sporadic incidences of thromboembolism were also reported). This hemostatic therapy eventually could not improve the patient's functional and survival outcomes. This therapy was also employed for a specific patient cohort of acute intracerebral hemorrhage (ICH) with a spot sign on CT angiography (marker for hemorrhage expansion), although the result reported non-significant changes in the clinical outcomes ("Spot Sign" Selection of Intracerebral Hemorrhage to Guide Hemostatic Therapy (SPOTLIGHT); The Spot Sign for Predicting and Treating ICH Growth Study (STOP-IT)).	1. Mayer SA, Brun NC, Begtrup K, Broderick J, Davis S, Diringer MN, Skolnick BE, Steiner T. (2005). Recombinant activated factor VII for acute intracerebral hemorrhage. *New Engl. J. Med.*, 352(8), 777–785. 2. Mayer SA, Brun NC, Begtrup K, et al.; FAST Trial Investigators. (2008). Efficacy and safety of recombinant activated factor VII for acute intracerebral hemorrhage. *N. Engl. J. Med.* 358(20):2127–2137. doi:10.1056/NEJMoa0707534 3. Gladstone DJ, Aviv RI, Demchuk AM, Hill MD, Thorpe KE, Khoury JC, Sucharew HJ, Al-Ajlan F, Butcher K, Dowlatshahi D, Gubitz G. (2019). Effect of recombinant activated coagulation factor VII on hemorrhage expansion among patients with spot sign–positive acute intracerebral hemorrhage: The spotlight and STOP-IT randomized clinical trials. *JAMA Neurol.*, 76(12), 1493–1501.
SM20302	It is a novel nonpeptide GP IIb/IIIa receptor (fibrinogen receptor of platelets) antagonist that hinders the platelet aggregation while maintain vascular integrity and patency.	It is reported to decrease the Incidence of tPA-induced hemorrhage in a large clot embolism model of rabbit at a dose range of 0.1–10 mg/kg.	1. Lapchak, PA. (2002). Hemorrhagic transformation following ischemic stroke: Significance, causes, and relationship to therapy and treatment. *Curr. Neurol. Neurosci. Rep.*, 2(1), 38–43. doi:10.1007/s11910-002-0051-0 2. Rebello SS, Huang J, Shiu WJ, et al. (1998). Pharmacokinetics and pharmacodynamics of SM-20302, a GPIIb/IIIa receptor antagonist, in anesthetized dogs. *J. Cardiovasc. Pharmacol.*, 32:485–494. 3. Lapchak PA, Araujo DM, Song D, Zivin JA. (2001). The nonpeptide GPIIb/IIIa platelet receptor antagonist SM-20302 reduces tissue plasminogen activator (tPA)-induced intracerebral hemorrhage following a thromboembolic stroke. *Stroke*, in press.

(*Continued*)

TABLE 8.1 ***(Continued)***
Established and Novel Therapies for Hemorrhagic Stroke

Therapy	Mechanism	Results	References
Apixaban	Direct inhibitor of factor Xa, thus acting as an anticoagulant	Apixaban has been reported to be used for treating subclinical atrial fibrillation, although the risk of major bleeding has been reported by many studies, such as in the ARISTOTLE Trial (Apixaban for Reduction in Stroke and Other Thromboembolic Events in Atrial Fibrillation). In the Apixaban After Anticoagulation-associated Intracerebral Hemorrhage in Patients With Atrial Fibrillation (APACHE-AF) trial, the drug yielded no significant difference between non-fatal stroke or vascular death among the control and intervention groups.	1. Hylek EM, Held C, Alexander JH, Lopes RD, De Caterina R, Wojdyla DM, Huber K, Jansky P, Steg PG, Hanna M, Thomas L. (2014). Major bleeding in patients with atrial fibrillation receiving apixaban or warfarin: The ARISTOTLE Trial (apixaban for reduction in stroke and other thromboembolic events in atrial fibrillation): Predictors, characteristics, and clinical outcomes. *J. Am. Coll. Cardiol.*, 63(20), 2141–2147. 2. Healey JS, Lopes RD, Granger CB, Alings M, Rivard L, McIntyre WF, Atar D, Birnie DH, Boriani G, Camm AJ, Conen D. (2024). Apixaban for stroke prevention in subclinical atrial fibrillation. *New Engl. J. Med.*, 390(2), 107–117. 3. Zille M, Farr TD, Keep RF, Römer C, Xi G, Boltze J. (2022). Novel targets, treatments, and advanced models for intracerebral haemorrhage. *EBioMedicine*, 76.

8.9 CONCLUSION

Hemorrhagic stroke being one of the most disproportionately lethal pathologies, requires an effective therapeutic strategy to manage and treat it. However, the available therapies are unable to effectively resolve the damage and reduce the mortality that follows post-stroke. This has led to the need for exploration of the disease pathology to find newer and effective therapeutic targets to limit the progression of this pathology. With the recent advancements in clinical trials, creating an effective adjuvant therapy will be possible to reduce the neurological impairment and mortality associated with hemorrhagic stroke.

REFERENCES

1. Feigin VL, Brainin M, Norrving B, et al. World Stroke Organization (WSO): Global Stroke fact sheet 2022. *International Journal of Stroke*. 2022;17(1):18–29.
2. Amarenco P, Bogousslavsky J, Caplan L, Donnan G, Hennerici M. Classification of stroke subtypes. *Cerebrovascular Diseases*. 2009;27(5):493–501.
3. Shi Z. Pathophysiology of hemorrhagic stroke. In P. A. Lapchak & G.-Y. Yang (Eds.), *Translational Research in Stroke*. 2017:77–96. https://doi.org/10.1007/978-981-10-5804-2_5
4. Ye F, Garton HJ, Hua Y, Keep RF, Xi G. The role of thrombin in brain injury after hemorrhagic and ischemic stroke. *Translational Stroke Research*. 2021;12:496–511.
5. Duan M, Xu Y, Li Y, Feng H, Chen Y. Targeting brain-peripheral immune responses for secondary brain injury after ischemic and hemorrhagic stroke. *Journal of Neuroinflammation*. 2024;21(1):102.
6. Thomas SE, Plumber N, Venkatapathappa P, Gorantla V. A review of risk factors and predictors for hemorrhagic transformation in patients with acute ischemic stroke. *International Journal of Vascular Medicine*. 2021;2021(1):4244267.
7. Montaño A, Hanley DF, Hemphill III JC. Hemorrhagic stroke. *Handbook of Clinical Neurology*. 2021;176:229–248.
8. Gruenbaum BF, Kutz R, Zlotnik A, Boyko M. Blood glutamate scavenging as a novel glutamate-based therapeutic approach for post-stroke depression. *Therapeutic Advances in Psychopharmacology*. 2020;10:2045125320903951.
9. Zhang C, Zhong J, Chen W-X, et al. Usage of angiotensin-converting enzyme inhibitor or angiotensin II receptor blocker in hypertension intracerebral hemorrhage. *Neuropsychiatric Disease and Treatment*. 2021:355–363.
10. Ohashi SN, DeLong JH, Kozberg MG, et al. Role of inflammatory processes in hemorrhagic stroke. *Stroke*. 2023;54(2):605–619.
11. Magid-Bernstein J, Girard R, Polster S, et al. Cerebral hemorrhage: Pathophysiology, treatment, and future directions. *Circulation Research*. 2022;130(8):1204–1229.
12. Lee S-H. Introduction on hemorrhagic stroke. *Stroke Revisited: Hemorrhagic Stroke*. 2018:3–6.
13. Badjatia N, Rosand J. Intracerebral hemorrhage. *The Neurologist*. 2005;11(6):311–324.
14. Ikram MA, Wieberdink RG, Koudstaal PJ. International epidemiology of intracerebral hemorrhage. *Current Atherosclerosis Reports*. 2012;14:300–306.
15. Abraham M, Marda M. Cerebrovascular disease. *Essentials of Neuroanesthesia*. 2017:345–366.
16. Gomes J, Wachsman AM. Types of strokes. *Handbook of Clinical Nutrition and Stroke*. 2013:15–31.
17. Abraham MK, Chang W-TW. Subarachnoid hemorrhage. *Emergency Medicine Clinics*. 2016;34(4):901–916.
18. Ballabh P. Pathogenesis and prevention of intraventricular hemorrhage. *Clinics in Perinatology*. 2014;41(1):47–67.
19. An SJ, Kim TJ, Yoon B-W. Epidemiology, risk factors, and clinical features of intracerebral hemorrhage: An update. *Journal of Stroke*. 2017;19(1):3.
20. Wang S, Zou X-L, Wu L-X, et al. Epidemiology of intracerebral hemorrhage: A systematic review and meta-analysis. *Frontiers in Neurology*. 2022;13:915813.
21. Chen S, Feng H, Sherchan P, et al. Controversies and evolving new mechanisms in subarachnoid hemorrhage. *Progress in Neurobiology*. 2014;115:64–91.
22. Chaudhary N, Pandey AS, Wang X, Xi G. Hemorrhagic stroke—Pathomechanisms of injury and therapeutic options. *CNS Neuroscience & Therapeutics*. 2019;25(10):1073.
23. Fang Y, Gao S, Wang X, et al. Programmed cell deaths and potential crosstalk with blood–brain barrier dysfunction after hemorrhagic stroke. *Frontiers in Cellular Neuroscience*. 2020;14:68.

24. Morais Filho AB, Rego TLH, Mendonca LL et al. The physiopathology of spontaneous hemorrhagic stroke: A systematic review. *Reviews in the Neurosciences*. 2021;32(6):631–658.
25. He C, Liu R, Fan Z, et al. Microglia in the pathophysiology of hemorrhagic stroke and the relationship between microglia and pain after stroke: A narrative review. *Pain and Therapy*. 2021;10:927–939.
26. Aronowski J, Zhao X. Molecular pathophysiology of cerebral hemorrhage: Secondary brain injury. *Stroke*. 2011;42(6):1781–1786.
27. Ma Q, Chen S, Hu Q, Feng H, Zhang JH, Tang J. NLRP3 inflammasome contributes to inflammation after intracerebral hemorrhage. *Annals of Neurology*. 2014;75(2):209–219.
28. Sanicola HW, Stewart CE, Luther P, et al. Pathophysiology, management, and therapeutics in subarachnoid hemorrhage and delayed cerebral ischemia: An overview. *Pathophysiology*. 2023;30(3):420–442.
29. Budohoski KP, Guilfoyle M, Helmy A, et al. The pathophysiology and treatment of delayed cerebral ischaemia following subarachnoid haemorrhage. *Journal of Neurology, Neurosurgery & Psychiatry*. 2014;85(12):1343–1353.
30. Provencio JJ, Vora N. *Subarachnoid hemorrhage and inflammation: Bench to bedside and back*. Copyright© 2005 by Thieme Medical Publishers, Inc.; 2005:435–444.
31. Andone S, Bajko Z, Motataianu A, Maier S, Barcutean L, Balasa R. Neuroprotection in stroke—Focus on the renin-angiotensin system: a systematic review. *International Journal of Molecular Sciences*. 2022;23(7):3876.
32. Vasudeva K, Balyan R, Munshi A. ACE-triggered hypertension incites stroke: Genetic, molecular, and therapeutic aspects. *Neuromolecular Medicine*. 2020;22:194–209.
33. Kangussu LM, Marzano LA, Souza CF, Dantas CC, Miranda AS, Simoes e Silva AC. The renin-angiotensin system and the cerebrovascular diseases: Experimental and clinical evidence. *Protein and Peptide Letters*. 2020;27(6):463–475.
34. Atkinson J. Stroke, high blood pressure and the renin–angiotensin–aldosterone system–new developments. *Frontiers in Pharmacology*. 2011;2:22.
35. Zhong P, Lu Z, Li Z, et al. Effect of renin-angiotensin-aldosterone system inhibitors on the rupture risk among hypertensive patients with intracranial aneurysms. *Hypertension*. 2022;79(7):1475–1486.
36. Fitzgerald S. RAAS inhibitors May help protect against brain aneurysm rupture. *Neurology Today*. 2022;22(14):5–6.
37. Xu W, Yang Y-m, Zhu J, et al. Impact of renin–angiotensin–aldosterone-system inhibitor drugs on mortality in patients with atrial fibrillation and hypertension. *BMC Cardiovascular Disorders*. 2022;22(1):141.
38. Zhong P, Lu Z, Li Z, et al. Impact of premorbid hypertension and renin-angiotensin-aldosterone system inhibitors on the severity of aneurysmal subarachnoid haemorrhage: A multicentre study. *Stroke and Vascular Neurology*. 2024; 10(2), e003052. https://doi.org/10.1136/svn-2023-003052.
39. Li W, Li S-G, Li L, et al. Soyasaponin I alleviates hypertensive intracerebral hemorrhage by inhibiting the renin–angiotensin–aldosterone system. *Clinical and Experimental Hypertension*. 2023;45(1):2177667.
40. Verduzco-Mendoza A, Carrillo-Mora P, Avila-Luna A, et al. Role of the dopaminergic system in the striatum and its association with functional recovery or rehabilitation after brain injury. *Frontiers in Neuroscience*. 2021;15:693404.
41. Li M, Zhou L, Sun X, et al. Dopamine, a co-regulatory component, bridges the central nervous system and the immune system. *Biomedicine & Pharmacotherapy*. 2022;145:112458.
42. Baik J-H. Stress and the dopaminergic reward system. *Experimental & Molecular Medicine*. 2020;52(12):1879–1890.
43. Gasiorowska A, Wydrych M, Drapich P, et al. The biology and pathobiology of glutamatergic, cholinergic, and dopaminergic signaling in the aging brain. *Frontiers in Aging Neuroscience*. 2021;13:654931.
44. Liu Y, Xue L, Zhang Y, Xie A. Association between stroke and Parkinson's disease: A meta-analysis. *Journal of Molecular Neuroscience*. 2020;70(8):1169–1176.
45. Joshi H, McIntyre WB, Kooner S, et al. Decreased expression of cerebral dopamine neurotrophic factor in platelets of stroke patients. *Journal of Stroke and Cerebrovascular Diseases*. 2020;29(1):104502.
46. Villa M, Martínez-Vega M, Del Pozo A, et al. The role of the dopamine system in post-stroke mood disorders in newborn rats. *International Journal of Molecular Sciences*. 2023;24(4):3229.
47. Park H-Y, Kim Y, Oh HM, Kim T-W, Park G-Y, Im S. Potential prognostic impact of dopamine receptor D1 (rs4532) polymorphism in post-stroke outcome in the elderly. *Frontiers in Neurology*. 2021;12:675060.
48. Tater P, Pandey S. Post-stroke movement disorders: Clinical spectrum, pathogenesis, and management. *Neurology India*. 2021;69(2):272–283.

49. Kaushik P, Ali M, Tabassum H, Parvez S. Post-ischemic administration of dopamine D2 receptor agonist reduces cell death by activating mitochondrial pathway following ischemic stroke. *Life Sciences*. 2020;261:118349.
50. Urmila A, Rashmi P, Nilam G, Subhash B. Recent advances in the endogenous brain renin-angiotensin system and drugs acting on it. *Journal of the Renin-Angiotensin-Aldosterone System*. 2021;2021:9293553.
51. Gupta S, Khan J, Ghosh S. Molecular mechanism of cognitive impairment associated with Parkinson's disease: A stroke perspective. *Life Science*. 2024;337:122358. doi:10.1016/j.lfs.2023.122358
52. Verduzco-Mendoza A, Carrillo-Mora P, Avila-Luna A, et al. Role of the dopaminergic system in the striatum and its association with functional recovery or rehabilitation after brain injury. *Frontiers in Neuroscience*. 2021;15:693404. doi:10.3389/fnins.2021.693404
53. Mendiola JMF, Arboix A, García-Eroles L, Sánchez-López MJ. Acute spontaneous lobar cerebral hemorrhages present a different clinical profile and a more severe early prognosis than deep subcortical intracerebral hemorrhages – A hospital-based stroke registry study. *Biomedicines*. 2023;11(1). doi:10.3390/biomedicines11010223
54. Chen XG, Shi SY, Hu L, et al. Longitudinal changes in the hypothalamic-pituitary-adrenal axis and sympathetic nervous system are related to the prognosis of stroke. *Frontiers in Neurology*. 2022;13:946593. doi:10.3389/fneur.2022.946593
55. Smith EE. Introduction to focused update vascular cognitive impairment and dementia: Better diagnosis, more avenues for prevention. *Stroke*. 2024;55(4):788–790. doi:10.1161/STROKEAHA.124.046673
56. Ueha R, Cotaoco C, Kondo K, Yamasoba T. Management and treatment for dysphagia in neurodegenerative disorders. *Journal of Clinical Medicine*. 2023;13(1). doi:10.3390/jcm13010156
57. Wu Y, Deng J, Ma J et al. Unraveling the pathogenesis of post-stroke depression in a hemorrhagic mouse model through frontal lobe circuitry and JAK-STAT signaling. *Advanced Science*. 2024:2402152.
58. Freeman ME, Kanyicska B, Lerant A, Nagy G. Prolactin: Structure, function, and regulation of secretion. *Physiological Reviews*. 2000;80(4):1523–631. doi:10.1152/physrev.2000.80.4.1523
59. Pérez-López FR, López-Baena MaT, Pérez-Roncero GR. Prolactin secretion in women: Narrative review. *Gynecology and Pelvic Medicine*. 2021;4:1–10.
60. Castillo X, Ortiz G, Arnold E, et al. The influence of the prolactin/vasoinhibin axis on post-stroke lesion volume, astrogliosis, and survival. *Journal of Neuroendocrinology*. 2024:e13415. doi:10.1111/jne.13415
61. Ramos-Martinez E, Ramos-Martínez I, Molina-Salinas G, Zepeda-Ruiz WA, Cerbon M. The role of prolactin in central nervous system inflammation. *Reviews in the Neuroscience*. 2021;32(3):323–340. doi:10.1515/revneuro-2020-0082
62. Çöven I, Kırcelli A, Duman E, Pınar HU, Basaran B. High prolactin level as a predictor of vasospasm in aneurysmal subarachnoidal hemorrhage. *Medical Science Monitor*. 2017;23:3831–3836. doi:10.12659/msm.906010
63. Ramos-Martinez E, Ramos-Martínez I, Molina-Salinas G, Zepeda-Ruiz WA, Cerbon M. The role of prolactin in central nervous system inflammation. *Reviews in the Neurosciences*. 2021;32(3):323–340. doi:10.1515/revneuro-2020-0082
64. Borba VV, Zandman-Goddard G, Shoenfeld Y. Prolactin and Autoimmunity. *Frontiers in Immunology*. 2018;9:73. doi:10.3389/fimmu.2018.00073
65. Papazoglou AS, Leite AR, Moysidis DV, et al. Serum prolactin levels and mortality in adults without prolactinoma: A meta-analysis. *Journal of Clinical Endocrinology & Metabolism*. 2024;109(8):e1652–e1664. doi:10.1210/clinem/dgae087
66. Yousefvand S, Hadjzadeh MA, Vafaee F, Dolatshad H. The protective effects of prolactin on brain injury. *Life Sci*. 2020;263:118547. doi:10.1016/j.lfs.2020.118547
67. Csikós V, Oláh S, Dóra F, Arrasz N, Cservenák M, Dobolyi A. Microglia depletion prevents lactation by inhibition of prolactin secretion. *iScience*. 2023;26(3):106264. doi:10.1016/j.isci.2023.106264
68. Yousefvand S, Hadjzadeh MA, Keshavarzi Z, et al. Effects of prolactin on movement disorders and APOE, GFAP, and PRL receptor gene expression following intracerebral hemorrhage in rats. *Iranian Journal of Basic Medical Science*. 2021;24(12):1709–1716. doi:10.22038/ijbms.2021.58176.12927
69. Costanza M, Pedotti R. Prolactin: Friend or foe in central nervous system autoimmune inflammation? *International Journal of Molecular Science*. 2016;17(12). doi:10.3390/ijms17122026
70. Suzuki H, Kawakita F, Asada R, Nakano F, Nishikawa H, Fujimoto M. Old but still hot target, glutamate-mediated neurotoxicity in stroke. *Translational Stroke Research*. 2021;10(28):13. doi:10.1007/s12975-021-00958-6
71. Gruenbaum BF, Zlotnik A, Frenkel A, Fleidervish I, Boyko M. Glutamate efflux across the blood-brain barrier: New perspectives on the relationship between depression and the glutamatergic system. *Metabolites*. 2022;12(5). doi:10.3390/metabo12050459

72. Pal MM. Glutamate: The master neurotransmitter and its implications in chronic stress and mood disorders. *Frontiers in Human Neuroscience*. 2021;15:722323. doi:10.3389/fnhum.2021.722323
73. Gruenbaum BF, Kutz R, Zlotnik A, Boyko M. Blood glutamate scavenging as a novel glutamate-based therapeutic approach for post-stroke depression. *Therapeutic Advances in Psychopharmacology*. 2020;10:2045125320903951. doi:10.1177/2045125320903951
74. Martín A, Domercq M, Matute C. Inflammation in stroke: The role of cholinergic, purinergic and glutamatergic signaling. *Therapeutic Advances in Neurological Disorders*. 2018;11:1756286418774267. doi:10.1177/1756286418774267
75. Bai W, Li W, Ning YL, et al. Blood glutamate levels are closely related to acute lung injury and prognosis after stroke. *Frontiers in Neurology*. 2017;8:755. doi:10.3389/fneur.2017.00755
76. Morales TI, Stearns-Yoder KA, Hoffberg AS, Khan TK, Wortzel H, Brenner LA. Interactions of glutamate and gamma amino butyric acid with the insulin-like growth factor system in traumatic brain injury (TBI) and/or cardiovascular accidents (CVA or stroke): A systematic review. *Heliyon*. 2022;8(3):e09037. doi:10.1016/j.heliyon.2022.e09037
77. Cha H, Choi JH, Jeon H, et al. Aquaporin-4 deficiency is associated with cognitive impairment and alterations in astrocyte-neuron lactate shuttle. *Molecular Neurobiology*. 2023;60(11):6212–6226. doi:10.1007/s12035-023-03475-9
78. Kaplan-Arabaci O, Acari A, Ciftci P, Gozuacik D. Glutamate scavenging as a neuroreparative strategy in ischemic stroke. *Frontiers in Pharmacology*. 2022;13:866738. doi:10.3389/fphar.2022.866738
79. Deng S, Chen X, Lei Q, Lu W. AQP2 promotes astrocyte activation by modulating the TLR4/NFκB-p65 pathway following intracerebral hemorrhage. *Frontiers in Immunology*. 2022;13:847360. doi:10.3389/fimmu.2022.847360
80. Visse R, Nagase H. Matrix metalloproteinases and tissue inhibitors of metalloproteinases: Structure, function, and biochemistry. *Circulation Research*. 2003;92(8):827–39. doi:10.1161/01.Res.0000070112.80711.3d
81. Liu Y, Mu Y, Li Z, Yong VW, Xue M. Extracellular matrix metalloproteinase inducer in brain ischemia and intracerebral hemorrhage. *Front Immunol*. 2022;13:986469. doi:10.3389/fimmu.2022.986469
82. Lattanzi S, Di Napoli M, Ricci S, Divani AA. Matrix metalloproteinases in acute intracerebral hemorrhage. *Neurotherapeutics*. 2020;17(2):484–496. doi:10.1007/s13311-020-00839-0
83. Morancho A, Rosell A, García-Bonilla L, Montaner J. Metalloproteinase and stroke infarct size: Role for anti-inflammatory treatment? *Annals of the New York Academy of Sciences*. 2010;1207:123–33. doi:10.1111/j.1749-6632.2010.05734.x
84. Lo EH, Wang X, Cuzner ML. Extracellular proteolysis in brain injury and inflammation: Role for plasminogen activators and matrix metalloproteinases. *Journal of Neuroscience Research*. 2002;69(1):1–9. doi:10.1002/jnr.10270
85. Yang Y, Estrada EY, Thompson JF, Liu W, Rosenberg GA. Matrix metalloproteinase-mediated disruption of tight junction proteins in cerebral vessels is reversed by synthetic matrix metalloproteinase inhibitor in focal ischemia in rat. *Journal of Cerebral Blood Flow & Metabolism*. 2007;27(4):697–709. doi:10.1038/sj.jcbfm.9600375
86. Lu J, Wang J, Lin Z, et al. MMP-9 as a biomarker for predicting hemorrhagic strokes in moyamoya disease. *Frontiers in Neurology*. 2021;12:721118. doi:10.3389/fneur.2021.721118
87. Mechtouff L, Bochaton T, Paccalet A, et al. Matrix metalloproteinase-9 relationship with infarct growth and hemorrhagic transformation in the era of thrombectomy. *Frontiers in Neurology*. 2020;11:473. doi:10.3389/fneur.2020.00473
88. Veeravalli KK. Implications of MMP-12 in the pathophysiology of ischaemic stroke. *Stroke and Vascular Neurology*. 2024;9(2):97–107. doi:10.1136/svn-2023-002363
89. Clément T, Rodriguez-Grande B, Badaut J. Aquaporins in brain edema. *Journal of Neuroscience Research*. 2020;98(1):9–18. doi:10.1002/jnr.24354
90. Badaut J, Fukuda AM, Jullienne A, Petry KG. Aquaporin and brain diseases. *Biochimica et Biophysica Acta*. 2014;1840(5):1554–65. doi:10.1016/j.bbagen.2013.10.032
91. Shi ZF, Fang Q, Chen Y et al. Methylene blue ameliorates brain edema in rats with experimental ischemic stroke via inhibiting aquaporin 4 expression. *Acta Pharmacologica Sinica*. 2021;42(3):382–392. doi:10.1038/s41401-020-0468-5
92. Bernardo-Castro S, Sousa JA, Brás A, et al. Pathophysiology of blood-brain barrier permeability throughout the different stages of ischemic stroke and its implication on hemorrhagic transformation and recovery. *Frontiers in Neurology*. 2020;11:594672. doi:10.3389/fneur.2020.594672

93. Ramiro L, Simats A, Penalba A, et al. Circulating aquaporin-4 as a biomarker of early neurological improvement in stroke patients: A pilot study. *Neuroscience Letters*. 2020;714:134580. doi:10.1016/j.neulet.2019.134580
94. Liu E, Peng X, Ma H, et al. The involvement of aquaporin-4 in the interstitial fluid drainage impairment following subarachnoid hemorrhage. *Frontiers in Aging Neuroscience*. 2020;12:611494. doi:10.3389/fnagi.2020.611494
95. Badaut J, Brunet JF, Grollimund L, et al. Aquaporin 1 and aquaporin 4 expression in human brain after subarachnoid hemorrhage and in peritumoral tissue. *Acta Neurochirurgica Supplement*. 2003;86:495–8. doi:10.1007/978-3-7091-0651-8_101
96. Wang Y, Huang C, Guo Q, Chu H. Aquaporin-4 and cognitive disorders. *Aging and Disease*. 2022;13(1):61–72. doi:10.14336/ad.2021.0731
97. Fang Y, Dai S, Jin C, et al. Aquaporin-4 polymorphisms are associated with cognitive performance in Parkinson's disease. *Frontiers in Aging Neuroscience*. 2021;13:740491. doi:10.3389/fnagi.2021.740491
98. Badaut J, Ashwal S, Obenaus A. Aquaporins in cerebrovascular disease: A target for treatment of brain edema. *Cerebrovascular Disease*. 2011;31(6):521–31. doi:10.1159/000324328
99. Ye F, Garton HJL, Hua Y, Keep RF, Xi G. The role of thrombin in brain injury after hemorrhagic and ischemic stroke. *Translational Stroke Research*. 2021;12(3):496–511. doi:10.1007/s12975-020-00855-4
100. Lee J, Mun S, Park A, et al. Proteomics reveals plasma biomarkers for ischemic stroke related to the coagulation cascade. *Journal of Molecular Neuroscience*. 2020;70(9):1321–1331. doi:10.1007/s12031-020-01545-4
101. Cognasse F, Duchez AC, Audoux E, et al. Platelets as key factors in inflammation: Focus on CD40L/CD40. *Frontiers in Immunology*. 2022;13:825892. doi:10.3389/fimmu.2022.825892
102. Cheng Y, Xi G, Jin H, Keep RF, Feng J, Hua Y. Thrombin-induced cerebral hemorrhage: Role of protease-activated receptor-1. *Translational Stroke Research*. 2014;5(4):472–5. doi:10.1007/s12975-013-0288-8
103. Rawish E, Nording H, Münte T, Langer HF. Platelets as mediators of neuroinflammation and thrombosis. *Frontiers in Immunology*. 2020;11:548631. doi:10.3389/fimmu.2020.548631
104. Krenzlin H, Gresser E, Jussen D, et al. The cerebral thrombin system is activated after intracerebral hemorrhage and contributes to secondary lesion growth and poor neurological outcome in C57Bl/6 mice. *Journal of Neurotrauma*. 2020;37(12):1481–1490. doi:10.1089/neu.2019.6582.
105. Pleşeru AM, Mihailă RG. The role of thrombin in central nervous system activity and stroke. *Clujul Medical*. 2018;91(4):368–371. doi:10.15386/cjmed-973
106. Shlobin NA, Har-Even M, Itsekson-Hayosh Z, Harnof S, Pick CG. Role of thrombin in central nervous system injury and disease. *Biomolecules*. 2021;11(4). doi:10.3390/biom11040562
107. Huang P, He XY, Xu M. Protease-activated receptor 1 inhibitor improves brain edema in rats with intracerebral hemorrhage. *Bratislavské Leárske Listy*. 2020;121(8):600–604. doi:10.4149/bll_2020_099
108. Magid-Bernstein J, Girard R, Polster S, et al. Cerebral hemorrhage: Pathophysiology, treatment, and future directions. *Circulation Research*. 2022;130(8):1204–1229. doi:10.1161/circresaha.121.319949
109. Tadi P, Lui F. Acute stroke. In *StatPearls*. StatPearls Publishing LLC.; 2024.
110. Vasudeva K, Balyan R, Munshi A. ACE-triggered hypertension incites stroke: Genetic, molecular, and therapeutic aspects. *Neuromolecular Medicine*. 2020;22(2):194–209. doi:10.1007/s12017-019-08583-1
111. Hernandez-Guillamon M, Delgado P, Penalba A, et al. Plasma β-amyloid levels in cerebral amyloid angiopathy-associated hemorrhagic stroke. *Neurodegenerative Disease*. 2012;10(1–4):320–3. doi:10.1159/000333811
112. Qureshi AI, Palesch YY, Martin R, et al. Effect of systolic blood pressure reduction on hematoma expansion, perihematomal edema, and 3-month outcome among patients with intracerebral hemorrhage: Results from the antihypertensive treatment of acute cerebral hemorrhage study. *Archives of Neurology*. 2010;67(5):570–6. doi:10.1001/archneurol.2010.61
113. Simats A, Liesz A. Systemic inflammation after stroke: Implications for post-stroke comorbidities. *EMBO Molecular Medicine*. 2022;14(9):e16269. doi:10.15252/emmm.202216269
114. Thomas SE, Plumber N, Venkatapathappa P, Gorantla V. A Review of risk factors and predictors for hemorrhagic transformation in patients with acute ischemic stroke. *International Journal of Vascular Medicine*. 2021;2021:4244267. doi:10.1155/2021/4244267
115. Wong YS, Tsai CF, Ong CT. Risk factors for stroke recurrence in patients with hemorrhagic stroke. *Scientific Reports*. 2022;12(1):17151. doi:10.1038/s41598-022-22090-7

116. Stankovic S, Majkic-Singh N. Genetic aspects of ischemic stroke: Coagulation, homocysteine, and lipoprotein metabolism as potential risk factors. *Critical Reviews in Clinical Laboratory Science.* 2010;47(2):72–123. doi:10.3109/10408361003791520
117. Jin X, Chen H, Shi H, et al. Lipid levels and the risk of hemorrhagic stroke: A dose-response meta-analysis. *Nutrition Metabolism and Cardiovasc Disease.* 2021;31(1):23–35. doi:10.1016/j.numecd.2020.10.014
118. Almuklass AM, Alawad YA, Alanazi AS, et al. Hyperlipidemia and hypertension are associated with intracerebral hemorrhage incidence: A retrospective study. *Cureus.* 2023;15(1):e33236. doi:10.7759/cureus.33236
119. Tschoe C, Bushnell CD, Duncan PW, Alexander-Miller MA, Wolfe SQ. Neuroinflammation after intracerebral hemorrhage and potential therapeutic targets. *Journal of Stroke.* 2020;22(1):29–46. doi:10.5853/jos.2019.02236
120. Alsbrook DL, Di Napoli M, Bhatia K, et al. Neuroinflammation in acute ischemic and hemorrhagic stroke. *Current Neurology and Neuroscience Reports.* 2023;23(8):407–431. doi:10.1007/s11910-023-01282-2
121. Jiang C, Wang Y, Hu Q, et al. Immune changes in peripheral blood and hematoma of patients with intracerebral hemorrhage. *The FASEB Journal.* 2020;34(2):2774–2791. doi:10.1096/fj.201902478R
122. Zhou Y, Wang Y, Wang J, Anne Stetler R, Yang QW. Inflammation in intracerebral hemorrhage: From mechanisms to clinical translation. *Progress in Neurobiology.* 2014;115:25–44. doi:10.1016/j.pneurobio.2013.11.003
123. Galea J, Ogungbenro K, Hulme S, et al. Reduction of inflammation after administration of interleukin-1 receptor antagonist following aneurysmal subarachnoid hemorrhage: Results of the subcutaneous interleukin-1Ra in SAH (SCIL-SAH) study. *Journal of Neurosurgery.* 2018;128(2):515–523. doi:10.3171/2016.9.Jns16615
124. Zhu H, Wang Z, Yu J, et al. Role and mechanisms of cytokines in the secondary brain injury after intracerebral hemorrhage. *Progress in Neurobiology.* 2019;178:101610. doi:10.1016/j.pneurobio.2019.03.003
125. Kinter KJ, Alfaro R, Kinter C, et al. The effects of body mass index on in-hospital mortality following first ischemic or hemorrhagic stroke events: Does the "obesity paradox" apply? *Annals of Medicine and Surgery (London).* 2021;70:102839. doi:10.1016/j.amsu.2021.102839
126. Forlivesi S, Cappellari M, Bonetti B. Obesity paradox and stroke: A narrative review. *Eating and Weight Disorders.* 2021;26(2):417–423. doi:10.1007/s40519-020-00876-w
127. Kumral E, Erdoğan CE, Arı A, Bayam FE, Saruhan G. Association of obesity with recurrent stroke and cardiovascular events. *Revue Neurologique (Paris).* 2021;177(4):414–421. doi:10.1016/j.neurol.2020.06.019
128. Shiozawa M, Kaneko H, Itoh H, et al. Association of body mass index with ischemic and hemorrhagic stroke. *Nutrients.* 2021;13(7). doi:10.3390/nu13072343
129. Chen CJ, Ding D, Derdeyn CP, et al. Brain arteriovenous malformations: A review of natural history, pathobiology, and interventions. *Neurology.* 2020;95(20):917–927. doi:10.1212/wnl.0000000000010968
130. Jäkel L, De Kort AM, Klijn CJM, Schreuder F, Verbeek MM. Prevalence of cerebral amyloid angiopathy: A systematic review and meta-analysis. *Alzheimer's & Dementia.* 2022;18(1):10–28. doi:10.1002/alz.12366
131. Sudarshan VK, Raghavendra U, Gudigar A, et al. Assessment of CT for the categorization of hemorrhagic stroke (HS) and cerebral amyloid angiopathy hemorrhage (CAAH): A review. *Biocybernetics and Biomedical Engineering.* 2022;42(3):888–901.
132. Seiffge DJ, Polymeris AA, Law ZK, et al. Cerebral amyloid angiopathy and the risk of hematoma expansion. *Annals of Neurology.* 2022;92(6):921–930. doi:10.1002/ana.26481
133. Smith EE, Charidimou A, Ayata C, Werring DJ, Greenberg SM. Cerebral amyloid angiopathy-related transient focal neurologic episodes. *Neurology.* 2021;97(5):231–238. doi:10.1212/wnl.0000000000012234
134. Lee S, Kim Y, Navi BB, et al. Risk of intracranial hemorrhage associated with pregnancy in women with cerebral arteriovenous malformations. *Journal of NeuroInterventional Surgery.* 2021;13(8):707–710. doi:10.1136/neurintsurg-2020-016838
135. El-Abtah ME, Petitt JC, Kashkoush A, Achey R, Bain MD, Moore NZ. Endovascular management of arteriovenous malformation–associated intracranial aneurysms: A systematic literature review. *World Neurosurgery.* 2022;164:257–269.
136. Shaligram SS, Winkler E, Cooke D, Su H. Risk factors for hemorrhage of brain arteriovenous malformation. *CNS Neuroscience & Therapeutics.* 2019;25(10):1085–1095. doi:10.1111/cns.13200
137. Rutledge C, Cooke DL, Hetts SW, Abla AA. Brain arteriovenous malformations. *Handbook of Clinical Neurology.* 2021;176:171–178. doi:10.1016/b978-0-444-64034-5.00020-1

138. Pan P, Weinsheimer S, Cooke D, et al. Review of treatment and therapeutic targets in brain arteriovenous malformation. *Journal Cerebral Blood Flow & Metabolism*. 2021;41(12):3141–3156. doi:10.1177/0271678x211026771
139. Hakimi R, Garg A. Imaging of hemorrhagic stroke. *CONTINUUM: Lifelong Learning in Neurology*. 2016;22(5):1424–1450.
140. Linn J, Brückmann H. Differential diagnosis of nontraumatic intracerebral hemorrhage. *Klinische Neuroradiologie*. 2009;19(1):45–61. doi:10.1007/s00062-009-8036-x
141. Wardlaw JM, Mair G, von Kummer R, et al. Accuracy of automated computer-aided diagnosis for stroke imaging: A critical evaluation of current evidence. *Stroke*. 2022;53(7):2393–2403. doi:10.1161/strokeaha.121.036204
142. Montaño A, Hanley DF, Hemphill JC, 3rd. Hemorrhagic stroke. *Handbook of Clinical Neurology*. 2021;176:229–248. doi:10.1016/b978-0-444-64034-5.00019-5
143. Kothari RU, Brott T, Broderick JP, et al. The ABCs of measuring intracerebral hemorrhage volumes. *Stroke*. 1996;27(8):1304–5. doi:10.1161/01.str.27.8.1304
144. Goldstein JN, Fazen LE, Snider R, et al. Contrast extravasation on CT angiography predicts hematoma expansion in intracerebral hemorrhage. *Neurology*. 2007;68(12):889–94. doi:10.1212/01.wnl.0000257087.22852.21
145. Hallevi H, Dar NS, Barreto AD, et al. The IVH score: A novel tool for estimating intraventricular hemorrhage volume: Clinical and research implications. *Critical Care Medicine*. 2009;37(3):969–74. doi:10.1097/CCM.0b013e318198683a
146. Delgado Almandoz JE, Schaefer PW, Goldstein JN, et al. Practical scoring system for the identification of patients with intracerebral hemorrhage at highest risk of harboring an underlying vascular etiology: The Secondary Intracerebral Hemorrhage Score. *AJNR Americal Journal Neuroradiology*. 2010;31(9):1653–60. doi:10.3174/ajnr.A2156
147. Morgan TC, Dawson J, Spengler D, et al. The Modified Graeb Score: An enhanced tool for intraventricular hemorrhage measurement and prediction of functional outcome. *Stroke*. 2013;44(3):635–41. doi:10.1161/strokeaha.112.670653
148. Wijman CA, Venkatasubramanian C, Bruins S, Fischbein N, Schwartz N. Utility of early MRI in the diagnosis and management of acute spontaneous intracerebral hemorrhage. *Cerebrovascular Disease*. 2010;30(5):456–63. doi:10.1159/000316892
149. Wada R, Aviv RI, Fox AJ, et al. CT angiography "spot sign" predicts hematoma expansion in acute intracerebral hemorrhage. *Stroke*. 2007;38(4):1257–62. doi:10.1161/01.STR.0000259633.59404.f3
150. Qureshi AI, Palesch YY. Antihypertensive treatment of acute cerebral hemorrhage (ATACH) II: Design, methods, and rationale. *Neurocritical Care*. 2011;15(3):559–576. doi:10.1007/s12028-011-9538-3
151. Anderson CS, Huang Y, Wang JG, et al. Intensive blood pressure reduction in acute cerebral haemorrhage trial (INTERACT): A randomised pilot trial. *The Lancet Neurology*. 2008;7(5):391–9. doi:10.1016/s1474-4422(08)70069-3
152. Mendelow AD, Gregson BA, Fernandes HM, et al. Early surgery versus initial conservative treatment in patients with spontaneous supratentorial intracerebral haematomas in the International Surgical Trial in Intracerebral Haemorrhage (STICH): A randomised trial. *The Lancet*. 2005;365(9457):387–397.
153. Hemphill IIIJC, Greenberg SM, Anderson CS, et al. Guidelines for the management of spontaneous intracerebral hemorrhage: A guideline for healthcare professionals from the American Heart Association/American Stroke Association. *Stroke*. 2015;46(7):2032–2060.
154. Mayer SA, Brun NC, Begtrup K, et al. Efficacy and safety of recombinant activated factor VII for acute intracerebral hemorrhage. *New England Journal of Medicine*. 2008;358(20):2127–2137.
155. Takeuchi S, Wada K, Nagatani K, Otani N, Mori K. Decompressive hemicraniectomy for spontaneous intracerebral hemorrhage. *Neurosurgical Focus*. 2013;34(5):E5. doi:10.3171/2013.2.Focus12424
156. Kellner CP, Connolly ES Jr. Neuroprotective strategies for intracerebral hemorrhage: Trials and translation. *Stroke*. 2010;41(10 Suppl):S99–102. doi:10.1161/strokeaha.110.597476.
157. Hattori N, Katayama Y, Maya Y, Gatherer A. Impact of stereotactic hematoma evacuation on activities of daily living during the chronic period following spontaneous putaminal hemorrhage: A randomized study. *Journal of Neurosurgery*. 2004;101(3):417–20. doi:10.3171/jns.2004.101.3.0417
158. Mould WA, Carhuapoma JR, Muschelli J, et al. Minimally invasive surgery plus recombinant tissue-type plasminogen activator for intracerebral hemorrhage evacuation decreases perihematomal edema. *Stroke*. 2013;44(3):627–634.

159. Morgan T, Awad I, Keyl P, Lane K, Hanley D. Preliminary report of the clot lysis evaluating accelerated resolution of intraventricular hemorrhage (CLEAR-IVH) clinical trial. *Acta Neurochirurgica Supplement*. 2008;105:217–20. doi:10.1007/978-3-211-09469-3_41
160. Moussa WM, Khedr W. Decompressive craniectomy and expansive duraplasty with evacuation of hypertensive intracerebral hematoma, a randomized controlled trial. *Neurosurgical Review*. 2017;40(1):115–127. doi:10.1007/s10143-016-0743-6
161. Laskowitz DT, Kolls BJ. Neuroprotection in subarachnoid hemorrhage. *Stroke*. 2010;41(10 Suppl):S79–S84. doi:10.1161/strokeaha.110.595090

9 Obesity and Its Impact on Brain Health

Irma Wati Ngadimon and Alina Arulsamy

9.1 INTRODUCTION

Obesity, classified by the World Health Organization as a body mass index (BMI) of 30 or higher, has emerged as a global health crisis impacting over one billion individuals as of 2023.[1] Its prevalence has surged in regions, such as the Middle East, Pacific Islands, and Southeast Asia,[2] fueled by urbanization, dietary shifts, and increasingly sedentary lifestyles. While obesity is widely recognized as a precursor to metabolic and cardiovascular diseases, its significant association with neurological disorders highlights its multidimensional threat to public health.[3] The economic implications of obesity are substantial, consuming 2.19% of global GDP in 2019, with costs projected to exceed $4 trillion annually by 2035 due to healthcare expenditures and productivity losses.[2]

Recent research has underscored obesity's critical role in increasing the pathogenesis of neurodegenerative diseases, such as AD, Parkinson's disease, and vascular dementia. These associations extend beyond general metabolic dysfunction, implicating chronic inflammation, insulin resistance, and oxidative stress originating from dysfunctional adipose tissue.[4,5] Obesity also contributes to a range of neuropsychological issues, including depression, anxiety, and cognitive decline, adding complexity to its mental health impact.[6,7] While earlier studies focused predominantly on obesity's metabolic consequences, recent work highlights its neurological impact, including neuroinflammation, hormonal dysregulation, and the role of the gut-brain axis in amplifying oxidative stress and neuroinflammatory responses.[8] These findings indicate that the effects of obesity on brain health are both profound and multifaceted, driven by underlying pathological mechanisms associated with adipose tissue dysfunction.

This chapter examines the critical mechanisms through which obesity contributes to neurological impairment. We will explore the mechanisms through which obesity drives neurological impairment, focusing on chronic inflammation, insulin resistance, and the gut-brain axis. Understanding these interactions is essential for developing targeted interventions to improve both the physical and mental health outcomes of individuals affected by this pervasive condition.

9.2 ADIPOSE TISSUE: AN INFLAMMATORY POWERHOUSE

9.2.1 Adipose Tissue as an Active Endocrine Organ

Adipose tissue, traditionally viewed as a passive fat storage site, is now recognized as a dynamic endocrine organ, particularly in the context of obesity. Obesity triggers significant hypertrophy (enlargement) and hyperplasia (increase in number) of fat cells, stimulating the secretion of pro-inflammatory cytokines, including tumor necrosis factor (TNF) and interleukin-6 (IL-6). These cytokines fuel a state of persistent, low-grade inflammation known as "metaflammation."[9] Importantly, this inflammatory response extends systemically, affecting multiple organ systems, including the central nervous system (CNS).[10] Visceral adipose tissue (VAT), surrounding internal organs, is particularly implicated in this process, producing inflammatory cytokines, such as C-reactive protein (CRP), IL-6, and TNF. The accumulation of VAT in obesity heightens widespread inflammation and impairs metabolic function, contributing to reduced insulin sensitivity and a cascade of adverse effects.[11] Dysfunctional adipose tissue also disrupts the balance of adipokines,

DOI: 10.1201/9781003500025-9

increasing leptin levels while decreasing adiponectin, an anti-inflammatory hormone, thus worsening the inflammatory environment.[12]

9.2.2 Chronic Inflammation and Brain Health

Chronic inflammation originating from dysfunctional adipose tissue has profound implications for brain health. Pro-inflammatory cytokines such as TNF and IL-6 can cross the blood-brain barrier (BBB) and initiate neuroinflammatory processes. Research consistently shows that systemic inflammation is often elevated in obesity, correlates with cognitive decline, and an increased risk of neurodegenerative diseases. Elevated systemic inflammatory markers like CRP and IL-6 have been directly linked to reductions in gray matter volume, particularly in brain regions critical for memory and executive functions, such as the hippocampus and prefrontal cortex.[13,14]

Studies emphasize that dysregulated white adipose tissue (WAT), characterized by hypertrophy and inflammation, significantly contributes to neuroinflammation and impaired brain perfusion. This is associated with heightened risks of cognitive impairment.[15] Visceral fat accumulation and elevated inflammatory markers like CRP and IL-6 further mediate the adverse effects of obesity on cognitive function.[16]

9.2.3 Disruption of the Blood-Brain Barrier

Chronic inflammation also compromises the integrity of the BBB, a protective layer regulating the passage of substances between the bloodstream and the brain. Studies show that individuals with obesity and heightened systemic inflammation display increased BBB permeability, particularly in areas like the hippocampus. This increased permeability allows pro-inflammatory molecules, immune cells, and reactive oxygen species (ROS) to invade the brain, exacerbating neuroinflammation and accelerating cognitive decline.[17,18]

The breakdown of the BBB is a key mechanism through which inflammation-induced adipose tissue dysfunction promotes the development of neurodegenerative diseases, such as Alzheimer's disease (AD). This disruption highlights the critical role of inflammation in linking obesity to impaired brain health.[18,19]

9.2.4 Preclinical and Clinical Studies on Adipose Tissue and Brain Health

Insights from both preclinical and clinical studies have significantly advanced our understanding of the relationship between adipose tissue dysfunction and brain health. Preclinical models have shown that targeting adipose tissue inflammation can reduce neuroinflammatory responses and improve cognitive outcomes. A 2022 systematic review identified various anti-inflammatory agents—both natural and synthetic—that effectively modulate inflammation in adipose tissue, offering promising therapeutic strategies to mitigate the neurological effects of obesity.[19]

Clinical studies provide evidence that targeting adipose tissue inflammation can positively influence brain health. For instance, weight loss interventions, such as bariatric surgery, have been associated with reductions in systemic inflammation and improvements in cognitive function. Patients undergoing bariatric surgery exhibited decreased levels of inflammatory markers and enhanced cognitive performance post-surgery. Additionally, pharmacological treatments aimed at reducing adipose tissue inflammation have shown promise.[20] Similarly, pharmacological treatments like GLP-1 receptor agonists, initially developed for diabetes management, have shown anti-inflammatory effects that extend to brain health by reducing systemic inflammation and improving metabolic parameters.[21] These findings underscore the potential of targeting adipose tissue inflammation as a viable approach to preserving brain health in the context of obesity.

Lifestyle interventions also play a critical role. Caloric restriction and physical exercise have been shown to reduce adipose tissue inflammation, subsequently decreasing neuroinflammation and enhancing cognitive function.[22] A 2019 systematic review investigating the impact of whole foods and dietary patterns on inflammation in overweight and obese adults found that certain dietary interventions can significantly decrease markers of subclinical inflammation, potentially improving cognitive outcomes.[23] Moreover, exercise has been shown to reduce inflammation in adipose tissue. Additionally, exercise has proven effective in reducing adipose tissue inflammation and improving insulin sensitivity, as evidenced in studies involving elderly populations.[24]

9.3 THE ROLE OF INSULIN RESISTANCE IN BRAIN DYSFUNCTION

9.3.1 Insulin Resistance: Bridging Peripheral and Brain Health

Insulin resistance, a hallmark of type 2 diabetes mellitus (T2DM), affects not only peripheral tissues like the liver and skeletal muscles but also impairs the brain's ability to regulate glucose. Given the brain's reliance on glucose as its primary energy source, disruptions in insulin signaling can lead to significant metabolic dysfunctions. Recent studies suggest that brain insulin resistance substantially contributes to cognitive decline, particularly in individuals with metabolic disorders like obesity and T2DM.[25,26]

While insulin facilitates glucose uptake in peripheral tissues, its role in the brain is more complex, involving the regulation of synaptic plasticity, neuronal survival, and memory processes. The occurrence of insulin resistance within the brain results in diminished glucose uptake, leading to energy deficits in neurons and subsequent cognitive impairments. Disrupted insulin signaling within the brain diminishes glucose uptake, leading to energy shortages that impair neuronal function. These metabolic dysfunctions are often observed in early stages of neurodegenerative diseases, particularly AD.[27] This disruption is a precursor to synaptic dysfunction, neuronal loss, and the eventual clinical manifestations of cognitive decline.

9.3.2 Impaired Neuronal Function and Synaptic Plasticity

Synaptic plasticity, the brain's ability of synapses to alter their strength in response to activity—is essential for the brain's information processing and memory retention. In the hippocampus, a region critical for learning and memory, insulin resistance disrupts synaptic plasticity by impairing long-term potentiation (LTP), a process fundamental for memory formation. Insulin signaling facilitates LTP through the activation of pathways, such as PI3K-Akt, which are crucial for neuronal survival and synapse formation.[28]

Experimental studies further underscore the link between insulin resistance and cognitive dysfunction.[29] Animal models of T2DM demonstrate significant impairments in hippocampal synaptic plasticity, accompanied by diminished neuronal resilience. In these models, insulin resistance weakened the protective effects of insulin on synapses, exacerbating learning and memory deficits.[30]

9.3.3 The Concept of "Type 3 Diabetes"

The term "Type 3 Diabetes" has been proposed to characterize AD as a neuroendocrine disorder marked by insulin resistance and impaired glucose metabolism in the brain. This concept highlights the overlap between T2DM and AD, underscoring the shared mechanisms of metabolic dysfunction and neurodegeneration.[31,32]

In AD, brain insulin resistance contributes to the accumulation of amyloid-beta and hyperphosphorylated tau proteins, which form neurotoxic plaques and tangles. Disrupted insulin signaling exacerbates Aβ production and tau phosphorylation, further accelerating neurodegeneration. Mitochondrial dysfunction, oxidative stress, and neuroinflammation exacerbate these pathological processes. This interplay initiates a self-reinforcing cycle, ultimately contributing to progressive neuronal damage and cognitive decline.[33] Disrupted insulin signaling exacerbates tau phosphorylation and increases amyloid-beta production, accelerating neurodegeneration. Additionally, insulin resistance may impair key pathways, such as PI3K-Akt, leading to downstream effects that promote cognitive decline.[25]

Preclinical research has demonstrated that restoring insulin signaling, for example, through intranasal insulin administration, can mitigate cognitive decline, reinforcing the link between insulin dysregulation and AD. The overlap between T2DM and AD suggests that therapeutic strategies targeting insulin signaling pathways could offer significant benefits for individuals at risk of both conditions.[34]

From impairing synaptic plasticity to driving neuroinflammation and protein aggregation, insulin resistance is central to the pathogenesis of AD and other neurodegenerative disorders. The concept of "Type 3 Diabetes" underscores the critical role insulin plays in brain health, linking metabolic disorders like T2DM to cognitive decline. Emerging therapies that target insulin signaling hold promise for mitigating the neurodegenerative processes associated with insulin resistance.

9.4 NEUROINFLAMMATION AND OXIDATIVE STRESS: THE DOUBLE THREAT

9.4.1 Systemic Inflammation Fueling Neuroinflammation

Systemic inflammation, often linked to obesity and metabolic imbalances, has far-reaching effects, extending into the CNS and triggering neuroinflammation. A key factor in this process is the overactivation of microglial cells, the brain's immune defenders. Under normal conditions, microglia play a critical role in clearing debris and supporting neuronal function. However, prolonged systemic inflammation can push microglia into an overactive state, leading to the release of inflammatory molecules like IL-6, TNF, and ROS. This persistent inflammatory state is a major contributor to neurodegenerative diseases like AD and Parkinson's disease (PD).[35,36]

When chronically activated, microglia lose their protective function and begin to exacerbate neuronal damage. In AD, for example, these overactive cells struggle to clear amyloid-beta plaques, intensifying neuroinflammation. Mitochondrial dysfunction further exacerbates this process by increasing the production of ROS, creating a vicious cycle of oxidative stress and inflammation. Consequently, this cascade accelerates neuronal death and cognitive decline.[37]

9.4.2 Oxidative Stress: A Compounding Factor

Oxidative stress, deeply tied to neuroinflammation, further damages neurons by overwhelming the brain's antioxidant defenses. Excessive ROS production leads to the oxidation of critical cellular components, such as DNA, proteins, and lipids. This is especially harmful in obesity and metabolic conditions, where excess fat tissue amplifies oxidative stress, leaving neurons more vulnerable to damage. The brain's high lipid content and metabolic demands make it particularly susceptible to these effects.[38]

Mitochondria are central to this process, serving as both the primary source of ROS and key targets of oxidative damage. When mitochondria are compromised, their capacity to produce ATP—the energy essential for cellular functions—is diminished, resulting in impaired neuronal function. In neurodegenerative diseases, mitochondrial dysfunction leads to an increased production of ROS, further damaging neurons.[39] Additionally, ROS function as signaling molecules that activate microglia, which perpetuates the inflammatory cycle and exacerbates the overall condition of the brain.[40]

9.4.3 The Cumulative Effect on Brain Aging

The combined effects of chronic neuroinflammation and oxidative stress significantly accelerate brain aging and contribute to the onset of neurodegenerative diseases, such as AD and PD.

Persistent oxidative damage and inflammation disrupt synaptic plasticity, impair neuronal function, and lead to the accumulation of misfolded proteins like amyloid-beta and tau. These processes, central to AD are worsened by mitochondrial dysfunction, which hampers the brain's ability to recover and repair.[41]

Over time, these factors lead to structural changes in the brain, such as thinning of the cortex and shrinkage of the hippocampus, both of which are linked to cognitive decline. However, research suggests that targeting oxidative stress and neuroinflammation may hold promise for slowing brain aging. Therapeutic approaches, including antioxidant and anti-inflammatory treatments, are being investigated to maintain cognitive performance and possibly slow the advancement of neurodegenerative conditions.[42]

9.5 GUT-BRAIN AXIS: A PATHWAY TO NEUROLOGICAL IMPACT

9.5.1 Understanding the Gut-Brain Connection

Obesity has a profound effect on gut health, disrupting the delicate balance of microorganisms in the gut, known as the gut microbiota. This disruption, called dysbiosis, weakens the intestinal barrier—a condition often referred to as "leaky gut." When this barrier becomes compromised, harmful substances like lipopolysaccharides (LPS) can escape into the bloodstream, triggering widespread inflammation. This inflammation doesn't just stay in the body; it can cross the BBB and contribute to neuroinflammation, which plays a significant role in conditions like AD and PD.[43,44]

The effects of gut dysbiosis in obesity go beyond inflammation. It can disrupt the communication between the gut and the brain through a network of immune, endocrine, and neural pathways known as the gut-brain axis (GBA). This disruption has been closely linked to cognitive decline and mental health issues. For instance, research shows that a weakened gut barrier in people with mild cognitive impairment (MCI) and AD is associated with higher levels of inflammation and noticeable cognitive challenges.[45]

Additionally, dysbiosis can affect mood and emotional well-being by altering the production of neurotransmitters, which may explain the higher prevalence of anxiety and depression among individuals with obesity.[46]

Encouragingly, there are ways to address these issues. Strategies like incorporating probiotics, prebiotics, and adopting a healthier diet have shown promise in restoring balance to the gut microbiota and strengthening the intestinal barrier. These interventions can help reduce inflammation and improve both cognitive function and mental health.[47] Understanding the link between obesity, gut health, and brain function highlights the importance of looking at the body as a whole when tackling the challenges of obesity-related brain health issues.

9.6 OBESITY AND RISK TOWARD NEURODEGENERATIVE DISEASE

As discussed previously, the link between obesity and neurological health involves multiple pathophysiological pathways. Obesity induces chronic inflammation in adipose tissue, insulin resistance, oxidative stress, BBB disruption, and gut-brain axis dysregulation, leading to neurological alterations. These changes manifest as cognitive decline, increased risk of neurodegenerative diseases such as Alzheimer's and Parkinson's, and heightened prevalence of mental health issues like anxiety and depression, as depicted in Figure 9.1. Recognizing these connections is essential for developing targeted interventions to mitigate obesity's impact on brain health.

9.6.1 Pathological Impact on Alzheimer's Disease

Obesity is a significant risk factor for AD, driving pathological mechanisms, such as insulin resistance, chronic neuroinflammation, and oxidative stress. Insulin resistance in obesity disrupts

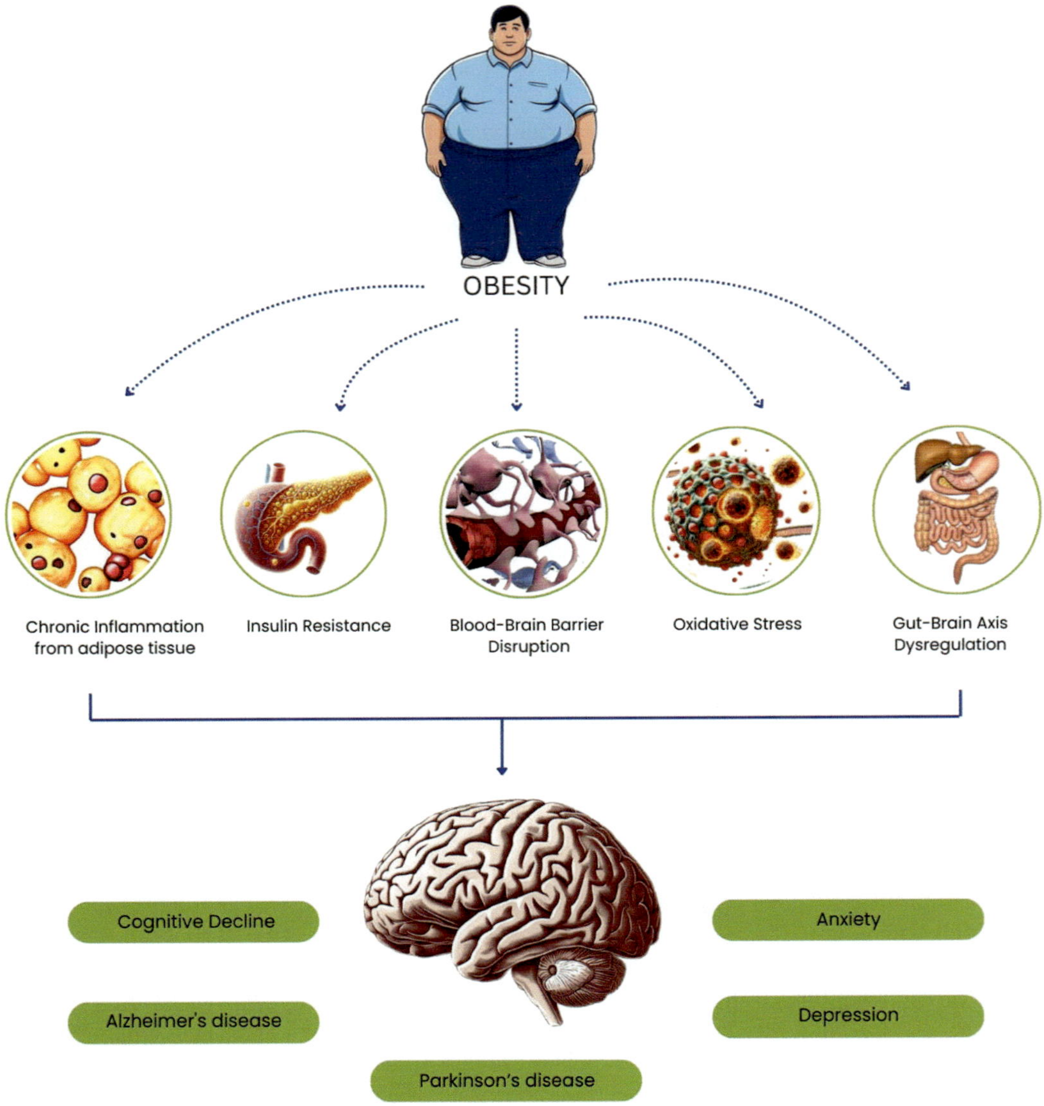

FIGURE 9.1 Pathophysiological pathways linking obesity to neurological outcomes. The multifaceted mechanisms through which obesity exacerbates neurologic disorders. Central Obesity triggers systemic changes, including chronic inflammation derived from adipose tissue, insulin resistance, disruption of the BBB, oxidative stress, and gut-brain axis dysregulation. These alterations lead to a spectrum of neurological consequences, ranging from cognitive decline and major neurodegenerative diseases such as Alzheimer's and Parkinson's disease to mental health disorders like anxiety and depression.

Source: Created using Canva Pro.

neuronal glucose metabolism, resulting in energy deficits that worsen the build-up of Aβ plaques and tau tangles, which are characteristic features of AD. These processes contribute to mitochondrial dysfunction, further impairing neuronal health.[48,49]

A recent study involving 1,736 Asian participants found that late-life obesity correlated with lower Aβ accumulation, while being underweight significantly increased the risk of Aβ burden. These findings highlight the complex metabolic interplay underlying AD pathology, indicating that both extremes of body weight can influence disease progression.[50]

Chronic inflammation caused by dysfunctional adipose tissue, activates the NLRP3 inflammasome, leading to the release of pro-inflammatory cytokines like IL-1β and IL-18. This inflammatory response disrupts insulin signaling, setting off a continuous cycle of neuroinflammation and neurodegeneration. This cycle progressively damages synapses, which accelerates the progression of AD.[51,52]

9.6.2 Cerebrovascular Impacts of Obesity on Cognitive Decline and Mental Health

Obesity significantly increases the risk of vascular dementia, largely due to its effects on cerebrovascular health. Excess adipose tissue promotes vascular dysfunction, including hypertension, atherosclerosis, and endothelial damage, all of which impair blood flow to the brain. This results in reduced oxygen and nutrient delivery, exacerbating oxidative stress and accelerating neuronal damage.[53]

Arterial stiffness, a hallmark of vascular dysfunction in obesity, further impairs cerebral autoregulation. This reduced vascular elasticity limits the brain's ability to adapt to metabolic demands, promoting ischemic injury and cognitive deficits.[54] Additionally, obesity is linked to cerebral small vessel disease (CSVD), a condition characterized by microvascular damage that amplifies risks for both vascular dementia and AD.[55]

9.6.3 Obesity and Neuropsychiatric Disorder

Obesity contributes to mental health disorders, such as depression and anxiety, largely driven by chronic inflammation and neurochemical imbalances. Elevated levels of pro-inflammatory cytokines, such as IL-6 and TNF, cross the BBB, disrupting emotional regulation and impairing brain regions like the hippocampus and amygdala.[56] This inflammation significantly impacts mood by affecting brain regions involved in emotional regulation.

Neurochemical changes in obesity, particularly reductions in serotonin and disruptions in dopamine regulation, further exacerbate mood disorders. Reduced serotonin levels are associated with depression, while altered dopamine signaling diminishes motivation and increases stress and anxiety.[57]

Interventions targeting inflammation and metabolic health, such as anti-inflammatory diets and physical activity, have shown promise in improving these mental health conditions.

9.6.4 Parkinson's Disease and Other Neurodegenerative Disorders

Although PD is distinct from AD in etiology, both disorders share overlapping mechanisms of inflammation and oxidative stress, particularly in the context of obesity. Obesity-induced oxidative stress and chronic inflammation activate microglial cells in the brain, leading to the release of reactive ROS and pro-inflammatory cytokines. These processes contribute to the degeneration of dopaminergic neurons in the substantia nigra, a hallmark of PD.[58]

Recent studies emphasize how obesity and metabolic conditions, including T2DM, activate the NF-κB/NLRP3 inflammasome pathway, promoting neuronal apoptosis and intensifying neuroinflammation.[59] Mitochondrial dysfunction further compounds oxidative damage, playing a critical role in neurodegeneration in both PD and other neurodegenerative diseases.[60]

9.7 CONCLUSION

Obesity drives neurodegenerative diseases such as Alzheimer's and Parkinson's through interconnected mechanisms, including insulin resistance, chronic inflammation, oxidative stress, and gut-brain axis dysfunction. These processes collectively damage neurons, disrupt synaptic function, and impair cognitive abilities, increasing susceptibility to conditions like vascular dementia.

Early interventions, such as healthy diets and physical activity, alongside emerging treatments like anti-inflammatory therapies, insulin-sensitizing agents, and gut microbiome interventions, show promise in mitigating these effects. Future research should prioritize personalized medicine and biomarker diagnostics to develop tailored strategies for managing obesity's impact on brain health.

REFERENCES

1. World Health Organization (WHO). *Obesity and overweight.* 2024. https://www.who.int/news-room/fact-sheets/detail/obesity-and-overweight
2. World Obesity Federation. *World Obesity Atlas.* 2023. https://data.worldobesity.org/publications/WOF-Obesity-Atlas-V5.pdf
3. Tahmi M, Palta P, Luchsinger JA. Metabolic syndrome and cognitive function. *Curr Cardiol Rep.* 2021;23(12):180. doi:10.1007/s11886-021-01615-y.
4. Morys F, Potvin O, Zeighami Y, et al. Obesity-associated neurodegeneration pattern mimics Alzheimer's disease in an observational cohort study. *J Alzheimer's Dis.* 2022;91(3):1059–1071. doi:10.3233/JAD-220535.
5. Hayes JP, Moody JN, Roca JG, Hayes SM, Initiative ADN. Body mass index is associated with smaller medial temporal lobe volume in those at risk for Alzheimer's disease. *NeuroImage Clin.* 2020;25:102156.
6. Russo C, Valle MS, Russo A, Malaguarnera L. The interplay between ghrelin and microglia in neuroinflammation: Implications for obesity and neurodegenerative diseases. *Int J Mol Sci.* 2022;23(21):13432.
7. Romain AJ, Marleau J, Baillot A. Association between physical multimorbidity, body mass index and mental health/disorders in a representative sample of people with obesity. *J Epidemiol Community Health.* 2019;73(9):874–880. doi:10.1136/jech-2018-211497.
8. Le Thuc O, García-Cáceres C. Obesity-induced inflammation: Connecting the periphery to the brain. *Nat Metab.* 2024;6(7):1237–1252. doi:10.1038/s42255-024-01079-8.
9. Torres S, Fabersani E, Marquez A, Gauffin-Cano P. Adipose tissue inflammation and metabolic syndrome. The proactive role of probiotics. *Eur J Nutr.* 2019;58(1):27–43. doi:10.1007/s00394-018-1790-2.
10. Schleh MW, Caslin HL, Garcia JN, et al. Metaflammation in obesity and its therapeutic targeting. *Sci Transl Med.* 2024;15(723):eadf9382. doi:10.1126/scitranslmed.adf9382.
11. Moreira VC, Silva CMS, Welker AF, da Silva ICR. Visceral adipose tissue influence on health problem development and its relationship with serum biochemical parameters in middle-aged and older adults: A literature review. *J Aging Res.* 2022;2022:8350527. doi:10.1155/2022/8350527.
12. Kolb H. Obese visceral fat tissue inflammation: From protective to detrimental? *BMC Med.* 2022;20(1):494. doi:10.1186/s12916-022-02672-y.
13. Karlsson IK, Zhan Y, Wang Y, et al. Adiposity and the risk of dementia: Mediating effects from inflammation and lipid levels. *Eur J Epidemiol.* 2022;37(12):1261–1271. doi:10.1007/s10654-022-00918-w.
14. Morys F, Dadar M, Dagher A. Association between midlife obesity and its metabolic consequences, cerebrovascular disease, and cognitive decline. *J Clin Endocrinol Metab.* 2021;106(10):e4260–e4274. doi:10.1210/clinem/dgab135.
15. Vreeken D, Seidel F, de La Roij G, et al. Impact of white adipose tissue on brain structure, perfusion, and cognitive function in patients with severe obesity. *Neurology.* 2023;100(7):e703–e718. doi:10.1212/WNL.0000000000201538.
16. Cannavale CN, Bailey M, Edwards CG, et al. Systemic inflammation mediates the negative relationship between visceral adiposity and cognitive control. *Int J Psychophysiol.* 2021;165:68–75. doi:10.1016/j.ijpsycho.2021.03.010.
17. Feng Z, Fang C, Ma Y, Chang J. Obesity-induced blood-brain barrier dysfunction: Phenotypes and mechanisms. *J Neuroinflammation.* 2024;21(1):110. doi:10.1186/s12974-024-03104-9.
18. Boleti AP de A, Cardoso PH de O, Frihling BEF, Silva PS e, Moraes LFRN de, Migliolo L. Adipose tissue, systematic inflammation, and neurodegenerative diseases. *Neural Regen Res.* 2023;18(1):38–46. https://journals.lww.com/nrronline/fulltext/2023/01000/adipose_tissue,_systematic_inflammation,_and.5.aspx
19. da Cruz Nascimento SS, Carvalho de Queiroz JL, Fernandes de Medeiros A, et al. Anti-inflammatory agents as modulators of the inflammation in adipose tissue: A systematic review. *PLoS One.* 2022;17(9):e0273942. https://doi.org/10.1371/journal.pone.0273942
20. Hathaway I, Momodu OE, Williams DM, Beamish AJ, Barry J, Stephens JW. Changes in cognitive function following bariatric surgery: An updated systematic review. *Obes Surg.* 2024;34(6):2216–2226. doi:10.1007/s11695-024-07235-z.

21. Chen B, Yu X, Horvath-Diano C, et al. GLP-1 programs the neurovascular landscape. *Cell Metab.* 2024;36(10):2173–2189. doi:10.1016/j.cmet.2024.09.003.
22. Henn RE, Elzinga SE, Glass E, et al. Obesity-induced neuroinflammation and cognitive impairment in young adult versus middle-aged mice. *Immun Ageing.* 2022;19(1):67. doi:10.1186/s12979-022-00323-7.
23. Cowan SF, Leeming ER, Sinclair A, Dordevic AL, Truby H, Gibson SJ. Effect of whole foods and dietary patterns on markers of subclinical inflammation in weight-stable overweight and obese adults: A systematic review. *Nutr Rev.* 2020;78(1):19–38. doi:10.1093/nutrit/nuz030.
24. Čížková T, Štěpán M, Daďová K, et al. Exercise training reduces inflammation of adipose tissue in the elderly: Cross-sectional and randomized interventional trial. *J Clin Endocrinol Metab.* 2020;105(12):e45 10–e4526. doi:10.1210/clinem/dgaa630.
25. Berlanga-Acosta J, Guillén-Nieto GE, Rodríguez-Rodríguez N, et al. Insulin resistance at the crossroad of Alzheimer disease pathology: A review. *Front Endocrinol (Lausanne).* 2020;11. https://api.semantic-scholar.org/CorpusID:226247232
26. Zhang S, Zhang Y, Wen Z, et al. Cognitive dysfunction in diabetes: Abnormal glucose metabolic regulation in the brain. *Front Endocrinol (Lausanne).* 2023;14. https://api.semanticscholar.org/CorpusID:259303056
27. Kumar V, Kim S-H, Bishayee K. Dysfunctional glucose metabolism in Alzheimer's disease onset and potential pharmacological interventions. *Int J Mol Sci.* 2022;23(17). doi:10.3390/ijms23179540.
28. Sharma M, Yadav Y, Dey CS. Neuronal insulin signaling and resistance: A balancing act of kinases and phosphatases. *J Endocrinol.* 2023. https://api.semanticscholar.org/CorpusID:265032585
29. Potenza MA, Sgarra L, Desantis V, Nacci C, Montagnani M. Diabetes and Alzheimer's disease: Might mitochondrial dysfunction help deciphering the common path? *Antioxidants.* 2021;10. https://api.semanticscholar.org/CorpusID:237321590
30. Franklin W, Krishnan B, Taglialatela G. Chronic synaptic insulin resistance after traumatic brain injury abolishes insulin protection from amyloid beta and tau oligomer-induced synaptic dysfunction. *Sci Rep.* 2019;9(1):8228. doi:10.1038/s41598-019-44635-z.
31. Cai Z, Xiao M, Chang L, Yan L-J. Role of insulin resistance in Alzheimer's disease. *Metab Brain Dis.* 2015;30(4):839–851. doi:10.1007/s11011-014-9631-3.
32. Wei Z, Koya J, Reznik SE. Insulin resistance exacerbates Alzheimer disease via multiple mechanisms. *Front Neurosci.* 2021;15:687157. doi:10.3389/fnins.2021.687157.
33. Singh DD, Shati AA, Alfaifi MY, et al. Development of dementia in type 2 diabetes patients: Mechanisms of insulin resistance and antidiabetic drug development. *Cells.* 2022;11(23). doi:10.3390/cells11233767.
34. Lanzillotta C, Tramutola A, Di Giacomo G, et al. Insulin resistance, oxidative stress and mitochondrial defects in Ts65dn mice brain: A harmful synergistic path in Down syndrome. *Free Radic Biol Med.* 2021;165:152–170. doi:10.1016/j.freeradbiomed.2021.01.042.
35. Afridi R, Lee W-H, Suk K. Microglia gone awry: Linking immunometabolism to neurodegeneration. *Front Cell Neurosci.* 2020;14:246.
36. Silva DF, Empadinhas N, Cardoso SM, Esteves AR. Neurodegenerative microbially-shaped diseases: Oxidative stress meets neuroinflammation. *Antioxidants.* 2022;11(11):2141.
37. Kaur K, Narang RK, Singh S. Role of Nrf2 in oxidative stress, neuroinflammation and autophagy in Alzheimer's disease: Regulation of Nrf2 by different signaling pathways. *Curr Mol Med.* 2025;25(4):372–387.
38. Simpson DSA, Oliver PL. ROS generation in microglia: Understanding oxidative stress and inflammation in neurodegenerative disease. *Antioxidants.* 2020;9(8):743.
39. Rey F, Ottolenghi S, Zuccotti GV, Samaja M, Carelli S. Mitochondrial dysfunctions in neurodegenerative diseases: Role in disease pathogenesis, strategies for analysis and therapeutic prospects. *Neural Regen Res.* 2022;17(4):754–758. doi:10.4103/1673-5374.322430.
40. Ragupathy H, Vukku M, Barodia SK. Cell-type-specific mitochondrial quality control in the brain: A plausible mechanism of neurodegeneration. *Int J Mol Sci.* 2023;24(19):14421.
41. Nasi M, De Gaetano A, Bianchini E, et al. Mitochondrial damage-associated molecular patterns stimulate reactive oxygen species production in human microglia. *Mol Cell Neurosci.* 2020;108:103538.
42. Teleanu DM, Niculescu A-G, Lungu II, et al. An overview of oxidative stress, neuroinflammation, and neurodegenerative diseases. *Int J Mol Sci.* 2022;23(11). doi:10.3390/ijms23115938.
43. Kesika P, Suganthy N, Sivamaruthi BS, Chaiyasut C. Role of gut-brain axis, gut microbial composition, and probiotic intervention in Alzheimer's disease. *Life Sci.* 2021;264:118627. doi:10.1016/j.lfs.2020.118627.
44. Das TK, Ganesh BP. Interlink between the gut microbiota and inflammation in the context of oxidative stress in Alzheimer's disease progression. *Gut Microbes.* 2023;15(1). doi:10.1080/19490976.2023.2206504.

45. Pei Y, Lu Y, Li HZ, Jiang CY, Wang L. Gut microbiota and intestinal barrier function in subjects with cognitive impairments: A cross-sectional study. *Front Aging Neurosci.* 2023;15(June):1–13. doi:10.3389/fnagi.2023.1174599.
46. Misiak B, Łoniewski I, Marlicz W, et al. The HPA axis dysregulation in severe mental illness: Can we shift the blame to gut microbiota? *Prog Neuro-Psychopharmacology Biol Psychiatry.* 2020;102:109951. doi:10.1016/j.pnpbp.2020.109951.
47. Wijdeveld M, Nieuwdorp M, IJzerman R. The interaction between microbiome and host central nervous system: The gut-brain axis as a potential new therapeutic target in the treatment of obesity and cardiometabolic disease. *Expert Opin Ther Targets.* 2020;24(7):639–653. doi:10.1080/14728222.2020.1761958.
48. Terzo S, Amato A, Mulè F. From obesity to Alzheimer's disease through insulin resistance. *J Diabetes Complications.* 2021;35(11):108026. doi:10.1016/j.jdiacomp.2021.108026.
49. Barone E, Di Domenico F, Perluigi M, Butterfield DA. The interplay among oxidative stress, brain insulin resistance and AMPK dysfunction contribute to neurodegeneration in type 2 diabetes and Alzheimer disease. *Free Radic Biol Med.* 2021;1760016–33. doi:10.1016/j.freeradbiomed.2021.09.006.
50. Lee EH, Yoo H, Kim YJ et al. Different associations between body mass index and Alzheimer's markers depending on metabolic health. *Alzheimers Res Ther.* 2024;16(1):194. doi:10.1186/s13195-024-01563-z.
51. Litwiniuk A, Bik W, Kalisz M, Baranowska-Bik A. Inflammasome NLRP3 potentially links obesity-associated low-grade systemic inflammation and insulin resistance with Alzheimer's disease. *Int J Mol Sci.* 2021;22(11). doi:10.3390/ijms22115603.
52. Helgudóttir SS, Mørkholt AS, Lichota J, et al. Rethinking neurodegenerative diseases: Neurometabolic concept linking lipid oxidation to diseases in the central nervous system. *Neural Regen Res.* 2024;19(7). https://journals.lww.com/nrronline/fulltext/2024/07000/rethinking_neurodegenerative_diseases_.21.aspx
53. Khan MSH, Hegde V. Obesity and diabetes mediated chronic inflammation: A potential biomarker in Alzheimer's disease. *J Pers Med.* 2020;10(2). doi:10.3390/jpm10020042.
54. Hayden MR. Overview and new insights into the metabolic syndrome: Risk factors and emerging variables in the development of type 2 diabetes and cerebrocardiovascular disease. *Med.* 2023;59(3). doi:10.3390/medicina59030561.
55. Mengr A, Strnadová V, Strnad Š, et al. Feeding high-fat diet accelerates development of peripheral and central insulin resistance and inflammation and worsens AD-like pathology in APP/PS1 mice. *Nutrients.* 2023;15(17). doi:10.3390/nu15173690.
56. Tasnim N, Khan N, Gupta A, et al. Exploring the effects of adiponectin and leptin in correlating obesity with cognitive decline: A systematic review. *Ann Med Surg.* 2023;85(6). https://journals.lww.com/annals-of-medicine-and-surgery/fulltext/2023/06000/exploring_the_effects_of_adiponectin_and_leptin_in.93.aspx
57. Humińska-Lisowska K. Dopamine in sports: A narrative review on the genetic and epigenetic factors shaping personality and athletic performance. *Int J Mol Sci.* 2024;25(21). doi:10.3390/ijms252111602.
58. Bittencourt A, Brum PO, Ribeiro CT, et al. High fat diet-induced obesity causes a reduction in brain tyrosine hydroxylase levels and non-motor features in rats through metabolic dysfunction, neuroinflammation and oxidative stress. *Nutr Neurosci.* 2022;25(5):1026–1040. doi:10.1080/1028415X.2020.1831261.
59. Alrouji M, Al-kuraishy HM, Al-Gareeb AI, et al. NF-κB/NLRP3 inflammasome axis and risk of Parkinson's disease in type 2 diabetes mellitus: A narrative review and new perspective. *J Cell Mol Med.* 2023;27(13):1775–1789. doi:10.1111/jcmm.17784.
60. Schmitt LO, Gaspar JM. Obesity-induced brain neuroinflammatory and mitochondrial changes. *Metabolites.* 2023;13(1). doi:10.3390/metabo13010086.

10 Metabolic Syndrome
A Gateway to Neurological Disorders

Kamlesh Kumar Pandey, Kamakshi Mehta, Bhaskar Sharma, Yogesh Sharma, Nikhil Maheshwari, and Prabhakar Singh

10.1 INTRODUCTION

MetS comprises a set of various interrelated risk factors that significantly increase the risk of developing chronic conditions like cardiovascular disease (CVD), type 2 diabetes, and stroke (Hayden 2023). Over the past 30 years, there has been an increase in the number of people suffering from neurological conditions. According to Ding et al., more than three billion people experienced neurological conditions in 2021 (Ding et al. 2022). The impact of neurological disorders is highest in Africa and most countries in Southeast Asia. The syndrome is diagnosed based on at least three out of five key factors: Abdominal obesity, insulin resistance, hypertension, hyperglycaemia, and dyslipidaemia (Shi et al. 2023). MetS reflects a systemic metabolic disturbance that affects numerous organs, and its rising prevalence (Figure 10.1) worldwide is a growing public health concern.

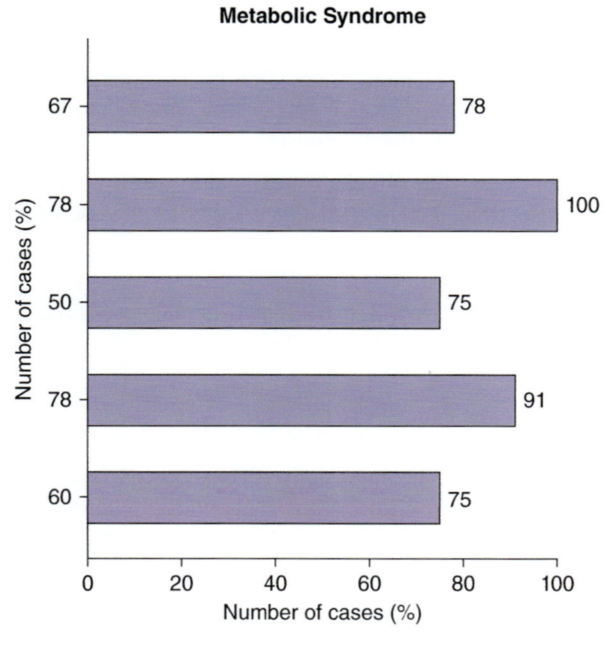

FIGURE 10.1 The bar graph highlights the prevalence of obesity (Riaz et al. 2024), hyperlipidaemia (Jamali et al. 2024; Eckel 2007), hypertension (Basu et al. 2023), and diabetes (Alawdi et al. 2024) reported among a total number of metabolic syndrome cases.

DOI: 10.1201/9781003500025-10

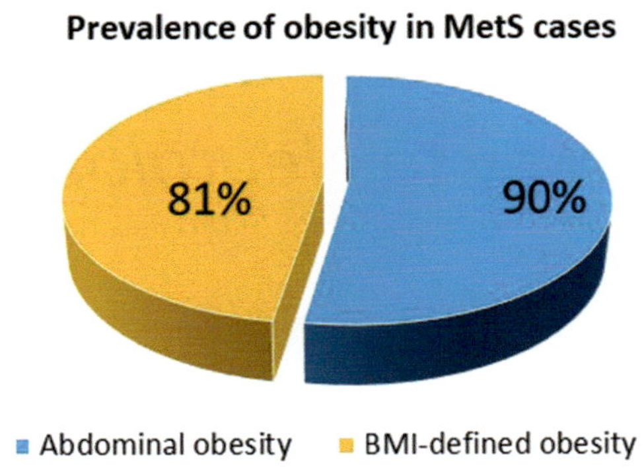

FIGURE 10.2 The graph shows the prevalence of abdominal (90%) and BMI-defined (81%) obesity, specifically within MetS cases reported in recent studies (Engin 2024; Swarup et al. 2025).

10.1.1 Criteria of MetS

The MetS is characterised by five clinical criteria used for diagnostic purposes, as explained below:

10.1.1.1 Abdominal Obesity

Central or visceral obesity, typically measured by waist circumference, is a critical marker for MetS. Unlike general obesity, central obesity is a stronger predictor of metabolic dysfunction due to its association with adipose tissue surrounding internal organs. This type of fat is particularly harmful as it promotes inflammation, oxidative stress, and hormone dysregulation, contributing to insulin resistance (Angelico et al. 2023). The prevalence of abdominal and BMI-defined obesity in individuals with MetS is given in Figure 10.2.

Criteria: Waist circumference thresholds vary by ethnicity and region, e.g.,

Men: ≥102 cm (40 inches) in the United States, ≥90 cm (35 inches) in Asian populations.
Women: ≥88 cm (35 inches) in the United States, ≥80 cm (31 inches) in Asian populations.

10.1.1.2 Insulin Resistance and Hyperglycaemia

Insulin resistance is the reduced ability of cells to respond to insulin, which is critical for glucose uptake and metabolism. Over time, insulin resistance leads to chronic hyperglycaemia (elevated blood sugar levels), which can progress to type 2 diabetes. Figure 10.3 shows the prevalence of insulin resistance and hyperglycaemia.

Criteria: Fasting glucose levels ≥100 mg/dL (5.6 mmol/L) or diagnosed type 2 diabetes. Alternatively, a glucose tolerance test or elevated insulin levels can be used to assess insulin resistance.

10.1.1.3 Hypertension

Elevated blood pressure contributes to the development of CVDs by placing excess strain on the heart and damaging blood vessels. Long-standing hypertension can lead to complications like atherosclerosis, kidney damage, and heart failure (Masenga and Kirabo 2023). Figure 10.4 depicts the prevalence of hypertension cases reported among individuals with MetS.

Criteria: Blood pressure ≥130/85 mmHg or current use of antihypertensive medication. Dyslipidaemia – elevated triglycerides.

10.1.1.4 Hyperlipidaemia

High levels of triglycerides, a type of fat found in the blood, are a significant risk factor for CVD. They contribute to the formation of atherosclerotic plaques, which can clog arteries and reduce

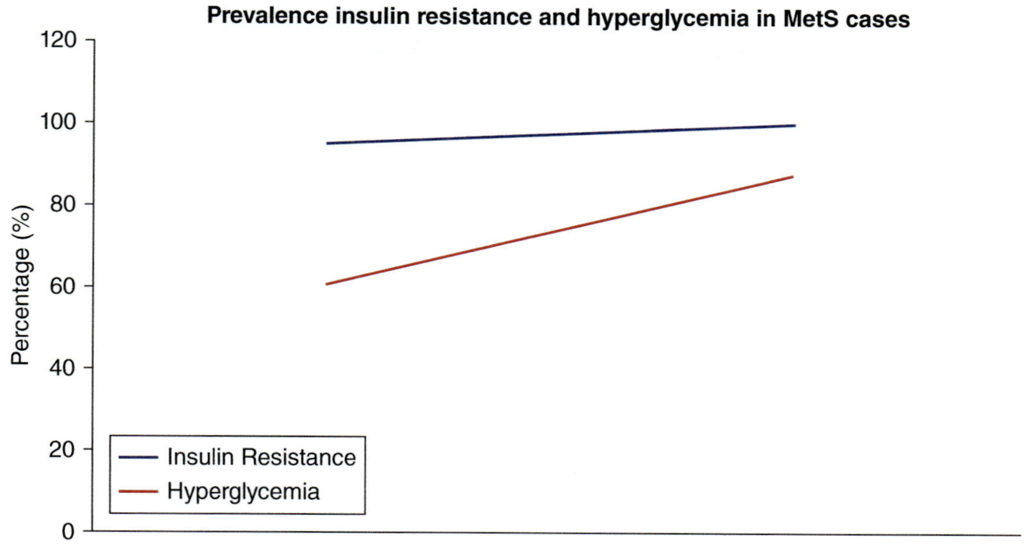

FIGURE 10.3 Recent studies reported that T2DM patients with MetS identified 95%–100% insulin resistance cases, and 61%–87.5% cases show hyperglycaemia (Alawdi et al. 2024).

blood flow (Nelson 2013). Figure 10.5 shows the prevalence of altered lipid profiles in individuals with MetS.

Criteria: Triglyceride levels ≥150 mg/dL (1.7 mmol/L) or use of medication to lower triglycerides. Dyslipidaemia – low high-density lipoproteins (HDL) cholesterol.

HDL cholesterol is often called "good" cholesterol because it helps eliminate excess cholesterol from the bloodstream, thereby preventing plaque formation in the arteries. Lower HDL levels are associated with an increased risk of atherosclerosis and CVD.

Criteria: Men: HDL levels <40 mg/dL (1.0 mmol/L). Women: HDL levels <50 mg/dL (1.3 mmol/L).

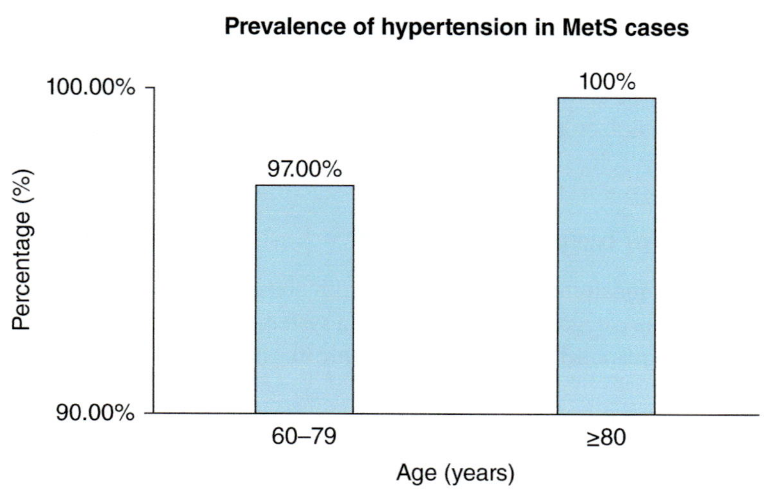

FIGURE 10.4 The bar graph shows the prevalence of hypertension in MetS cases. In elderly patients (≥80 years), the prevalence was 100%, while in the age group 60–79 years, it was 97% (Van Jaarsveldt et al. 2024).

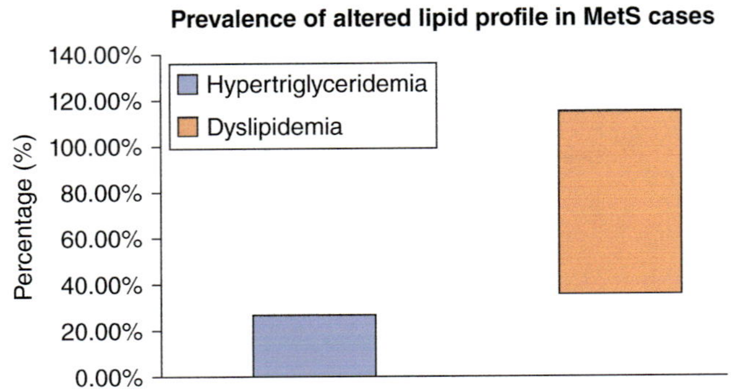

FIGURE 10.5 The graph shows the prevalence of hypertriglyceridaemia (20%) and dyslipidaemia (33%–95%) in MetS cases (Eckel 2007; Jamali et al. 2024).

10.1.2 Epidemiology of MetS

The prevalence of MetS is increasing globally, particularly in developed and rapidly urbanising regions. It is estimated that over 20%–30% of the adult population in countries like the United States and Europe meet the diagnostic criteria for MetS. The prevalence of MetS is also increasing in Asia and Latin America due to the changes in diet and lifestyle, such as greater consumption of processed foods, refined sugars, and sedentary behaviour (Ranasinghe et al. 2017).

10.1.3 The Demographic Trends

Ageing Populations: The elderly are more prone to develop MetS due to the physiological changes, reduced muscle mass and the metabolic slowdown. The main causative factor is the lifestyle risk factors, like poor diet and physical inactivity over time.

Urbanisation and Lifestyle Shifts: The global development, economic transitions, and rapid urbanisation have decreased physical activity in people and led to the adoption of Western dietary patterns, which contain high levels of fats, sugars, and processed foods. These factors, coupled with the increased obesity rates, are responsible for the growth of MetS in both developed and developing nations (Zong et al. 2024).

Obesity Epidemic: The main factor reported to be associated with the MetS prevalence is the global obesity epidemic. Currently, the developed and developing nations are experiencing alarmingly high rises in obesity rates, which is caused by high-calorie diets and a lack of exercise. Obesity is considered the main risk factor for insulin resistance and other components of MetS (Anderson and Durstine 2019).

10.1.4 Systemic Health Implications of MetS

Unfortunately, MetS has much more complicated health consequences outside its association with CVD and diabetes. We can categorise this disorder as a systemic condition affecting multiple organs simultaneously, which is responsible for various chronic diseases.

10.1.4.1 Cardiovascular Disease (CVD)

The most severe health condition driven by MetS is cardiovascular disorders. MetS significantly increases the risk of coronary artery disease, heart attacks, and strokes due to the combined effects of hypertension, dyslipidaemia, and insulin resistance. We are aware that these factors are responsible for atherosclerosis, the buildup of plaques in the arteries, which impairs the blood flow and can cause severe cardiovascular consequences.

10.1.4.2 Type 2 Diabetes

It is not surprising that insulin resistance is a hallmark of MetS. Insulin resistance is responsible for type 2 diabetes. The individuals with MetS are noticeably more prone to diabetes, which further exacerbates the already present risk of cardiovascular and kidney diseases (Hayden 2023).

10.1.4.3 Non-Alcoholic Fatty Liver Disease (NAFLD)

MetS also gives rise to NAFLD, a health condition diagnosed by excessive accumulation of fat in the liver. If we don't treat NAFLD, it can result in more severe liver diseases, like non-alcoholic steatohepatitis (NASH) and cirrhosis (Cotter and Rinella 2020).

10.1.4.4 Chronic Kidney Disease (CKD)

This is another severe implication of MetS. The hypertension, hyperglycaemia, and dyslipidaemia associated with MetS collectively contribute to kidney damage. This is responsible for the development of CKD, which can lead to kidney failure if not treated on time (Prasad 2014).

10.1.4.5 Cognitive Decline and Dementia

There is numerous evidence which links MetS to cognitive decline, with insulin resistance and inflammation. This is noting that MetS can result in the development of AD and other forms of dementia (Frisardi et al. 2010).

10.1.4.6 Cancer

By affecting the metabolism, the MetS may increase the risk of certain cancers, including breast, colorectal, and pancreatic cancers. The exact mechanisms are still under investigation, but chronic inflammation and insulin resistance are kind of playing an important role (Doyle et al. 2012).

10.2 PATHOPHYSIOLOGICAL MECHANISMS CONNECTING MetS AND NEUROLOGICAL DISORDERS

MetS is found to severely impact the central nervous system (CNS). It causes metabolic disturbances with neurological conditions, which involve neurodegenerative diseases, cognitive decline, and cerebrovascular disorders. There are multiple interconnected mechanisms, like chronic low-grade inflammation, oxidative stress, mitochondrial dysfunction, and vascular abnormalities, which collectively associate the MetS with brain dysfunction (Farooqui et al. 2012). The brain is one of the most sensitive organs of the human body. This is to noting that these phenomena disrupt the neural homeostasis, contributing to neuronal damage, and thus increasing the risk of neurological diseases, such as Alzheimer's Disease (AD), Parkinson's Disease (PD), and stroke.

10.2.1 Chronic Inflammation and Neuroinflammation

10.2.1.1 The Systemic Inflammation in MetS

MetS is also known to be associated with a chronic state of low-grade, systemic inflammation, mainly caused by visceral adipose tissue, which acts as an endocrine organ. In obese people, the excess adipose tissue secretes a lot of pro-inflammatory cytokines and adipokines, for example, tumour necrosis factor-alpha (TNF-α), interleukin-6 (IL-6), interleukin-1 beta (IL-1β), and C-reactive protein (CRP). These molecules contribute to insulin resistance, in addition to being responsible for endothelial dysfunction and immune dysregulation, which can pose a negative impact on the CNS (Virtue and Vidal-Puig 2010). Excessive reactive oxygen species (ROS) production results in damage to endothelial cells and the basement membrane, damaging the blood-brain barrier (BBB) (Maiuolo et al. 2021).

10.2.1.2 Link to Neuroinflammation

a. *Blood-Brain Barrier (BBB) Breakdown*

This is surprising that the chronic inflammation in MetS individuals can even disrupt the integrity of the BBB. We know that this selective barrier shields the brain from harmful substances present in the bloodstream. When the BBB is compromised, various pro-inflammatory cytokines, immune cells, and potentially neurotoxic substances can enter the brain, which triggers local inflammation and collectively contributes to multiple neurodegenerative changes (Mou et al. 2022). The infiltration of inflammatory mediators upregulates the metalloproteinase enzymes. This causes the degradation of claudin-5 and occluding proteins, which ultimately leads to damage to the tight junction. Astrocyte dysregulation, pericyte dysfunction, and tight junction damage result in the structurally compromised BBB (Mauro et al. 2015; Van Dyken and Lacoste 2018).

b. *Microglial Activation*

When they are inside the brain, the pro-inflammatory cytokines cause activation of microglial cells, the main immune cells of the CNS. These activated microglial cells produce additional cytokines, nitric oxide (NO), and ROS, creating a vicious cycle of neuroinflammation. This chronic microglial activation is implicated in the pathogenesis of severe neurodegenerative disorders like AD and PD (Amor et al. 2010).

c. *Neurodegeneration*

In MetS patients, the insulin resistance and brain glucose dysregulation, chronic inflammation, metabolic and oxidative stress, result in the decrease in cognitive function, protein misfolding, and neuroinflammation (Maiuolo et al. 2021; Capucho et al. 2022). If it stays for a longer time, the neuroinflammation and BBB leakage can lead to neuronal dysfunction and even death. As an example, in AD, inflammation enhances the production and aggregation of amyloid-beta (Aβ) plaques. It also increases the hyperphosphorylated tau tangles, the primary pathological hallmarks of the disease. Similarly, chronic inflammation contributes to the degeneration of dopaminergic neurons in the substantia nigra in PD, which leads to the motor deficits.

d. *Cognitive Impairment and Dementia*

Chronic inflammation is found to have a role in cognitive decline and the development of dementia. There are numerous reports documenting that the increased systemic inflammatory markers, such as CRP and IL-6, can be correlated with poor cognitive performance, declined memory, loss of executive function, and decreased processing speed. The cognitive deficits noticed in AD and people with other dementias are attributed to the prolonged neuroinflammation (Cunningham and Hennessy 2015).

10.2.2 Oxidative Stress and Neural Damage

In MetS, oxidative stress arises through interconnected pathways that generate substantial ROS and impair antioxidant status. The primary ROS sources in MetS patients include the overexpression of the nicotinamide adenine dinucleotide phosphate (NADPH) oxidase enzyme (NOX) and mitochondrial dysfunction. AGEs activate NOX pathways, and NOX2 generates superoxide radicals (O_2^-) under pathological conditions. Excessive carbohydrate and lipid consumption trigger mitochondrial dysfunction and overactivity, enhancing electron leakage during oxidative phosphorylation (Francesca et al. 2015; Martemucci et al. 2023; Masenga and Kirabo 2023). Multiple factors exacerbate redox imbalance, including hyperglycaemia, dyslipidaemia, adipose tissue dysfunction, antioxidant depletion, the initiation of a lipid peroxidation cascade, and inflammation. Activation of polyol pathways in Met S patients depletes the major intracellular antioxidant reduced glutathione (GSH) (Roberts and Sindhu 2009). Decreasing GSH level further impairs thioredoxin function, preventing the neutralisation of lipid peroxides and H_2O_2 (Vona et al. 2019). Adipose hypertrophy leads to hypoxia and the infiltration of macrophages and necrotic adipocytes, which activate proinflammatory cytokines (TNF-α, IL-6) through NF-κB signalling and M1 macrophages to release ROS (Francisqueti et al. 2017). Peroxidation of free fatty acids in the membrane produces active aldehydes that cause oxidative damage (Hutcheson and Rocic 2012).

Excessive ROS generation leads to the oxidative modification of lipids, proteins, and DNA, resulting in widespread cellular dysfunction. Oxidative stress significantly contributes to the development of various neurological disorders, including AD, PD, and stroke, leading to neuronal damage, disrupted neurochemical signalling, and disease progression.

a. *Neural Vulnerability*
   The brain is particularly susceptible to oxidative stress due to its high oxygen consumption, lipid-rich content (especially in neuronal membranes), and relatively low levels of antioxidant defences compared to other tissues. ROS can damage the delicate lipids in neuronal cell membranes, disrupting cellular communication and synaptic function.
b. *Aβ and Tau*
   In AD, oxidative stress accelerates the aggregation of Aβ peptides into toxic plaques and the formation of hyperphosphorylated tau protein, which forms neurofibrillary tangles. Both pathologies disrupt neural networks and are closely associated with cognitive decline and memory loss (Rajmohan and Reddy 2017).
c. *PD*
   The main key factor in the selective degeneration of dopaminergic neurons in the substantia nigra is the oxidative stress in patients with PD. The main feature of oxidative stress is mitochondrial dysfunction, which contributes to the depletion of dopamine. Dopamine is a neurotransmitter required for motor control, and its decrease leads to the hallmark motor symptoms of the disease (Dias et al. 2013).
d. *Cerebral Ischaemia and Stroke*
   Oxidative stress also plays a critical role in ischaemic stroke. During ischaemia, the lack of oxygen leads to a surge in ROS production upon reperfusion (restoration of blood flow), which damages the surrounding brain tissue, a process known as ischaemia-reperfusion injury. This oxidative damage exacerbates neuronal loss and worsens the neurological outcomes after stroke (Allen and Bayraktutan 2009).

10.2.3 Mitochondrial Dysfunction and Neurodegeneration

10.2.3.1 Mitochondrial Dysfunction in MetS

Mitochondria are the primary organelles responsible for generating ATP through oxidative phosphorylation. In MetS, mitochondrial dysfunction occurs through several interconnected pathways. The key factors include the oxidative phosphorylation overload, dysregulated mitochondrial dynamics, mitochondrial DNA (mtDNA) abnormalities, biogenesis failure, and inflammation cross-talk. Overconsumption of glucose and lipids forces mitochondria to overproduce ATP, causing the leakage of electrons from the electron transport chain (ETC). ETC disruption and AGEs production in chronic hyperglycaemia (Bhatti et al. 2017) generate ROS that damage the mtDNA, lipids, and proteins. Damaged mtDNA increases ROS production, creating a self-reinforcing cycle (Alves et al. 2017). Downregulation of Peroxisome proliferator-activated receptor gamma coactivator (PGC)-1α, the master regulator of mitochondrial production, causes a failure in mitochondrial biogenesis, leading to compromised oxidative capacity and worsening lipid and glucose handling (Annayya et al. 2012). Consequently, cells struggle to generate energy from glucose, and elevated free fatty acids overload mitochondria, releasing ROS and further impairing energy production and metabolism (Prasun 2020). Adipose tissue macrophages release TNF-α/IL-6, which inhibit ETC activity and promote ceramide synthesis. Ceramides directly damage the mitochondrial membranes (Annayya et al. 2012). Sometimes, nutrient overload dysregulates mitochondrial dynamics by activating

DRP1, which fragments mitochondria into dysfunctional organelles, reduces MFN1/MFN2 and OPA1 expression, functionally declines, and prevents damaged mitochondria from sharing components (Alves et al. 2017).

Overall, these processes generate a vicious cycle of metabolic dysfunction by

- ROS damages mitochondria
- Dysfunctional mitochondria generate more ROS
- Worsening insulin resistance and metabolic dysregulation

10.2.3.2 Impact on the Brain

a. *Neuronal Energy Demand*

Neurons have a remarkably high energy demand because of their constant activity and synaptic transmission. Mitochondrial dysfunction in neurons results in energy deficits, impairing synaptic function, neurotransmitter release, and, ultimately, neuronal survival (Nicholls and Budd 2000).

b. *AD*

In AD, mitochondrial dysfunction has been linked to reduced cerebral glucose metabolism, a characteristic disease feature. Impaired mitochondrial function worsens oxidative stress and causes a decline in neuronal function, contributing to cognitive decline.

c. *PD*

The pathogenesis mechanism in PD is mitochondrial dysfunction, which leads to the cellular death of dopaminergic neurons in the substantia nigra. This damage is associated with the mutations in genes responsible for mitochondrial maintenance, such as PINK1 and Parkin. Their impaired function affects the energy production and results in increased oxidative stress in the affected neurons (Federico et al. 2012).

10.3 ROLE OF METABOLIC FACTORS IN NEUROLOGICAL DISORDERS

The neurological disorders possess a complicated pathogenesis. MetS significantly affects brain function and contributes to neurodegenerative diseases via complex metabolic pathways. There are various neurological impacts posed by metabolic factors like insulin resistance and dyslipidaemia, inflammation and oxidative stress, hormonal dysregulation, endothelial dysfunction, and autophagy impairment. These factors negatively affect the cognitive decline, neuronal dysfunction, neurodegeneration, BBB disruption, and neuroinflammation. The following are the key metabolic factors responsible for the neurological disorders, such as AD, PD, and stroke.

10.3.1 Adipokines and Neuroinflammation

10.3.1.1 Dysregulated Adipokines in MetS

Adipokines are signalling proteins secreted by adipose tissue and are very important for the energy balance, inflammation, and insulin sensitivity. There is an excessive fat accumulation, especially visceral fat, in MetS. This creates an imbalance between pro-inflammatory and anti-inflammatory adipokines, triggering the systemic inflammation that influences the function of the CNS (Richard et al. 2020).

10.3.1.2 Leptin Resistance and Cognitive Impairment

Leptin regulates the appetite and energy expenditure in healthy people, and it can cross the BBB to negatively impact brain regions, such as the hypothalamus, hippocampus, and cerebral cortex. However, in MetS and obesity, the brain develops leptin resistance, leading to dysfunctional signalling (Burguera et al. 2000).

High leptin levels, coupled with resistance, have been implicated in neuroinflammation by promoting the production of pro-inflammatory cytokines within the CNS. Studies indicate that leptin resistance exacerbates hippocampal dysfunction, contributing to cognitive deficits and the early onset of dementia. Leptin's effect on neurogenesis and synaptic plasticity is diminished in MetS, resulting in problems related to learning and memory and decreased cognitive performance.

10.3.1.3 Reduced Adiponectin and Neurodegeneration

Adiponectin is an adipokine with anti-inflammatory and insulin-sensitising properties. It plays a protective role in our brain by reducing oxidative stress and inflammation. In MetS, the adiponectin levels are decreased, which imposes a risk of neurodegenerative diseases. We can find lots of studies advocating that lower adiponectin levels are linked to an increase in neuronal oxidative damage and thus a decline in the synaptic function, particularly in the hippocampus region of our brain. In AD models, adiponectin deficiency is found to be associated with enhanced Aβ accumulation and impaired neuronal survival (Verdile et al. 2015).

10.3.1.4 Pro-inflammatory Adipokines (TNF-α, IL-6, Resistin)

Adipose tissue secretes pro-inflammatory cytokines, such as TNF-α, IL-6, and resistin in patients with MetS. These cytokines are known to cross the BBB and impact microglia. The Microglia are the brain's resident immune cells; their damage can lead to chronic neuroinflammation. TNF-α and IL-6 promote the formation of Aβ plaques and phosphorylate the tau protein; both their processed forms are critical in AD. Additionally, prolonged exposure to these cytokines disrupts the synaptic function and induces neuronal apoptosis, resulting in neurotoxicity (Wang et al. 2015).

10.3.2 Mechanisms of Neuroinflammation

There are various mechanisms of neuroinflammation. In MetS patients, the primary mechanisms include chronic activation of microglia, BBB disruption, activation of proinflammatory pathways, signalling pathways involved, molecular and cellular consequences, and cytokine and chemokine release.

10.3.2.1 Chronic Activation of Microglia

The microglial cells maintain homeostasis by clearing cellular debris and pathogens in our brain. However, in MetS patients, the sustained activation of microglia caused by systemic inflammation results in a pro-inflammatory state in our CNS. This overactivation releases multiple neurotoxic chemicals, like NO and ROS, which damage neurons.

10.3.2.2 Blood-Brain Barrier (BBB) Dysfunction

The chronic systemic inflammation in MetS conditions disrupts the integrity of the BBB, which allows the peripheral immune cells and inflammatory mediators to invade the CNS. This process even complicates the neuroinflammation and disrupts the ability of our brain to eliminate the toxic proteins, hastening neurodegeneration (Saraiva et al. 2016).

10.3.2.3 Link to Specific Neurological Disorders

a. *AD*

In MetS, the increase in pro-inflammatory adipokines correlated with enhanced Aβ deposition and increased tau tangle formation. This results in synaptic dysfunction and neuronal death. The leptin resistance and decrease in the adiponectin levels exacerbate memory deficits and thus accelerate the progression of AD.

b. *PD*

Notably, the pro-inflammatory cytokines are known to play an important role in the degeneration of dopaminergic neurons in the substantia nigra; which is a characteristic

TABLE 10.1
The Mechanism and Description of Factors Involved in Neuroinflammation During Metabolic Syndrome

Mechanism Aspect	Description
BBB Disruption	Increased permeability, adhesion molecule upregulation, and astrocyte dysfunction
NF-κB Pathway Activation	Central mediator triggered by metabolic stress, oxidative, and ER stress
TLR4 Signalling	Activates NF-κB, linking overnutrition to brain inflammation
Cytokine Release	TNF-α, IL-1β, and IL-6 promote immune cell recruitment and sustain inflammation.
Cellular Damage	Neuronal/glial dysfunction, hormonal imbalance, cognitive impairment
Additional Pathways	VEGF, HIFs, PKC, Rho/ROCK, eNOS, and Wnt/β-catenin modulate inflammation and BBB integrity

feature of PD. The chronic neuroinflammation resulting from the increase in adipokine levels leads to increased oxidative stress and impaired mitochondrial function. These are the critical factors in the deterioration of our nerve cells.

The above neuroinflammatory cascade creates a vicious cycle where metabolic disturbances perpetuate the inflammation in our brain, contributing to neurodegenerative diseases and cognitive decline in MetS. The key features and pathways are summarised in Table 10.1.

10.3.3 Insulin Resistance and Brain Health

10.3.3.1 Peripheral and Brain Insulin Resistance

a. *Peripheral Insulin Resistance in MetS*

Generally, we observed in MetS that insulin resistance occurs in peripheral tissues, such as muscle, liver, and adipose tissue. This condition is characterised by the impaired uptake of glucose, which causes hyperglycaemia and compensatory hyperinsulinemia. Over time, this insulin resistance leads to systemic inflammation and imposes metabolic stress.

b. *Brain Insulin Resistance*

The hippocampus and cortex of our brain rely on insulin for proper synaptic function, memory, and neurogenesis. The neurons become less responsive to insulin signalling in brain insulin resistance, impacting the glucose metabolism, synaptic plasticity, and cognitive functions. This process is further worsened by hyperinsulinemia, oxidative stress, and neuroinflammation.

10.3.3.2 Mechanisms of Brain Insulin Resistance

a. *Impaired Glucose Metabolism*

Our neurons take glucose in influence with the insulin function, serving as the primary energy source. In cases of brain insulin resistance, neurons are not able to efficiently metabolise glucose, which results in energy deficits and impairs synaptic transmission. This phenomenon leads to loss of memory and cognitive decline, which are characteristic of the early onset of AD.

b. *Oxidative Stress*

Mitochondrial dysfunction causes excessive ROS generation, triggering oxidative stress if we have brain insulin resistance. This process damages crucial cellular components, including DNA, proteins, and lipids, accelerating neuronal degeneration and cell death.

c. *Neuroinflammation*

Insulin resistance in the brain triggers pro-inflammatory signalling pathways, especially through microglia activation. Chronic neuroinflammation additionally disrupts insulin signalling and worsens synaptic dysfunction.

10.3.3.3 The "Type 3 Diabetes" Hypothesis in Alzheimer's Disease (AD)

a. *Insulin and Aβ Accumulation*

Insulin-degrading enzyme (IDE) plays a crucial role in clearing Aβ, a toxic protein that builds up in AD. In insulin-resistant brains, IDE activity decreases, resulting in the accumulation of Aβ plaques (Messier and Teutenberg 2005).

b. *Tau Pathology*

Insulin resistance leads to the hyperphosphorylation of tau proteins, resulting in their aggregation into neurofibrillary tangles, a significant characteristic of AD. Disruption of insulin signalling undermines neuronal stability and facilitates the propagation of tau pathology across the brain (Akhtar and Sah 2020).

10.3.3.4 Cognitive Decline and Dementia:

Insulin resistance is closely linked to cognitive decline. Individuals with type 2 diabetes face a 50%–100% higher risk of developing AD. Brain insulin resistance disrupts synaptic plasticity, which impairs learning and memory processes. Additionally, hyperinsulinemia contributes to vascular damage, elevating the risk of stroke and vascular dementia.

10.3.4 Glucotoxicity and Neuronal Health

10.3.4.1 Chronic Hyperglycaemia in MetS

Prolonged high blood sugar levels in MetS cause glucotoxicity, a condition in which neurons are exposed to excessive glucose levels, leading to cellular damage. Glucotoxicity leads to the generation of advanced glycation end-products (AGEs). This further contributes to the increase in oxidative stress and chronic inflammation, which ultimately harms neuronal function (Giri et al. 2018). Therefore, the insulin resistance causes Chronic hyperglycaemia in MetS patients. The other factors involve pancreatic compensation failure and metabolic disturbances caused by obesity and inflammation, causing an increase in glucose level (Table 10.2).

10.3.4.2 Mechanisms of Glucotoxicity

a. *Advanced Glycation End-products (AGEs)*

AGEs are formed when glucose reacts with proteins, lipids, and nucleic acids, creating stable cross-linked molecules in our cells. AGEs accumulate in neurons and disrupt cellular functions by altering the three-dimensional structures of proteins and impairing ligand-receptor signalling. On top of that, they promote inflammation by activating the receptor for advanced glycation end-products (RAGE), which amplifies neuroinflammation and oxidative damage (Singh et al. 2014).

TABLE 10.2
The Key Factors That Govern Hyperglycaemia in MetS Patients

Factor	Role of Hyperglycaemia in Metabolic Syndrome
Insulin resistance	Reduces cellular glucose uptake
Hyperinsulinemia	Initial compensatory response by the pancreas
Pancreatic β-cell fatigue	Leads to inadequate insulin production over time
Central obesity	Increases free fatty acids and inflammation, worsening insulin resistance

b. *Oxidative Stress*

Hyperglycaemia causes mitochondrial dysfunction, producing excessive ROS in our brain. These reactive molecules result in peroxidation of lipids, DNA damage, and protein oxidation. These factors all contribute to neuronal apoptosis and synaptic dysfunction.

c. *Neuroinflammation*

AGEs and oxidative stress activate the inflammatory pathways, especially the NF-κB pathway, which drives the production of pro-inflammatory cytokines in our brain. This ongoing inflammation contributes to neurodegeneration and impairs synaptic plasticity in the CNS (Wang et al. 2015).

10.3.4.3 Impact on Neurodegenerative Diseases

a. *AD*

We know that the glucotoxicity increases the rate of the accumulation of Aβ plaques and tau tangles in the brain. AGEs have been observed in amyloid plaques and neurofibrillary tangles, and their presence is linked to increased neurotoxicity and decline in cognitive function (Bennett et al. 2004).

b. *PD*

In PD, there is limited information about the glucotoxicity, but we know that the oxidative stress induced by hyperglycaemia contributes to the degeneration of dopaminergic neurons in the substantia nigra. This elevated glucose level may also impact the function of mitochondria in our neurons, further worsening the neurodegeneration (Patergnani et al. 2021).

10.3.4.4 Neuronal Apoptosis and Synaptic Dysfunction

a. *Neuronal Apoptosis*

The mechanism of neuronal apoptosis in MetS patients is affected by multiple interconnected mechanisms. For example, mitochondria play a central role in energy metabolism and the regulation of apoptosis in general. We can find that under the glucotoxic conditions, mitochondrial function is compromised. This leads to the energy deficits in our neurons. Furthermore, an increase in ROS production activates apoptotic pathways, which leads to programmed cell death (Greene et al. 1999). An increase in glucose level, hyperglycaemia, disrupts the autophagy, and causes the accumulation of damaged proteins in our brain and organelles that trigger neuronal death (Maiuolo et al. 2021). Persistent inflammation and ongoing stress in the endoplasmic reticulum (ER) trigger the onset of pro-apoptotic pathways, which promote neurodegeneration in MetS patients (Cai and Liu 2012; Maiuolo et al. 2021). The negatively affected antioxidation pathways and caspase activation further facilitate neuronal cell apoptosis (Wilson and Kumar 2018; Zhou et al. 2019).

b. *Synaptic Damage*

The MetS patients have compromised synaptic functions due to an increase in hyperglycaemia, insulin resistance, dysfunction of lysosomes, and oxidative stress. The impaired insulin signalling impacts plasticity in our synapses by decreasing the phosphorylation of the insulin receptor and its downstream effector proteins, such as Akt (Agrawal and Gomez-Pinilla 2012). Insulin resistance is also known to disrupt the release of glutamate and negatively affect the postsynaptic signalling by downregulating NMDA and α-amino-3-hydroxy-5-methyl-4-isoxazolepropionic acid (AMPA) receptors, leading to impaired long-term potentiation (LTP) and reduced synaptic strength (Soda et al. 2024). Synaptic transmission is also affected by hyperglycaemic conditions through the mechanism of altering the phosphorylation of synaptic proteins. This impairs the neurotransmitter release and reduces the synaptic plasticity. This all leads to cognitive deficits and memory loss in neurodegenerative conditions (Duarte 2014). Elevation in ROS levels and lipid peroxidation harms the synaptic proteins and membranes, which further undermines the synaptic function (Agrawal and Gomez-Pinilla 2012). Table 10.3 summarises the factors involved in synaptic damage in MetS patients.

TABLE 10.3
The Key Factors Involved in Neuronal Apoptosis and Synaptic Dysfunction in MetS Patients.

Mechanism	Neuronal Apoptosis	Synaptic Dysfunction
Oxidative stress	Mitochondrial damage, apoptosis	Damages synaptic proteins
Impaired autophagy	Accumulation of damaged organelles	Disrupts neurotransmission
Insulin resistance	Reduces neuroprotection	Impairs synaptic plasticity and signalling
Inflammation/ER stress	Activates pro-apoptotic pathways	Contributes to synaptic loss
Glutamate signalling	–	Reduces LTP, synaptic strength

10.4 CEREBROVASCULAR HEALTH AND NEUROLOGICAL DISORDERS

Cerebrovascular health is also affected by the complexities of MetS, which significantly increases the risk of various neurological disorders and a decline in cognition (Farooqui et al. 2012; Rojas et al. 2021b). We can characterise the MetS as a cluster of conditions, including obesity and insulin resistance, hypertension, and dyslipidaemia. This can lead to impaired cerebral blood flow and endothelial dysfunction (Mohamed et al. 2023). These alterations are responsible for the cerebrovascular accidents, including strokes, which subsequently increase the risk of developing neurodegenerative disorders like AD and vascular dementia. We can see that the MetS-associated chronic inflammation and oxidative stress (Luque-Contreras et al. 2014; Li et al. 2023) contribute to neuronal damage, worsening cognitive decline, and other neurological deficits. This is an alarming situation. We need to understand the relationship between MetS and cerebrovascular health to develop targeted interventions that may reduce the onset and progression of neurological disorders. There is an importance of a holistic health approach that prioritises metabolic and neurological well-being.

The intricate relationship between cerebrovascular health and neurological disorders sets the stage for a deeper exploration of specific risk factors within MetS. One of the most critical components is hypertension, which significantly contributes to the onset and progression of cerebrovascular diseases, particularly stroke. While the detrimental effects of high blood pressure on cerebrovascular events are well-established, the nuances of how these conditions intertwine to precipitate cognitive decline warrant further investigation. Following hypertension, dyslipidaemia emerges as another pivotal factor, characterised by abnormal lipid profiles that exacerbate cerebrovascular and neurodegenerative processes. Together, these components of MetS create a multifaceted risk landscape that influences brain health, necessitating a closer look at their individual and combined impacts.

10.4.1 Hypertension and Stroke

Hypertension, often referred to as a "silent killer," is the leading global risk factor for death and the second for morbidity, affecting over a quarter of the world's population in 2000 and projected to reach about 29% by 2025. It is a crucial component of MetS that plays an essential role in the progression of cerebrovascular diseases, particularly strokes. While the link between hypertension and stroke is well-established, with high blood pressure contributing to both ischaemic and haemorrhagic strokes, there remains a limited understanding of how these two conditions together lead to cognitive decline. Ischaemic strokes, which result from obstructed blood flow, are the most common type, while haemorrhagic strokes happen due to the rupture of weakened blood vessels. The potential mechanisms driving cognitive decline are still not fully understood. Decline includes neuroinflammation, hypoperfusion, and subcortical ischaemic vascular disease.

10.4.2 Dyslipidaemia and Brain Health

Dyslipidaemia is characterised by abnormal lipid levels, particularly elevated low-density lipoprotein (LDL) cholesterol, elevated triglyceride levels, and reduced HDL cholesterol (Hedayatnia et al. 2020). This imbalance in lipid content is critical to developing cerebrovascular disease and other neurological disorders. In MetS, altered lipid profiles can significantly contribute to cerebrovascular diseases and neurodegeneration.

MetS encompasses several interrelated metabolic risk factors, including central obesity, insulin resistance, hypertension, and dyslipidaemia. The link between these factors is well-documented; for example, elevated triglyceride levels are often associated with insulin resistance, a hallmark of MetS (Reitz 2013a). The presence of dyslipidaemia within this syndrome exacerbates the risks of CVD. However, emerging evidence suggests it also has profound effects on brain health.

10.4.2.1 Impact on Brain Health

a. *Cognitive Decline and Dementia*

   MetS is associated with cognitive impairment and an increased risk of all-cause dementia, including vascular dementia. Patients with MetS have a 24% higher risk of early-onset dementia and exhibit impairments in memory, psychomotor coordination, processing speed, executive function, and reasoning (Koutsonida et al. 2022; Alsuwaidi et al. 2023). MetS disrupts cognition by inducing oxidative stress, insulin resistance (IR), and chronic inflammation, which impair glucose metabolism. IR disturbs neuronal glucose metabolism and accelerates tau hyperphosphorylation. Compromised insulin signalling disrupts brain energy metabolism and accelerates Aβ pathology associated with AD (Ezkurdia et al. 2023). Furthermore, meta-analyses have demonstrated that higher triglyceride levels and glucose index are associated with an increased risk of cognitive impairment and dementia in Asian individuals (Wu et al. 2019; Wang et al. 2023). Excess free radicals and reactive species production in MetS patients oxidatively modify lipids, proteins, and DNA in brain cells (Yates et al. 2012; Koutsonida et al. 2022). This redox imbalance diminishes antioxidant defences, alters glycolytic pathways, and harms neurons. Dyslipidaemia has been linked to cognitive decline and the onset of neurodegenerative diseases, including AD. Research suggests that high LDL and low HDL levels may contribute to the development of AD by encouraging amyloid plaque accumulation (Reitz 2013b). Additionally, MetS causes cerebrovascular damage, compromising BBB function and Aβ clearance, which worsens cognitive decline (Rojas-Gutierrez et al. 2017). These factors show how metabolic dysfunction speeds up neurodegeneration and cognitive decline.

b. *Neuroinflammation*

   MetS exerts neuroinflammatory effects through systemic and local inflammation, cytokine dysregulation, adipokine and hormonal imbalance, or BBB disruption. Chronic dyslipidaemia leads to neuroinflammation, a critical factor in developing neurological diseases. The accumulation of lipids in the brain can trigger inflammatory pathways, potentially contributing to neurodegeneration. In particular, the roles of oxidative stress and inflammatory cytokines within MetS and dyslipidaemia garner increasing interest (Reitz 2013b; Rojas-Gutierrez et al. 2017). In MetS patients, secretion of proinflammatory adipokines from the adipose tissue sometimes exacerbates neuroinflammation (Esposito and Giugliano 2004). In MetS cases, elevated levels of proinflammatory cytokines like TNF-α and IL-6 cause insulin resistance (Tylutka et al. 2023). Insulin resistance (IR) in the brain impairs insulin signalling that is critical for memory and learning, while chronic low-grade inflammation accelerates neurodegeneration through pro-inflammatory cytokines and oxidative stress (Kshirsagar et al. 2021; Rojas et al. 2021a). MetS elicits chronic systemic inflammation that may compromise the BBB and increase its permeability, allowing the infiltration of inflammatory mediators, toxins, pathogens, and immune cells

(Van Dyken and Lacoste 2018). This inflammation and disrupted glucose uptake lead to tau hyperphosphorylation and Aβ accumulation, both hallmark features of AD.

c. *Vascular Contributions*

MetS significantly increases cerebrovascular risks like ischaemic stroke driven by a complex array of pathophysiological mechanisms and the clustering of various risk factors. MetS accelerates the development and progression of atherosclerotic lesions in brain-supplying arteries. Dyslipidaemia is closely linked to atherosclerosis, which can impair cerebral blood flow and lead to vascular dementia. Conditions associated with MetS, such as hypertension and diabetes, further exacerbate pro-atherothrombotic effects, endothelial dysfunction, and chronic vascular inflammation and damage contributing to cognitive decline (Reitz 2013a, 2013b; He et al. 2020). Increased prevalence of carotid and intracranial atherosclerosis in MetS is well-established, with insulin resistance being a key factor in plaque formation and growth. MetS patients face a threefold increased risk of developing AD, as MetS also adversely affects cerebrovascular health by compromising the integrity of the BBB and hindering the clearance of Aβ. This impairment results in Aβ accumulation and subsequent cognitive decline. The MetS patients are at a 24% higher risk of developing early-onset dementia before the age of 65 (Lee et al. 2025). Hypertension and hyperlipidaemia lead to vascular damage, hypoperfusion, cerebral hypoxia, and ischaemic changes, all linked to the onset of AD. Metabolic disturbances like dyslipidaemia and hyperglycaemia, along with chronic inflammation, further exacerbate neuronal dysfunction and enhance amyloid and tau pathology (Ezkurdia et al. 2023).

Therefore, the vascular component of MetS should not be overlooked in discussions of brain health. Metabolic impairment related to MetS disrupts the walls of cerebral arteries and promotes a hypercoagulable state, further increasing the risk of stroke. Chronic damage to the endothelium, impaired fibrinolysis, and a pro-inflammatory state amplify ischaemic damage and may worsen outcomes from stroke (Arenillas et al. 2007; Yan et al. 2024).

d. *Effect on Mood and Mental Health*

Beyond cognitive decline, dyslipidaemia has also been linked to mood disorders. Some studies indicate that individuals with dyslipidaemia may face a higher risk of depression and anxiety, potentially due to shared pathophysiological mechanisms involving inflammation and oxidative stress. Chronic inflammation can affect brain function, contributing to neurodegenerative changes that influence mood regulation. For instance, research shows that elevated levels of inflammatory markers like IL-6 and CRP are prevalent in individuals experiencing both CVD and mood disorders, such as depression. Recent meta-analyses have bolstered this hypothesis by demonstrating that patients with depression display significantly higher peripheral levels of inflammatory markers, including CRP, IL-6, IL-10, IL-12, TNF-α, soluble IL-2 receptor (sIL-2R), and IL-1 receptor antagonist, compared to healthy individuals.

10.5 CLINICAL AND THERAPEUTIC IMPLICATIONS

MetS significantly elevates the risk of neurological disorders, making early identification and management crucial.

10.5.1 Early Intervention

Early detection and management of the key factors of MetS are crucial for preventing or delaying the onset of neurological disorders. Identifying risk factors such as obesity, hypertension, dyslipidaemia, and insulin resistance at an early stage allows for timely interventions. Clinical screenings and routine assessments can facilitate monitoring these components. Interventions initiated at this stage can significantly lower the risk of cognitive decline and neurodegenerative diseases, as studies indicate that addressing MetS can improve both metabolic health and brain function. Educating patients about recognising early symptoms and engaging in preventive measures is essential for enhancing long-term neurological outcomes.

10.5.2 Lifestyle Modifications

Lifestyle changes are crucial in reducing the risk of MetS and related neurological disorders. Key strategies involve adopting a balanced diet that includes whole foods such as fruits, vegetables, whole grains, and lean proteins while minimising processed foods high in sugar and unhealthy fats (Hoyas and Leon-Sanz 2019). Regular physical activity, targeting at least 150 minutes of moderate exercise weekly, can aid in weight management, improve insulin sensitivity, and lower blood pressure (Ortega et al. 2022; Kobiec et al. 2023). Stress management techniques, such as environmental enrichment, mindfulness, meditation, and adequate sleep, also contribute to improved metabolic health (Ortega et al. 2022). By emphasising these lifestyle changes, healthcare providers can empower patients to take control of their health, thus reducing the risk of MetS and its neurological consequences.

10.5.3 Pharmacological Approaches

Pharmacological interventions targeting the components of MetS can significantly mitigate neurological risks. Amphetamines, initially introduced as weight loss drugs in the 1930s, originated from their early use for narcolepsy and Parkinsonism, quickly gaining popularity for their appetite-suppressing effects. This popularity led to the development of various combinations with other substances to alleviate side effects, including methamphetamine derivatives, which were also marketed for weight loss (Stemmer et al. 2019). Antihypertensive medications can help manage high blood pressure, reducing the risk of stroke and cognitive impairment (Kuang 2020). Renin-angiotensin-aldosterone system (RAAS) inhibitors, angiotensin receptor blockers (ARBs), and angiotensin-converting enzyme (ACE) inhibitors are generally used to treat arterial hypertension (Rask Larsen et al. 2018). Statins may control dyslipidaemia by improving lipid profiles, reducing cholesterol levels, and potentially lowering the risk of atherosclerosis, which can impact cerebral blood flow and cognitive health (Kuang 2020). Statins and insulin sensitisers impart anti-inflammatory effects during cardiovascular complications in MetS patients. The first-line pharmacological approach, like insulin-sensitising agents, sodium-glucose cotransporter (SGLT)-2 inhibitors, and glucagon-like peptide (GLP)-1 receptor agonists, is generally used to treat glucose intolerance (Binesh Marvasti and Adeli 2010; Rask Larsen et al. 2018). GLP-1 receptor agonists and anti-obesity drugs are generally used to prevent obesity, and in severe cases, bariatric surgery is recommended (Lillich et al. 2021). Insulin sensitisers like metformin can enhance insulin resistance and support weight loss, reducing MetS risk factors. By employing a tailored combination of these medications based on individual patient needs, healthcare providers can significantly affect both metabolic health and neurological outcomes.

10.5.4 Use of Antioxidant Drugs

The management of oxidative stress in MetS through medication has been widely explored, with various antioxidant strategies showing potential in mitigating conditions like diabetic neuropathy. Among the antioxidants studied, vitamins A, C, and E, flavonoids, alpha-lipoic acid (ALA), lutein, and docosahexaenoic acid (DHA) have demonstrated neuroprotective properties (Etchegoyen et al. 2018). During antioxidant therapy, natural antioxidants like Ginsenosides increase glucose uptake via GLUT4 translocation, decrease proinflammatory cytokines, and inhibit NF-κB signalling (Lin and Sun 2024). Natural polyphenols like hibiscus and oleuropein scavenges ROS and improve lipid and carbohydrate metabolism (Abdulghani and Al-Fayyadh 2024). Other compounds, such as taurine, acetyl L-carnitine, and N-acetylcysteine, along with synthetic ROS scavengers like tempol and superoxide dismutase (SOD) mimetics, have also shown promise. Additionally, antioxidants like coenzyme Q10 and mitochondrial uncoupling proteins (UCPs) protect against oxidative damage. At the same time, anti-inflammatory agents such as TNF-α inhibitors and IL-10 can alleviate oxidative stress-induced damage, particularly in neurological contexts like brain stroke. Further advancements in targeting oxidative pathways include aldose reductase inhibitors (ARIs), protein kinase C (PKC) inhibitors, and anti-AGEs agents, which address specific biochemical processes involved in oxidative stress-related neuropathy. Statins and supplementation in high doses may cause adverse effects, so the emerging trends include the targeted delivery of nanoparticle-based antioxidants and supplementation of multi-mechanistic agents, including aloe vera and berberine, which counter the oxidative stress, insulin resistance, and inflammation simultaneously (Mohammadian et al. 2024).

10.5.5 Multidisciplinary Approach

Addressing the complex interplay between MetS and neurological disorders necessitates a multidisciplinary approach. Integrated care that includes endocrinology, neurology, and cardiovascular medicine specialists is crucial for comprehensive management. This collaboration facilitates the creation of personalised treatment plans that consider all aspects of a patient's health. Regular communication among healthcare providers helps ensure that interventions are coordinated and effective, enhancing the potential for improved outcomes. Furthermore, educating patients on the significance of this integrated care model can empower them to engage actively in their health journey, ultimately lowering the risk of metabolic and neurological disorders.

Concentrating on these clinical and therapeutic implications can lay the groundwork for improved prevention and management strategies that tackle the complex challenges posed by MetS and its effects on neurological health.

10.6 CONCLUSION

In conclusion, the intricate relationship between metabolic health and neurological function underscores the importance of maintaining a balanced metabolism to support cognitive well-being. As emerging research reveals, MetS not only impacts physical health but also serves as a significant risk factor for various neurological disorders. This connection highlights the need for a holistic health approach considering metabolic and neurological dimensions.

Given the shared mechanisms that drive MetS and neurological disorders, there is an urgent call for further research in this area. Understanding these underlying pathways could unlock new therapeutic strategies and interventions that target metabolic dysfunction and cognitive decline, ultimately improving patient outcomes.

Finally, advocating for comprehensive and proactive strategies to manage MetS is imperative. By addressing lifestyle factors such as diet, physical activity, and stress management, we can reduce the global burden of neurological diseases. Prioritising metabolic health benefits individuals and has far-reaching implications for public health, emphasising the necessity for collaborative efforts among researchers, healthcare providers, and policymakers. Together, we can work towards a future where both metabolic and neurological health are prioritised, leading to healthier lives.

REFERENCES

MF Abdulghani, S Al-Fayyadh (2024). Natural products for managing metabolic syndrome: A scoping review. Front Pharmacol 15:1366946. https://doi.org/10.3389/fphar.2024.1366946

R Agrawal, F Gomez-Pinilla (2012). 'Metabolic syndrome' in the brain: Deficiency in omega-3 fatty acid exacerbates dysfunctions in insulin receptor signalling and cognition. J Physiol 590(10):2485–2499. https://doi.org/10.1113/jphysiol.2012.230078

A Akhtar, SP Sah (2020). Insulin signaling pathway and related molecules: Role in neurodegeneration and Alzheimer's disease. Neurochem Int 135:104707. https://doi.org/10.1016/j.neuint.2020.104707

SH Alawdi, M Al-Dholae, S Al-Shawky (2024). Metabolic syndrome and pharmacotherapy outcomes in patients with type 2 diabetes mellitus. Front Clin Diabetes Healthc 5:1380244. https://doi.org/10.3389/fcdhc.2024.1380244

CL Allen, U Bayraktutan (2009). Oxidative stress and its role in the pathogenesis of ischaemic stroke. Int J Stroke Off J Int Stroke Soc 4(6):461–470. https://doi.org/10.1111/j.1747-4949.2009.00387.x

HN Alsuwaidi, AI Ahmed, HA Alkorbi, SM Ali, LN Altarawneh, SI Uddin, SR Roueentan, AA Alhitmi, L Djouhri, T Chivese (2023). Association between metabolic syndrome and decline in cognitive function: A cross-sectional study. Diab Metab Syndr Obes Targ Ther 16:849–859. https://doi.org/10.2147/DMSO.S393282

D Alves, ME Calmeiro, C Macário, R Silva. (2017). Family phenotypic heterogeneity caused by mitochondrial DNA mutation A3243G. Acta Med Port 30:581–585. https://doi.org/10.20344/amp.8638

S Amor, F Puentes, D Baker, P Van Der Valk (2010). Inflammation in neurodegenerative diseases. Immunology 129(2):154–169. https://doi.org/10.1111/j.1365-2567.2009.03225.x

E Anderson, JL Durstine (2019). Physical activity, exercise, and chronic diseases: A brief review. Sports Med Health Sci 1(1):3–10. https://doi.org/10.1016/j.smhs.2019.08.006

F Angelico, F Baratta, M Coronati, D Ferro, M Del Ben (2023). Diet and metabolic syndrome: A narrative review. Intern Emerg Med 18(4):1007–1017. https://doi.org/10.1007/s11739-023-03226-7

RA Annayya, C Mandavia, J Ren, JR Sowers, L Pulakat (2012). Mitochondria and oxidative stress in the cardiorenal metabolic syndrome. Cardiorenal Med 2(2):87–109. https://doi.org/10.1159/000335675

JF Arenillas, MA Moro, A Dávalos. (2007). The metabolic syndrome and stroke. Stroke 38(7):2196–2203. https://doi.org/10.1161/STROKEAHA.106.480004

S Basu, AJ Thirunavukarasu, V Maheshwari, M Zode, R Hassan-Burden (2023). Determinants and treatment status of metabolic syndrome among older adults in India: A nationally representative, community-based cross-sectional survey. BMJ Public Health 1:e000389.

DA Bennett, JA Schneider, RS Wilson, JL Bienias, SE Arnold (2004). Neurofibrillary tangles mediate the association of amyloid load with clinical Alzheimer disease and level of cognitive function. Arch Neurol 61(3):378–384. https://doi.org/10.1001/archneur.61.3.378

JS Bhatti, GK Bhatti, PH Reddy (2017). Mitochondrial dysfunction and oxidative stress in metabolic disorders – A step towards mitochondria based therapeutic strategies. Biochim Biophys Acta Mol Basis Dis 1863(5):1066–1077. https://doi.org/10.1016/j.bbadis.2016.11.010

T Binesh Marvasti, K Adeli (2010). Pharmacological management of metabolic syndrome and its lipid complications. Daru: Journal of faculty of pharmacy. Tehran Univ Med Sci 18(3):146–154.

B Francesca, FR Luigi, R Rita (2015). Metabolic syndrome, aging and involvement of oxidative stress. Aging Dis 6(2):109–120 https://doi.org/10.14336/AD.2014.0305

B Burguera, ME Couce, GL Curran, MD Jensen, RV Lloyd, MP Cleary, JF Poduslo (2000). Obesity is associated with a decreased leptin transport across the blood-brain barrier in rats. Diabetes 49(7):1219–1223. https://doi.org/10.2337/diabetes.49.7.1219

D Cai, T Liu (2012). Inflammatory cause of metabolic syndrome via brain stress and NF-κB. Aging 4(2):98–115. https://doi.org/10.18632/aging.100431

AM Capucho, A Chegão, FO Martins, H Vicente Miranda, SV Conde (2022). Dysmetabolism and neurodegeneration: Trick or treat? Nutrients 14(7):1425. https://doi.org/10.3390/nu14071425

TG Cotter, M Rinella (2020). Nonalcoholic fatty liver disease 2020: The state of the disease. Gastroenterology 158(7):1851–1864. https://doi.org/10.1053/j.gastro.2020.01.052

C Cunningham, E Hennessy (2015). Co-morbidity and systemic inflammation as drivers of cognitive decline: New experimental models adopting a broader paradigm in dementia research. Alzheimers Res Ther 7(1):33. https://doi.org/10.1186/s13195-015-0117-2

V Dias, E Junn, MM Mouradian (2013). The role of oxidative stress in Parkinson's disease. J Park Dis 3(4):461–491. https://doi.org/10.3233/JPD-130230.

C Ding, Y Wu, X Chen, Y Chen, Z Wu, Z Lin, D Kang, W Fang, F Chen (2022). global, regional, and national burden and attributable risk factors of neurological disorders: The global burden of disease study 1990–2019. Front Public Health 10:952161. https://doi.org/10.3389/fpubh.2022.952161

SL Doyle, CL Donohoe, J Lysaght, JV Reynolds (2012). Visceral obesity, metabolic syndrome, insulin resistance and cancer. Proc Nutr Soc 71(1):181–189. https://doi.org/10.1017/S002966511100320X

J Duarte (2014). Metabolic alterations associated to brain dysfunction in diabetes. Aging Dis. https://doi.org/10.14336/ad.2014.1104

R Eckel (2007. Treating dyslipidemia of the metabolic syndrome: Where's the evidence? Nat Rev Endocrinol 3:437. https://doi.org/10.1038/ncpendmet0514

A Engin (2024). The definition and prevalence of obesity and metabolic syndrome: Correlative clinical evaluation based on phenotypes. Adv Exp Med Biol 1460:1–25. https://doi.org/10.1007/978-3-031-63657-8_1

K Esposito, D Giugliano (2004). The metabolic syndrome and inflammation: Association or causation? Nutr Metab Cardiovasc Dis 14(5):228–232. https://doi.org/10.1016/s0939-4753(04)80048-6

M Etchegoyen, MH Nobile, F Baez, B Posesorski, J González, N Lago, J Milei, M Otero-Losada (2018). Metabolic syndrome and neuroprotection. Front Neurosci 12:196. https://doi.org/10.3389/fnins.2018.00196

A Ezkurdia, MJ Ramírez, M Solas (2023). Metabolic syndrome as a risk factor for Alzheimer's Disease: A focus on insulin resistance. Int J Mol Sci 24(5):4354. https://doi.org/10.3390/ijms24054354

AA Farooqui, T Farooqui, F Panza, V Frisardi (2012). Metabolic syndrome as a risk factor for neurological disorders. Cell Mol Life Sci 69(5):741–762. https://doi.org/10.1007/s00018-011-0840-1

A Federico, E Cardaioli, P Da Pozzo, P Formichi, GN Gallus, E Radi (2012). Mitochondria, oxidative stress and neurodegeneration. J Neurol Sci 322(1–2):254–262. https://doi.org/10.1016/j.jns.2012.05.030

FV Francisqueti, LC Chiaverini, KC Santos, IO Minatel, CB Ronchi, AJ Ferron, AL Ferreira, CR Corrêa (2017). The role of oxidative stress on the pathophysiology of metabolic syndrome. Rev Assoc Med Brasil (1992) 63(1):85–91. https://doi.org/10.1590/1806-9282.63.01.85

V Frisardi, V Solfrizzi, D Seripa, C Capurso, A Santamato, D Sancarlo, G Vendemiale, A Pilotto, F Panza (2010). Metabolic-cognitive syndrome: A cross-talk between metabolic syndrome and Alzheimer's disease. Ageing Res Rev 9(4):399–417. https://doi.org/10.1016/j.arr.2010.04.007

B Giri, S Dey, T Das, M Sarkar, J Banerjee, SK Dash (2018). Chronic hyperglycemia mediated physiological alteration and metabolic distortion leads to organ dysfunction, infection, cancer progression and other pathophysiological consequences: An update on glucose toxicity. Biomed Pharmacother Biomedecine Pharmacother 107:306–328. https://doi.org/10.1016/j.biopha.2018.07.157.

DA Greene, MJ Stevens, I Obrosova, EL Feldman (1999). Glucose-induced oxidative stress and programmed cell death in diabetic neuropathy. Eur J Pharmacol 375(1–3):217–223. https://doi.org/10.1016/s0014-2999(99)00356-8

MR Hayden (2023). Overview and new insights into the metabolic syndrome: Risk factors and emerging variables in the development of type 2 diabetes and cerebrocardiovascular disease. Medicina (Mex) 59(3):561. https://doi.org/10.3390/medicina59030561

J-T He, X Zhao, L Xu, C-Y Mao (2020). Vascular risk factors and Alzheimer's disease: Blood-brain barrier disruption, metabolic syndromes, and molecular links. J Alzheimers Dis 73(1):39–58. https://doi.org/10.3233/JAD-190764

M Hedayatnia, Z Asadi, R Zare-Feyzabadi, M Yaghooti-Khorasani, H Ghazizadeh, R Ghaffarian-Zirak, A Nosrati-Tirkani, M Mohammadi-Bajgiran, M Rohban, F Sadabadi, H-R Rahimi, M Ghalandari, M-S Ghaffari, A Yousefi, E Pouresmaeili, M-R Besharatlou, M Moohebati, GA Ferns, H Esmaily, M Ghayour-Mobarhan (2020). Dyslipidemia and cardiovascular disease risk among the MASHAD study population. Lipids Health Dis 19(1):42. https://doi.org/10.1186/s12944-020-01204-y

I Hoyas, M Leon-Sanz (2019). Nutritional challenges in metabolic syndrome. J Clin Med 8(9):1301. https://doi.org/10.3390/jcm8091301

R Hutcheson, P Rocic (2012). The metabolic syndrome, oxidative stress, environment, and cardiovascular disease: The great exploration. J Diab Res 271028, 13. https://doi.org/10.1155/2012/271028

Z Jamali, F Ayoobi, Z Jalali et al. (2024). Metabolic syndrome: A population-based study of prevalence and risk factors. Sci Rep 14:3987. https://doi.org/10.1038/s41598-024-54367-4

T Kobiec, C Mardaraz, N Toro-Urrego, R Kölliker-Frers, F Capani, M Otero-Losada (2023). Neuroprotection in metabolic syndrome by environmental enrichment. A lifespan perspective. Front Neurosci 17:1214468. https://doi.org/10.3389/fnins.2023.1214468

M Koutsonida, G Markozannes, E Bouras, E Aretouli, KK Tsilidis (2022). Metabolic syndrome and cognition: A systematic review across cognitive domains and a bibliometric analysis. Front Psychol 13:981379. https://doi.org/10.3389/fpsyg.2022.981379

V Kshirsagar, C Thingore, A Juvekar (2021). Insulin resistance : A connecting link between Alzheimer's disease and metabolic disorder. Metab Brain Dis 36:67–83.

Z-M Kuang (2020). Effect of combined antihypertensive and lipid-lowering therapies on cognitive function: A new treatment strategy? Cardiol Res Pract 2020:1–10. https://doi.org/10.1155/2020/1484357

JY Lee, K Han, J Kim, JS Lim, DY Cheon, M Lee (2025). Association between metabolic syndrome and young-onset dementia: A nationwide population-based study. Neurology 104(10):e213599. https://doi.org/10.1212/WNL.0000000000213599

H Li, J Ren, Y Li, Q Wu, J Wei (2023). Oxidative stress: The nexus of obesity and cognitive dysfunction in diabetes. Front Endocrinol 14:1134025. https://doi.org/10.3389/fendo.2023.1134025

FF Lillich, JD Imig, E Proschak (2021). Multi-target approaches in metabolic syndrome. Front Pharmacol 11:554961. https://doi.org/10.3389/fphar.2020.554961

Z Lin, L Sun (2024). Research advances in the therapy of metabolic syndrome. Front Pharmacol 15:1364881. https://doi.org/10.3389/fphar.2024.1364881

D Luque-Contreras, K Carvajal, D Toral-Rios, D Franco-Bocanegra, V Campos-Peña (2014). Oxidative stress and metabolic syndrome: Cause or consequence of Alzheimer's disease? Oxid Med Cell Longev 2014:1–11. https://doi.org/10.1155/2014/497802

J Maiuolo, M Gliozzi, V Musolino, C Carresi, F Scarano, S Nucera, M Scicchitano, F Bosco, S Ruga, MC Zito, R Macri, R Bulotta, C Muscoli, V Mollace (2021). From metabolic syndrome to neurological diseases: Role of autophagy. Front Cell Dev Biol 9:651021. https://doi.org/10.3389/fcell.2021.651021

J Maiuolo, M Gliozzi, V Musolino, C Carresi, F Scarano, S Nucera, M Scicchitano, F Bosco, S Ruga, MC Zito, R Macri, R Bulotta, C Muscoli, V Mollace (2021). From metabolic syndrome to neurological diseases: Role of autophagy. Front Cell Develop Biol 9:651021. https://doi.org/10.3389/fcell.2021.651021

G Martemucci, G Fracchiolla, M Muraglia, R Tardugno, RS Dibenedetto, AG D'Alessandro (2023). Metabolic syndrome: A narrative review from the oxidative stress to the management of related diseases. Antioxidants 12(12):2091. https://doi.org/10.3390/antiox12122091

SK Masenga, A Kirabo (2023). Hypertensive heart disease: Risk factors, complications and mechanisms. Front Cardiovasc Med 10:1205475. https://doi.org/10.3389/fcvm.2023.1205475

C Mauro, V De Rosa, F Marelli-Berg, E Solito (2015). Metabolic syndrome and the immunological affair with the blood–brain barrier. Front Immunol 5:677. https://doi.org/10.3389/fimmu.2014.00677

C Messier, K Teutenberg (2005). The role of insulin, insulin growth factor, and insulin-Degrading enzyme in brain aging and Alzheimer's disease. Neural Plast 12(4):311–328. https://doi.org/10.1155/NP.2005.311

SM Mohamed, MA Shalaby, RA El-Shiekh, HA El-Banna, SR Emam, AF Bakr (2023). Metabolic syndrome: Risk factors, diagnosis, pathogenesis, and management with natural approaches. Food Chem Adv 3:100335. https://doi.org/10.1016/j.focha.2023.100335

K Mohammadian, F Fakhar, S Keramat, A Stanek. (2024). The role of antioxidants in the treatment of metabolic dysfunction-associated fatty liver disease: A systematic review. Antioxidants 13(7):797. https://doi.org/10.3390/antiox13070797

Y Mou, Y Du, L Zhou, J Yue, X Hu, Y Liu, S Chen, X Lin, G Zhang, H Xiao, B Dong (2022). Gut Microbiota interact with the brain through systemic chronic inflammation: Implications on neuroinflammation. Neurodegener Aging Front Immunol 13:796288. https://doi.org/10.3389/fimmu.2022.796288

RH Nelson (2013). Hyperlipidemia as a risk factor for cardiovascular disease. Prim Care Clin Off Pract 40(1):195–211. https://doi.org/10.1016/j.pop.2012.11.003

DG Nicholls, SL Budd (2000). Mitochondria and neuronal survival. Physiol Rev 80(1):315–360. https://doi.org/10.1152/physrev.2000.80.1.315.

FB Ortega, J Mora-Gonzalez, C Cadenas-Sanchez, I Esteban-Cornejo, JH Migueles, P Solis-Urra, J Verdejo-Román, M Rodriguez-Ayllon, P Molina-Garcia, JR Ruiz, V Martinez-Vizcaino, CH Hillman, KI Erickson, AF Kramer, I Labayen, A Catena (2022). Effects of an exercise program on brain health outcomes for children with overweight or obesity: The active brains randomized clinical trial. JAMA Netw Open 5(8):e2227893. https://doi.org/10.1001/jamanetworkopen.2022.27893

S Patergnani, E Bouhamida, S Leo, P Pinton, A Rimessi (2021). Mitochondrial oxidative stress and "Mito-Inflammation": Actors in the diseases. Biomedicines 9(2):216. https://doi.org/10.3390/biomedicines9020216

GR Prasad (2014). Metabolic syndrome and chronic kidney disease: Current status and future directions. World J Nephrol 3(4):210. https://doi.org/10.5527/wjn.v3.i4.210

P Prasun (2020). Mitochondrial dysfunction in metabolic syndrome. Biochim Biophys Acta BBA – Mol Basis Dis 1866(10):165838. https://doi.org/10.1016/j.bbadis.2020.165838

R Rajmohan, PH Reddy (2017). Amyloid-Beta and phosphorylated Tau accumulations cause abnormalities at synapses of Alzheimer's disease neurons. J Alzheimers Dis 57(4):975–999. https://doi.org/10.3233/JAD-160612

P Ranasinghe, Y Mathangasinghe, R Jayawardena, AP Hills, A Misra (2017). Prevalence and trends of metabolic syndrome among adults in the Asia-pacific region: A systematic review. BMC Public Health 17(1):101. https://doi.org/10.1186/s12889-017-4041-1

J Rask Larsen, L Dima, CU Correll, P Manu (2018). The pharmacological management of metabolic syndrome. Expert Rev Clin Pharmacol 11(4):397–410. https://doi.org/10.1080/17512433.2018.1429910

C Reitz (2013a). Dyslipidemia and the risk of Alzheimer's disease. Curr Atheroscler Rep 15(3):307. https://doi.org/10.1007/s11883-012-0307-3

C Reitz (2013b). Dyslipidemia and the risk of Alzheimer's disease. Curr Atheroscler Rep 15(3):307. https://doi.org/10.1007/s11883-012-0307-3

A Riaz, S Asghar, S Shahid, H Tanvir, MH Ejaz, M Akram (2024). Prevalence of metabolic syndrome and its risk factors influence on microvascular complications in patients with type 1 and type 2 diabetes mellitus. Cureus 16(3):e55478. https://doi.org/10.7759/cureus.55478

AJ Richard, U White, CM Elks, JM Stephens (2020). Adipose Tissue: Physiology to Metabolic Dysfunction. In Feingold KR, Anawalt B, Blackman MR, Boyce A, Chrousos G, Corpas E, de Herder WW, Dhatariya K, Dungan K, Hofland J, Kalra S, Kaltsas G, Kapoor N, Koch C, Kopp P, Korbonits M, Kovacs CS, Kuohung W, Laferrère B, Levy M, McGee EA, McLachlan R, New M, Purnell J, Sahay R, Shah AS, Singer F, Sperling MA, Stratakis CA, Trence DL, Wilson DP (eds) Endotext. South Dartmouth, MA: MDText.com, Inc.

CK Roberts, KK Sindhu (2009). Oxidative stress and metabolic syndrome. Life Sci 84(21–22):705–712. https://doi.org/10.1016/j.lfs.2009.02.026

M Rojas, M Chávez-Castillo, D Pirela, H Parra, M Nava, M Chacín, L Angarita, R Añez, J Salazar, R Ortiz, S Durán Agüero, M Gravini-Donado, V Bermúdez, E Díaz-Camargo (2021b). Metabolic syndrome: Is it time to add the Central nervous system? Nutrients 13(7):2254. https://doi.org/10.3390/nu13072254

M Rojas, D Pirela, H Parra, M Nava, M Chac, L Angarita, R Añez, J Salazar, R Ortiz, S Dur, M Gravini-donado, V Berm, D Edgar (2021a). Metabolic syndrome : Is it time to add the Central nervous system? Nutrients 13(2254):1–22. https://doi.org/10.3390/nu13072254

E Rojas-Gutierrez, G Muñoz-Arenas, S Treviño, B Espinosa, R Chavez, K Rojas, G Flores, A Díaz, J Guevara (2017). Alzheimer's disease and metabolic syndrome: A link from oxidative stress and inflammation to neurodegeneration. Synapse 71(10):e21990. https://doi.org/10.1002/syn.21990

C Saraiva, C Praça, R Ferreira, T Santos, L Ferreira, L Bernardino (2016). Nanoparticle-mediated brain drug delivery: Overcoming blood-brain barrier to treat neurodegenerative diseases. J Control Release Off J Control Release Soc 235:34–47. https://doi.org/10.1016/j.jconrel.2016.05.044

M Shi, S Han, K Klier, G Fobo, C Montrone, S Yu, M Harada, A-K Henning, N Friedrich, M Bahls, M Dörr, M Nauck, H Völzke, G Homuth, HJ Grabe, C Prehn, J Adamski, K Suhre, W Rathmann, A Ruepp, J Hertel, A Peters, R Wang-Sattler (2023). Identification of candidate metabolite biomarkers for metabolic syndrome and its five components in population-based human cohorts. Cardiovasc Diabetol 22(1):141. https://doi.org/10.1186/s12933-023-01862-z.

VP Singh, A Bali, N Singh, AS Jaggi (2014). Advanced glycation end products and diabetic complications. Korean J Physiol Pharmacol 18(1):1. https://doi.org/10.4196/kjpp.2014.18.1.1

T Soda, T Pasqua, G De Sarro, F Moccia (2024). Cognitive impairment and synaptic dysfunction in cardiovascular disorders: The new frontiers of the Heart–Brain axis. Biomedicines 12(10):2387. https://doi.org/10.3390/biomedicines12102387

K Stemmer, TD Müller, RD DiMarchi, PT Pfluger, MH Tschöp (2019). CNS-targeting pharmacological interventions for the metabolic syndrome. J Clin Invest 129(10):4058–4071. https://doi.org/10.1172/JCI129195

S Swarup, I Ahmed, Y Grigorova, et al. (2025). Metabolic syndrome. [Updated 2024 Mar 7]. In StatPearls [Internet]. Treasure Island, FL: StatPearls Publishing. Available from: https://www.ncbi.nlm.nih.gov/books/NBK459248/

A Tylutka, B Morawin, Ł Walas. et al. (2023). Assessment of metabolic syndrome predictors in relation to inflammation and visceral fat tissue in older adults. Sci Rep 13:89. https://doi.org/10.1038/s41598-022-27269-6

P Van Dyken, B Lacoste (2018). Impact of metabolic syndrome on neuroinflammation and the blood-brain barrier. Front Neurosci 12:930. https://doi.org/10.3389/fnins.2018.00930

P Van Dyken, B Lacoste (2018). Impact of metabolic syndrome on neuroinflammation and the blood-brain barrier. Front Neurosci 12:930. https://doi.org/10.3389/fnins.2018.00930.

C Van Jaarsveldt, T Jabari, E Zwarts, S Färber, Y Sikuza, H Schilling, S Pauw, E Klein, C Van Rooyen, G Joubert, CC Van der Bijl (2024). Prevalence of metabolic syndrome among adults treated at a district hospital outpatient department. SAFP Off J South African Acad Fam Prac/Prim Care 66(1):e1–e7. https://doi.org/10.4102/safp.v66i1.5959

G Verdile, KN Keane, VF Cruzat, S Medic, M Sabale, J Rowles, N Wijesekara, RN Martins, PE Fraser, P Newsholme (2015). Inflammation and oxidative stress: The molecular connectivity between insulin resistance, obesity, and Alzheimer's disease. Mediat Inflam 2015(1):105828. https://doi.org/10.1155/2015/105828

S Virtue, A Vidal-Puig (2010). Adipose tissue expandability, lipotoxicity and the metabolic syndrome – An allostatic perspective. Biochim Biophys Acta BBA – Mol Cell Biol Lipids 1801(3):338–349. https://doi.org/10.1016/j.bbalip.2009.12.006

R Vona, L Gambardella, C Cittadini, E Straface, D Pietraforte (2019). Biomarkers of oxidative stress in metabolic syndrome and associated diseases. Oxidat Med Cell Long 8267234, 19. https://doi.org/10.1155/2019/8267234

H Wang, Q Ling, Y Wu, M Zhang (2023). Association between the triglyceride glucose index and cognitive impairment and dementia: A meta-analysis. Front Aging Neurosci 15. https://doi.org/10.3389/fnagi.2023.1278730

W-Y Wang, M-S Tan, J-T Yu, L Tan (2015). Role of pro-inflammatory cytokines released from microglia in Alzheimer's disease. Ann Transl Med 3(10):136. https://doi.org/10.3978/j.issn.2305-5839.2015.03.49

CH Wilson, S Kumar (2018). Caspases in metabolic disease and their therapeutic potential. Cell Death Differ 25:1010–1024. https://doi.org/10.1038/s41418-018-0111-x

Y Wu, Z Wang, X Jia, H Zhang, H Zhang, J Li, K Zhang (2019). Prediction of Alzheimer's disease with serum lipid levels in Asian individuals: A meta-analysis. Biomarkers 24(4):341–351. https://doi.org/10.1080/1354750X.2019.1571633

G Yan et al (2024). Integrative insights into cerebrometabolic disease: Understanding, management, and future prospects. J Neurorestoratol 12(2):100107.

KF Yates, V Sweat, PL Yau, MM Turchiano, A Convit (2012). Impact of metabolic syndrome on cognition and brain: A selected review of the literature. Arterioscler Thromb Vasc Biol 32(9):2060–2067. https://doi.org/10.1161/ATVBAHA.112.252759

W Zhou, Y Yao, J Li, D Wu, M Zhao, Z Yan, A Pang, L Kong (2019). TIGAR attenuates high glucose-induced neuronal apoptosis *via* an autophagy pathway. Front Mol Neurosci 12:193. https://doi.org/10.3389/fnmol.2019.00193

X Zong, R Kelishadi, HS Kim, P Schwandt, TE Matsha, JG Mill, CA Caserta, CCM Medeiros, A Kollias, PH Whincup, L Pacifico, A López-Bermejo, M Zhao, M Zheng, B Xi (2024). A proposed simplified definition of metabolic syndrome in children and adolescents: A global perspective. BMC Med 22(1):190. https://doi.org/10.1186/s12916-024-03406-y

11 Molecular and Cellular Responses to Spinal Cord Injury

Anjana Sinha, Shampa Ghosh, Krishna Kumar Singh, Amit Kumar Pandey, and Jitendra Kumar Sinha

11.1 INTRODUCTION

The most debilitating type of neurological damage is spinal cord injury (SCI) among other neurological disorders, which usually frequently leads to irreversible motor, sensory, and autonomic dysfunction. Beyond just physical impairment, SCI has major psychological, social, and economic repercussions for both individuals and our entire society.[1,2] Approximately, there are 15–40 occurrences of SCI per million people worldwide each year, with greater rates in low- and middle-income nations because of higher risk factors such violence, falls, and traffic accidents.[3,4] There are major two stages of SCI the primary injury, which is the initial mechanical disruption of spinal cord tissue, and the secondary injury, which is a series of cellular and biochemical processes that worsen the damage over the course of hours to weeks following the insult.[5,6] Although primary damage cannot be reversed, there is still window for therapeutic intervention during the secondary phase. This has led to a significant change in SCI research, with a focus on comprehending the cellular and molecular processes that underlie secondary injury. Spinal cord damage triggers a cascade inflammatory response at the molecular level that includes free radicals, chemokines, proteases, and cytokines. These factors work together to cause neuronal death and demyelination.[7,8] Concomitantly a wide range of cell types, including as oligodendrocytes, astrocytes, microglia, and infiltrating immune cells, experience notable phenotypic changes, some of which worsen neurodegeneration and others of which are neuroprotective.[9] In addition to determining the degree of tissue preservation and functional recovery, these overlapping and temporally dynamic processes may also be targets for therapeutic modulation. There are still few viable treatments for SCI despite decades of research. The complexity of the injury environment has made clinical translation difficult, despite current strategies that aim to reduce secondary damage, improve axonal regeneration, and encourage remyelination.[10] Thus, a better comprehension of the cellular and molecular reactions to SCI is essential for creating combinatorial and targeted treatments. The molecular and cellular reactions to SCI, such as inflammation, apoptosis, glial scarring, and neuronal regeneration, are thoroughly covered in this chapter. Additionally, it looks at how these processes interact to affect recovery and talks about new and existing therapy approaches that try to control or harness these reactions for better results.

11.2 PATHOPHYSIOLOGY OF SPINAL CORD INJURY

The two distinct but interconnected stages of SCI pathophysiology are the first injury brought on by mechanical trauma and the secondary injury (Figure 11.1), which is a complex series of cellular and molecular processes that worsen tissue damage and impede healing.[11]

DOI: 10.1201/9781003500025-11

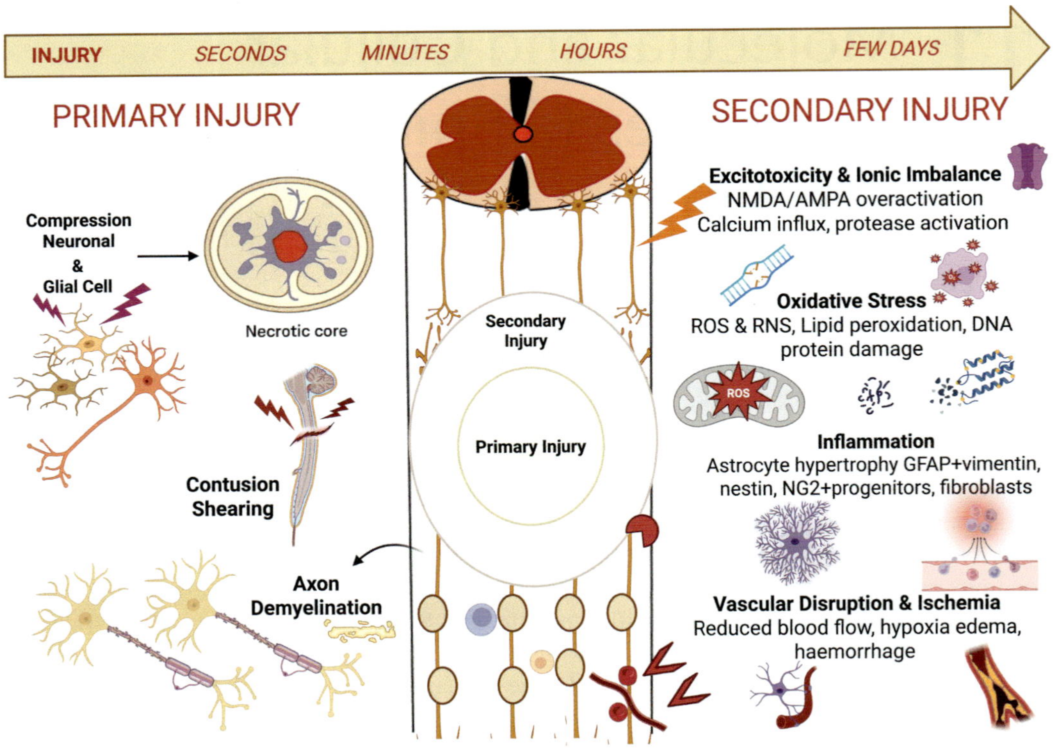

FIGURE 11.1 Primary and secondary injury cascade after central nervous system trauma. The initial primary injury occurs within seconds and is produced by mechanical forces (compression, contusion, and shearing) that cause immediate neuronal and glial cell damage, axonal disruption and demyelination, and formation of a necrotic core. Over minutes to hours and into days, a secondary injury expands outward from the primary focus. Key mechanisms include excitotoxicity and ionic imbalance (NMDA/AMPA overactivation, Ca^{2}+ influx, protease activation), oxidative stress (ROS/RNS, lipid peroxidation, DNA/protein damage), inflammation (reactive astrocytosis with GFAP, vimentin and nestin expression; recruitment of NG2+ progenitors and fibroblasts), and vascular disruption/ischemia (reduced blood flow, hypoxia, oedema, haemorrhage). Together, these processes propagate tissue loss and further demyelination, enlarging the lesion beyond the primary site over time. The timeline at the top denotes the progression from seconds to days.

11.2.1 Primary Injury

The primary injury occurs following trauma and can be defined by direct structural harm to the neuronal and vascular components of the spinal cord. Compression, contusion, laceration, or transection of the cord can cause the immediate loss of nerve cells, glial cells, and blood vessels.[12] The biomechanical forces involved trigger tissue deformation, disruption of axonal pathways, haemorrhage, and microvascular rupture, leading to a necrotic core at the exact location of injury (Figure 11.1).

11.2.2 Secondary Injury

The secondary injury phase begins within minutes of the main insult and can last for weeks or more. This phase involves a series of biochemical and cellular processes that worsen the initial injury. Secondary damage is characterised by ischaemia, inflammation, excitotoxicity, ionic imbalance, oxidative stress, and apoptosis, which all lead to increasing cell death and demyelination.[13]

11.2.3 Vascular Disruption and Ischemia

Disruption of the spinal vasculature causes reduced perfusion, ischaemia, and hypoxia in the spinal cord parenchyma. Breakdown of the blood-spinal cord barrier causes oedema, infiltration of immune cells, and haemorrhage, aggravating the local injury.[14] Reduced oxygen and glucose availability contribute to energy failure and metabolic dysfunction in neurons and glia.

11.2.4 Excitotoxicity and Ionic Imbalance

The following damage, excessive excitatory neurotransmitter release, such as glutamate, causes excitotoxicity, predominantly by overstimulation of N-methyl-D-aspartate (NMDA) and α-amino-3-hydroxy-5-methyl-4-isoxazolepropionic acid (AMPA) receptors. This results in tremendous calcium influx into neurons and glia, activating calpains and other calcium-dependent proteases that degrade cytoskeletal and membrane proteins.[15] Potassium efflux and sodium buildup both destabilise membrane potential and cellular equilibrium.

11.2.5 Oxidative Stress and Lipid Peroxidation

The secondary damage generates huge amounts of reactive oxygen species (ROS) and reactive nitrogen species (RNS), causing oxidative stress. This causes lipid peroxidation, DNA damage, protein oxidation, and mitochondrial dysfunction.[8] Free radicals reduce cell survival and promote inflammation, resulting in a self-sustaining cycle of harm.

11.2.6 Inflammation

The SCI activates a strong inflammatory response that includes resident microglia and astrocytes, as well as infiltrating neutrophils, macrophages, and T lymphocytes.[16] While initially protective, chronic inflammation can lead to bystander damage via pro-inflammatory cytokines (examples include tumour necrosis factor-alpha (TNF-α), interleukin-1 beta (IL-1β)), chemokines, and matrix metalloproteinases (MMP).[17] The polarisation of macrophages into neurotoxic (M1) or neuroprotective (M2) phenotypes is an important predictor of outcome.

11.2.7 Apoptosis and Cell Death

During the subacute phase of SCI, neurones and oligodendrocytes experience widespread apoptosis. Both internal (mitochondrial) and extrinsic (death receptor-mediated) apoptotic mechanisms are engaged, which involve Bcl-2 family proteins, cytochrome C release, caspase-3 activation, and DNA fragmentation.[18] Apoptosis promotes demyelination and further restricts axonal conduction and repair.

11.2.8 Glial Scar and Inhibitory Environment

As a late response, stimulated astrocytes proliferate and create a glial scar, which has both protective and inhibitory functions. It helps seal the injury site and prevent further inflammation, but it also inhibits axonal regeneration by expressing inhibitory chemicals, such as chondroitin sulphate proteoglycans (CSPGs). Understanding this evolving pathophysiological landscape is essential for identifying therapeutic targets aimed at reducing secondary damage, promoting neuroprotection, and facilitating regeneration after SCI.

11.2.9 Astrocyte Reactivity and Glial Scar Composition

After SCI, a process known as reactive astrogliosis occurs in astrocytes, the main constituent of the glial scar. This includes vimentin, nestin, and glial fibrillary acidic protein (GFAP) hypertrophy,

TABLE 11.1
The Timeline of Pathological Events That Follows Spinal Cord Injury

Time Post-Injury	Pathological Events	Key Features/Molecules
Immediate (seconds to minutes)	Primary mechanical injuries, vascular disruption, haemorrhage, cell membrane rupture	Mechanical compression, haemorrhage, axonal shearing
Early acute (minutes to hours)	Excitotoxicity, ionic imbalance, oxidative stress, oedema	↑ Glutamate, ↑ Ca^{2}+ influx, ROS/RNS production
Acute (hours to days)	Inflammation onset, immune cell infiltration, necrosis, apoptosis initiation	TNF-α, IL-1β, neutrophils, and microglial activation
Subacute (days to weeks)	Glial scar formation, sustained inflammation, apoptosis, demyelination, Wallerian degeneration	GFAP↑, caspase-3 activation, CSPGs, MMPs
Chronic (weeks to months)	Stabilised glial/fibrotic scar, cavity formation, axonal dieback, regeneration failure	Persisting CSPGs, axonal retraction, and ECM remodelling

proliferation, and overexpression.[19] Numerous signals, such as transforming growth factor-β (TGF-β), TNF-α, and IL-1β, are increased in the injury microenvironment and induce astrocyte activation.[20] In order to physically isolate the injured area, reactive astrocytes move towards the lesion site and create a dense network of processes. Astrocytes are not the only cells that make up the glial scar. Microglia, NG2+ oligodendrocyte progenitor cells, and fibroblasts derived from the meninges and perivascular spaces are further cellular contributions.

Astrocytes are not the only cells that make up the glial scar. Microglia, NG2+ oligodendrocyte progenitor cells, and fibroblasts derived from the meninges and perivascular spaces are further cellular contributions.[21] The inhibitory landscape is further amplified by fibroblasts, in particular, which help build a fibrotic scar inside the lesion core, frequently combined with infiltrating macrophages (Table 11.1).

11.3 MOLECULAR RESPONSES TO SPINAL CORD INJURY

The molecular response to SCI is a complex and dynamic network of signalling channels that govern both degenerative and reparative processes. Following the initial mechanical insult, an intricate interplay of pro-inflammatory cytokines, proteolytic enzymes, neurotrophic factors, apoptotic regulators, and epigenetic changes shapes the course of tissue damage and neural plasticity. In order to understand these molecular events, it is critical for designing pharmacological or gene-based therapies that preserve brain architecture while encouraging functional recovery.

11.3.1 Inflammatory Mediators

11.3.1.1 Cytokines: TNF-α, IL-1β, and IL-6

The pro-inflammatory cytokines such as TNF-α, IL-1β, and interleukin-6 (IL-6) play a key role in the early and long-lasting response to SCI.[22] After the activation of the microglia, infiltrating macrophages, and reactive astrocytes release these cytokines within minutes to hours of injury. TNF-α increases inflammation and death by activating the NF-κB and caspase-8 pathways, but it also helps with tissue healing in further stages. Increased IL-1β levels are linked to neuronal death and glial activation. The blocking of IL-1 signalling can reduce the damage to the tissue.[23] The IL-6 has both pro- and anti-inflammatory actions, influencing astrocyte activation, B cell maturation, and acute-phase response regulation.[24] The cytokine milieu influences not just the local immune response but also the severity of secondary damage.

11.3.1.2 Chemokines and Recruitment of Immune Cells

Chemokines such as CXCL1, CCL2 (MCP-1), and CX3CL1 (fractalkine) direct the recruitment and accumulation of neutrophils, monocytes, and T-cells into the damaged spinal cord.[25] These molecules are upregulated at the lesion site and act on certain receptors (e.g., CCR2, CX3CR1) in order to regulate immune cell trafficking. While peripheral immune cells are required for debris clearing and wound stabilisation, their excessive or chronic recruitment exacerbates tissue damage and scar formation.

11.3.2 Matrix Metalloproteinases (MMPs)

The MMPs are zinc-dependent endopeptidases that modify the extracellular matrix (ECM). In SCI, MMPs, especially MMP-2 and MMP-9, are highly elevated and contribute to both blood-spinal cord barrier (BSCB) disintegration and neuroinflammation.[26] The MMP-9 is rapidly produced within hours post-injury and plays a critical role in the degradation of tight junction proteins such as claudin-5 and occludin, hence increasing vascular permeability and immune cell infiltration. MMP-2, on the other hand, shows a more delayed expression pattern and is involved in long-term scar maturation and axonal remodelling.[27] Although MMP inhibition has shown neuroprotective effects in preclinical studies, their multifaceted activities demand context-specific targeted techniques.

11.3.3 Growth Factors and Neurotrophins

The growth factors and neurotrophins influence neuronal survival, axonal regeneration, synaptic plasticity, and glial behaviour following SCI. Among the most extensively studied are brain-derived neurotrophic factor (BDNF), nerve growth factor (NGF), and glial cell line-derived neurotrophic factor (GDNF). BDNF, via its receptor TrkB, promotes neuronal survival, synaptic effectiveness, and axonal sprouting. In experimental SCI models, administering BDNF after damage improves motor function and lowers lesion size. NGF acts on TrkA receptors to induce the regeneration of sensory neurons and alter pain signalling pathways. The GDNF, which signals through the RET receptor complex, supports the survival of motor neurons and has been shown to decrease apoptosis and promote tissue sparing. Despite their regeneration potential, neurotrophins have limited therapeutic applications due to their short half-lives, poor tissue penetration, and dose-dependent adverse effects. Advances in gene therapy and biomaterial delivery systems attempt to break down these barriers and allow controlled, localised release of these molecules.

11.3.4 Apoptosis-Related Genes and Proteins

Apoptosis is a hallmark of the secondary injury phase in SCI, contributing significantly to oligodendrocyte loss, myelin degeneration, and reduced axonal conductivity. Both intrinsic (mitochondrial) and extrinsic (death receptor) mechanisms culminate on effector caspase activation, specifically caspase-3.[27] Activated caspase-3 is a strong indicator of apoptotic death in neurones and oligodendrocytes, especially in white matter regions days after injury.[27] The Bcl-2 family of proteins regulates cytochrome c release and the permeability of the outer membrane of the mitochondria. These proteins include pro-apoptotic members (Bax, Bak) and anti-apoptotic regulators (Bcl-2, Bcl-xL). Additionally, the tumour suppressor p53 is essential for detecting DNA damage and initiating apoptosis, especially in neurones and glial progenitor cells.[28] In experimental settings, strategies that repress p53, overexpress Bcl-2, or decrease caspase activity have shown neuroprotective effects. Targeting apoptosis alone is challenging, though, because it is intricately linked to inflammation, necroptosis, and autophagy.

11.3.5 Epigenetic Modifications

After SCI, epigenetic alterations including DNA methylation, histone modifications, and non-coding RNAs have become significant modulators of gene expression. Without changing the underlying DNA sequence, these changes have an impact on decisions about cell destiny, inflammation, and regeneration. The spinal cord tissues have been shown to exhibit both selective hypermethylation of neuroprotective gene promoters and post-injury global hypomethylation. Following SCI, immune gene expression is impacted by the dysregulation of enzymes such as DNA methyltransferases (DNMTs) and ten-eleven translocation (TET) enzymes. The histone acetylation alters transcriptional accessibility and chromatin structure and is controlled by histone acetyltransferases (HATs) and histone deacetylases (HDACs).

Following SCI, HDAC inhibitors such as valproic acid have demonstrated promise in lowering inflammation and fostering neurogenesis.[29] Additionally, the damaged spinal cord exhibits variable expression of non-coding RNAs, such as long non-coding RNAs (lncRNAs) and microRNAs (miRNAs). For instance, miR-124 encourages neuronal differentiation and lowers inflammation, whereas miR-21 is increased after SCI and contributes to astrocyte reactivity and scar formation.[30,31]

11.4 CELLULAR RESPONSES TO SPINAL CORD INJURY

Central nervous system cells coordinate a complex and dynamic response that influences inflammation, healing, and regeneration following SCI. The main cellular participants—astrocytes, neurones, oligodendrocyte-lineage cells, microglia/macrophages, and invading immune cells—are reviewed in this section with a focus on their polarisation states, temporal roles, and contributions to both pathology and recovery.

11.4.1 Microglia and Macrophage Activation

The central nervous system's (CNS) resident immune cells, known as microglia, are activated minutes after SCI. A variety of phenotypes are displayed by activated microglia and infiltrating monocyte-derived macrophages, which are traditionally classified as M1 (pro-inflammatory) and M2 (anti-inflammatory and pro-repair). M1-type cells worsen neuronal damage by secreting cytokines like TNF-α and IL-1β. On the other hand, M2-type macrophages aid in tissue repair and axonal regeneration by producing IL-10 and arginase-1.[32] The persistence of M1 phenotypes after the acute phase can result in chronic neuroinflammation and demyelination, despite their advantageous roles.[8] Consequently, methods that alter macrophage polarisation are being researched in an effort to tip the scales in favour of a regenerative M2 response, overall summarised in Table 11.2.

11.4.2 Astrocyte Reactivity and Glial Scar Formation

A characteristic of reactive gliosis, astrocytes undergo hypertrophy, proliferate, and upregulate GFAP in response to injury.[33] The reactive astrocytes produce inhibitory chemicals, such as CSPGs, which build the glial scar, but they also initially aid in neuroprotection by isolating the damage site and repairing the blood–spinal cord barrier (BSCB). Although new research indicates

TABLE 11.2
Microglia and Macrophage Polarisation After SCI

Subtype	Markers	Functions	Impact on Recovery
M1	iNOS, TNF-α, IL-1β	Pro-inflammatory	Worsens secondary damage
M2	Arg-1, IL-10, CD206	Anti-inflammatory, tissue repair	Promotes regeneration

TABLE 11.3
Comparison of Major Glial Cell Responses

Cell Type	Marker	Response to SCI	Dual Roles
Microglia	Iba1, CD68	Activation, phagocytosis	Damage and repair
Astrocytes	GFAP	Scar formation, cytokine release	Protection and inhibition
Oligodendrocytes	MBP, CNPase	Apoptosis, demyelination	Limited remyelination

that astrocytes may promote regeneration in specific circumstances, contingent on the extracellular matrix composition and signalling signals, the glial scar serves as a physical and physiological barrier to axonal regrowth.[34,35] As a result, the astrocytic response has two aspects, and controlling it may have therapeutic benefits, which are summarised in Table 11.3.

11.4.3 Oligodendrocyte Death and Demyelination

As the cells that myelinate central nervous system axons, oligodendrocytes are particularly vulnerable to secondary damage mechanisms, such as oxidative stress and excitotoxicity. Demyelination from their loss affects axonal conduction and leads to functional impairments.[35] After an injury, endogenous oligodendrocyte precursor cells (OPCs) multiply and, given the right circumstances, can develop into adult oligodendrocytes. However, inflammatory mediators and inhibitory extracellular cues frequently cause this remyelination process to be incomplete.[36] Ongoing regeneration methods aim to improve OPC differentiation and survival.

11.4.4 Neuronal Apoptosis and Survival

One of the main characteristics of SCI is the death of neurones, which can happen through extrinsic (death receptor-mediated) or intrinsic (mitochondrial-mediated) apoptotic mechanisms. As early as four hours after injury, caspase-3 activation has been found in neurones, suggesting that programmed cell death has begun quickly.[37] Additionally, the Bcl-2 family proteins and the p53 pathway are important regulators of neuronal death.[38] In some situations, partial functional recovery is caused by surviving neurones exhibiting a degree of neuroplasticity, such as sprouting and synaptic reorganisation, despite significant neuronal loss.[39] Therefore, neuronal survival and synaptic plasticity-promoting interventions are essential for SCI rehabilitation.

11.4.5 Infiltrating Immune Cells

The peripheral immune cells can infiltrate the lesion site because SCI interferes with the BSCB. Among the first to arrive are neutrophils, which cause additional tissue damage by producing ROS and proteases.[40] When T-cells, especially CD4+ and CD8+ subsets, enter the damaged cord, they can either promote healing through regulatory roles or worsen injury through cytotoxic actions.[41] The adaptive immune response is shaped by dendritic cells (DCs), which also penetrate the spinal cord and function as antigen-presenting cells.[42] They might be both potential allies and foes in SCI pathology due to their complicated and context-dependent roles, as summarised in Table 11.4.

11.5 THERAPEUTIC IMPLICATIONS

Therapeutic approaches for SCI must take into account the complicated pathological landscape of the condition and the multiple nature of the main and subsequent injury cascades. Key pathogenic pathways such as inflammation, apoptosis, and regeneration failure are the focus of current research

TABLE 11.4
Types of Infiltrating Immune Cells and Their Functions

Cell Type	Arrival Time Post-Injury	Key Functions	Harmful or Protective
Neutrophils	Within hours	ROS release, proteases	Harmful
T-cells	Days	Modulate inflammation	Mixed
Dendritic Cells	Days–weeks	Antigen presentation	Mostly harmful

and clinical application methods. A number of significant treatment approaches with encouraging translational potential are outlined in this section.

11.5.1 Anti-inflammatory Agents

One of the main causes of subsequent damage after SCI is inflammation. The goal of therapeutic inflammation targeting is to lessen immune cell recruitment and activation as well as the generation of reactive species and pro-inflammatory cytokines. Ibuprofen and indomethacin are examples of non-steroidal anti-inflammatory medications (NSAIDs) that have shown some neuroprotective effects by lowering prostaglandin synthesis and blocking cyclooxygenase enzymes.[43] The acute treatment of SCI has traditionally involved the administration of corticosteroids, especially methylprednisolone. They work by inhibiting lipid peroxidation, suppressing the transcription of inflammatory genes, and maintaining the integrity of the blood-spinal cord barrier.[44] However, because of serious side effects such immunosuppression and gastrointestinal issues, their use is still debatable.[45] A tetracycline derivative called minocycline has shown promise as an anti-inflammatory and anti-apoptotic drug in preclinical models of SCI. It lessens neuronal loss and enhances functional recovery by blocking matrix metalloproteinase expression, microglial activation, and caspase-1 and -3 activity.[26]

11.5.2 Neuroprotective and Regenerative Strategies

The cell-based treatments have become more popular as regenerative and neuroprotective approaches. Mesenchymal stem cells (MSCs), neural stem/progenitor cells (NSPCs), and induced pluripotent stem cells (iPSCs) are among the stem cell transplants that have the potential to replace neurones, provide trophic support, and alter the inflammatory microenvironment.[46]

The delivery of neurotrophic factors is another compelling tactic. Neuronal survival, axonal regeneration, and synaptic plasticity are all facilitated by substances like GDNF, BDNF, and NGF. Systemic effects have been reduced, and therapeutic efficacy has increased with controlled, localised delivery using viral vectors or biomaterial scaffolds.

11.5.3 Targeting Molecular Pathways

Modulation of specific molecular pathways involved in cell death and matrix remodelling represents a rational therapeutic approach. Caspase inhibitors have shown promise in preclinical models by blocking apoptotic cascades. For instance, inhibition of caspase-3 activation has led to reduced neuronal apoptosis and improved behavioural outcome.[27] MMPs, especially MMP-9 and MMP-12, are implicated in tissue degradation and breakdown of the blood-spinal cord barrier. Better histological preservation and functional results have been linked to their suppression by pharmacological agents or gene silencing.[25,26] The precise modification of genes implicated in the pathophysiology of SCI is being investigated using gene editing technologies, such as clustered regularly interspaced short palindromic repeats (CRISPR)-Cas9. Targeting genes that control neurotrophic signalling, apoptosis, or inflammation may allow for customised therapeutic interventions; however, clinical translation is still in its early phases.[47]

11.5.4 Biomaterial and Scaffold Approaches

A supporting matrix for cellular infiltration, axonal regeneration, and the regulated release of medicinal drugs is offered by scaffolds made of biomaterials. Hydrogels made of synthetic (like PEG, PLGA) or natural (like collagen, hyaluronic acid) polymers have been developed to carry medications, neurotrophic agents, or stem cells straight to the site of injury.[48] These scaffolds can be loaded with nanoparticles or functionalised with ligands to accomplish immunomodulation, targeted delivery, and sustained release. For instance, in experimental SCI, hydrogel-based BDNF or GDNF delivery has demonstrated improved axonal regeneration and decreased glial scar formation.[49] When taken as a whole, these therapeutic approaches offer a multifaceted strategy for treating SCI. In order to optimise neuroprotection and functional recovery, future efforts will focus on optimising the time, dose, and combinatorial administration of these therapies.

11.6 CHALLENGES AND FUTURE DIRECTIONS

The pathophysiology of SCI has become better understood, but its clinical translation is still a challenge. The preclinical models used to investigate SCI frequently fall short of reproducing the complexity and variability seen in actual injuries, which is one of the main drawbacks. The majority of animal models, especially rodent models, use controlled contusion or transection injuries in young, healthy subjects. These settings are very different from the age-related reactions, comorbidities, and varied processes in human patients.[46] The elevated prevalence of translational failure of experimental medicines in clinical trials is largely caused by these discrepancies.

The existence of translational barriers, such as immunological incompatibilities among species, variations in pharmacodynamics and pharmacokinetics, and moral dilemmas related to the application of gene-editing or stem cell-based treatments in humans, is another significant obstacle. Additionally, administering neuroprotective and regenerative medicines is made more difficult by the blood–spinal cord barrier (BSCB), which restricts drug transport to the central nervous system.[50]

A substantial functional recovery is unlikely with monotherapies that focus on a single biochemical route due to the complex nature of SCI pathophysiology. The necessity of customised or combinatorial treatments that concurrently address inflammation, cell death, demyelination, and axonal development inhibition is becoming more widely acknowledged. For instance, anti-inflammatory drugs may be more effective when combined with growth-promoting substances or stem cell transplants.[9]

In the future, new technologies present exciting opportunities for creativity. The use of CRISPR-Cas9 gene-editing technology has made it possible to precisely modify genes related to axonal regeneration, death, or inflammation.[51] In mouse models, for example, it has been demonstrated that specific deletion of the phosphatase and tensin homolog (PTEN) gene promotes axonal regeneration.[52]

Comparably, single-cell RNA sequencing, or scRNA-seq, is transforming our knowledge of the temporal dynamics and cellular heterogeneity of SCI. High-resolution mapping of transcriptome alterations specific to cell types is made possible by this method, providing information on the molecular conditions of neurones, glia, and immune cells that infiltrate throughout time.[53] The creation of stage-specific therapies could be guided by this kind of thorough characterisation.

The application of machine learning and artificial intelligence (AI) to injury classification, prognosis, and therapy optimisation is another frontier. To forecast recovery trajectories and choose the best times for interventions, AI can analyse complicated information, such as MRI imaging, electrophysiological, and multi-omics profiles.[54] Precision medicine strategies catered to unique damage patterns and biological reactions may be made possible by this.

Therefore, even though SCI research has advanced significantly, interdisciplinary and integrative work is still needed to address current issues. Overcoming translational barriers and bringing next-generation treatments from the laboratory to the bedside will need sustained investment in model improvement, biomarker identification, and computational tools. Some of the future directions could be as follows:

1. *Humanised Animal Models to Improve Immune Translational Relevance*

   Humanised mouse models that include elements of the human immune system have advanced. These can aid in bridging the gap between preclinical results in rodents and clinical outcomes in humans by accurately reflecting human immunological responses, particularly those to injury and medication.[55,56]
2. *Combination Therapies and Multifaceted Interventions*

   Combinatorial approaches—combining cell transplantation, biomaterials, growth factors, immunomodulation, nanotechnology, etc.—have garnered a lot of support. In preclinical conditions, these work better than single-agent treatments.[57,58]

   One potential synergy between stem/progenitor cells and nanoparticles (NPs) is the combination of cell grafts and biodegradable polymers, such as PLGA, that deliver trophic or anti-inflammatory cargos.[58]
3. *Biomaterials, Scaffolds, and Microenvironment Modulation*

   The development of bioactive scaffolds that modify the glial scar, supply neurotrophic factors, or stimulate angiogenesis is ongoing. These aid by guiding regenerating fibres, promoting axonal growth, modifying the hostile post-injury milieu, and assisting newly differentiated cells.[59] The more focus is being placed on physical regulatory techniques (such as fibre alignment, scaffold matrix stiffness, and controlled release of biomolecules).[59]
4. *Neurostimulation and Rehabilitation Innovations*

   Though they are not usually cures, methods like robotic exoskeletons, transcutaneous spinal cord stimulation, epidural stimulation, and more rigorous rehabilitative training are showing promise as adjuncts that can significantly enhance function.[60,61] Time is crucial: Early haemodynamic optimisation (maintaining spinal perfusion pressure) and early surgical decompression are increasingly recognised as significant controllable factors.
5. *Single-Cell Omics and Molecular Mapping*

   As you've already mentioned, scRNA-seq is mapping cellular heterogeneity following SCI. More recent reviews have emphasised spatial transcriptomics and multispectral imaging to better understand how immune cells, astrocytes, microglia, and neurones behave in time and place. This helps identify biomarkers and possibly "windows" when therapies could be more successful.[59]
6. *Clinical Trials and Translation Efforts*

   A number of recent or ongoing clinical trials of cell therapies for human SCI have been conducted; for instance, the "Stem cell therapies for spinal cord injury in humans" review (2025) summarises the results of human trials, the endpoints used, and the gaps that still exist. Some trials combine therapies (e.g., immunomodulation + rehabilitation, scaffolds + stem cells), but many use small sample sizes, have heterogeneity in injury level/severity, and limited long-term follow-up, all of which make it difficult to identify clinically meaningful improvements.[62]

11.7 CONCLUSION

A debilitating neurological disorder, SCI is typified by intricate and dynamic pathogenic cascades that change over time. The main molecular and cellular processes that contribute to secondary injury have been covered in detail in this chapter. These include inflammation, glial activation, demyelination, neuronal death, and disruption of the blood–spinal cord barrier (BSCB). Every biological reaction, from oligodendrocyte loss and immune cell infiltration to microglial polarisation and astrocyte reactivity, has a dual function in aggravating injury and starting the healing process.[63,64] Crucially, complex molecular signalling pathways, including as cytokines (like TNF-α and IL-1β), MMP, neurotrophic factors (like BDNF and GDNF), and apoptotic regulators like caspases and the Bcl-2 family, control these cellular reactions.[8] Developments in neuroimmunology and molecular biology have started to clarify how these pathways interact to dictate post-injury cell fate and regenerative potential. These results highlight the critical need for multi-targeted and integrative therapeutic approaches from a clinical perspective. While standard pharmacological approaches such as corticosteroids and NSAIDs provide little neuroprotection, novel therapies using stem cells, biomaterials, neurotrophic factor administration, and gene-editing technologies like CRISPR give renewed promise.[65,66] Furthermore, AI-guided diagnostics and single-cell transcriptomics-enabled precision medicine techniques could lead to customised treatment plans based on each patient's unique biological profile.

Integrating knowledge from advanced biomaterials, artificial intelligence, and systems biology with clinical neurology will be crucial as we proceed. In addition to stopping the progression of secondary damage, such interdisciplinary and integrative treatments have the potential to significantly improve the long-term quality of life for people with SCI and enable real neurological rehabilitation. In summary, knowing the cellular and molecular landscape of SCI is critical for both improving clinical outcomes and creating successful therapies. For SCI patients, an integrative strategy that incorporates immunomodulation, neuroprotection, and regenerative techniques has the potential to greatly enhance rehabilitation and quality of life.

REFERENCES

1. Oyinbo CA. Secondary injury mechanisms in traumatic spinal cord injury: A nugget of this multiply cascade. *Acta Neurobiol Exp (Wars)*. 2011;71(2):281–299. doi:10.55782/ane-2011-1848.
2. Fehlings M, Singh A, Tetreault L, Kalsi-Ryan S, Nouri A. Global prevalence and incidence of traumatic spinal cord injury. *CLEP*. 2014:309. doi:10.2147/CLEP.S68889
3. Alizadeh A, Dyck SM, Karimi-Abdolrezaee S. Traumatic spinal cord injury: an overview of pathophysiology, models and acute injury mechanisms. *Front Neurol*. 2019;10:282. doi:10.3389/fneur.2019.00282
4. Lee BB, Cripps RA, Fitzharris M, Wing PC. The global map for traumatic spinal cord injury epidemiology: Update 2011, global incidence rate. *Spinal Cord*. 2014;52(2):110–116. doi:10.1038/sc.2012.158
5. Ahuja CS, Martin AR, Fehlings MG. Recent advances in managing a spinal cord injury secondary to trauma. *F1000Res*. 2016;5:1017. doi:10.12688/f1000research.7586.1
6. Tran AP, Warren PM, Silver J. The biology of regeneration failure and success after spinal cord injury. *Physiol Rev*. 2018;98(2):881–917. doi:10.1152/physrev.00017.2017
7. Ahuja CS, Nori S, Tetreault L, et al. Traumatic spinal cord injury—repair and regeneration. *Neurosurgery*. 2017;80(3S):S9–S22. doi:10.1093/neuros/nyw080
8. David S, Kroner A. Repertoire of microglial and macrophage responses after spinal cord injury. *Nat Rev Neurosci*. 2011;12(7):388–399. doi:10.1038/nrn3053
9. Silva NA, Sousa N, Reis RL, Salgado AJ. From basics to clinical: a comprehensive review on spinal cord injury. *Prog Neurobiol*. 2014;114:25–57. doi:10.1016/j.pneurobio.2013.11.002
10. Rowland JW, Hawryluk GWJ, Kwon B, Fehlings MG. Current status of acute spinal cord injury pathophysiology and emerging therapies: Promise on the horizon. *FOC*. 2008;25(5):E2. doi:10.3171/FOC.2008.25.11.E2
11. Sekhon LHS, Fehlings MG. Epidemiology, demographics, and pathophysiology of acute spinal cord injury. *Spine*. 2001;26(Suppl):S2–S12. doi:10.1097/00007632-200112151-00002
12. Tator CH, Koyanagi I. Vascular mechanisms in the pathophysiology of human spinal cord injury. *J Neurosurg*. 1997;86(3):483–492. doi:10.3171/jns.1997.86.3.0483
13. Casella GTB, Marcillo A, Bunge MB, Wood PM. New vascular tissue rapidly replaces neural parenchyma and vessels destroyed by a contusion injury to the rat spinal cord. *Exper Neurol*. 2002;173(1):63–76. doi:10.1006/exnr.2001.7827

14. Park E, Velumian AA, Fehlings MG. The role of excitotoxicity in secondary mechanisms of spinal cord injury: a review with an emphasis on the implications for white matter degeneration. *J Neurotrauma.* 2004;21(6):754–774. doi:10.1089/0897715041269641
15. Liu N, Zhang YP, Titsworth WL, et al. A novel role of phospholipase A_2 in mediating spinal cord secondary injury. *Ann Neurol.* 2006;59(4):606–619. doi:10.1002/ana.20798
16. Shechter R, Miller O, Yovel G, et al. Recruitment of beneficial M2 macrophages to injured spinal cord is orchestrated by remote brain choroid plexus. *Immunity.* 2013;38(3):555–569. doi:10.1016/j.immuni.2013.02.012
17. Beattie MS, Farooqui AA, Bresnahan JC. Review of current evidence for apoptosis after spinal cord injury. *J Neurotrauma.* 2000;17(10):915–925. doi:10.1089/neu.2000.17.915
18. Silver J, Miller JH. Regeneration beyond the glial scar. *Nat Rev Neurosci.* 2004;5(2):146–156. doi:10.1038/nrn1326
19. Okada S, Nakamura M, Katoh H, et al. Conditional ablation of Stat3 or Socs3 discloses a dual role for reactive astrocytes after spinal cord injury. *Nat Med.* 2006;12(7):829–834. doi:10.1038/nm1425
20. Barnabé-Heider F, Göritz C, Sabelström H, et al. Origin of new glial cells in intact and injured adult spinal cord. *Cell Stem Cell.* 2010;7(4):470–482. doi:10.1016/j.stem.2010.07.014
21. Allan SM, Rothwell NJ. Cytokines and acute neurodegeneration. *Nat Rev Neurosci.* 2001;2(10):734–744. doi:10.1038/35094583
22. Pineau I, Lacroix S. Proinflammatory cytokine synthesis in the injured mouse spinal cord: multiphasic expression pattern and identification of the cell types involved. *J Compar Neurol.* 2007;500(2):267–285. doi:10.1002/cne.21149
23. Kwon B. Pathophysiology and pharmacologic treatment of acute spinal cord injury. *Spine J.* 2004;4(4):451–464. doi:10.1016/j.spinee.2003.07.007
24. Donnelly DJ, Popovich PG. Inflammation and its role in neuroprotection, axonal regeneration and functional recovery after spinal cord injury. *Exper Neurol.* 2008;209(2):378–388. doi:10.1016/j.expneurol.2007.06.009
25. Noble LJ, Donovan F, Igarashi T, Goussev S, Werb Z. Matrix metalloproteinases limit functional recovery after spinal cord injury by modulation of early vascular events. *J Neurosci.* 2002;22(17):7526–7535. doi:10.1523/JNEUROSCI.22-17-07526.2002
26. Wells JEA, Rice TK, Nuttall RK, et al. An adverse role for matrix metalloproteinase 12 after spinal cord injury in mice. *J Neurosci.* 2003;23(31):10107–10115. doi:10.1523/JNEUROSCI.23-31-10107.2003
27. Springer JE, Azbill RD, Knapp PE. Activation of the caspase-3 apoptotic cascade in traumatic spinal cord injury. *Nat Med.* 1999;5(8):943–946. doi:10.1038/11387
28. Takagi T, Takayasu M, Mizuno M, Yoshimoto M, Yoshida J. Caspase activation in neuronal and glial apoptosis following spinal cord injury in mice. *Neurol Med Chir(Tokyo).* 2003;43(1):20–30. doi:10.2176/nmc.43.20
29. Wadan AHS, Shaaban AH, El-Sadek MZ, et al. Mitochondrial-based therapies for neurodegenerative diseases: A review of the current literature. *Naunyn-Schmiedeberg's Arch Pharmacol.* 2025;398(9):11357–11386. doi:10.1007/s00210-025-04014-0
30. Wang Y, Lin ZP, Zheng HZ, et al. Abnormal DNA methylation in the lumbar spinal cord following chronic constriction injury in rats. *Neuroscience Letters.* 2016;610:1–5. doi:10.1016/j.neulet.2015.10.048
31. Deguchi K, Miyazaki K, Tian F, et al. Erratum to "Modifying neurorepair and neuroregenerative factors with tPA and edaravone after transient middle cerebral artery occlusion in rat brain" [Brain Research 1436 (2012) 168–178]. *Brain Res.* 2012;1444:112–113. doi:10.1016/j.brainres.2012.01.042
32. Shechter R, London A, Varol C, et al. Infiltrating blood-derived macrophages are vital cells playing an anti-inflammatory role in recovery from spinal cord injury in mice. *PLoS Med.* 2009;6(7):e1000113. doi:10.1371/journal.pmed.1000113
33. Burda JE, Sofroniew MV. Reactive gliosis and the multicellular response to CNS damage and disease. *Neuron.* 2014;81(2):229–248. doi:10.1016/j.neuron.2013.12.034
34. Anderson MA, Burda JE, Ren Y, et al. Astrocyte scar formation aids central nervous system axon regeneration. *Nature.* 2016;532(7598):195–200. doi:10.1038/nature17623
35. Crowe MJ, Bresnahan JC, Shuman SL, Masters JN, Beattie MS. Apoptosis and delayed degeneration after spinal cord injury in rats and monkeys. *Nat Med.* 1997;3(1):73–76. doi:10.1038/nm0197-73
36. Tripathi R, McTigue DM. Prominent oligodendrocyte genesis along the border of spinal contusion lesions. *Glia.* 2007;55(7):698–711. doi:10.1002/glia.20491
37. McEwen ML, Springer JE. A mapping study of caspase-3 activation following acute spinal cord contusion in rats. *J Histochem Cytochem.* 2005;53(7):809–819. doi:10.1369/jhc.4A6467.2005

38. Kotipatruni RR, Dasari VR, Veeravalli KK, Dinh DH, Fassett D, Rao JS. p53- and Bax-mediated apoptosis in injured rat spinal cord. *Neurochem Res.* 2011;36(11):2063–2074. doi:10.1007/s11064-011-0530-2
39. Raineteau O, Schwab ME. Plasticity of motor systems after incomplete spinal cord injury. *Nat Rev Neurosci.* 2001;2(4):263–273. doi:10.1038/35067570
40. Sterner RC, Sterner RM. Immune response following traumatic spinal cord injury: Pathophysiology and therapies. *Front Immunol.* 2023;13:1084101. doi:10.3389/fimmu.2022.1084101
41. Jones TB, Basso DM, Sodhi A, et al. Pathological CNS autoimmune disease triggered by traumatic spinal cord injury: Implications for autoimmune vaccine therapy. *J Neurosci.* 2002;22(7):2690–2700. doi:10.1523/JNEUROSCI.22-07-02690.2002
42. Ankeny DP, Guan Z, Popovich PG. B cells produce pathogenic antibodies and impair recovery after spinal cord injury in mice. *J Clin Invest.* 2009;119(10):2990–2999. doi:10.1172/JCI39780
43. Profyris C, Cheema SS, Zang D, Azari MF, Boyle K, Petratos S. Degenerative and regenerative mechanisms governing spinal cord injury. *Neurobiol Dis.* 2004;15(3):415–436. doi:10.1016/j.nbd.2003.11.015
44. Evaniew N, Noonan VK, Fallah N, et al. Methylprednisolone for the treatment of patients with acute spinal cord injuries: A propensity score-matched cohort study from a Canadian multi-center spinal cord injury registry. *J Neurotrauma.* 2015;32(21):1674–1683. doi:10.1089/neu.2015.3963
45. Hurlbert RJ. Methylprednisolone for acute spinal cord injury: An inappropriate standard of care. *J Neurosurg Spine.* 2000;93(1):1–7. doi:10.3171/spi.2000.93.1.0001
46. Tetzlaff W, Okon EB, Karimi-Abdolrezaee S, et al. A systematic review of cellular transplantation therapies for spinal cord injury. *J Neurotrauma.* 2011;28(8):1611–1682. doi:10.1089/neu.2009.1177
47. Bennett JP, Onyango IG. Energy, entropy and quantum tunneling of protons and electrons in brain mitochondria: Relation to mitochondrial impairment in aging-related human brain diseases and therapeutic measures. *Biomedicines.* 2021;9(2):225. doi:10.3390/biomedicines9020225
48. Yunta M, Nieto-Díaz M, Esteban FJ, et al. MicroRNA dysregulation in the spinal cord following traumatic injury. *PLoS ONE.* 2012;7(4):e34534. doi:10.1371/journal.pone.0034534
49. Schwartz CM. *Derivation, Enrichment and Characterization of Dopaminergic Neurons from Pluripotent Stem Cells.* Karolinska Institutet (Sweden); 2010.
50. Stenudd M, Sabelström H, Frisén J. Role of endogenous neural stem cells in spinal cord injury and repair. *JAMA Neurol.* 2015;72(2):235. doi:10.1001/jamaneurol.2014.2927
51. Danilov CA, Steward O. Conditional genetic deletion of PTEN after a spinal cord injury enhances regenerative growth of CST axons and motor function recovery in mice. *Exper Neurol.* 2015;266: 147–160. doi:10.1016/j.expneurol.2015.02.012
52. Liu K, Lu Y, Lee JK, et al. PTEN deletion enhances the regenerative ability of adult corticospinal neurons. *Nat Neurosci.* 2010;13(9):1075–1081. doi:10.1038/nn.2603
53. Milich LM, Ryan CB, Lee JK. The origin, fate, and contribution of macrophages to spinal cord injury pathology. *Acta Neuropathol.* 2019;137(5):785–797. doi:10.1007/s00401-019-01992-3
54. Cui Y, Zhu J, Duan Z, Liao Z, Wang S, Liu W. Artificial intelligence in spinal imaging: Current status and future directions. *IJERPH.* 2022;19(18):11708. doi:10.3390/ijerph191811708
55. Kim JT, Bresson-Tan G, Zack JA. Current advances in humanized mouse models for studying NK cells and HIV infection. *Microorganisms.* 2023;11(8):1984. doi:10.3390/microorganisms11081984
56. Karnik I, Her Z, Neo SH, Liu WN, Chen Q. Emerging preclinical applications of humanized mouse models in the discovery and validation of novel immunotherapeutics and their mechanisms of action for improved cancer treatment. *Pharmaceutics.* 2023;15(6):1600. doi:10.3390/pharmaceutics15061600
57. Griffin JM, Bradke F. Therapeutic repair for spinal cord injury: Combinatory approaches to address a multifaceted problem. *EMBO Mol Med.* 2020;12(3):e11505. doi:10.15252/emmm.201911505
58. García E, Sánchez-Noriega S, González-Pacheco G, González-Vázquez AN, Ibarra A, Rodríguez-Barrera R. Recent advances in the combination of cellular therapy with stem cells and nanoparticles after a spinal cord injury. *Front Neurol.* 2023;14:1127878. doi:10.3389/fneur.2023.1127878
59. Hu X, Xu W, Ren Y, et al. Spinal cord injury: Molecular mechanisms and therapeutic interventions. *Sig Transduct Target Ther.* 2023;8(1):245. doi:10.1038/s41392-023-01477-6
60. Montoto-Marqués A, Benito-Penalva J, Ferreiro-Velasco ME, et al. Advances and new therapies in traumatic spinal cord injury. *JCM.* 2025;14(7):2203. doi:10.3390/jcm14072203
61. Chopra M, Kumar H. Navigating the complexities of spinal cord injury: An overview of pathology, treatment strategies and clinical trials. *Drug Discov Today.* 2025;30(6):104387. doi:10.1016/j.drudis.2025.104387
62. Sugai K, Nakamura M, Okano H, Nagoshi N. Stem cell therapies for spinal cord injury in humans: A review of recent clinical research. *Brain and Spine.* 2025;5:104207. doi:10.1016/j.bas.2025.104207

63. Hohlfeld R, Beltran E, Gerdes LA, Dornmair K. Tissue-resident CD8+ memory T cells in multiple sclerosis. *Brain*. 2021;144(1):e7–e7. doi:10.1093/brain/awaa352
64. Bradbury EJ, Burnside ER. Moving beyond the glial scar for spinal cord repair. *Nat Commun*. 2019;10(1):3879. doi:10.1038/s41467-019-11707-7
65. Fan H, Gao J, Chen Q, et al. Emerging regenerative strategies for spinal cord injury: Exosome-derived mechanisms and therapeutic insights. *Front Neurosci*. 2025;19:1652196. doi:10.3389/fnins.2025.1652196
66. Ashammakhi N, Kim HJ, Ehsanipour A, et al. Regenerative therapies for spinal cord injury. *Tissue Eng Part B Rev*. 2019;25(6):471–491. doi:10.1089/ten.TEB.2019.0182

12 Rehabilitation Strategies and Neuroplasticity in Spinal Cord Injury

Shreya Sridhar, Tejasvani Khanna, and Akash Gautam

12.1 OVERVIEW OF SPINAL CORD INJURY

Spinal cord injury (SCI) pertains to the impairment of functionality and structure of the spinal cord due to external trauma or internal pathogenic factors. The spinal cord conveys sensory and motor information between the brain and different body parts. SCI is described as a devastating neurological syndrome because of the resultant morbidity and lifetime disability in addition to acute and chronic sensorimotor dysfunction (1, 2). Along with a partial or total loss of sensory and motor abilities, individuals with SCI may also have autonomic dysfunction (3). At present, the yearly occurrence of SCI is believed to range from 40 to 80 cases per million individuals, with most cases happening among young adults. Approximately 80% of these cases are in individuals under 40 (4). The incidence rate of SCI is higher in males, with a male-to-female ratio between 1.6:1 and 8:1 (4).

As per the International Standards for the Neurological Classification of SCI1 and the American Spinal Injuries Association Impairment Scale (AIS), SCI is classified into complete and incomplete. Complete lesions are placed under the category AIS A, and incomplete lesions are placed under AIS B, AIS C, AIS D or AIS E (5). Incomplete SCI (ISCI) comprises Central Cord Syndrome, Brown-Séquard Syndrome, Anterior Cord Syndrome, and Posterior Cord Syndrome, each exhibiting distinct patterns of symptoms and functional impairments, of which Central Cord Syndrome is the most prevalent form of ISCI. It primarily occurs due to damage to the central portion of the spinal cord, often as a result of neck hyperextension injuries, such as those sustained in severe whiplash from car accidents. The clinical symptoms of complete and ISCI are listed in Table 12.1.

Based on their cause, SCI can be categorized into two types: Traumatic and non-traumatic SCI. The most common etiology of traumatic SCI (TSCI) is trauma, such as traffic accidents, falls, or acts of violence (6). Among patients with TSCI, cervical injuries constitute the largest proportion (7). TSCI can be further divided into primary and secondary SCI depending upon the impact and progression of the injury. Primary SCI occurs due to the immediate harm inflicted on the spinal cord during traumatic incidents like accidents, falls, and sports injuries. The mechanisms involved in the pathogenesis of primary SCI include mechanical forces such as compression, stretching, or torsion applied to the spinal cord, resulting in the rupture of nerve fibers, injury to spinal cord cells, and damage to blood vessels that can lead to vascular rupture. Secondary SCI develops from the ongoing injury and functional impairments that follow the initial injury. This process includes inflammation, cellular swelling, and oxidation of lipids, which can worsen the primary injury, further contributing to neuronal death and functional loss (8, 9).

Non-traumatic SCI can arise from a variety of etiologies, including congenital defects, degenerative diseases, vascular issues, inflammatory conditions, tumors, infections, and metabolic disorders. These causes can be broadly categorized into three major groups: Congenital, genetic, and acquired, with acquired conditions being the most prevalent (10). Congenital conditions include type 2 Chiari malformations and spinal dysraphism. Acquired causes include degenerative conditions like

DOI: 10.1201/9781003500025-12

TABLE 12.1
Clinical Symptoms of Complete and Incomplete Spinal Cord Injury (SCI)

Features	Complete SCI	Incomplete SCI
Anatomic symptoms	Complete loss of motor and sensory function below the level of lesion	Some amount of preserved motor and sensory function below the level of lesion
Type of injury	Tetraplegia, Paraplegia	Central Cord Syndrome, Brown Sequard Syndrome, Anterior Cord Syndrome
Motor and sensory function	Complete loss of voluntary movement, loss of pain, temperature, or sensation of touch below the level of injury	Symptoms may vary based on the tracts injured For example, Brown-Séquard Syndrome typically presents with ipsilateral motor loss and contralateral loss of pain and temperature sensation
Other clinical symptoms	Autonomic dysregulation results in urinary retention, incontinence, or bowel dysfunction, such as constipation, sweating, and high blood pressure	Spasticity and autonomic dysregulation
Prognosis	Unlikely to regain any functional abilities below the level of injury	Based on the type and severity of injury, there is a potential for more significant recovery over time

spondylosis, amyotrophic lateral sclerosis, and vascular causes, such as hemorrhage or infarction, autoimmune conditions, such as multiple sclerosis and transverse myelitis, benign or malignant tumors, spinal cord abscesses, meningitis, metabolic disorders, and toxins (11–16). Unlike traumatic spinal cord injuries, which are often associated with accidents or physical impacts, neurotrauma spinal cord injury (NTSCI) reflects a broader spectrum of conditions, making it an essential focus of research and clinical care due to its complex nature and varying prognosis for affected individuals (10, 12).

12.2 DIAGNOSIS OF SPINAL CORD INJURY

For the appropriate treatment, prognosis assessment, and management of neurological complications, health professionals must essentially diagnose SCI. The diagnosis of SCI can be broadly divided into three strategies: Neurological examination, imaging, and electrophysiological examination.

12.2.1 Neurological Examination

The first step in assessing the severity and level of SCI is to conduct a detailed neurological examination. The AIS is an established neurological assessment that includes a motor examination based on myotomes, a sensory examination based on dermatomes, and an examination of the anorectal area. SCI is classified based on the American Spinal Injury AIS, which is a modification of the previously used Frankel Scale, which is a 5-point severity scale, which is listed in Table 12.2 (17, 18). According to the results of these assessments, the severity grade and level of the injury are determined (17).

The motor examination comprises testing ten myotomes, C5 to T1, corresponding to bilateral upper limbs and L2-S1, corresponding to bilateral lower limbs. Motor strength is assessed based on a standardized six-point scale (graded 0–5), more details have been listed in

TABLE 12.2
American Spinal Injury Association Impairment Scale

Grading	Type of Injury	Motor and Sensory functions
A	Complete	No motor, sensory, and sacral sparing
B	Incomplete	Sensory sparing but motor functions lost below the neurological level, including sacral segments
C	Incomplete	Motor functions spared below the neurological level, ≥50% of muscles – Grade less than 3
D	Incomplete	Motor functions spared below the neurological level, ≥50% of muscles – Grade more than 3
E	Normal	Motor and sensory functions are normal

Tables 12.3 and 12.4A. Sensory examination encompasses testing bilateral pinprick and light touch in 28 dermatomes. The dermatomes are considered from level C2 to S5, where S4 and S5 are taken as one single dermatome. Each dermatome and laterality are graded for sensations. Anorectal examination aids in determining the completeness of the injury and assessing spinal shock. The external anal sphincter is evaluated through digital examination to assess both motor and sensory capabilities. These functions are rated in a binary manner, where 0 indicates absence, and 1 indicates presence. The bulbocavernosus reflex is also evaluated through a digital rectal examination, during which both the internal and external anal sphincters contract in response to squeezing the glans penis or clitoris, which can be palpable upon examination, more details are listed in Table 12.3 and 12.4B (17).

TABLE 12.3
Dermatomes and Myotomes for Neurological Assessment of Upper and Lower Extremities in the AIS Scale.

Root	Muscle Group	Muscles	Dermatome
Upper Limb			
C5	Elbow Flexors	Biceps, brachialis	Lateral shoulder
C6	Wrist Extensors	Extensor carpi radialis longus, Extensor carpi radialis brevis	Lateral forearm, dorsum of thumb and index finger
C7	Elbow Extensors	Triceps	Dorsum of the third finger
C8	Finger Flexors	Flexor digitorum profundus	Dorsum of the ring and little finger
T1	Finger Abductors	Abductor digiti minimi (little finger)	Medial forearm
Lower Limb			
L2	Hip Flexors	Iliopsoas	Anterior mid-thigh
L3	Knee Extensors	Quadriceps	Anterior knee
L4	Dorsiflexors of ankle	Tibialis Anterior	The medial side of the leg and medial malleolus
L5	Extensors of long-toe	Extensor hallucis longus	The lateral side of the leg, medial dorsum of the foot
S1	Plantar flexors of the ankle	Gastrocnemius and Soleus complex	Distal calf, lateral plantar foot

TABLE 12.4
Motor and Sensory Grading

3A. Motor Strength Grading

Grade	Muscle Strength
0	Complete paralysis
1	Palpable or visible contraction
2	Active movement on the elimination of gravity
3	Active movement against gravity
4	Active movement against gravity and some resistance in a specific position
5	Active movement against gravity and full resistance
5*	The active movement against gravity and adequate resistance would be accepted as normal provided no inhibiting factors (like pain, disuse) were present
NT	Not testable

3B. Sensory Grading

Grade	Sensory Function
0	Absent
1	Altered (can either be increased or decreased)
2	Normal
NT	Not testable

12.2.2 Imaging

Though different imaging techniques like X-rays, CT scans, and MRI are available for the diagnosis of SCI, MRI is the gold standard investigation of choice in SCI (18–20). MRI in acute SCI is warranted in any patient who has a neurological deficit that persists post-trauma. The characteristic features of an SCI lesion on MRI include a spindle-shaped lesion with a central hyperintensity associated with surrounding oedema. Many literature reviews indicate that MRI doesn't have a prognostic value in SCI (18–20). However, if an SCI is detected on a CT scan in the form of bleeding or swelling and a neurologic examination isn't possible, it may have some prognostic value. X-rays can assist healthcare providers in identifying fractures, dislocations, or other bony abnormalities in patients.

12.2.3 Electrophysiological Examination

Somatosensory evoked potentials (SSEP), nerve conduction studies (NCS), and motor evoked potentials (MEP) may be used to diagnose SCI (18). Research suggests that electromyography (EMG) might be the most reliable technique for diagnosing SCI. According to existing literature, after experiencing SCI, both the amplitude of motor-evoked potentials (MEPs) and reflex activity can be undetectable for a period of 14 weeks (21). Following this, during rehabilitation, the amplitude of MEPs may slowly rise over the next 4 weeks and stabilize by week 22 in instances of mild injury where paraparesis remains intact (18).

12.3 PATHOPHYSIOLOGY OF SCI

The pathophysiology of SCI is divided into acute, subacute and chronic phases (22).

The acute phase begins with neutrophilic infiltration and cellular processes like hemorrhage, necrosis, and demyelination. Neutrophils produce metalloproteinase-9, which breaks down the extracellular matrix and disrupts the blood–spinal cord barrier. Proinflammatory cytokines are also upregulated, which include tumor necrosis factor-alpha (TNFα), interleukin-1 beta (IL1β), and interleukin-6 (IL-6). They contribute to the release of glutamate and nitric oxide by activating resident microglia, which further release pro-inflammatory cytokines responsible for this action that affects cell survival. This phase occurs within 48 h of injury (22).

Glial fibrillary acidic protein (GFAP) upregulation and heightened proliferation of astrocytes are features of the subacute phase. The blood–spinal cord barrier and glutamate reuptake, mediated by signal transducer and transcription 3 activation (STAT3), are compromised by loss of aquaporin-4 (AQP4) activity in astrocytes. Between 2 and 14 days after the injury, ependymal cells travel to the injury site and contribute to the formation of astrocytes at the central area of the damage. Additionally, there is a rise in the production of chondroitin sulfate proteoglycans (CSPGs) by resident astrocytes, which continues to occur alongside persistent ischemia (22–26).

The chronic phases are marked by the consolidation of perilesional scar and the development of a fibrotic core by type A pericytes, which inhibits neural growth and extension. The fibrotic center is encircled by activated endogenous microglia and macrophages derived from monocytes, CSPGs, and irregular blood vessels and is enclosed by border-forming reactive astrocytes that inhibit the dispersion of immune cells beyond the core of the lesion (22–26). Schematic representation of pathological processes involved in SCI are shown in Figure 12.1 (8, 22).

12.4 THERAPEUTIC STRATEGIES FOR SPINAL CORD REPAIR

To restore the lost sensory or motor functions and improve the patient's quality of life, rehabilitation strategies for the repair of the spinal cord are essentially required. The current clinical treatment for SCI is divided into acute and chronic phases (23). The acute phase is aimed at early surgical decompression and stabilization, augmentation of spinal cord perfusion, intravenous high-dose corticosteroids, and anti-inflammatory therapy. In contrast, the chronic phase involves neurological rehabilitation training (22). These rehabilitation strategies include physical therapy (including passive, active, and neuroprostheses techniques), occupational therapy (OT), speech therapy, and psychological therapy.

Technological advancements have provided us with various non-invasive neuromodulation techniques for SCI rehabilitation. This neuromodulation involves either brain stimulation by techniques like transcranial magnetic stimulation (TMS), peripheral nerve stimulation (PNS), vagus nerve stimulation (VNS), deep brain stimulation (DBS), transcranial direct current stimulation (tDCS), or spinal cord stimulation by techniques like epidural spinal cord stimulation (ESCS), transcutaneous spinal cord stimulation (TSCS), and paired corticospinal–motoneuronal stimulation (PCMS) (24–26).

Cellular regenerative strategies, while still in the experimental stages, offer hope for potential functional recovery and improved quality of life for individuals with SCI. These strategies include treatment with endogenous stem cells, induced pluripotent stem cells (IPSCs) derived neural progenitor cells (NPCs), neural stem cells (NSCs), mesenchymal stem cells (MSCs), oligodendrocyte progenitor cells (OPCs), tissue-engineered scaffolds, direct cell reprograming (22, 27). Non-cellular therapies for SCI consist of various pharmacological treatments, neuromodulation techniques, and innovative drug delivery systems.

One of the most widely studied pharmacological interventions for acute TSCI is methylprednisolone sodium succinate (MPSS), a synthetic glucocorticoid (28). Riluzole, minocycline, intravenous immunoglobulin (IVIG), Imatinib, Hepatocyte growth factor, anti-Nogo-A antibody, and anti-RGMa antibody are some other pharmacotherapies that have been used (22, 28). Anticonvulsants have been used to manage neuropathic pain post-SCI (29). Botulinum toxin and phenol blocks have also been used to reduce mixed pain post-SCI (29). Researchers are also exploring the use of lipid nanovesicles and extracellular vesicles (EVs) for targeted drug delivery (30). These systems can improve the accumulation of therapeutic agents at the SCI site and shift inflammatory responses from the pro-inflammatory M1 macrophage phenotype to the neuroprotective M2 phenotype, thereby fostering a more conducive environment for recovery (30).

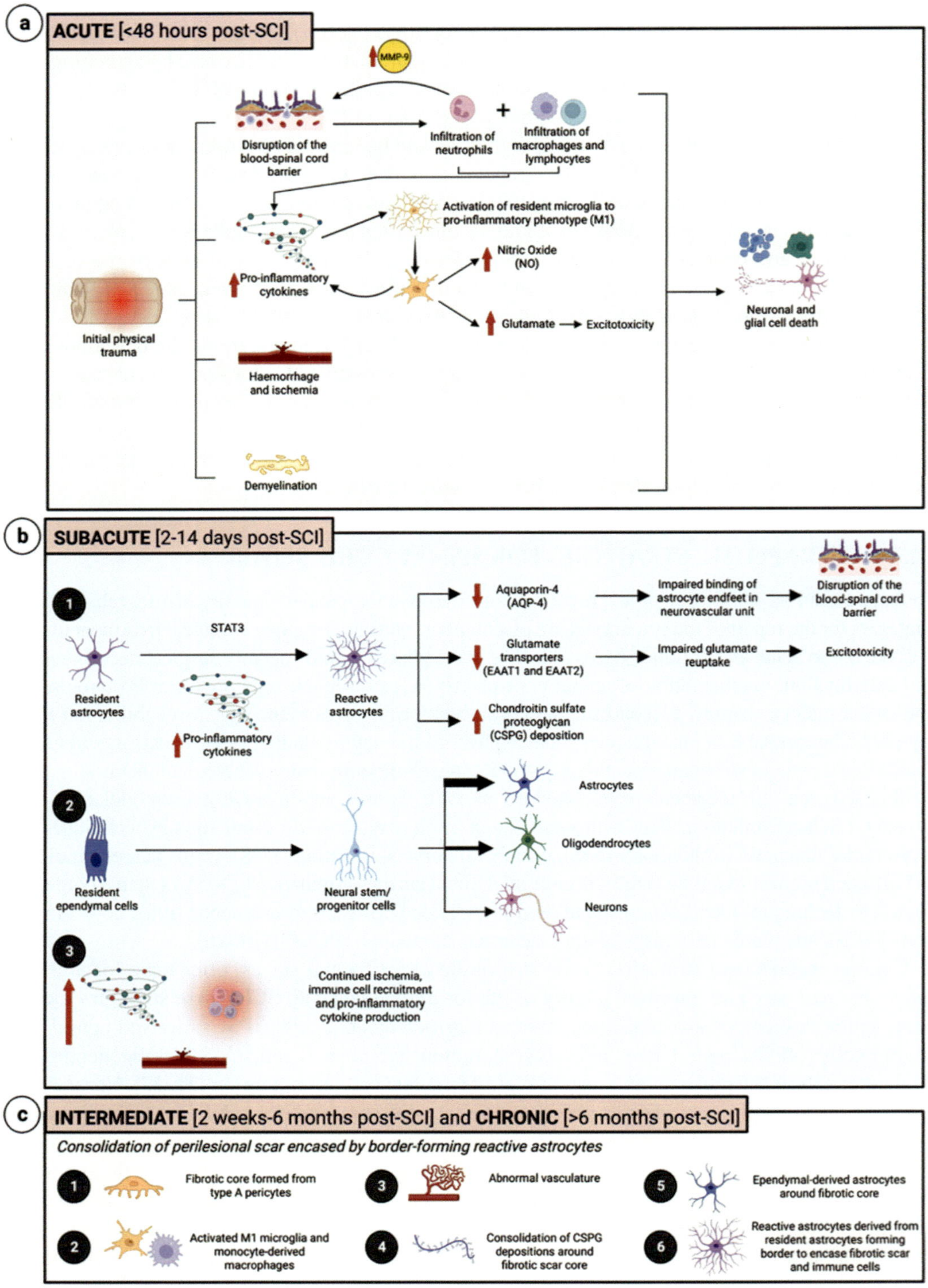

FIGURE 12.1 Illustration of the pathophysiological processes involved in acute, subacute, intermediate, and chronic phases of spinal cord injury (22).

Rehabilitation focuses not only on restoring physical function but also on addressing the psychological aspects of recovery. A multidisciplinary approach involves physical therapists, occupational therapists, psychologists, and other specialists to support the individual's physical and emotional needs (31).

12.5 NEUROPLASTICITY IN SPINAL CORD INJURY

Neuroplasticity refers to the brain's ability to change, adapt, and reorganize. It includes the central nervous system's capability to strengthen and reshape neural connections in reaction to alterations in sensory inputs or motor demands (24). Recent research highlights the role of neuroplasticity in recovery from SCI. Following SCI, spinal interneurons exhibit neuroplastic circuit rearrangements that can either facilitate or hinder functional recovery over weeks and months (22).

Neuroplasticity is characterized by both adaptive and maladaptive processes. Adaptive plasticity refers to beneficial changes that promote recovery, while maladaptive plasticity may hinder recovery, such as when compensatory mechanisms lead to inefficient movement patterns or chronic pain. This adaptive capacity of the nervous system emphasizes the importance of rehabilitation programs that can enhance recovery through targeted therapies. Research indicates that TSCI leads to an immediate cessation of communication among neuronal circuits both upstream and downstream from the injury site. The loss of supraspinal control triggers a compensatory restructuring of sensory and interneuronal circuits, which plays a role in adaptive and maladaptive neuroplastic adaptations within the networks located caudal to the lesion (22, 32).

Neuroplasticity is categorized into structural and functional plasticity. Functional plasticity refers to the brain's capacity to shift functions from injured regions to intact ones, particularly after injury. For instance, in the case of SCI, spared neural pathways may allow for some recovery of lost functions, depending on the extent of the injury (32). Structural Plasticity refers to the physical changes in the brain's neural networks, consisting of the growth of new connections or the strengthening of existing ones. Following SCI, individuals may experience changes in their brain structure that correlate with improved motor function, illustrating the brain's capacity to adapt physically to enhance function (33).

The following factors affect the extent and efficacy of neuroplasticity following SCI:

a. Timing of Rehabilitation: Early intervention is often linked with more significant recovery. Engaging in rehabilitation shortly after the injury can lead to heightened levels of neuroplasticity following SCI (22, 31).
b. Age: The capacity for neuroplasticity is often greater in younger individuals, particularly during critical periods of development. However, adults also retain a degree of neuroplasticity that can be channeled through effective rehabilitation techniques (34).
c. Intensity and Specificity of Training: Intensive and targeted training is essential to promote neuroplasticity. This approach ensures that the spinal cord perceives the demand for specific functions, facilitating the necessary neuroadaptive changes (22, 35).

12.6 MECHANISMS OF NEUROPLASTICITY IN SPINAL CORD INJURY

Neuroplasticity involves various mechanisms, including synaptic plasticity, which facilitates the reorganization of neural circuits following injury. This process includes axonal sprouting, synaptic remodeling, and axon regeneration (36). Electrical stimulation techniques, such as functional electrical stimulation (FES), can induce neuroplastic changes at spinal synapses, enhancing synaptic efficiency and promoting the strengthening of connections between neurons. Electrical stimulation and voluntary motor training have been shown to improve motor function by enhancing spinal excitability (22, 36).

The reorganization of descending propriospinal circuits is crucial in restoring locomotion directed by supraspinal centers following ISCI. It has been noted that the severed corticospinal axons produce collateral sprouts and establish new synaptic links with interneuron networks located above the injury site. Although this sprouting is not specific to interneurons immediately after the injury, subsequent spontaneous locomotor recovery results in preferential maintenance of the corticospinal synaptic connections on the distant descending propriospinal neurons (22).

Neurotrophic factors are crucial in promoting regeneration and synaptic plasticity after SCI. These proteins act as growth factors for neurons, stimulating their survival, growth, and differentiation. Brain-derived neurotrophic factor (BDNF) is a significant player in promoting neuroplasticity following SCI (36). It has been found that exercise-induced neuroplasticity produces BDNF, which is responsible for structural and functional transformations within the brain that influence synaptogenesis. BDNF not only aids in the survival of injured neurons but also exhibits anti-inflammatory properties that help mitigate secondary damage following SCI. The timing and dosage of BDNF administration are critical for maximizing its therapeutic effects, with early intervention correlating with better functional outcomes. BDNF enhances neuronal survival, axonal growth, and the formation of new synapses through various signaling pathways, including the Protein kinase B (AKT) and extracellular signal-regulated kinase (ERK) pathways (32, 36).

Other neurotrophins involved in neuroplasticity include nerve growth factor (NGF) and neurotrophin-3 (NT3) (37). TrkA is the receptor for NGF, while TrkB is the receptor for NT-3. The mechanisms through which NT3 exerts its effects on neuroplasticity involve multiple signaling pathways. Research indicates that NT3 facilitates repair processes via the mitogen-activated protein kinase (MAPK) signaling pathway, essential for cellular proliferation and differentiation following injury (38). Additionally, NT3 has been linked to the transforming growth factor-beta (TGF-β) signaling pathway, suggesting that it may enhance axonal regeneration and myelination in a supportive environment (39). NGF also plays a pivotal role in neuroplasticity following SCI. It exerts protective effects against apoptosis, especially induced by endoplasmic reticulum (ER) stress (33, 37, 40). Studies have also indicated that the genes related to axon regeneration, postsynaptic densities, and neurotransmitter receptors have been significantly overexpressed in Shox2+ V2d interneurons and neurons of the ascending spinocerebellar tract. It has also been observed that post-SCI therapy significantly increases the recruitment of Hoxa10+ V2a interneurons, which play a role in lumbar interneuron plasticity (22). Various factors affecting neuroplasticity and regeneration are mentioned in Figure 12.2 (33). The balance between these stimulators and inhibitors determines whether the nervous system prioritizes regeneration or stability.

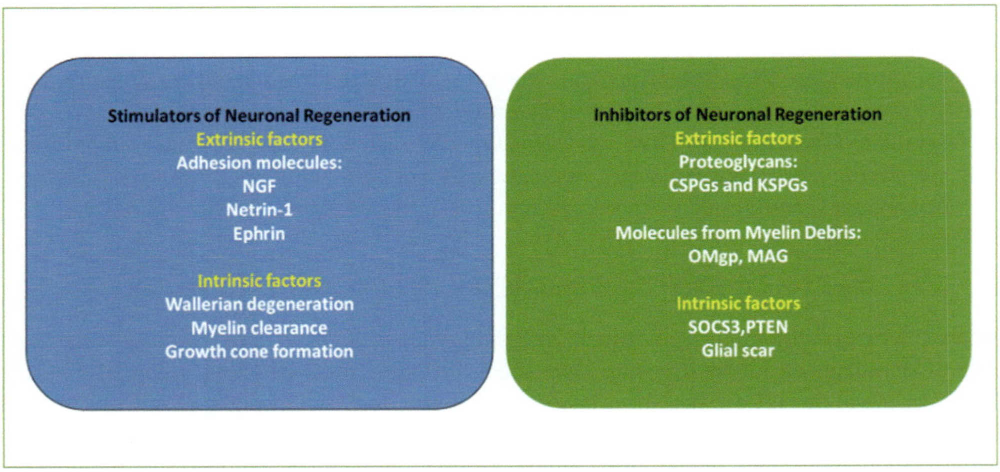

FIGURE 12.2 Illustration of factors influencing stimulation and inhibition of regeneration of neurons in the nervous system. Key components include NGF (nerve growth factor), CSPGs (chondroitin sulfate proteoglycans), KSPGs (keratan sulfate proteoglycans), OMgp (oligodendrocyte myelin glycoprotein), MAG (myelin-associated glycoprotein), SOCS3 (suppressor of cytokine signaling 3), PTEN (phosphatase and tensin homolog). These intrinsic and extrinsic factors collectively promote or inhibit neural repair and recovery.

12.7 REHABILITATION STRATEGIES IN SCI

Effective rehabilitation programs provide the necessary stimuli that guide the maturation of newly formed neural circuits, enhancing the likelihood of functional recovery. Rehabilitation must occur promptly following an injury, as the timing of interventions can significantly impact the outcomes of neuroplastic changes (22, 31). Rehabilitation plays a critical role in ensuring that the potential generated by plasticity and neuro-regeneration is channeled into restoring the functions lost due to SCI. Studies have shown that rehabilitation after plasticity induction enhances the recovery of the rehabilitated function. Recent advancements in rehabilitation approaches emphasize the importance of specific, repetitive, and intense training regimens designed to stimulate these neuroadaptive changes (33). A study by Wagner et al. showed that patterned electronic stimulation, when combined with rehabilitation, helps provide a synergistic effect for functional recovery by guiding plasticity in the spinal cord (41). Therefore, combining rehabilitation with pharmacologic intervention and electrical stimulation can promote functional recovery in SCI.

Choosing the type of rehabilitation strategy becomes crucial in individuals with SCI. It involves consideration of a diverse range of factors, including but not limited to the type and extent of injury, age of the individual, occupation, time since injury, effectiveness, and feasibility. Rehabilitation strategies include activity-based therapy, including physical therapy, occupational and speech therapy, psychotherapy, brain stimulation, and spinal cord stimulation. The goal of rehabilitation is to help the patient to lead a productive and satisfactory life. It involves a multi-disciplinary and patient-centered approach (5).

12.7.1 Physical Therapy – Passive, Active, and Neuro-prosthesis

The World Federation of Neurological Societies Committee has recommended various physical therapy interventions for SCI (42) to increase strength, treat and prevent contractures, improve performance of motor tasks, improve respiratory function, improve joint mobility, improve cardiorespiratory fitness, and interventions to manage pain, spasticity, and swelling. Physiotherapy includes passive and active interventions and neuroprosthesis. A brief description of each type is given below.

12.7.1.1 Passive Interventions

Passive interventions comprise passive movements, passive stretches, tilt table standing, splinting, motorized passive bicycle training, hydrotherapy, massage, cryotherapy, and laser therapy (42–44). These aim to maintain the extensibility of soft tissues overlapping joints and improve joint mobility and range of motion. They also reduce the risk of cartilage deterioration (45). Passive movements minimize contractures, adhesion formation, soreness in muscles, muscle atrophy, and increase muscle strength and power (45). Repetitions of the passive movement ranging from 3 to 30 repetitions for 7–10 min showed an increase in muscle power in one of the studies conducted by Alaparthi et al. (46). Passive stretches are typically extended-duration stretches designed to prevent and address joint mobility limitations and the development of contractures. The elbow, wrist, hip and ankle are the most common sites of contracture development (47). No randomized controlled trials have been found that have studied the effects of stretch administered for long durations, i.e., >3 months, making the added benefit of stretch appear to be clinically of lesser significance. Nevertheless, contractures in SCI are disabling, stretches shouldn't be discarded altogether.

Motorized passive bicycle training is aimed at improving motor performance and cardiorespiratory fitness. FES cycling can be used in people who have higher level tetraplegia and are unable to use their arms to exercise. Some studies have concluded that FES cycling decreases spasticity, while some results have been inconclusive (44). Hydrotherapy is the therapeutic external use of fresh water through showers and baths. Hydrokinesitherapy has been demonstrated to enhance mental functions. Its primary methods involve water-based exercise therapy, such as swimming, walking training, aquatic gymnastics, underwater traction, and breathing exercises. Contraindications encompass open wounds, trophic ulcers, postoperative fistulas, as well as both acute and chronic diseases of the skin, eyes, ears, nose, and throat (ENT). Additionally, conditions occurring after previous infectious diseases and chronic infectious diseases are also contraindications to hydrokinesitherapy (42).

Similarly, short massage sessions (ranging from a few minutes) of the triceps muscle of the lower leg led to a decrease in H reflexes, but this effect lasted only for a few minutes. Cryotherapy has been found to decrease muscle spasticity for up to an hour following the cessation of cold application.

12.7.1.2 Active Interventions

These are comprised of strength training, Gait-related training, bodyweight-supported treadmill training (BWSTT), mobility training, and electric stimulation. Active interventions aim to improve motor performance, joint mobility, cardiorespiratory fitness, increase strength, and improve respiratory function. The primary goal of strength training is to address weakness in partially paralyzed and non-paralyzed muscles. Limited strength in SCI may be due to decreased muscle mass due to disuse atrophy or may be neurologically induced (48). Progressive resistance training involving electrical stimulation as well as without electrical stimulation is part of the regimen. It has been demonstrated that an 8-week strength training program that combines electrical stimulation and progressive resistance training is beneficial for SCI patients. Many studies have been inconclusive in establishing a guideline for the type and duration of strength training and its efficacy. Thus, the frequently employed method may consist of progressive resistance training and repetitive practice with lighter loads and higher repetitions. Electrical stimulation can be combined with maximum voluntary exertion and high resistance (5). A detailed description of various electrical stimulation strategies for modulating neuroplasticity in SCI has been provided in the subsequent paragraphs. Task-oriented training and cumulative resistance training are utilized in clinical settings to enhance the strength of partially paralyzed muscles.

Gait-related training includes robotic-assisted gait training, treadmill training with or without overhead suspension, standing with assistance, and electrical and somatosensory stimulation (44). Many studies incorporating gait training in clinical practice included increased repetition rather than increased load and progressive resistance. The strength of partially paralyzed muscles was seen to increase in some studies. BWSTT also aims to improve gait in SCI. Mobility training consists of manual wheelchair mobility training, and seated mobility training, Arm cranking, hand cycling, and circuit training, with and without electric stimulation. These are aimed at improving motor performance and cardiopulmonary fitness. Wheelchair mobility training was effective in SCI, while seated mobility training was effective in another study. Determining when to stop enhancing neurological recovery is a challenge for both the physiotherapist and the individual suffering from SCI. In certain instances, helping a person regain movement patterns similar to those before the injury can be less challenging than instructing someone with SCI on entirely new movement techniques. Respiratory muscle training with hand biking, interval training, and arm pedaling are aimed at improving cardiopulmonary fitness. Due to the very low certainty of the evidence, certain suggestions put forth by various studies were deemed weak.

12.7.1.3 Neuro-Prosthetics

Neuroprosthetics have emerged as a transformative approach in rehabilitating individuals with SCIs, aiming to restore lost motor and sensory functions through advanced technologies that interface with the nervous system (49). Neuroprosthetics can be broadly classified into implantable and non-implantable devices. They exert their mechanism of action by stimulating neural pathways and promoting the reorganization of neural connections (50). Implantable devices are surgically placed within the body to directly interface with the nervous system (51). They aim to bypass damaged areas of the spinal cord and facilitate communication between the brain and muscles or sensory systems. Various electrical stimulation techniques include FES, epidural, and intraspinal microstimulation, bladder neuroprosthesis, and cortical stimulation (51).

FES sequentially activates paralyzed muscles to restore functional movements and is particularly effective in gait restoration (51). Epidural and intraspinal microstimulation involve the implantation of electrodes along the spinal cord to deliver precise electrical impulses, re-establishing communication pathways between the brain and the lower body. This technique has been shown to facilitate voluntary movements and even enable standing and walking with assistance in individuals with SCI (52). Bladder neuroprosthetics are designed to restore bladder control in individuals with SCI by stimulating the relevant neural pathways (52).

Non-implantable devices include transcutaneous electrical stimulation, brain–computer interfaces (BCIs), and bidirectional interfaces (53). TENS is a non-invasive method that uses electrical currents applied to the skin to activate peripheral nerves and spinal pathways, promoting functional recovery without surgical intervention (53). BCIs allow the translation of neuronal activity into control signals for external devices. These interfaces operate through non-invasive or invasive methods and significantly enhance the potential for rehabilitation and functional recovery in individuals with SCI (54). BCIs utilize signals from the brain to control external devices, such as computers, robotic systems, and assistive technologies. BCIs can interpret neural activity through non-invasive methods like electroencephalography (EEG) to provide command signals for device operation (54). Bidirectional interfaces integrate recording and stimulation, allowing real-time interaction between the brain and neuroprosthetic devices. This technology aims to create a feedback loop that enhances user control and performance (50).

12.7.2 Occupational Therapy

OT is one of the major components of rehabilitation after SCI. The primary goals of OT in SCI are maximizing functional independence by focusing on self-care and mobility, promoting participation in meaningful activities, improving quality of life, and empowering self-advocacy and adjustment to disability. OT interventions include but are not limited to activities of daily living (ADLs), mobility exercises, community reintegration, adaptive equipment training, splinting, strengthening and endurance, airway management, home management, assistive technology and environmental control, and therapeutic activities (55, 56). These interventions are designed based on a person's social and cultural features, level of education, personality traits, interests, values, attitudes, and behaviors (57).

ADLs can be further divided into basic activities of daily living (BADLs) and instrumental activities of daily living (IADLs). BADLs include tasks like bathing, grooming, feeding, clothing management, and personal hygiene, whereas IADLs include more complex activities necessary for independent living, like meal preparation, household chores, and managing finances. Mobility exercises include wheelchair mobility, education about transfers from a wheelchair to other surfaces (e.g., bed, toilet), training in adaptive equipment like transfer boards and lift systems, propulsion techniques, and car transfers for community access. Community reintegration involves preparing people with SCIs to participate in work-related events, social events, and other community-related activities.

Adaptive equipment training includes feeding (like adaptive utensils, plate guards, and drinking aids to foster independence in feeding), dressing (like Button hooks, zipper pulls, and dressing sticks can simplify the dressing process), and mobility aids (like transfer boards, specialized wheelchairs, and power-assist devices). Strengthening and endurance training includes sitting tolerance, strengthening/no equipment, manual resistance, basic/low-tech equipment, gym machines, overhead slings/mobile arm support, neuromuscular re-education, arm ergometer, motomed, participation sport, yoga, pilates, standing frame/tilt table (55). On the other hand, assistive technology is often necessary to facilitate communication, especially for people with high tetraplegia. Therapeutic interventions target dexterity enhancement, manual therapy, respiratory exercises for airway control, tenodesis practice, vestibular exercises, swelling management, Visual/perceptual training, cognitive training, assisted devices, and desensitization.

12.7.3 Speech Therapy

Early initiation of vocational rehabilitation in patients with SCI is very effective. It has been found that engaging patients in an early vocational rehabilitation program, especially in people still in the hospital setting, has good potential for improving post-injury labor-force participation (58). Speech therapy aims to improve communication, ensure safe swallowing, and improve articulation.

Speech therapy interventions comprise respiratory therapy, articulation therapy, swallowing therapy, and cognitive-communication therapy. Speech-language pathologists (SLPs) play a major role in SCI. They assess patients' speech, swallowing, and respiratory abilities and design targeted rehabilitation plans. SLPs also educate patients and their families on communication strategies, safe swallowing techniques, and exercises (59).

12.7.3.1 Respiratory Therapy

Breath control exercises are essential for SCI patients with respiratory impairments, especially in high cervical injuries. Techniques, such as controlled exhalation and incentive spirometry improve breath support, critical for sustained speech, and voice projection (60). Strengthening the diaphragm through breathing exercises enhances respiratory function, which helps improve voice intensity and quality. SLPs also use techniques like the Lee Silverman Voice Treatment (LSVT) to amplify vocal intensity in patients with weak vocal cords (American Speech-Language-Hearing Association, ASHA). In the voice modulation exercises, patients work on achieving clearer, more audible speech by practicing pitch control, volume modulation, and sustained phonation. These exercises help SCI patients project their voices and communicate effectively in social settings (61).

12.7.3.2 Articulation Therapy

SCI patients often require exercises that target the lips, tongue, and jaw to improve speech clarity. SLPs may use repetitive sound drills, exaggerated enunciation, and pacing exercises to increase articulation strength and coordination. Speech drills for rhythm and pace incorporate drills that enhance rhythm, speed, and accuracy, helping patients regain control over their speech rate (57,62).

12.7.3.3 Swallowing Therapy (Dysphagia Therapy)

To reduce the risk of aspiration, SCI patients with dysphagia learn compensatory strategies, such as chin tucks or head tilts. SLPs also teach patients positioning and pacing techniques to make swallowing safer and reduce the risk of pneumonia (63). SLPs work with nutritionists to suggest modified food textures for patients at risk of choking, offering softer or thicker foods that are easier to swallow safely. This individualized diet plan helps SCI patients meet their nutritional needs safely (63). SLPs may use sensory stimulation methods, like thermal-tactile stimulation, to strengthen the swallow reflex and promote safe and effective swallowing (64).

12.7.3.4 Cognitive-Communication Therapy

SCI patients display a marked reduction in attention, memory, processing speed, and executive functions. Moreover, they have impaired social communication due to speech and language processing defects. SCI can also lead to emotional and psychological challenges, affecting mood, motivation, and self-esteem. Targeted exercises in the form of cognitive communication therapy can enhance attention, memory, processing speed, and executive function. Memory and recall exercises with various memory tasks, including sequence recall and story retelling, help improve short-and long-term memory (62). Similarly, problem-solving and social skills training exercises help to improve social skills. Conversational appropriateness is key for reintegrating patients into their communities. Techniques may include role-playing and real-life scenario simulations to build confidence and social rapport (62).

12.7.4 Psychological Therapy

SCI poses a significant medical and psychological challenge. It can lead to secondary physical and emotional effects, such as a higher likelihood of depression and anxiety, as well as a diminished quality of life (65). Individuals with higher levels of injury are more vulnerable and experience greater disability-related challenges and problems with accessibility, which further diminish life satisfaction and increase hospitalization frequency due to complications (66). Given the multifaceted nature of the recovery process in SCI, rehabilitation practitioners need to adopt a holistic approach that addresses both physical and psychological needs (67). Various types of psychological therapy include cognitive behavioral therapy, mindfulness-based interventions, and technology-based interventions.

12.7.4.1 Cognitive Behavioral Therapy (CBT)

CBT is a widely used therapeutic approach that modifies negative thought patterns and maladaptive behaviors. It includes a variety of techniques to facilitate emotional and behavioral change in an individual. Several widely used techniques in CBT include but are not limited to problem-solving approaches, assertiveness training, cognitive restructuring, enhancing motivation to engage in rewarding activities, and relaxation practices (68). A systematic review analysis by Mehta et al. supported the effectiveness of CBT to be moderate in the reduction of depressive symptoms after SCI. A conflicting level 2 evidence was found for the effectiveness of CBT in diminishing symptoms of anxiety (68). The levels of evidence were drawn following the modified Sackett scale (69).

12.7.4.2 Mindfulness-Based Interventions (MBI)

Unlike CBT, which aims to confront fundamental beliefs and alter negative thoughts and unhelpful behaviors, MBIs seek to promote mindfulness of the current moment and help cultivate acceptance (38). A systematic review analysis by Hearn et al. (65) presented mixed evidence regarding the effectiveness of mindfulness in enhancing outcomes after SCI. Although the review pointed out the need for more thorough research in this area, it did suggest that mindfulness practices may be beneficial in reducing the symptoms of anxiety and depression (65).

12.7.4.3 Technology-Based Interventions

Several digital psychosocial tools (such as internet- and mobile-based interventions, smartphone applications, and virtual reality (VR) applications) have been created specifically for individuals with SCI. A scoping review by Armstrong et al. examined three categories of interventions: Mobile applications for smartphones, internet- and mobile-based interventions (IMIs), and VR applications (70). IMIs are self-directed interventions and include online educational programs that individuals can use to obtain health-related information. These interventions have been shown to produce a desirable effect on the mental health, e.g., depression and quality of life of users (70). Mobile health apps (like mHealth) convey therapeutic information via smartphones. Initial studies suggest these tools are effective in treating various mental health conditions, such as phobias, post-traumatic stress disorder, and stress. Additionally, they show promise in alleviating symptoms of depression, anxiety, and pain (70).

Recent advancements in technology have introduced robotic devices into rehabilitation practices. These devices are designed to assist therapists in delivering care and tracking patients' progress by accurately recording biomechanical parameters during therapy sessions. However, they have not consistently demonstrated superior outcomes compared to conventional methods (33). Another emerging tool in rehabilitation is VR, which has been integrated to enhance patient engagement and motivation. VR applications create an interactive, three-dimensional world for users. While some reviews suggest that VR may improve motor and cognitive functions, results are mixed regarding its effectiveness compared to traditional rehabilitation methods. As research in this area evolves, VR is increasingly viewed as an adjunct to conventional therapy, potentially increasing overall therapy time and enriching the rehabilitation experience (33).

12.8 ELECTRICAL STIMULATION TECHNIQUES TARGETING NEUROPLASTICITY

Rehabilitation approaches for people with SCI have increasingly focused on enhancing neuroplasticity to enable recovery and improve function (24). Neuromodulation techniques can be broadly categorized into brain Stimulation and spinal cord stimulation. A combination of non-invasive brain stimulation and rehabilitation has proven effective in hastening recovery for individuals with SCI.

12.8.1 Brain Stimulation

Non-invasive brain stimulation techniques include TMS, tDCS, transcranial alternative current stimulation (tACS), transcranial random noise stimulation (tRNS), laser, ultrasound, virus, and other vectors. TMS and tDCS are the most widely used in clinical practice (71).

TMS consists of a wire coil positioned over the scalp, where electric current passes through it, inducing an electromagnetic field that further induces the current. This modulates cortical excitability and synaptic plasticity. Repetitive TMS (rTMS) modulates neuronal activity in the targeted cortical regions (71, 72). TMS affects long-term depression (LTD) and long-term potentiation (LTP) (73). LTD is a long-term decrease in synaptic potency and can be stimulated by low-frequency synaptic activation (74). This is inhibitory and can weaken or eliminate connections between neurons that are not in use anymore. LTP is a long-term increase in the strength of a synapse, is stimulated by high-frequency stimulation of the synapse, and has a role in Hebbian learning and long-term memory formation. TMS also plays a significant role in pain modulation, improving spasticity, recovery of muscle strength, and motor function (75).

Individuals with complete SCI with injuries rostral to T9 have been shown to have limited restoration of lower limb function as opposed to patients with incomplete injury who generally have a good prognosis in regaining locomotor function within a year of injury. However, TMS has improved spasticity, which may be utilized in patients with complete SCI (24). Research has demonstrated a transient decrease in the activity of inhibitory neurons in the cerebral cortex and a short-term decrease in cortical inhibition with rTMS (72). Many studies have shown that combining spinal and brain stimulation can magnify the benefits of locomotor training (75, 76).

tDCS is another widely used non-invasive brain stimulation modality, which involves the application of electric currents to the brain with the help of electrode pads placed on the scalp (71). Compared with rTMS, tDCS is more affordable, available, easier to apply, and can be utilized for home use (77). The formation of a weak electric field modifies neuronal activity by modifying the resting membrane potentials. Anodal tDCS applies a positive current and enhances cortical activity, while cathodal stimulation reduces it (71). The application of tDCS in SCI aims to improve motor functionality, improve neuropathic pain, and reduce spasticity.

A systematic review and meta-analysis were conducted by De Araujo et al. to observe the potential effects of anodal tDCS on enhancing muscle strength and motor function following an ISCI (78). Six randomized controls were included in the study. The results showed a significant improvement in motor functionality with tDCS compared to sham tDCS stimulation with a small ES and a CI crossing zero. However, tDCS had no statistically significant effect on muscle compared to sham tDCS (78). Another study showed that tDCS may provide relief in patients with neuropathic pain by altering pain-processing pathways within the brain and targeting the motor cortex with anodal tDCS demonstrated pain-relieving effects (79).

12.8.2 Spinal Cord Stimulation

Spinal cord stimulation (SCS) is another important neuromodulation intervention in SCI (51). ESCS, TSCS, and PCMS are some of the major therapeutic methods designed to enhance movement and sensory abilities in people with SCI (80).

12.8.2.1 Epidural Spinal Cord Stimulation

ESCS involves the implantation of a device that delivers electrical impulses to the spinal cord via electrodes placed in the epidural space (81). It engages spinal reflex pathways that consist of monosynaptic pathways (Ia, muscle spindle fibers, and Ib, Golgi tendon organ fibers) as well as polysynaptic pathways (II), which stimulate extensors and flexors accordingly (82). Long-latency muscle response is evoked by Group II afferent fibers that produce stepping patterns in volitional locomotor control. Additionally, ESCS activates the lumbar spinal cord's Central Pattern Generators (CPGs), which are responsible for the rhythmic contraction of muscles of the flexor and extensor compartments when walking.

It has been proposed that electrical stimulation of proprioceptive circuits can decrease threshold potentials and increase the excitability of neuronal networks. This priming enables these networks to be ready for activation from supraspinal or propriospinal inputs, facilitating sensory information's role in guiding voluntary movement. Many animal models have been studied, and they have shown that sensory feedback has played a part in the restoration of rhythmic, coordinated hindlimb movement and has facilitated stepping patterns in rodents (82).

The mechanism of autonomic control through ESCS involves stimulating spinal circuits that govern sympathetic and parasympathetic responses. ESCS activates propriospinal networks, which help generate apt respiratory muscle pattern activity via stimulation and inhibition of motor neurons involved in respiration. ESCS recruits the diaphragm and inspiratory intercostal muscles, enabling breathing patterns closer to spontaneous respiration. ESCS also centrally controls motor neuron activation, allowing asynchronous and physiologic recruitment of respiratory muscles.

Recent studies have indicated that ESCS can aid in restoring autonomic function by improving the interplay between the sympathetic and parasympathetic nervous systems. These effects have been observed as improvements in cardiovascular stability, including increases in blood pressure ranging from 10 to 40 mmHg during stimulation, alongside enhanced heart rate variability and temperature regulation (83).

Investigating how ESCS affects the genitourinary system requires understanding the topographical arrangement of neural networks in the lumbosacral region. Afferent fibers that prevent bladder activity during filling are activated by excitation in the sympathetic networks located in the L1–L2 region, which are involved in the bladder's storage function. The L3–L4 region regulates sexual function and governs the external urethral sphincter and detrusor muscle (L3–L4 in rodents, L3–L5 in humans). Stimulation of the sacral cord recruits pelvic and pudendal nerve axons, influencing micturition, and sacral nerve stimulation in individuals with SCI has also shown that spino–bulbo–spinal pathways play a role in bladder control (82).

In cases of SCI, hemodynamic instability may present as orthostatic hypotension, syncope, or autonomic dysreflexia due to disrupted supraspinal control of sympathetic ganglia. Studies have shown that ESCS activated autonomic responses and synchronized baroreceptor signaling to stabilize blood pressure in cervical SCI in animal models (82).

To summarize, the role of ESCS in SCI lies in improving locomotor control, autonomic regulation, and restoration of sensorimotor functions. ESCS has shown promise in enhancing motor function, particularly in individuals with incomplete motor lesions, who may experience improved neurological function and voluntary control over lower extremity muscles over time, even after the cessation of stimulation. Additionally, clinical studies have highlighted the benefits of ESCS in autonomic function improvement (82). It works by blocking the transmission of pain signals through nociceptive nerve fibers entering the dorsal horn (DH) and helps improve neuropathic pain (82).

12.8.2.2 Transcutaneous Spinal Cord Stimulation

It is a non-invasive neuromodulation technique used in SCI in which electrodes are used to stimulate spinal circuits through an electric current. This technique involves the stimulation of spinal circuits via transcutaneous input instead of epidural implantation. According to several studies, the vertebral levels most frequently stimulated to target motor activity in the lower limbs are T11–T12 and/or L1–L2. The position of the cathode has been variable across studies. Most of the studies placed the cathode dorsally over the vertebrae, while others placed it dorsally in the paravertebral position. In the majority of other studies, the electrode was positioned along the midline of the vertebral column, while the anode was placed either on the iliac crest, anterior superior iliac spine (ASIS), or near the paraumbilical region over the anterior abdomen (84).

TSCS functions through a complex interplay of neuronal circuits and plasticity mechanisms that modulate sensory and motor pathways in individuals with SCI. The DH's excitatory and inhibitory interneurons play a pivotal role in processing nociceptive and non-nociceptive signals. These interneurons contribute to pain modulation by balancing excitatory and inhibitory inputs within the spinal circuitry, particularly in laminae I–III of the DH.

TSCS, when applied, activates large-diameter Aβ fibers, which play a role in non-nociceptive sensory signaling. These fibers, when activated, modulate wide dynamic range (WDR) and nociceptive-specific (NS) neurons within the DH. The WDR neurons respond to both noxious and non-noxious stimuli and increase firing rates in response to T-SCS, thereby facilitating pain relief and enhancing motor control through their projections to higher brain centers. Conversely, TSCS inhibits NS neurons, which only respond to noxious stimuli, thus reducing the perception of pain (85, 86).

TSCS plays a role in enhancing motor recovery, pain modulation, and reduction of spasticity. Studies, including animal models and in individuals with SCI, have shown that TSCS can lead to reorganizing motor pathways, enabling voluntary control of paralyzed muscles when combined with activity-based training. The engagement of cortico-reticulo-spinal circuits is believed to play a role in this mechanism. These circuits are responsible for voluntary movement commands that reach the spinal cord after injury (84, 87, 88).

TSCS has been utilized in various clinical contexts to improve upper extremity function and overall mobility in patients with motor ISCI. The combination of TSCS with exercise has shown benefits in improving motor activity across various studies. A study by Seanez et al. documented sustained recovery of arm and hand movements following tSCS when combined with activity-based training (88). A study by McHugh et al. showed that TSCS, when used as an adjunct to walking therapy, is both clinically feasible and useful in individuals with motor ISCI (89).

12.8.2.3 Paired Corticospinal-Motoneuronal Stimulation

PCMS is an emerging neuromodulation intervention to enhance motor recovery in individuals with SCI. The idea of spike-time-dependant plasticity, which postulates that the relative timing of synaptic inputs can influence the strength of synaptic connections, is the foundation of PCMS (90). PCMS is a relatively newer technique compared to TSCS and ESCS (90, 91). It involves a combination of TMS, which targets peripheral nerves, with TMS applied to the primary motor cortex. This improves synaptic transmission by causing plasticity at corticospinal–motoneuronal synapses (91). It works on the principle of synchronizing the timing of corticospinal volleys generated by TMS to reach the junction between corticospinal pathways and motor neurons. with the antidromic potentials produced by electrical stimulation of peripheral nerves (91).

The enhancement of corticospinal excitability through PCMS has been linked to significant improvements in motor output and functional capabilities of leg muscles, suggesting its potential as a valuable adjunct to conventional rehabilitation strategies (92). A correlation was found between motor output and the intensity of the induced plasticity in the study by Bunday and Perez et al. in humans with incomplete cervical SCI. This study showed that plasticity can be induced at residual corticospinal–motoneuronal synapses. It demonstrated that when presynaptic volleys reached the spinal cord just before motoneuronal volleys, there was a rise in corticospinal responsiveness in the first dorsal interosseous (FDI) muscle. This enhancement was evident in hand mobility, measured by the time taken to complete the Nine-Hole-Peg Test, which assesses fine motor skills. The improvement in corticospinal communication was positively linked to enhancement in voluntary motor performance (92).

Another study by Grover et al. demonstrated that increasing the number of paired stimuli during PCMS can significantly enhance MEPs and improve motor performance in tasks, such as the nine-hole peg test. It was found that 360 paired pulses (PCMS-360) led to greater improvements in MEP size and functional outcomes compared to 180 paired pulses (PCMS-180) in individuals with chronic incomplete cervical SCI, thus demonstrating that increasing the number of PCMS paired pulses can improve corticospinal excitability and voluntary motor output, which can effectively increase functional capabilities in individuals with SCI (93). Furthermore, the potentiation of corticospinal transmission is more pronounced when PCMS is applied during voluntary activity as opposed to its application during rest. Thus, combining activity and PCMS may lead to a more effective rehabilitation process (94).

12.9 MOLECULAR BASIS OF RECOVERY IN SCI

The growth and survival of neurons as well as synaptic plasticity, are dependent on neurotrophic growth factors. Thus, it is no surprise that these growth factors are being studied as potential treatment modalities for SCI (95). These factors include glial cell-derived neurotrophic factor (GDNF), neurotrophin 4/5 (NT-4/5), BDNF, NGF, neurotrophin 3 (NT3), fibroblast growth factor (FGF), hepatocyte growth factor (HGF), insulin-like growth factor-1 (IGF-1), epidermal growth factor (EGF), and basic fibroblast growth factor (bFGF or FGF2) (95). BDNF RNA therapy was observed to improve motor recovery after SCI in animal models (96). Retrograde neurotrophin 3 (NT3) transport can promote lumbar motor circuit reorganization and improved recovery (97). IGF-1 overexpression and administration of FGF2 have been found to improve myelination and functional improvement (98, 99). Animal studies have also revealed a beneficial role of VEGF in SCI recovery (100).

Granulocyte colony-stimulating factor (G-CSF) promotes the movement of bone marrow-derived stem cells to the injury site. It appears capable of reducing inflammation, protecting myelin, and stimulating axonal regeneration and can improve motor and sensory function after SCI (101). Additionally, various biomolecules influence recovery after SCI. Blockage of IL-7 receptors and administration of IL-4 and IL-10 can enhance recovery (102). Downregulating the mTOR pathway activated by SCI can promote improvement (103). A recently discovered circular RNA, CircPlek, increases after SCI, and its role is being studied (104).

Transcription factors, including Krüppel-like factors (Klfs), Sox11, Stat3, cAMP-response element binding protein (Creb), and hypoxia-inducible factor-1a (Hif-1a) can potentially contribute to regeneration (105). Epigenetic modifications may enable the transcription of certain genomic regions and influence neuronal plasticity (106). Regeneration of axons is ostensibly opposed by CSPGs and myelin-associated inhibitors (MAIs) (107). The generation of reactive oxygen species and microglia activation following SCI leads to axon degeneration, secondary injury, and consequently permanent damage to neurons. In an animal model, IrFPHtz, an iridium complex, suppressed inflammation and enhanced recovery (108). The CSF1R inhibitor GW2580 has been found to reduce the proliferation of microglia and inflammatory damage (109).

Axon regeneration requires microtubule stabilization, which may be augmented by drugs, such as taxol and epothilone B (110). Knockdown of *Rhoa* is associated with improved axonal regeneration (111). Deletion of *Snph*, a mitochondrial anchor protein, has been found to enhance metabolism and, consequently, axonal regeneration (112). Additionally, animal models have exhibited better regeneration of axons after drug-induced blockade of $\alpha 2\delta 2$ subunit of voltage-gated calcium channels (113). Axon regeneration depends not only on intrinsic growth capacity but also on the external environment induced by SCI. Various factors act as inhibitors to recovery. High concentration of CSPGs prevents axonal growth, but these can be degraded by enzymes to improve SCI recovery (114). Antibody neutralization, receptor antagonists and genetic inhibition can be used to reduce inhibition medicated by MAI. Anti-Nogo A has been found to improve bladder function after SCI (115).

12.10 CHALLENGES AND PROSPECTS OF NEUROPLASTICITY AND REHABILITATION STRATEGIES IN SCI

Despite its remarkable potential, neuroplasticity has some limitations. Certain brain regions responsible for critical functions may be less capable of recovery following damage, and factors such as substance abuse, age or disease can negatively impact the brain's plasticity. Despite the potential for neuroplasticity and regeneration, several challenges hinder recovery post-SCI. These include inhibitory factors in the extracellular matrix, loss of trophic support, and glial scar tissue development at the injury site, which restricts axonal growth and hinders functional reconnection.

The field of SCI is plagued by various challenges. One of the main limitations to testing interventions is that rodent models differ significantly from primates regarding spinal cord anatomy, size, repair, and motor circuit responses. Thus, the results of animal models often do not translate into human trials (44). Conventional anti-inflammatory drugs have limited efficacy in SCI (44). For instance, G-CSF showed encouraging results in preliminary trials but not in the Phase-3 trial (116).

With the advancements in state-of-the-art facilities, emerging technologies like the usage of stem cells, the production of exoskeletons and VR have the potential to overcome many of these challenges. Considering the extensive cell death that follows SCI, stem cell replacement has become a vital area of research (13) and has largely proven safe (95). It is believed that the ability for multipotent regeneration in the spinal cord mainly originates from the radial and tanycyte ependymal cells located in the central canal (115). After SCI, these cells upregulate progenitor markers like Cd13, prominin, and nestin and express transcription factors by migrating to the surrounding parenchyma (115). Various in vitro studies have supported the role of Neurogenin2 in differentiating into neurons and Mash1 in differentiating into oligodendrocytes (117).

Autologous somatic cells can be developed into IPSCs, which can be transplanted after SCI (118). One limitation of transplanting stem cells is that the inflammatory environment of SCI induces disproportionately high differentiation into astrocytes. However, recovery demands the generation of oligodendrocytes and neurons as well. Using caudalizing agents, ventralizing agents, and platelet-derived growth factors before transplantation can bias stem cells to develop into oligodendrocytes. Meanwhile, using NPC's that express GDNF can induce differentiation into neurons (119). Studies have demonstrated improved motor recovery using these biasing methods (120).

Another novel treatment modality for SCI is the utilization of tissue engineering and scaffolds (95). Various biomaterial structures such as channels, fibers, and scaffolds have been employed (121). They act as fillers to reconstitute the extracellular matrix, facilitate transduction, and bridge gaps for axon sprouting (102). Meanwhile, hydrogels can influence the appropriate delivery of immunomodulators (122). The mechanical properties of the scaffold play a crucial role in influencing their efficacy. Neuronal branching is increased on soft constructs, and astrocyte growth on a stiff architecture (123). Increased scaffold and natural extracellular matrix (ECM) similarity facilitate better recovery (124). Scaffolds can also incorporate electroconductive and electroconductive components to promote cellular differentiation, proliferation, and migration (125).

3D bioprinting has emerged as a novel method of nervous system reconstruction by printing cells and essential cytokines (126). This can replicate complex native neuronal structures (126). Various cell types, including Schwann cells, NSCs, precursors to oligodendrocytes, etc., have been employed for printing and incorporated into scaffolds (127). However, the field is plagued by limitations, such as low cell density, high viscosity, unsuitable mechanical characteristics, and inflammatory environments (128).

There is some hope for neuromodulatory techniques to improve prognosis and support daily activities. Brain machine interface helped a tetraplegic patient regain upper limb movement (129). However, these techniques have several concerns and limitations, including cost, availability, safety and durability (129). Moreover, SCI demands a combinatorial treatment plan. Robust multidisciplinary collaboration is essential to determine the safety and efficacy of such treatments.

Thus, it is the need of the hour to design randomized, large-scale clinical trials for SCI treatment. This is all the more important considering the possible side effects of treatment and the limited reliability of animal models.

REFERENCES

1. Brown, AR, Martinez, M. From cortex to cord: motor circuit plasticity after spinal cord injury. Neu Reg Res. 2019;14(12):2054–2062. doi:10.4103/1673-5374.262572
2. Kazim SF, Bowers CA, Lloyd RA, Cole CD, Schmidt MH. Letter: frailty and outcomes of traumatic spinal cord injury. Neurosurgery. 2021;89(3):E185–E186. doi:10.1093/neuros/nyab210
3. Tarvonen-Schröder S, Tenovuo O, Kaljonen A, Laimi K. Comparing disability between traumatic brain injury and spinal cord injury using the 12-item WHODAS 2.0 and the WHO minimal generic data set covering functioning and health. Clin Rehab. 2018;32(12):1676–1683. doi:10.1177/0269215518785945
4. Ding W, Hu S, Wang P, Kang H, Peng R, Dong Y, Li F. Spinal cord injury: the global incidence, prevalence, and disability from the global burden of disease study 2019. Spine. 2022;47(21):1532–1540. doi:10.1097/BRS.0000000000004417
5. Harvey LA. Physiotherapy rehabilitation for people with spinal cord injuries. J Physiother. 2016;62(1):4–11.
6. Safdarian M, Trinka E, Rahimi-Movaghar V, Thomschewski A, Aali A, Abady GG, et al. Global, regional, and national Burden of spinal cord injury, 1990–2019: a systematic analysis for the Global Burden of Disease Study 2019. Lancet Neurol. 2023;22(11):1026–47.
7. Lu Y, Shang Z, Zhang W, Pang M, Hu X, Dai Y, et al. Global incidence and characteristics of spinal cord injury since 2000–2021: a systematic review and meta-analysis. BMC Med. 2024;22(1):285.
8. Li K, Chen Z, Chang X, Xue R, Wang H, Guo W. Wnt signaling pathway in spinal cord injury: from mechanisms to potential applications. Front Mol Neurosci. 2024;17:1427054.
9. Sterner RC, Sterner RM. Immune response following traumatic spinal cord injury: pathophysiology and therapies. Front Immunol. 2023;13:1084101.
10. Müller-Jensen L, Ploner CJ, Kroneberg D, Schmidt WU. Clinical presentation and causes of non-traumatic spinal cord injury: an observational study in emergency patients. Front Neurol. 2021;12:701927.
11. Papageorgiou NA, Papageorgiou P, Kotroni A, Vasiliadis E. A comprehensive review of the importance of the main comorbidities in developing cognitive disorders in patients with spinal cord injuries. Cureus. 2024;16(9):e70071. doi:10.7759/cureus.70071
12. Lena E, Timelli L, DI Fonzo S, Tonini A, Pisani V, Garcovich C, Covella E, Tamburella F, Scivoletto G. Unveiling the mosaic: comparing demographics and outcomes in traumatic vs. non-traumatic spinal cord injuries. Eur J Phys Rehabil Med. 2024. doi:10.23736/S1973-9087.24.08554-X
13. Choi Y, Leigh JH. Epidemiology of traumatic and non-traumatic spinal cord injury in Korea: a narrative review. Korean J Neurotrauma. 2023;19(4):434–445. doi:10.13004/kjnt.2023.19.e54
14. Khadour FA, Khadour YA, Meng L, XinLi C, Xu T. Epidemiology features of traumatic and non-traumatic spinal cord injury in China, Wuhan. Sci Rep. 2024;14(1):1640. doi:10.1038/s41598-024-52210-4
15. Musubire AK, Meya DB, Katabira ET, Meyer ACL, Bohjanen PR, Boulware DR, Minja F. Epidemiology of non-traumatic spinal cord injury in Uganda: a single center, prospective study with MRI evaluation. BMC Neurol. 2019;19(1):10. doi:10.1186/s12883-019-1236-3
16. Smith É, Fitzpatrick P, Lyons F, Morris S, Synnott K. Epidemiology of non-traumatic spinal cord injury in Ireland – A prospective population-based study. J Spinal Cord Med. 2022;45(1):76–81. doi:10.1080/10790268.2020.1762829
17. Roberts TT, Leonard GR, Cepela DJ. Classifications in brief: American Spinal Injury Association (ASIA) Impairment Scale. Clin Orthop. 2017;475(5):1499–504.
18. van Middendorp JJ, Goss B, Urquhart S, Atresh S, Williams RP, Schuetz M. Diagnosis and prognosis of traumatic spinal cord injury. Glob Spine J. 2011;1(1):001–007. doi:10.1055/s-0031-1296049

19. Parizel PM, Van Der Zijden T, Gaudino S, Spaepen M, Voormolen MHJ, Venstermans C, et al. Trauma of the spine and spinal cord: imaging strategies. Eur Spine J. 2010;19(S1):8–17.
20. Bozzo A, Marcoux J, Radhakrishna M, Pelletier J, Goulet B. The role of magnetic resonance imaging in the management of acute spinal cord injury. J Neurotrauma. 2011;28(8):1401–11.
21. Shigapova RR, Mukhamedshina YO. Electrophysiology methods for assessing of neurodegenerative and post-traumatic processes as applied to translational research. Life. 2024;14(6):737.
22. Punjani N, Deska-Gauthier D, Hachem LD, Abramian M, Fehlings MG. Neuroplasticity and regeneration after spinal cord injury. North Am Spine Soc J. 2023;15:100235.
23. Tian T, Zhang S, Yang M. Recent progress and challenges in the treatment of spinal cord injury. Protein Cell. 2023;14(9):635–52.
24. Evancho A, Tyler WJ, McGregor K. A review of combined neuromodulation and physical therapy interventions for enhanced neurorehabilitation. Front Hum Neurosci. 2023;17:1151218.
25. Sanches E, Ho D, Van De Looij Y, Aebi Toulotte A, Baud L, Bouteldja F, et al. Early intensive rehabilitation reverses locomotor disruption, decrease brain inflammation and induces neuroplasticity following experimental Cerebral Palsy. Brain Behav Immun. 2024;121:303–16.
26. Sandrow-Feinberg HR, Houlé JD. Exercise after spinal cord injury as an agent for neuroprotection, regeneration and rehabilitation. Brain Res. 2015;1619:12–21.
27. Bartlett RD, Burley S, Ip M, Phillips JB, Choi D. Cell therapies for spinal cord injury: trends and challenges of current clinical trials. Neurosurgery. 2020;87(4):E456–E472. doi:10.1093/neuros/nyaa149
28. Fischer G, Bättig L, Stienen MN, Curt A, Fehlings MG, Hejrati N. Advancements in neuroregenerative and neuroprotective therapies for traumatic spinal cord injury. Front Neurosci. https://doi.org/10.3389/fnins.2024.1372920
29. Mehta S, McIntyre A, Janzen S, Loh E, Teasell R. Systematic review of pharmacologic treatments of pain after spinal cord injury: an update. Arch Phys Med Rehabil. 2016;97(8):1381–1391.e1.
30. Zeng CW. Advancing spinal cord injury treatment through stem cell therapy: a comprehensive review of cell types, challenges, and emerging technologies in regenerative medicine. Int J Mol Sci. 2023;24(18):14349. doi:10.3390/ijms241814349
31. Behrman AL, Bowden MG, Nair PM. Neuroplasticity after spinal cord injury and training: an emerging paradigm shift in rehabilitation and walking recovery. Phys Ther. 2006;86(10):1406–25.
32. Martin JH. Neuroplasticity of spinal cord injury and repair. Handb Clin Neurol. 2022;184:317–30.
33. Nagappan PG, Chen H, Wang DY. Neuroregeneration and plasticity: a review of the physiological mechanisms for achieving functional recovery postinjury. Mil Med Res. 2020;7(1):30.
34. Cramer SC, Sur M, Dobkin BH, O'Brien C, Sanger TD, Trojanowski JQ, et al. Harnessing neuroplasticity for clinical applications. Brain. 2011;134(6):1591–609.
35. Calderone A, Cardile D, De Luca R, Quartarone A, Corallo F, Calabrò RS. Brain plasticity in patients with spinal cord injuries: a systematic review. Int J Mol Sci. 2024;25(4):2224.
36. Burak Yalçın M. Neuromodulatory Effect of BDNF in Spinal Cord Injury [Internet]. Biochemistry. IntechOpen; 2023. doi:10.5772/intechopen.111849
37. Li C, Luo Y, Li S. The roles of neural stem cells in myelin regeneration and repair therapy after spinal cord injury. Stem Cell Res Ther. 2024;15(1):204.
38. Ye J, Xue R, Ji ZY, Zou CJ, Chen YQ, Wang JJ, Cheng XD. Effect of NT-3 on repair of spinal cord injury through the MAPK signaling pathway. Eur Rev Med Pharmacol Sci. 2020;24(5):2165–2172. doi:10.26355/eurrev_202003_20481
39. Chen T, He X, Wang J, Du D, Xu Y. NT-3 combined with TGF-β signaling pathway enhance the repair of spinal cord injury by inhibiting glial scar formation and promoting axonal regeneration. Mol Biotechnol. 2024;66(6):1484–1495. doi:10.1007/s12033-023-00781-4
40. Zhang H, Wu F, Kong X, Yang J, Chen H, Deng L, et al. Nerve growth factor improves functional recovery by inhibiting endoplasmic reticulum stress-induced neuronal apoptosis in rats with spinal cord injury. J Transl Med. 2014;12(1):130.
41. Wagner FB, Mignardot JB, Le Goff-Mignardot CG, Demesmaeker R, Komi S, Capogrosso M, et al. Targeted neurotechnology restores walking in humans with spinal cord injury. Nature. 2018;563(7729):65–71.
42. Peev N, Komarov A, Osorio-Fonseca E, Zileli M. Rehabilitation of spinal cord injury: WFNS spine committee recommendations. Neurospine. 2020;17(4):820–32.
43. Glinsky JV, Harvey LA. Physiotherapy management of people with spinal cord injuries: an update. J Physiother. 2024;70(4):256–64.
44. Barbosa PHFA, Glinsky JV, Fachin-Martins E, Harvey LA. Physiotherapy interventions for the treatment of spasticity in people with spinal cord injury: a systematic review. Spinal Cord. 2021;59(3):236–47.

45. Afrin R. *Practice of Passive Movement and Stretching Exercise of the Limbs by the Rehabilitation Professionals for Spinal Cord Injury Patients at CRP.* Bangladesh Health Professions Institute (BHPI).
46. Alaparthi GK, Raigangar V, Chakravarthy Bairapareddy K, Gatty A, Mohammad S, Alzarooni A, et al. A national survey in United Arab Emirates on practice of passive range of motion by physiotherapists in intensive care unit. PLOS ONE. 2021;16(8):e0256453.
47. Diong J, Harvey LA, Kwah LK, Eyles J, Ling MJ, Ben M, et al. Incidence and predictors of contracture after spinal cord injury—A prospective cohort study. Spinal Cord. 2012;50(8):579–84.
48. Harvey LA, Glinsky JV, Bowden JL. The effectiveness of 22 commonly administered physiotherapy interventions for people with spinal cord injury: a systematic review. Spinal Cord. 2016;54(11):914–23.
49. Rothschild RM. Neuroengineering tools/applications for bidirectional interfaces, brain-computer interfaces, and neuroprosthetic implants – A review of recent progress. Front Neuroeng. 2010;3:112. doi:10.3389/fneng.2010.00112
50. Collinger JL, Foldes S, Bruns TM, Wodlinger B, Gaunt R, Weber DJ. Neuroprosthetic technology for individuals with spinal cord injury. J Spinal Cord Med. 2013;36(4):258–72. doi:10.1179/2045772313Y.0000000128
51. Karamian BA, Siegel N, Nourie B, Serruya MD, Heary RF, Harrop JS, et al. The role of electrical stimulation for rehabilitation and regeneration after spinal cord injury. J Orthop Traumatol. 2022;23(1):2.
52. Cho N, Squair JW, Bloch J, Courtine G. Neurorestorative interventions involving bioelectronic implants after spinal cord injury. Bioelectron Med. 2019;5(1):10.
53. Bonizzato M. Neuroprosthetics: an outlook on active challenges toward clinical adoption. J Neurophysiol. 2021;125(1):105–109. doi:10.1152/jn.00496.2020
54. Leach JB, Achyuta AK, Murthy SK. Bridging the divide between neuroprosthetic design, tissue engineering and neurobiology. Front Neuroeng. 2010;2:18. doi:10.3389/neuro.16.018.2009
55. Foy T, Perritt G, Thimmaiah D, Heisler L, Offutt JL, Cantoni K, et al. Occupational therapy treatment time during inpatient spinal cord injury rehabilitation. J Spinal Cord Med. 2011;34(2):162–75.
56. Zbogar D, Eng JJ, Miller WC, Krassioukov AV, Verrier MC. Movement repetitions in physical and occupational therapy during spinal cord injury rehabilitation. Spinal Cord. 2017;55(2):172–9.
57. Nas K. Rehabilitation of spinal cord injuries. World J Orthop. 2015;6(1):8.
58. Middleton JW, Johnston D, Murphy G, Ramakrishnan K, Savage N, Harper R, et al. Early access to vocational rehabilitation for spinal cord injury inpatients. J Rehabil Med. 2015;47(7):626–31.
59. Gordan W, Gerber D, David DS, Adornato V, Brougham R, Gassaway J, et al. Relationship of speech-language pathology inpatient rehabilitation interventions and patient characteristics to outcomes following spinal cord injury: the SCIRehab Project. J Spinal Cord Med. 2012;35(6):565–77.
60. Rodrigues FO, Frois CA, Sarmet M, Mangilli LD. Vocal parameters in individuals with traumatic spinal cord injury: a systematic review. J Voice. 2021;35(4):545–53.
61. Rodrigues FO, Sarmet M, Maldaner V, Yamasaki R, Behlau M, Davison Mangilli L. Traumatic spinal injury: preliminary results of respiratory function, voice and quality of life. J Voice. 2023;37(3): 469.e1–469.e10.
62. Brougham R, David DS, Adornato V, Gordan W, Dale B, Georgeadis AC, et al. Speech-language pathology treatment time during inpatient spinal cord injury rehabilitation: the SCIRehab project. J Spinal Cord Med. 2011;34(2):186–95.
63. Dick S, Thomas J, McMillan J, Davis K, Miles A. Swallowing rehabilitation following spinal injury: a case series. J Spinal Cord Med. 2022;45(1):65–75.
64. Wolf C, Meiners TH. Dysphagia in patients with acute cervical spinal cord injury. Spinal Cord. 2003;41(6):347–53.
65. Hearn JH, Cross A. Mindfulness for pain, depression, anxiety, and quality of life in people with spinal cord injury: a systematic review. BMC Neurol. 2020;20(1):32.
66. Bhattarai M, Smedema SM, Hoyt WT, Bishop M. The role of mindfulness in quality of life of persons with spinal cord injury: a cross-sectional study. Health Qual Life Outcomes. 2022;20(1):148.
67. Post MWM, Van Leeuwen CMC. Psychosocial issues in spinal cord injury: a review. Spinal Cord. 2012;50(5):382–9.
68. Mehta S, Orenczuk S, Hansen KT, Aubut JAL, Hitzig SL, Legassic M, et al. An evidence-based review of the effectiveness of cognitive behavioral therapy for psychosocial issues post-spinal cord injury. Rehabil Psychol. 2011;56(1):15–25.
69. Schulz R, Czaja SJ, Lustig A, Zdaniuk B, Martire LM, Perdomo D. Improving the quality of life of caregivers of persons with spinal cord injury: a randomized controlled trial. Rehabil Psychol. 2009;54(1):1–15.

70. Armstrong A, Oetinger K, Weimer K, Hönig K. Digital psychosocial interventions for individuals with spinal cord injury: a scoping review. Front Psychiatry. 2024;15:1289138.
71. Kesikburun S. Non-invasive brain stimulation in rehabilitation. Turk J Phys Med Rehabil. 2022;68(1):1–8.
72. Belci M, Catley M, Husain M, Frankel HL, Davey NJ. Magnetic brain stimulation can improve clinical outcome in incomplete spinal cord injured patients. Spinal Cord. 2004;42(7):417–9.
73. Gersner R, Kravetz E, Feil J, Pell G, Zangen A. Long-term effects of repetitive transcranial magnetic stimulation on markers for neuroplasticity: differential outcomes in anesthetized and awake animals. J Neurosci. 2011;31(20):7521–7526.
74. González JC, Lignani G, Maroto M, Baldelli P, Hernández-Guijo JM. Presynaptic muscarinic receptors reduce synaptic depression and facilitate its recovery at hippocampal GABAergic synapses. Cereb Cortex. 2014;24(7):1818–31.
75. Chen YT, Su YC, Lin YC. Effect of repetitive transcranial magnetic stimulation on gait function and strength among patients with spinal cord injury: a meta-analysis. Rehab Pract Sci. 2023;2023(2):10.
76. Pulverenti TS, Zaaya M, Grabowski E, Grabowski M, Knikou M. Brain and spinal cord paired stimulation coupled with locomotor training facilitates motor output in human spinal cord injury. Front Neurol. 2022;13:1000940.
77. Li C, Jirachaipitak S, Wrigley P, Xu H, Euasobhon P. Transcranial direct current stimulation for spinal cord injury-associated neuropathic pain. Korean J Pain. 2021;34(2):156–64.
78. De Araújo AVL, Ribeiro FPG, Massetti T, Potter-Baker KA, Cortes M, Plow EB, et al. Effectiveness of anodal transcranial direct current stimulation to improve muscle strength and motor functionality after incomplete spinal cord injury: a systematic review and meta-analysis. Spinal Cord. 2020;58(6):635–46.
79. Antal A, Terney D, Kühnl S, Paulus W. Anodal transcranial direct current stimulation of the motor cortex ameliorates chronic pain and reduces short intracortical inhibition. J Pain Symptom Manage. 2010;39(5):890–903.
80. Inanici F, Brighton LN, Samejima S, Hofstetter CP, Moritz CT. Transcutaneous spinal cord stimulation restores hand and arm function after spinal cord injury. IEEE Trans Neural Syst Rehabil Eng. 2021;29:310–9.
81. Choi E, Gattas S, Brown N, Hong J, Limbo J, Chan A, et al. Epidural electrical stimulation for spinal cord injury. Neural Regen Res. 2021;16(12):2367.
82. Chalif JI, Chavarro VS, Mensah E, Johnston B, Fields DP, Chalif EJ, et al. Epidural spinal cord stimulation for spinal cord injury in humans: a systematic review. J Clin Med. 2024;13(4):1090.
83. Luz A, Rupp R, Ahmadi R, Weidner N. Beyond treatment of chronic pain: a scoping review about epidural electrical spinal cord stimulation to restore sensorimotor and autonomic function after spinal cord injury. Neurol Res Pract. 2023;5(1):14.
84. Taylor C, McHugh C, Mockler D, Minogue C, Reilly RB, Fleming N. Transcutaneous spinal cord stimulation and motor responses in individuals with spinal cord injury: a methodological review. PLoS ONE. 2021;16(11):e0260166.
85. Martin R. Utility and feasibility of transcutaneous spinal cord stimulation for patients with incomplete SCI in therapeutic settings: a review of topic. Front Rehabil Sci. 2021;2:724003.
86. Caylor J, Reddy R, Yin S, Cui C, Huang M, Huang C, et al. Spinal cord stimulation in chronic pain: evidence and theory for mechanisms of action. Bioelectron Med. 2019;5(1):12.
87. Tefertiller C, Rozwod M, VandeGriend E, Bartelt P, Sevigny M, Smith AC. Transcutaneous electrical spinal cord stimulation to promote recovery in chronic spinal cord injury. Front Rehabil Sci. 2022;2:740307.
88. Seáñez I, Capogrosso M. Motor improvements enabled by spinal cord stimulation combined with physical training after spinal cord injury: review of experimental evidence in animals and humans. Bioelectron Med. 2021;7(1):16.
89. McHugh LV, Miller AA, Leech KA, Salorio C, Martin RH. Feasibility and utility of transcutaneous spinal cord stimulation combined with walking-based therapy for people with motor incomplete spinal cord injury. Spinal Cord Ser Cases. 2020;6(1):104.
90. Urbin MA, Ozdemir RA, Tazoe T, Perez MA. Spike-timing-dependent plasticity in lower-limb motoneurons after human spinal cord injury. J Neurophysiol. 2017;118(4):2171–80.
91. Jo HJ, Richardson MSA, Oudega M, Perez MA. Paired corticospinal-motoneuronal stimulation and exercise after spinal cord injury. J Spinal Cord Med. 44(Suppl 1):S23–7.
92. Bunday KL, Perez MA. Motor recovery after spinal cord injury enhanced by strengthening corticospinal synaptic transmission. Curr Biol. 2012;22(24):2355–61.

93. Grover FM, Chen B, Perez MA. Increased paired stimuli enhance corticospinal-motoneuronal plasticity in humans with spinal cord injury. J Neurophysiol. 2023;129(6):1414.
94. Bunday KL, Urbin MA, Perez MA. Potentiating paired corticospinal-motoneuronal plasticity after spinal cord injury. Brain Stimul Basic Transl Clin Res Neuromodulation. 2018;11(5):1083–92.
95. Yari D, Saberi A, Salmasi Z, Ghoreishi SA, Etemad L, Movaffagh J, et al. Recent advances in the treatment of spinal cord injury. Arch Bone Jt Surg. 2024;12(6):380–99.
96. Crowley ST, Fukushima Y, Uchida S, et al. Enhancement of motor, function recovery after spinal cord injury in mice by delivery of, brain-derived neurotrophic factor mRNA. Mol Ther Nucleic Acids, 2019;17:465–476.
97. Han Q, Ordaz JD, Liu NK, et al. Descending motor circuitry required, for NT-3 mediated locomotor recovery after spinal cord injury, in mice. Nat Commun 2019;10:5815.
98. Goldshmit Y, Frisca F, Pinto AR, et al. Fgf2 improves functional recovery-decreasing gliosis and increasing radial glia and neural progenitor cells after spinal cord injury. Brain Behav 2014;4:187–200.
99. Allahdadi KJ, de Santana TA, Santos GC, et al. IGF-1 overexpression, improves mesenchymal stem cell survival and promotes neurological recovery after spinal cord injury. Stem Cell Res Ther, 2019;10:146.
100. Lutton C, Young YW, Williams R, et al. Combined VEGF and PDGF, treatment reduces secondary degeneration after spinal cord, injury. J Neurotrauma 2012;29:957–970.
101. Aschauer-Wallner S, Leis S, Bogdahn U, Johannesen S, Couillard-Despres S, Aigner L. Granulocyte colony-stimulating factor in traumatic spinal cord injury. Drug Discov Today. 2021;26(7):1642–1655. doi:10.1016/j.drudis.2021.03.014.
102. Ashammakhi N, Kim HJ, Ehsanipour A, et al. Regenerative therapies for spinal cord injury. Tissue Eng Part B Rev. 2019;25(6):471–491. doi:10.1089/ten.teb.2019.0182.
103. Ding Y, Chen Q. mTOR pathway: a potential therapeutic target for spinal cord injury. Biomed Pharmacother. 2022;145:112430. doi:10.1016/j.biopha.2021.112430.
104. Wang W, He D, Chen J, et al. Circular RNA Plek promotes fibrogenic activation by regulating the miR-135b-5p/TGF-betaR1 axis after spinal cord injury. Aging (Albany NY). 2021;13(9):13211–13224. doi:10.18632/aging.203002.
105. Cho Y, Shin Jung E, Ewan EE, et al. Activating injury-responsive, genes with hypoxia enhances axon regeneration through neuronal HIF-1α. Neuron 2015;88:720–734.
106. Puttagunta R, Tedeschi A, Sória MG, et al. PCAF-dependent epigenetic, changes promote axonal regeneration in the central nervous, system. Nat Commun 2014;5:3527.
107. Geoffroy CG, Zheng B. Myelin-associated inhibitors in axonal growth, after CNS injury. Curr Opin Neurobiol 2014;27:31–38.
108. Ji Z, Gao GB, Ma YM, et al. Highly bioactive iridium metal-complex, alleviates spinal cord injury via ROS scavenging and inflammation reduction. Biomaterials 2022;284:121481.
109. Poulen G, Aloy E, Bringuier CM, et al. Inhibiting microglia proliferation, after spinal cord injury improves recovery in mice and nonhuman primates. Theranostics 2021;11:8640–8659.
110. Hellal F, Hurtado A, Ruschel J, et al. Microtubule stabilization reduces, scarring and causes axon regeneration after spinal cord injury. Science 2011;331:928–931.
111. Gwak SJ, Macks C, Jeong DU, et al. RhoA knockdown by cationic, amphiphilic copolymer/siRhoA polyplexes enhances axonal, regeneration in rat spinal cord injury model. Biomaterials 2017;121:155–166.
112. Han Q, Xie Y, Ordaz JD, et al. Restoring cellular energetics promotes, axonal regeneration and functional recovery after spinal cord, injury. Cell Metab. 2020;31:623–641.e8.
113. Tedeschi A, Dupraz S, Laskowski CJ, et al. The calcium channel subunit Alpha2delta2 suppresses axon regeneration in the adult.
114. Bradbury EJ, Burnside ER. Moving beyond the glial scar for spinal, cord repair. Nat Commun. 2019;10:3879.
115. Hachem LD, Mothe AJ, Tator CH. Unlocking the paradoxical endogenous stem cell response after spinal cord injury. Stem Cells. 2020;38:187–194. doi:10.1002/stem.3107
116. Koda M, Hanaoka H, Fujii Y, et al. Randomized trial of granulocyte colony-stimulating factor for spinal cord injury. Brain 2021;144:789–799.
117. Weiss S, Dunne C, Hewson J, et al. Multipotent CNS stem cells are present in the adult mammalian spinal cord and ventricular neuroaxis. J Neurosci. 1996;16:7599–7609. doi:10.1523/JNEUROSCI.16-23-07599.1996
118. Khazaei M, Ahuja CS, Fehlings MG. Induced pluripotent stem cells for traumatic spinal cord injury. Front Cell Dev Biol. 2017;4:152. doi:10.3389/fcell.2016.00152

119. Khazaei M, Ahuja CS, Nakashima H, et al. GDNF rescues the fate of neural progenitor grafts by attenuating Notch signals in the injured spinal cord in rodents. Sci Transl Med. 2020;12:eaau3538. doi:10.1126/scitranslmed.aau3538
120. Nagoshi N, Khazaei M, Ahlfors J-E, et al. Human spinal oligodendrogenic neural progenitor cells promote functional recovery after spinal cord injury by axonal remyelination and tissue sparing. Stem Cells Transl Med. 2018;7:806–818. doi:10.1002/sctm.17-0269
121. Martinez-Ramos C, Doblado LR, Mocholi EL, et al. Biohybrids for spinal cord injury repair. J Tissue Eng Regen Med. 2019;13(3):509–521. doi:10.1002/term.2816
122. Walsh CM, Wychowaniec JK, Brougham DF, Dooley D Functional hydrogels as therapeutic tools for spinal cord injury: new perspectives on immunopharmacological interventions. Pharmacol Ther. 2022;234:108043. doi:10.1016/j.pharmthera.2021.108043
123. Kubinová Š. Chapter 7 – Soft and rigid scaffolds for spinal cord injury regeneration. In: Perale G, editor. *Spinal Cord Injury (SCI) Repair Strategies*. Woodhead Publishing; 2020.
124. Yari D, Movaffagh J, Ebrahimzadeh MH, Saberi A, Qujeq D, Moradi A. Biomimetic ECM-based hybrid scaffold for cartilage tissue engineering applications. J Polym Environ. 2024:1–9.
125. Omidinia-Anarkoli A, Boesveld S, Tuvshindorj U, Rose JC, Haraszti T, De Laporte L. An injectable hybrid hydrogel with oriented short fibers induces unidirectional growth of functional nerve cells. Small. 2017;13:36.
126. Liu X, Hao M, Chen Z, et al. 3D bioprinted neural tissue constructs for spinal cord injury repair. Biomaterials. 2021;272:120771. doi:10.1016/j.biomaterials.2021.120771
127. Yu X, Zhang T, Li Y. 3D printing and bioprinting nerve conduits for neural tissue engineering. Polymers (Basel). 2020;12:8. doi:10.3390/polym12081637
128. Yuan TY, Zhang J, Yu T, Wu JP, Liu QY. 3D bioprinting for spinal cord injury repair. Front Bioeng Biotechnol. 2022;10:847344. doi:10.3389/fbioe.2022.847344
129. Ajiboye AB, Willett FR, Young DR, et al. Restoration of reaching and grasping movements through brain-controlled muscle stimulation in a person with tetraplegia: a proof-of-concept demonstration. Lancet 2017;389:1821–1830.

13 Micronutrients and Brain Health

Sakshi Singh, Bhaskar Sharma, Srishti Jain, Kulsum Jan, and Prabhakar Singh

13.1 INTRODUCTION

Brain health is dynamic for maintaining cognitive function, emotional stability, and global well-being. It is supported by factors such as a stable diet, regular physical activities, and mental stimulation, which help to avoid cognitive decline and slow down the risk of neurodegenerative diseases like Alzheimer's and Parkinson's. Nutrition plays a key role, with essential fatty acids, antioxidants, and micronutrients like B vitamins supporting neuronal function, reducing oxidative stress, and enhancing neuroplasticity (Parletta et al., 2013). Regular exercise boosts brain function by improving blood flow and supporting hippocampal health, reducing dementia risk (Livingston et al., 2017), while mental stimulation promotes cognitive resilience. Conversely, chronic stress negatively impacts brain health, leading to inflammation and mood disorders. The brain, as the central control system, regulates memory, motor functions, and emotions, influencing mood, stress, and overall bodily functions (Conkbayir, 2023). Micronutrients like omega-3 fatty acids, vitamins C, E, B6, and B12, and minerals like magnesium and zinc are essential for brain health, aiding neurotransmitter production, neurogenesis, and protection against cognitive decline and mood disorders (Lahoda Brodska et al., 2023). Maintaining a nutrient-rich diet and active lifestyle is critical for long-term cognitive health and well-being.

13.2 KEY MICRONUTRIENTS FOR BRAIN HEALTH

13.2.1 Vitamin A

Vitamin A is essential for vision and brain development, playing a key role in maintaining normal vision, as it is involved in the production of rhodopsin, which is present in retina rod cells (Sajovic et al., 2022). It helps to convert light into electrical signals for the brain, and its deficiency leads to night blindness and eye conditions like xerophthalmia. Beyond vision, vitamin A, in the form of retinoic acid, is vital for brain development by regulating gene expression for neurogenesis, synaptic plasticity, and neural tube formation during fetal growth (Wołoszynowska-Fraser et al., 2020). It enhances learning and memory through its effects on synaptic strength, especially in the hippocampus, and disruptions in its signaling are linked to cognitive deficits and neurodevelopmental disorders. Thus, adequate vitamin A intake is essential for both visual and cognitive health. Vitamin A role in brain health is presented in Figure 13.1.

13.2.2 Vitamin B

The B vitamins, comprising thiamine (B1), riboflavin (B2), niacin (B3), pantothenic acid (B5), pyridoxine (B6), biotin (B7), folate (B9), and cobalamin (B12), are vital water-soluble nutrients needed for energy metabolism, neurotransmitter synthesis, and cognitive performance. Acting as coenzymes, they bind to enzymes to facilitate a wide range of biochemical reactions. For example, vitamin B6 is a cofactor for over 140 enzymes involved in amino acid metabolism, while pantothenic

DOI: 10.1201/9781003500025-13

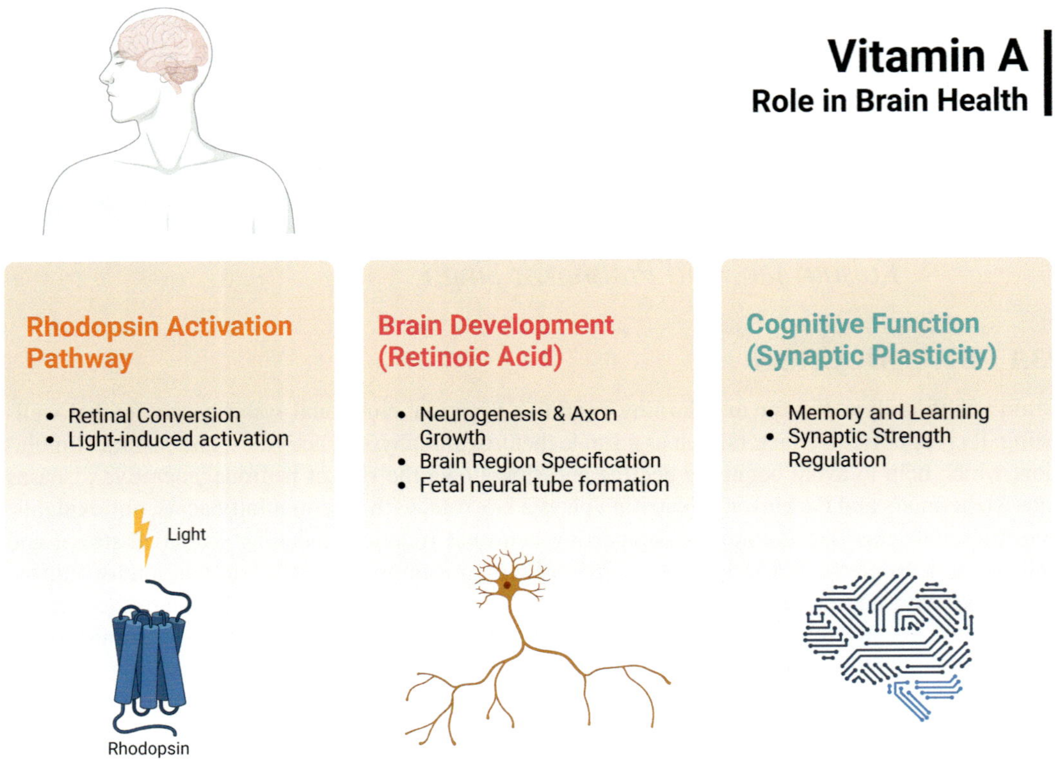

FIGURE 13.1 Schematic illustration depicting the influence of Vitamin A on brain health.

acid, through coenzyme A (CoA), supports energy production and biosynthesis of bioactive compounds (Zempleni et al., 2013). B vitamins play a significant role in the brain, including maintaining the blood-brain barrier. The brain has a higher concentration of important B vitamins than the rest of the body does, including methyltetrahydrofolate, biotin, and pantothenic acid.

Thiamine (B1) controls the cholinergic system and neuronal membranes in addition to acting as a coenzyme in the pentose phosphate pathway, which is critical for the synthesis of compounds necessary for brain function (Kerns et al., 2015). Through FMN and FAD, riboflavin (B2) enhances the function of enzymes involved in the metabolism of niacin, folate, and B6 as well as providing antioxidant protection (Sinigaglia-Coimbra et al., 2011). Niacin (B3) is necessary for the synthesis of energy, DNA repair, and the regulation of neurotransmitters and also affects the immunological responses related to Parkinson's and Schizophrenia. The production of neurotransmitters and the structure of brain cells depend on CoA, which is a precursor of pantothenic acid (B5) (10). Vitamin B6 (pyridoxine) is essential for the synthesis of neurotransmitters, influencing the production of dopamine, GABA, and serotonin as well as controlling inflammation and the immune system. Diseases like diabetes are associated with biotin (B7), which promotes glucose metabolism. Folate (B9) and vitamin B12 (cobalamin) participate in the folate and methionine cycles, which are necessary for the synthesis of DNA, the differentiation of neurons, and the synthesis of neurotransmitters. Literature shows that deficiencies in B vitamins play a major role in illnesses including dementia, neurodevelopmental abnormalities, and cognitive decline. The synthesis of neurotransmitters that control behavior, emotion, and thinking processes requires B vitamins. The production of serotonin, dopamine, norepinephrine, and GABA, which are essential for mood regulation and cognitive function, depends on pyridoxine (B6) as a cofactor. During one-carbon metabolism, folate (B9) transforms homocysteine into methionine, which is needed to make S-adenosylmethionine (SAM), an amino acid that helps in neurotransmitter production. Myelin integrity, neurotransmitter synthesis, and the synthesis of SAM depend on the methylation processes that cobalamin (B12) and folate carry

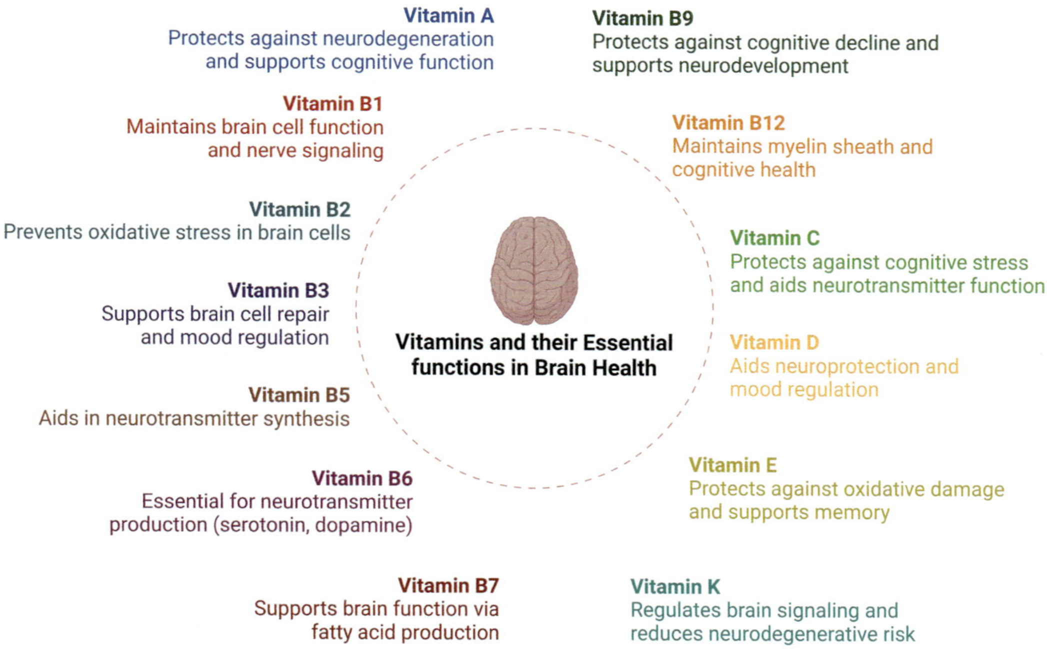

FIGURE 13.2 Schematic illustration highlighting various vitamins and their vital roles in maintaining brain health.

on. Cognitive impairments and Wernicke-Korsakoff syndrome can result from thiamine (B1) deficiency, which is essential for the metabolism of glucose. Adequate amounts of folate and B12 help to reduce homocysteine levels, which lowers the risk of dementia and cognitive decline. Memory problems have been associated with pyridoxine (B6) insufficiency, niacin (B3) protects against oxidative stress and cognitive dysfunction, and dementia may result from pellagra (Figure 13.2).

13.2.3 Vitamin D

Vitamin D is essential for healthy bones as well as brain function. Brain areas like the hippocampus, where it interacts with vitamin D receptors (VDRs), its active form, calcitriol, control the synthesis of neurotransmitters, neurotrophic factors, and antioxidant enzymes. Combined with its anti-inflammatory properties that lessen neuroinflammation connected to mood disorders and neurodegenerative diseases, vitamin D also aids in controlling calcium levels within neurons, which is critical for synaptic plasticity and neuroprotection. Because vitamin D regulates serotonin, reduces inflammation, and controls the HPA axis, low levels of the vitamin are linked to an increased risk of depression. Vitamin D also plays a protective role against cognitive decline, with low levels linked to increased risks of dementia and Alzheimer's disease (AD). Its regulation of calcium homeostasis prevents neuronal damage, while its anti-inflammatory effects help combat oxidative stress. Studies like the "InCHIANTI" research and meta-analyses highlight the association between vitamin D deficiency and cognitive decline, though clinical trials, such as the "Vital-D" trial, are still needed to confirm the vitamin's preventive potential in cognitive disorders. Vitamin D's role in brain health is presented in Figure 13.3.

13.2.4 Vitamin C

Ascorbic acid, or vitamin C, is a potent water-soluble antioxidant crucial for maintaining physiological functions, particularly in the nervous system, where it aids neurotransmitter production and

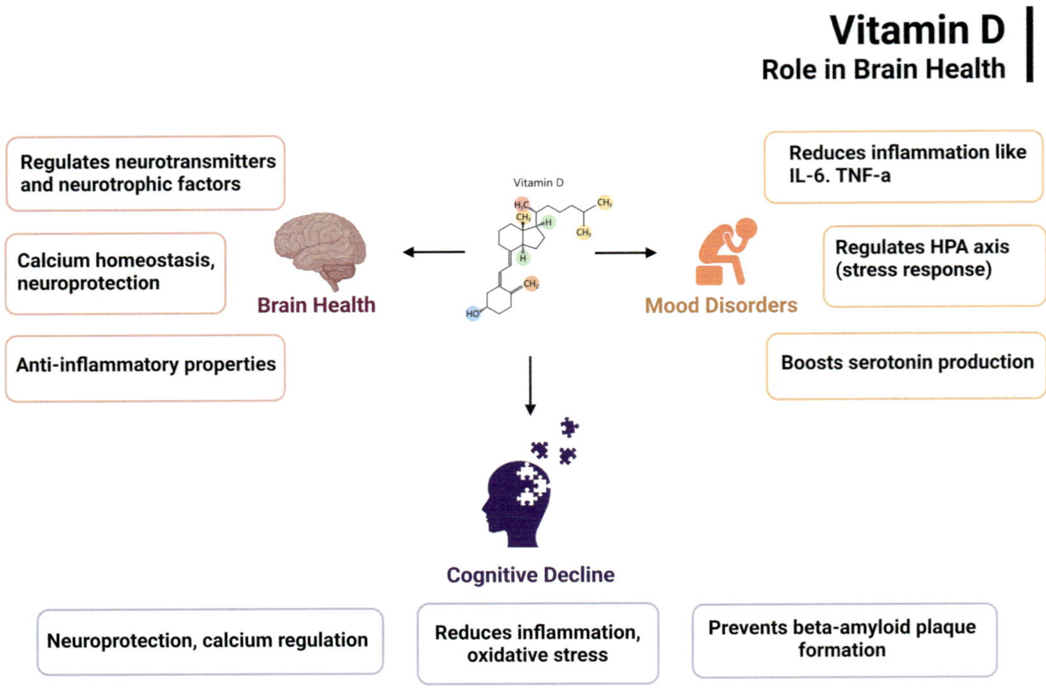

FIGURE 13.3 Schematic illustration depicting the role of Vitamin D in promoting and maintaining brain health.

antioxidant defense (Carr & Maggini, 2017). It neutralizes reactive oxygen and nitrogen species (ROS and RNS), safeguarding cells from oxidative stress that can harm lipids, proteins, and nucleic acids. In the brain, which is highly susceptible to oxidative damage due to its metabolic activity, vitamin C accumulates in neuronal and glial cells, providing neuroprotection against conditions like Alzheimer's and Parkinson's. Vitamin C helps in the synthesis of neurotransmitters by controlling tetrahydrobiopterin, a cofactor for tryptophan hydroxylase, which in turn influences the synthesis of serotonin. It also serves as a cofactor in the transformation of dopamine to norepinephrine. Additionally, it supports the production of neuropeptides, which maintain appropriate nervous system signaling. Furthermore, by restoring oxidized versions of other antioxidants like vitamin E, vitamin C strengthens the antioxidant system and protects essential biomolecules including lipids, proteins, and DNA. This antioxidant capacity is essential for halting oxidative stress-related neurodegeneration and cognitive loss. Anxiety, sadness, and cognitive impairment have all been related to vitamin C deficiency. Vitamin C is essential for brain function (Ballaz & Rebec, 2019). Its involvement in the synthesis of neurotransmitters, neuroprotection, and antioxidant protection. Vitamin C supplements reduce the symptoms of oxidative stress related to mood disorders and neurodegenerative illnesses (Plevin et al., 2020). Vitamin C plays two roles: It protects neurons and promotes neuroplasticity. Insufficient amounts of vitamin C can affect cognitive function, memory, and learning. Vitamin C supports neurotransmitter generation, antioxidant protection, and neuropeptide regulation, it is essential for maintaining cognitive function and delaying the development of neurological disorders.

13.2.5 Vitamin E

Vitamin E is a fat-soluble antioxidant comprising eight compounds, which is essential to prevent oxidative damage to cell membranes. α-tocopherol is the most physiologically active form and is extensively stored in human tissues among other forms (Eerman & Brodaty, 2004). It functions as a chain-breaking antioxidant, preventing lipid peroxidation driven by reactive oxygen species (ROS) in lipid-rich environments like cell membranes. It is especially important for the brain, which is vulnerable to oxidative stress due to its high oxygen demand and lipid content. Vitamin E's antioxidant properties play a significant role in reducing oxidative stress, which contributes to neurodegenerative diseases like Alzheimer's, Parkinson's, and Huntington's diseases (Teleanu et al., 2022). In Alzheimer's, vitamin E helps to reduce amyloid-β-induced lipid peroxidation, potentially slowing cognitive decline. Its neuroprotective effects may also shield dopaminergic neurons in Parkinson's, though clinical trial results are mixed. In Huntington's disease, vitamin E shows promise in mitigating oxidative stress-related neuronal damage. Vitamin E provides neuroprotection through several mechanisms, including inhibiting lipid peroxidation in cell membranes, which preserves neuronal structural integrity. By stopping free radical chain reactions, it helps to prevent the oxidation of polyunsaturated fatty acids (PUFAs). Additionally, vitamin E modulates the inflammatory response by inhibiting oxidative stress pathways and reducing the production of pro-inflammatory cytokines, which are linked to neurodegenerative diseases (Wallert et al., 2021). It also preserves mitochondrial function by protecting mitochondrial lipids and proteins from oxidative damage, supporting energy production, and reducing ROS. Vitamin E further safeguards synaptic function by protecting membrane phospholipids, ensuring proper neurotransmission (Tönnies & Trushina, 2017). While epidemiological studies suggest a protective role, particularly in AD, clinical trials show mixed results, likely due to differences in dosage and patient characteristics. Although supplementation may benefit those with mild to moderate Alzheimer's, high doses could interfere with other fat-soluble vitamins and increase the risk of hemorrhagic stroke, necessitating careful dosage considerations.

13.3 MINERALS

13.3.1 Iron

Iron is an essential trace mineral crucial for various physiological functions that support human health, particularly in oxygen transport, myelination, and neurotransmitter production, which significantly influence brain development, functionality, and cognitive abilities (Muñoz & Humeres, 2012). It is a main component of hemoglobin, which is present in red blood cells and carries oxygen; each hemoglobin molecule contains four iron atoms, each capable of binding with an oxygen molecule, facilitating effective oxygen delivery from the lungs to tissues of the body. Insufficient iron levels can disrupt this process, potentially leading to anemia characterized by fatigue, weakness, and cognitive difficulties (Muñoz & Humeres, 2012). Additionally, iron is critical in brain development, especially in myelination, the process where myelin, the fatty layer coating nerve fiber, enhances the speed of electrical signal transmission between neurons, which is critical for smooth neuronal communication and cognitive functions, such as learning and memory. Iron is necessary for the production of cholesterol and fatty acids, both essential for myelin production by oligodendrocytes in the central nervous system (CNS) (Affrald & Narayan, 2024). Lack of iron during crucial periods of brain development can impede myelination, leading to slowed neural processing and cognitive impairments. Iron deficiency leads to alterations in brain structure and function, particularly in critical areas like the hippocampus and prefrontal cortex, which are pivotal for memory and executive functions.

Iron is integral to the synthesis of various neurotransmitters, including dopamine, norepinephrine, and serotonin, serving as a cofactor for essential enzymes like tyrosine hydroxylase, which initiates dopamine production, a neurotransmitter vital for reward, motivation, and executive functions. Additionally, iron helps in the activity of monoamine oxidase, which breaks down neurotransmitters like serotonin and norepinephrine. An iron deficiency can lead to neurotransmitter imbalances, adversely affecting mood, attention, and cognitive capabilities, and has been associated with the development of neuropsychiatric conditions like attention-deficit hyperactivity disorder (ADHD). Iron deficiency, particularly during critical periods such as early childhood and adolescence, can have profound and lasting impacts on cognitive development; even mild iron deficiency without anemia has been linked to impairments in attention, memory, and executive function. The mechanisms of cognitive impairment include disrupted myelination, as iron is essential for forming myelin, with deficiencies leading to hypo-myelination that slows signal transmission in the brain, thereby hindering physical development, language acquisition, and overall cognitive function. Furthermore, altered neurotransmitter metabolism occurs with low iron levels, resulting in decreased dopamine in regions like the prefrontal cortex and striatum, which can lead to mood disorders and learning difficulties due to diminished serotonin and norepinephrine. Chronic iron deficiency anemia can also lead to hypoxia, or reduced oxygen supply in brain tissues, adversely affecting neural activity and energy metabolism, ultimately impairing cognitive growth and function (Figure 13.4).

Evidence from human studies underscores the critical role of iron in cognitive and behavioral development across various age groups. In infants and toddlers, iron deficiency has been associated with significant and lasting deficits in motor and cognitive abilities, as demonstrated by Lozoff et al. (2006), who found that iron-deficient infants exhibited slower development compared to their adequately nourished peers, with effects persisting into adolescence and suggesting lasting neural changes. In school-aged children, iron deficiency correlates with diminished academic performance, lower IQ scores, and difficulties in problem-solving tasks; research by Zimmermann & Boelaert, (2015) indicates that iron supplementation can enhance attention and memory in these children, although the extent of improvement is influenced by the severity and duration of the deficiency.

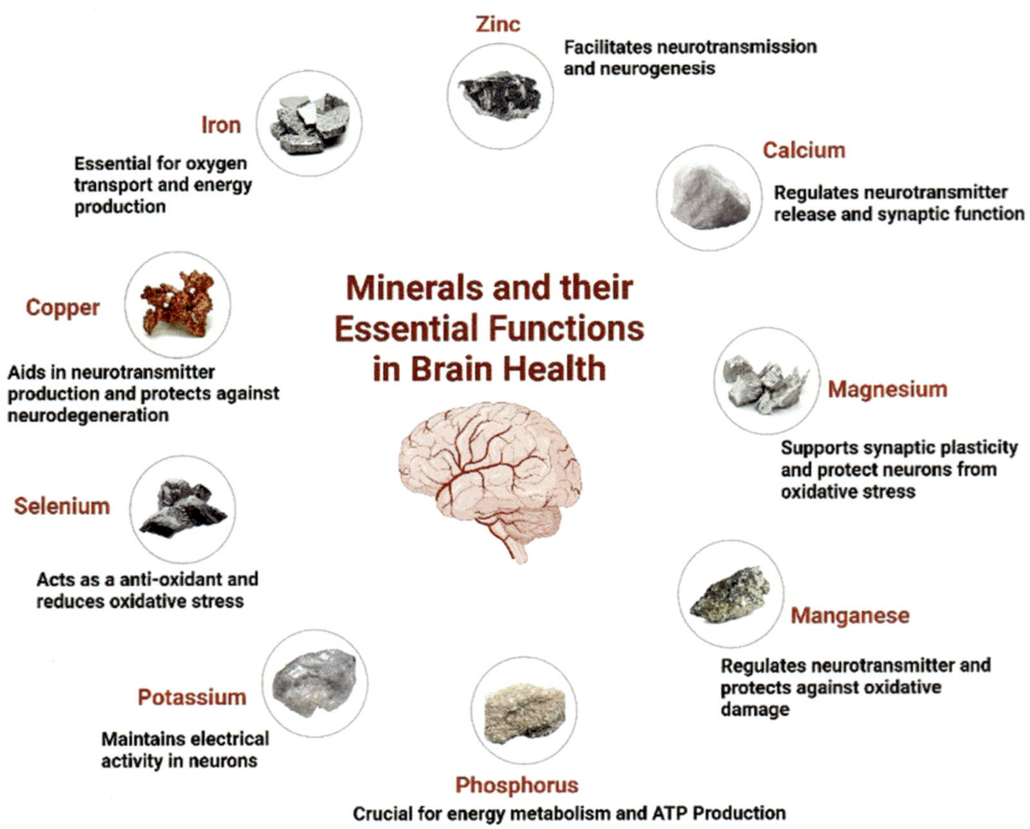

FIGURE 13.4 Representative illustration highlighting key minerals and their vital roles in maintaining brain health.

During adolescence, iron deficiency has been linked to impairments in cognitive flexibility and learning abilities.

13.3.2 Zinc

Zinc, an essential trace element, is vital for neurogenesis, especially in the hippocampus, which is involved in learning and memory. It is concentrated in synaptic vesicles and plays a crucial role in regulating neural stem cell proliferation, differentiation, and survival. Research indicates that zinc ions (Zn^{2}+) influence various signaling pathways related to neurogenesis, especially the brain-derived neurotrophic factor (BDNF) pathway, which is essential for neuron survival and synaptic plasticity; zinc deficiency can lead to decreased BDNF levels and hinder neurogenesis. Additionally, zinc serves as a neuromodulator in synaptic plasticity by affecting glutamatergic synapses and the N-methyl-D-aspartate (NMDA) receptor, critical for long-term potentiation (LTP), a process integral to memory consolidation. Both excess and deficiency of zinc can disrupt synaptic function, with overload inhibiting the NMDA receptor and shortage hindering LTP, leading to cognitive deficits. Proper zinc levels are necessary for effective memory formation and cognitive performance, and studies have shown that supplementation can enhance cognitive function, particularly in those with deficiencies. Furthermore, zinc is increasingly recognized for its role in mood regulation, as individuals with depression often exhibit lower zinc levels. It modulates neurotransmitters, such as serotonin, enhancing receptor effectiveness, and regulating the hypothalamic-pituitary-adrenal (HPA) axis, crucial for stress response. Zinc deficiency may exacerbate mood disorders by increasing glucocorticoid secretion, and supplementation has shown promise in alleviating depressive symptoms, particularly in those resistant to traditional antidepressants. Clinical studies have highlighted zinc's potential as an adjunct treatment for depression, with mechanisms likely involving neurotransmitter modulation, oxidative stress reduction, and inflammatory pathway management, as depressed individuals often exhibit elevated pro-inflammatory cytokines.

13.3.3 Copper

Copper is an essential trace element that plays an important role in various physiological processes vital for maintaining brain function, including energy production, neurotransmitter synthesis, and antioxidant defense. Its involvement in cellular energy production is particularly important, as copper is a key component of cytochrome c oxidase (COX), an enzyme crucial for the final stage of mitochondrial oxidative phosphorylation, facilitating ATP synthesis necessary for energy-demanding organs like the brain. Proper copper levels are also essential for the synthesis and regulation of neurotransmitters, including dopamine and norepinephrine, with copper-dependent enzymes like dopamine β-hydroxylase playing a significant role; imbalances in copper can disrupt these processes and lead to cognitive impairments and mood disorders. Additionally, copper acts as a cofactor for superoxide dismutase (SOD), a crucial enzyme in the brain that neutralizes ROS and protects against oxidative stress, thereby maintaining neuronal health. However, both copper deficiency and toxicity can lead to severe neurological issues, such as Menkes disease associated with deficiency, characterized by cognitive impairments and motor coordination difficulties, and Wilson's disease linked to toxicity, resulting in excessive copper accumulation in the brain, oxidative damage, and cognitive decline. Abnormal copper levels are implicated in various neurodegenerative diseases, such as Alzheimer's, where they can exacerbate amyloid-β plaque formation and tau protein tangles, and in prion diseases and Parkinson's disease (PD) through their effects on protein misfolding and oxidative stress. Current therapeutic strategies focusing on copper metabolism, including chelation therapy and supplementation, are being explored for their potential to restore balanced copper levels, reduce oxidative damage, and promote proper neuronal function, highlighting the importance of maintaining optimal copper levels for cognitive health.

13.3.4 Magnesium

Magnesium comes fourth in terms of abundance in the human body. It is essential for various physiological functions, including energy metabolism, protein synthesis, and DNA replication, with about 60% found in bones and the rest in muscles, soft tissues, and extracellular fluids. It is crucial for the proper functioning of the nervous system and neuromuscular transmission, acting as a cofactor in over 300 enzymatic reactions that are vital for neurotransmission and muscle function. In neurotransmission, magnesium preserves electrical potential across nerve membranes and regulates ion channels critical for signaling by acting as an antagonist to NMDA receptors and calcium channels, thereby preventing excessive neuronal excitation and neurotoxicity associated with excitotoxicity, which is linked to neurodegenerative diseases like Alzheimer's and amyotrophic lateral sclerosis (ALS). Furthermore, magnesium is essential for neuromuscular function, balancing calcium's role in muscle excitation and regulating acetylcholine release at the neuromuscular junction; a deficiency can lead to increased nerve excitability, resulting in muscle cramps and spasms. The mineral also plays a significant role in mood regulation, as its deficiency is associated with anxiety and depression, affecting the HPA axis and lowering serotonin levels, which are vital for mood stability; supplementation has shown potential antidepressant-like effects and can improve symptoms of anxiety and depression. Additionally, magnesium exhibits neuroprotective properties by managing oxidative stress and reducing excitotoxicity, safeguarding neurons from damage through its role in blocking NMDA receptors and regulating microglial cell activity to prevent chronic neuroinflammation. Its influence extends to cognitive health, with research suggesting that magnesium supplementation may enhance cognitive performance and synaptic plasticity, especially in individuals with cognitive impairment or those at risk for neurodegenerative diseases, making adequate magnesium levels crucial for overall brain health and function.

13.3.5 Selenium

Selenium (Se) is a trace element vital for various biological functions, particularly in antioxidant defense and cognitive health, with its significance largely stemming from selenoproteins that support redox balance, DNA synthesis, thyroid hormone metabolism, and immune system function. Given the brain's vulnerability to oxidative damage due to its high metabolic activity and lipid-rich composition, selenium-dependent antioxidant systems play a critical role in managing oxidative stress and maintaining cognitive function. The antioxidant properties of selenium are primarily facilitated by its integration into selenoproteins, including glutathione peroxidases (GPx), thioredoxin reductases (TrxR), and selenoprotein P (SelP), which counteract reactive oxygen and nitrogen species produced during metabolic processes. For instance, GPx enzymes protect neuronal membranes from oxidative stress by converting harmful lipid hydroperoxides into less damaging substances, while TrxR maintains a reducing environment that aids in cellular repair. SelP functions as both an antioxidant and a transporter, ensuring adequate selenium levels in the brain, which is crucial for optimal selenoprotein activity and regulation of oxidative stress associated with neurodegenerative conditions like AD and PD. Selenium deficiency is linked to cognitive decline and a higher risk of neurodegenerative diseases; low selenium levels in older adults are associated with memory loss and diminished executive function, likely due to compromised antioxidant defenses that heighten oxidative stress and neuronal damage. In AD, selenium may mitigate oxidative damage and slow disease progression by reducing amyloid-beta plaque formation, as evidenced by lower selenium levels in AD patients and improved cognitive function in animal models with selenium supplementation. In PD, selenium's antioxidant properties help preserve dopaminergic neurons by reducing oxidative stress and improving motor function in preclinical studies. The brain's susceptibility to oxidative damage, caused by its high oxygen demand and the presence of PUFAs, is countered by selenium through mechanisms such as preventing lipid peroxidation, reducing inflammatory processes associated with neuroinflammation, and regulating redox signaling via transcription factors that enhance the brain's antioxidant capacity. While selenium deficiency is rare, suboptimal levels can impair cognitive function, particularly in older adults; thus, studies indicate that selenium supplementation can improve cognitive performance, especially in memory and executive function tasks among the elderly, with evidence from studies like the EVA study highlighting the correlation between low plasma selenium levels and cognitive decline. Additionally, selenium may exhibit synergistic effects with other antioxidants, such as vitamins E and C, enhancing neuroprotection and improving cognitive outcomes in both animal studies and human trials.

13.4 MECHANISMS OF MICRONUTRIENT ACTION IN THE BRAIN

13.4.1 Antioxidants

The brain is particularly vulnerable to oxidative stress owing to its elevated oxygen requirements, substantial lipid composition, and comparatively inadequate antioxidant defenses, resulting in a disparity between the generation of ROS and the brain's capacity to mitigate them. This oxidative imbalance can damage essential cellular components, including lipids, proteins, and DNA, contributing to neurodegenerative disorders such as Alzheimer's and PDs, as well as age-related cognitive decline. Micronutrients, especially vitamins and minerals, are essential for bolstering the brain's antioxidant defenses, thus reducing free radical damage and promoting cognitive health. For instance, vitamin C, a potent water-soluble antioxidant, effectively neutralizes ROS and regenerates other antioxidants, such as vitamin E, in neurons, safeguarding them from oxidative stress while also facilitating dopamine production and improving iron absorption to minimize oxidative damage (Harrison & May, 2009). Vitamin E, being lipid-soluble, is essential for protecting cell membranes from lipid peroxidation, thereby preserving neuronal integrity and preventing the spread of oxidative stress, which is vital for preventing age-related neurodegenerative conditions. Selenium acts as a cofactor for selenoenzymes like GPx, which are instrumental in degrading hydrogen peroxide and organic hydroperoxides, thus thwarting ROS formation and protecting neurons from oxidative damage, particularly in critical brain regions associated with cognitive health, such as the hippocampus and cerebral cortex (Zoidis et al., 2018). Zinc plays a significant role in antioxidant defense by stabilizing cellular membranes, reducing free radical formation, and modulating the activity of SOD, an essential enzyme for maintaining antioxidant balance, while also regulating synaptic plasticity and neurotransmitter activity to guard against cognitive decline (Prakash et al., 2015). Copper, an essential trace element, serves as a cofactor for antioxidant enzymes like copper-zinc SOD (Cu/Zn-SOD), but its excess can induce oxidative stress by facilitating free radical production, making it crucial to maintain balanced copper levels for optimal brain health (Mezzaroba et al., 2019). Finally, while iron is essential for oxygen transport and neurotransmitter production, excess free iron can lead to harmful ROS generation through Fenton reactions, underscoring the importance of maintaining iron homeostasis, with antioxidants like vitamin C playing a key role in regulating iron bioavailability to reduce the risk of oxidative damage in the brain.

13.4.1.1 Mechanisms of Action

Antioxidant micronutrients, including vitamins C and E, play a pivotal role in scavenging free radicals by donating electrons, thus preventing damage to lipids, proteins, and nucleic acids; selenium and zinc further enhance antioxidant defenses by facilitating the breakdown of ROS through key enzymes such as GPx and SOD, respectively (Yuan et al., 2024). Moreover, micronutrients like zinc and copper help to mitigate the pro-oxidant activities of metals by chelating harmful metal ions, reducing the generation of hydroxyl radicals during Fenton reactions. The synergy among these antioxidants is evident in vitamin C's ability to regenerate oxidized vitamin E, bolstering the overall antioxidant defense system crucial for combating oxidative stress linked to neurodegenerative diseases like Alzheimer's and Parkinson's (Bule et al., 2019). Furthermore, various micronutrients serve essential functions in neurotransmitter synthesis, with vitamin B6 acting as a coenzyme for the conversion of amino acids into critical neurotransmitters such as serotonin and dopamine, while vitamin C assists in the biosynthesis of catecholamines. Vitamin D is important for calcium homeostasis, influencing neurotransmitter release, while iron acts as a cofactor in synthesizing dopamine and serotonin, with deficiencies leading to cognitive and emotional disorders. Zinc modulates neurotransmitter release and synaptic plasticity, particularly impacting glutamate and GABA systems essential for learning and memory, and magnesium supports neurotransmitter regulation, preventing excitotoxicity(Dakshinamurti et al., 2019). Copper acts as a cofactor for dopamine β-hydroxylase and regulates enzymes involved in neurotransmitter metabolism, while selenium's antioxidant properties protect neurotransmitter systems from oxidative damage. Lastly, folate is necessary for synthesizing neurotransmitters like serotonin and dopamine through its role in converting homocysteine to methionine, emphasizing the critical interconnectedness of these micronutrients in maintaining optimal brain health and function.

13.4.2 Neuroprotection and Neurogenesis

Neurogenesis, the formation of new neurons in the brain, is crucial for maintaining cognitive function, brain flexibility, and repair mechanisms, primarily occurring in the hippocampus and subventricular zone. Micronutrients, including vitamins, minerals, and trace elements, significantly influence neurogenesis, neuronal survival, and synaptic plasticity, processes vital for cognitive health. For instance, vitamin D promotes the growth and differentiation of neural stem cells and is linked to calcium balance and inflammation reduction, while B vitamins (B6, B9, and B12) are essential for DNA production and managing homocysteine levels, with deficiencies associated with cognitive decline and neurodegenerative diseases. Zinc plays a critical role in synaptic plasticity and neurogenesis, with deficiencies leading to increased oxidative stress susceptibility and cognitive impairments. Additionally, vitamin E protects neuronal membranes from oxidative damage, enhancing neuronal survival, while iron is vital for neurotransmitter production, myelin synthesis, and energy metabolism, although excess iron can induce oxidative stress. Magnesium regulates NMDA receptor activity and helps to prevent excitotoxicity, thus supporting synaptic plasticity. Lastly, copper is important for neurotransmitter synthesis and antioxidant defense, with both deficiency and toxicity negatively impacting neurogenesis. Overall, a balanced intake of essential micronutrients is crucial for fostering neurogenesis, supporting neuronal health, and preventing cognitive decline and mood disorders, highlighting the need for dietary strategies to enhance brain health.

13.4.3 Myelination and Brain Structure

Myelination is a crucial process in the CNS involving oligodendrocytes and Schwann cells that produce myelin sheaths, which enhance the speed and efficiency of electrical signal transmission along axons, essential for effective neuronal communication. Disruptions in this process can lead to neurological disorders, including multiple sclerosis, leukodystrophies, and peripheral neuropathies. Micronutrients play a vital role in the development and maintenance of myelin. Vitamin B12 is essential for synthesizing and repairing myelin through the methylation of myelin basic protein and phospholipid production; deficiencies can lead to demyelination and neurodegeneration, manifesting as neurological symptoms. Folate, working synergistically with B12, is crucial for nucleic acid synthesis in oligodendrocytes, and its deficiency can impair DNA synthesis and myelination, leading to developmental brain disorders during pregnancy (Parletta et al., 2013). Iron is important for oligodendrocyte development and myelin protein production, but deficiencies during critical developmental periods can cause impaired myelination and long-term cognitive deficits, while excess iron can lead to oxidative stress (Radlowski & Johnson, 2013). Zinc regulates enzymes involved in lipid metabolism critical for myelin composition and promotes the maturation of myelinating cells; its deficiency can impair cognitive function and memory due to defective myelination. Copper is vital for energy production and antioxidant defense in myelinating cells; deficiency can lead to severe neurodegeneration and demyelination, as seen in Menkes disease (Fontes et al., 2024). Magnesium stabilizes axonal membranes and supports oligodendrocyte function, with deficiencies linked to increased neuronal damage and impaired myelin repair (Yamanaka et al., 2019). Selenium acts as an antioxidant, protecting oligodendrocytes from oxidative stress, and its deficiency can exacerbate myelin damage (Cardoso et al., 2017). Vitamin D promotes myelination by regulating neurotrophic factors and cytokines that influence oligodendrocyte function; low levels are associated with increased risks of demyelinating conditions (Shirazi et al., 2015). Overall, adequate intake of these micronutrients is essential for the formation, maintenance, and repair of myelin sheaths, highlighting the importance of dietary approaches in preventing demyelination and supporting CNS health, particularly in neurodegenerative and demyelinating disorders.

13.4.4 Gene Expression and Epigenetic Modulation

Gene expression and epigenetic control are essential for brain development, neuroplasticity, and cognitive functioning, with epigenetic modifications including DNA methylation, histone modifications, and non-coding RNA regulation affecting gene activity without altering the genetic code. Micronutrients significantly contribute to these processes by acting as cofactors and regulators for the enzymes that facilitate epigenetic modifications, thereby impacting the gene expression pathways essential for brain function and neurodevelopment (Deibel et al., 2015). For instance, folate plays a pivotal role in one-carbon metabolism, providing methyl groups necessary for DNA methylation, and its deficiency is linked to neurodevelopmental disorders (Barua et al., 2014). Similarly, vitamin B12 is critical for the methylation cycle, and its deficiency can lead to abnormal gene expression associated with neurological disorders (Guéant et al., 2022). Choline influences both DNA methylation and histone acetylation, while zinc regulates gene expression through interactions with transcription factors, impacting neurogenesis and cognitive functions. Iron, crucial for neurotransmitter synthesis, affects gene expression related to neuroprotection, whereas selenium's antioxidant properties protect against oxidative stress, maintaining epigenetic integrity (Shepard & Nugent, 2020). Vitamin D, by binding to its receptor, regulates genes involved in neuroinflammation and neuroprotection, while magnesium supports DNA replication and repair, thus stabilizing the epigenetic landscape. Copper's role in epigenetic regulation is highlighted through its influence on histone acetylation, and omega-3 fatty acids, particularly DHA, modulate microRNA expression, affecting inflammation and cognitive performance. Overall, a deficiency or imbalance in these micronutrients can lead to aberrant gene expression, resulting in cognitive impairments and an increased risk of neurodegenerative diseases, emphasizing the importance of micronutrients in maintaining neurological health and their potential role in developing therapeutic strategies for neurological disorders.

13.5 MICRONUTRIENT DEFICIENCY AND COGNITIVE IMPAIRMENT

13.5.1 Iron

Iron plays a crucial role in cognitive development and neural functioning, particularly during infancy and childhood when the brain undergoes rapid growth and development. Observational studies indicate that iron-deficiency anemia in children is associated with poor cognitive outcomes, diminished school performance, and behavioral issues, although isolating the effects of anemia from other factors can be challenging. Proposed mechanisms for this association include reduced exploration and developmental delays, slower nerve conduction due to impaired myelination, and compromised neurotransmitter synthesis. Ensuring adequate iron levels may help protect intellectual development, while excessive iron, often resulting from genetic disorders, is linked to neurodegenerative diseases like Alzheimer's and Parkinson's due to increased oxidative stress and neuronal damage (Fleming & Ponka, 2012). Throughout life, iron requirements in the brain vary, with the highest concentrations found in the basal ganglia, reflecting its necessity for oxidation-reduction reactions, neurotransmitter synthesis, and myelin production. Iron uptake peaks during critical periods of brain growth, particularly in the first two years of life when cognitive development is particularly vulnerable. Prenatal factors, such as maternal iron deficiency or conditions affecting placental iron transport, can predispose infants to early iron deficiency even without anemia, leading to long-lasting impacts on brain structure and function. Research shows that early iron deficiency may result in deficits in brain iron levels, hypomyelination, and altered neurotransmitter function, which are often not reversible through supplementation. Studies in both animal models and children highlight that iron is essential for energy production, mitochondrial function, and nucleic acid metabolism, all crucial for cognitive processes, especially in the hippocampus, where rapid maturation occurs during early development. Furthermore, while older children and adolescents face a lower prevalence of anemia, iron deficiency remains a significant health concern among various populations, including women and the elderly. Meta-analyses suggest that iron supplementation can improve cognitive domains such as attention and concentration, particularly in those who are anemic; however, the long-term cognitive deficits that persist even after iron levels are restored underscore the complexity of iron's role in brain health (Falkingham et al., 2010). Understanding the optimal timing and dosage for iron supplementation, particularly in vulnerable populations, is essential, as excessive iron can lead to toxicity and oxidative damage.

13.5.2 Manganese

Manganese (Mn) is an essential nutrient and the fifth most abundant metal in the Earth's crust, playing a vital role in the proper functioning of the human CNS; however, excessive accumulation can lead to neurotoxic effects, particularly in the brain, when levels exceed the homeostatic range. Mn enters the environment through industrial processes, the application of Mn-containing pesticides, and gasoline additives like methylcyclopentadienyl manganese tricarbonyl (MMT). Most studies on Mn neurotoxicity have focused on occupational exposure, primarily through inhalation in industries such as smelting and welding, where toxicity has been linked to cognitive, neurological, and psychological impairments in adults. Recently, concerns have emerged regarding children's vulnerability to Mn overexposure and its neuropsychological impacts, as children are generally more susceptible to environmental contaminants. While dietary intake usually maintains systemic Mn homeostasis, excessive exposure can occur from airborne Mn, contaminated drinking water, or Mn-rich foods. Furthermore, infants consuming milk formula or receiving total parenteral nutrition may face increased Mn levels in the brain, especially given their immature blood-brain barrier and excretory systems. Factors contributing to heightened risk in children include higher breathing rates, greater intestinal absorption (potentially reaching 80% in neonates compared to 1-5% in adults), and lower excretion rates due to underdeveloped biliary systems, leading to greater Mn accumulation. Historically, the neurotoxic effects of Mn were first identified in 1837 among Scottish laborers, with symptoms of manganism including memory deficits, fatigue, balance issues, and tremors. Subsequent research has shown Mn intoxication emphasized the importance of understanding Mn's neurotoxic potential, particularly in vulnerable populations.

13.5.3 Cadmium

Cadmium is a neurotoxic heavy metal that poses significant risks to the CNS of fetuses and infants, primarily by disrupting placental functions, affecting enzyme activity, and altering the availability of nutrients essential for CNS development. During pregnancy, cadmium exposure can lead to detrimental effects on neurotransmitter levels, including norepinephrine, dopamine, serotonin, and acetylcholine, which are critical for proper brain function. Moreover, cadmium is associated with increased production of free radicals, causing cell membrane damage and resulting in various physiological changes. The neurotoxic effects on the developing fetus stem from impaired placental function and metabolic disturbances that disrupt essential trace elements in the CNS. Interestingly, cadmium can promote DNA synthesis and cell proliferation at low concentrations, while at higher levels, it can induce apoptosis and chromosomal abnormalities, indicating its effects are concentration-dependent. The challenge in assessing cadmium's neurotoxic effects is further complicated by co-exposure to lead (Pb), as both metals can penetrate the immature blood-brain barrier and accumulate in the developing brain. Studies suggest a complex, dose-dependent interaction between prenatal exposures to these metals, especially at low concentrations, indicating potential antagonistic effects. This interplay underscores the biological complexities in studying the neurodevelopmental consequences of co-exposure to multiple toxic substances, highlighting the need for further research in this area.

13.6 MICRONUTRIENT DEFICIENCIES IN AGING

As individuals age, they experience an increase in cellular senescence among neurons and microglia, leading to heightened apoptosis, protein aggregation, mitochondrial dysfunction, and the generation of ROS, which contributes to oxidative stress and damage to proteins and lipids, as well as DNA accumulation (Boccardi et al., 2017). The brain is particularly vulnerable to oxidative damage due to its relatively low levels of antioxidant defenses compared to other tissues, facilitating the accumulation of neurotoxic peptides such as amyloid-beta, closely associated with AD development (Morley, 2015). Aging also results in the reduction of neuronal glucose transporters, impairing glucose uptake, which correlates with cognitive decline. This decline is further exacerbated by the negative impact of aging on the oxidative phosphorylation system in mitochondria, leading to increased ROS production (Zapico & Ubelaker, 2013). Consequently, damaged mitochondrial DNA impairs the functionality of oxidative phosphorylation, initiating a damaging cycle. Cognitive decline is often linked to neuroinflammation; misfolded proteins activate Toll-like receptors on microglia, triggering a potentially harmful inflammatory response that releases excessive ROS and reactive nitrogen species (RNS), contributing to neurodegeneration (Dominguez & Barbagallo, 2017). Neurodegenerative diseases like Alzheimer's and Parkinson's significantly affect the elderly population, with AD characterized by cognitive decline and language skill deterioration, while PD presents pronounced motor impairments (Recchia et al., 2004; Van Giau et al., 2018). Despite distinct molecular features such as amyloid-beta production in AD and dopaminergic neuron degeneration in PD the underlying mechanisms remain unclear, though aging-related oxidative stress and mitochondrial dysfunction may play a role in their pathogenesis. Nutrition, sleep, and exercise significantly influence cognitive health in aging, as malnutrition is associated with cognitive impairment and functional decline, prompting the exploration of dietary interventions and their efficacy in preventing cognitive frailty and dementia (Dominguez & Barbagallo, 2017; Morley, 2018). Recent findings suggest that maintaining a balanced diet and a healthy body mass index (BMI) can preserve cognitive function, while sleep disturbances are recognized as risk factors for neurodegenerative diseases like AD (Spira et al., 2014). Although the protective role of exercise remains debated, some pieces of evidence indicate that it may reduce the risk of cognitive decline.

13.7 MEDITERRANEAN DIET (MEDI) AND OTHER DIETS

The Mediterranean diet (MeDi) has emerged as a promising dietary model for protecting against dementia, primarily due to its emphasis on high consumption of vegetables, fruits, legumes, whole grains, and nuts, alongside moderate intakes of fish, poultry, and red meat, with olive oil as the primary source of unsaturated fats (Panza et al., 2014). Numerous population studies have demonstrated that adherence to the MeDi is linked to slower cognitive decline, reduced risk of AD, and a lower likelihood of progression from mild cognitive impairment (MCI) to AD, as well as decreased mortality rates in AD patients (Solfrizzi et al., 2011). Systematic reviews and meta-analyses consistently support these findings, indicating that following a MeDi is associated with a lower risk of cognitive decline, MCI, and the progression of MCI to AD (Psaltopoulou et al., 2013; Singh et al., 2014). For instance, a six-year study in Italy revealed that adherence to the MeDi reduced the onset of frailty in elderly individuals Talegawkar et al., (2012), and the FINGER study showed that combining the MeDi with physical exercise, social engagement, cognitive training, and cardiovascular risk management slowed cognitive decline (Ngandu et al., 2015). A recent randomized controlled trial indicated that an enhanced MeDi, incorporating extra virgin olive oil or nuts, improved overall cognitive function over 6.5 years (Martínez-Lapiscina et al., 2013). These findings suggest that the MeDi, characterized by high intakes of monounsaturated fatty acids (MUFA), omega-3 PUFAs, fish, antioxidants, and moderate alcohol consumption, may confer protective benefits against dementia and AD. The effectiveness of the MeDi is often attributed to extra virgin olive oil, which has shown the ability to enhance memory and reduce oxidative damage in animal models of AD. Additionally, olive oil polyphenols can inhibit amyloid-beta aggregation and induce epigenetic changes (Yuda et al., 2020). Similarly, the Nordic diet, which emphasizes polyphenol-rich berries and whole grains, has also been associated with cognitive protection (Bauer & Morley, 2018), while diets in Asia rich in fruits, nuts, seeds, and fish have been inversely correlated with frailty. Moreover, fruits and vegetables, crucial components of such diets, are known for their high levels of vitamins, minerals, and antioxidants, along with low energy density and dietary fiber content, all of which contribute to cognitive health (Angelino et al., 2019; Loef & Walach, 2012). Higher vegetable consumption has been linked to reduced risk of dementia and slower cognitive decline, while nuts, rich in MUFA and PUFAs, may enhance cognitive function by lowering cholesterol levels and reducing risks of cardiovascular disease and type 2 diabetes, both of which are associated with cognitive decline. Coffee's effects on cognitive health are mixed, yet recent meta-analyses suggest that higher coffee intake may correlate with a reduced risk of AD, warranting further research for confirmation (Liu et al., 2016).

13.8 CASE STUDIES AND RESEARCH FINDINGS

Micronutrient deficiencies are increasingly recognized as significant contributors to cognitive impairments, particularly deficiencies in nutrients like iron and vitamin B12, which are crucial for essential processes such as oxygen transport, neurotransmitter synthesis, and myelination, ultimately leading to cognitive dysfunction. Iron deficiency, the most prevalent micronutrient deficiency globally, affects around 2 billion people, especially preschool-aged children and pregnant women. Early-life iron deficiency can cause lasting neurocognitive and behavioral issues, as evidenced by studies showing that prenatal iron supplementation positively impacts brain development and cognitive abilities in children. For instance, research in Nepal found that children whose mothers received iron and folic acid during pregnancy exhibited better cognitive outcomes compared to those whose mothers did not. Similarly, studies in Vietnam and China highlighted the benefits of iron supplementation on children's growth and motor skills, while longitudinal studies linked chronic iron deficiency to cognitive and socioemotional impairments throughout life. On the other hand, vitamin B12, predominantly found in animal products, is critical for cognitive function, with deficiencies posing risks for infants breastfed by mothers with inadequate B12 intake and for children following vegan diets. Research has shown that children with B12 deficiency experience developmental delays and poorer cognitive performance. Observational studies from Guatemala and the Netherlands indicate that these deficiencies can lead to slower reaction times and academic challenges. The potential dual impact of micronutrient deficiencies on cognitive development and caregiver interactions, termed "functional isolation," underscores the need for early intervention and preventive measures. Given the strong links between these deficiencies and adverse developmental outcomes, ongoing research is vital to address the complexities of timing, severity, and the interaction of multiple deficiencies, ultimately informing effective prevention strategies for at-risk populations.

13.9 MICRONUTRIENT SUPPLEMENTATION AND BRAIN HEALTH

A comprehensive analysis of the efficacy of various nutritional supplements on cognitive function encompassed 49 trials categorized by supplementation type, including proteins and amino acids, vitamins, polyphenols, PUFAs, and nutrient combinations. Within the protein and amino acid supplements, a 12-week study involving 25 participants with MCI showed improved global Clinical Dementia Rating scores with anserine and carnosine supplementation; however, other cognitive tests yielded no significant benefits, though APOE4+ individuals displayed some improvements (Masuoka et al., 2019). Similarly, a 24-week trial of a casein-derived peptide demonstrated enhanced performance on an Alzheimer's scale but not on total cognitive scores (Yuda et al., 2020). In contrast, trials involving taurine and lactotripeptides yielded no cognitive benefits. Notably, whey peptide supplementation led to increased verbal fluency, especially in fatigued participants. Among vitamin supplements, while three trials on vitamin D failed to show significant cognitive improvements, a study on B-group vitamins found that combined folic acid and vitamin B12 significantly enhanced cognitive performance in MCI participants. Polyphenol supplements, such as cocoa flavanols and anthocyanin-rich extracts, demonstrated variable effects on cognitive performance across studies, with some showing significant improvements in memory and cognitive tasks. In terms of PUFAs, studies using omega-3 supplements failed to demonstrate significant cognitive benefits compared to placebo, although a combination of omega-3s, ginsenosides, and catechins did show some positive outcomes. Lastly, combinations of nutrients highlighted mixed results; certain combinations led to improved cognitive performance, while others resulted in no significant benefits. Overall, the findings underscore the complexity of cognitive supplementation, suggesting that while some interventions show promise, many do not yield substantial cognitive enhancements across diverse populations and supplement types.

13.10 TARGETED SUPPLEMENTATION FOR VULNERABLE POPULATION

Micronutrients play a crucial role in supporting brain health and cognitive function by influencing essential processes, such as neurotransmission, neuroprotection, synaptic plasticity, and neurogenesis. A deficiency in these nutrients can impair cognitive abilities, increase susceptibility to neurodegenerative diseases, and lead to emotional instability. Specific populations, including the elderly, pregnant women, and vegetarians, are particularly vulnerable to micronutrient deficiencies due to various factors like physiological changes, dietary restrictions, or increased nutrient requirements. For elderly individuals, factors like reduced gastric acid production and decreased nutrient absorption exacerbate the risk of deficiencies in vital micronutrients such as vitamin B12, which is essential for neurotransmitter production and myelin sheath maintenance, and whose deficiency is linked to cognitive decline and dementia (Smith et al., 2018); vitamin D, where inadequate levels can lead to cognitive decline and mood disorders; and omega-3 fatty acids, crucial for neuronal membrane integrity, with insufficient intake increasing AD risk (Annweiler et al., 2015). Targeted supplementation can enhance cognitive function and lower neurodegenerative disease risk in this demographic. Pregnant women face heightened micronutrient needs essential for fetal brain development, making supplementation critical. Key nutrients include folic acid, which prevents neural tube defects; iron, necessary for oxygen transport and neurotransmitter production, with deficiencies linked to lower cognitive development and IQ in offspring; and iodine, vital for thyroid hormone production and brain development, where insufficiency can lead to developmental delays. Additionally, vegetarians and vegans are at risk for deficiencies in micronutrients typically obtained from animal products. Vitamin B12 is critical for neurotransmitter synthesis, and its absence can lead to neurological damage and cognitive decline. Iron, while present in plant foods, is less bioavailable, making deficiencies common and impairing cognitive function. Lastly, omega-3 fatty acids, primarily found in fish, are essential for brain health, and vegetarians and vegans may require algal-based DHA supplements to meet their needs. Supplementing these vital nutrients is essential for maintaining cognitive function and overall brain health across these vulnerable populations.

13.11 POTENTIAL RISKS OF OVER-SUPPLEMENTATION

While micronutrients are essential for maintaining brain health, excessive consumption through supplementation can lead to serious health risks, including neurotoxic effects, cognitive issues, and an increased risk of neurological disorders. Over-supplementation may result in toxicity, exemplified by vitamin A, where both acute and chronic excess can cause symptoms like increased intracranial pressure, dizziness, and cognitive impairments, as well as severe teratogenic effects during pregnancy effects (Penniston & Tanumihardjo, 2006). Similarly, excessive vitamin D can lead to hypercalcemia, negatively affecting brain function and causing confusion, cognitive impairment, and potential coma. Excessive intake of vitamin B6 can cause sensory neuropathy and impair cognitive function, while iron overload, particularly in cases like hemochromatosis, can accumulate in the brain, promoting oxidative stress and contributing to neurodegenerative diseases like Parkinson's and Alzheimer's. Zinc toxicity can disrupt homeostasis, impair memory and learning, and induce mood disorders. Magnesium over-supplementation can result in confusion and coma, especially in individuals with kidney issues, while excessive selenium intake may lead to cognitive issues and increase the risk of neurological diseases. Copper, necessary for neurotransmitter synthesis, can cause neurotoxicity when over-supplemented, potentially leading to cognitive decline and psychiatric symptoms (Kaler, 2011). Neurotoxic mechanisms associated with excessive micronutrient intake include oxidative stress and neuroinflammation, resulting from elevated levels of iron, copper, and zinc that produce free radicals, leading to neuronal damage and conditions like Alzheimer's and Parkinson's. Additionally, high doses of certain vitamins and minerals can disrupt neurotransmitter synthesis, leading to imbalances that affect mood and cognition, and calcium dysregulation due to vitamin D toxicity or excessive magnesium can impair synaptic function and increase susceptibility to cognitive dysfunction and neurodegenerative diseases. Understanding safe upper intake levels and the potential risks associated with excessive micronutrient consumption is critical, especially in populations that may use supplements without medical supervision.

13.12 DIET AND MICRONUTRIENT INTAKE

Maintaining adequate levels of essential micronutrients is vital for cognitive development and brain health, as a nutrient-dense diet rich in vitamins and minerals supports crucial neurological processes, including neurotransmitter production, myelination, and neurogenesis. Iron, present in both heme (found in red meat, poultry, and fish) and non-heme forms (found in legumes, leafy greens, fortified cereals, and nuts), plays a significant role in cognitive functions like memory and attention, and its absorption can be enhanced by pairing it with vitamin C-rich foods. Iodine is necessary for thyroid hormone production and brain development, with rich sources including seafood, dairy products, eggs, and iodized salt. Zinc, essential for neurotransmitter function and DNA synthesis, can be found in high levels in meat, shellfish, legumes, nuts, seeds, and whole grains, although its absorption can be improved through techniques like soaking or fermenting. Vitamin B12 is crucial for maintaining myelin sheath and neurotransmitter production, primarily sourced from animal products, fish, dairy, eggs, and fortified foods for those following plant-based diets, who may require supplementation. Folate is important for DNA synthesis and repair, found in leafy greens, legumes, fruits, fortified grains, and nuts. Additionally, other micronutrients that promote brain health include vitamin D, omega-3 fatty acids (DHA and EPA), and antioxidants like vitamins C

TABLE 13.1
Dietary Sources of Key Micronutrients for Cognitive Development

Micronutrients	Primary Food Sources	RDI (Adults)	Functions
Iron	Red meat, poultry, fish, lentils, spinach, fortified cereals	8 mg (men), 18 mg (women)	Vitamin C enhances non-heme iron absorption.
Iodine	Seafood, dairy products, iodized salt, eggs	150 µg	Excess intake can affect thyroid function.
Zinc	Meat, shellfish, legumes, nuts, seeds, whole grains	11 mg (men), 8 mg (women)	Phytates inhibit absorption; consider food preparation methods.
Vitamin B12	Meat, fish, dairy, eggs, fortified foods	2.4 µg	Vegans and vegetarians may need supplements.
Folate	Leafy greens, legumes, fruits, fortified grains	400 µg	Critical during pregnancy for neural tube development
Vitamin D	Fatty fish, fortified dairy, sunlight	600 IU (15 µg)	Deficiency linked to cognitive decline
Vitamin C	Citrus fruits, strawberries, capsicum, broccoli	90 mg (men), 75 mg (women)	Antioxidant that protects brain cells
Vitamin E	Nuts, seeds, spinach, broccoli	15 mg	Protects against oxidative stress in the brain

and E, all of which can be integrated into daily meals to prevent deficiencies and enhance cognitive function (Table 13.1).

13.13 DIETARY PATTERNS FOR OPTIMAL BRAIN HEALTH

The MeDi, rooted in the traditional eating patterns of populations along the Mediterranean Sea, such as those in Greece, France, Spain, and parts of North Africa, emphasizes unrefined carbohydrates like bread, pasta, and grains, alongside abundant dairy products, fruits, and vegetables (Katsiardanis et al., 2013; Zbeida et al., 2014). Animal protein sources such as chicken, fish, and eggs are taken infrequently, and red meat is restricted to several occasions each month. This diet, including 28–40% of total caloric intake from fat, is low in saturated and trans fats, primarily due to the predominant use of olive oil, which is abundant in oleic acid, a favorable monounsaturated fat. Investigations into the influence of the MeDi on cognitive function yield inconclusive results; numerous cross-sectional studies suggest a favorable association between compliance with this dietary regimen and cognitive performance, as demonstrated by significant outcomes from the Boston Puerto Rican Health Study and an analysis of participants from the National Health and Nutrition Examination Survey, both indicating elevated cognitive scores linked to increased adherence to the MeDi. In contrast, further investigations, including one with Australian participants and another from Hong Kong, revealed no significant correlation between the MeDi and cognitive function. Moreover, randomized controlled trials, including a clinical trial in Barcelona, demonstrated that participants adhering to a MeDi enriched with olive oil showed enhanced performance in short-term memory assessments compared to those following a low-fat diet, thereby substantiating the diet's cognitive advantages. Overall, while the MeDi is recognized for its nutritional advantages, the relationship between its adherence and cognitive function remains an area of ongoing research, warranting further exploration to better understand its implications for brain health (Table 13.2).

TABLE 13.2
Summary of Research Exploring the Link Between the Mediterranean Diet and Cognitive Function: Findings and Limitations

Study	Result	Limitation
Chan et al. (2013)	The Mediterranean Diet (MD) was not associated with cognitive performance.	Cross-sectional design limits the ability to determine causality; cognitive status was self-reported; and participants had higher education levels.
Crichton et al. (2013)	MD showed no association with cognitive function.	Cross-sectional design restricts the ability to determine causality; cognitive status was self-reported; inclusion and exclusion criteria were unclear.
Corley et al. (2013)	The Mediterranean diet was initially associated with enhanced cognitive performance; however, this association became insignificant when controlling for childhood IQ and socio-economic status.	Self-selecting sample; unclear inclusion and exclusion criteria; response rate not disclosed.
Zbeida et al. (2014)	MD was associated with enhanced cognitive performance in both the NHANES and MABAT-ZAHAV cohorts ($p < 0.001$ and 0.008, respectively).	Cross-sectional design limits causality determination; 24-hour recall may not represent typical consumption; unclear inclusion and exclusion criteria; funding sources not disclosed.
Samieri et al. (2013)	MD was associated with enhanced cognitive performance, including TICS scores, global cognition, and verbal memory ($p = 0.004$, 0.002, and <0.001, respectively). No impact of MD on cognitive function over time was detected.	Study included only females; participants were highly educated; the cognitive assessment was conducted via telephone; reliability was unclear.
Samieri et al. (2013)	MD exhibited no correlation with cognitive function.	Study included only females; participants were highly educated; the cognitive assessment was conducted via telephone.
Qin et al. (2015)	MD correlated with a decelerated cognitive decline ($\beta = 0.042$, 95% CI 0.002–0.081).	24-hour recall does not accurately reflect typical intake.
Feart et al. (2009)	MD was associated with improved cognition, evidenced by a reduction in MMSE errors ($\beta = -0.006$, 95% CI -0.01 to -0.0003 for each 1-unit rise in MD score).	MD was not associated with other cognitive assessments (IST, BVRT, or FCSRT) or dementia risk; selection bias due to disparities between participants with and without complete data.
Tangney et al. (2011)	MD correlated with enhanced cognition and reduced cognitive decline ($\beta = 0.0014$ per 1-point increment, $p = 0.0004$).	24-hour recall does not reflect usual intake.

13.14 PROSPECTS AND CONCLUSION

Emerging micronutrients of interest, such as choline, lutein, and selenium, have gained attention for their potential roles in brain health. Choline, essential for acetylcholine synthesis, supports cognitive function and memory. Lutein, a carotenoid, is being studied for its neuroprotective effects, particularly in reducing inflammation and oxidative stress. Selenium, known for its antioxidant properties, may help to prevent cognitive decline by protecting neurons from oxidative damage. Although these micronutrients are less studied than traditional vitamins and minerals, their growing recognition underscores their potential significance in maintaining brain health and preventing neurodegenerative diseases. Personalized nutrition and genomics explore how individual genetic differences influence micronutrient needs and brain health. Variations in genes related to nutrient metabolism, such as those affecting vitamin D, B vitamins, or omega-3 fatty acids, can impact cognitive function, memory, and mood regulation. This emerging field modifies dietary interventions based on genetic profiles to optimize brain health. Integrative approaches combine micronutrient supplementation with lifestyle factors like physical activity, sleep, and stress management, creating a holistic strategy for cognitive enhancement. This synergy between nutrition and lifestyle promotes long-term brain health, potentially reducing the risk of cognitive decline and neurodegenerative diseases. Micronutrients play essential roles in brain health by supporting cognitive function, neurotransmitter synthesis, and neuroprotection. Key vitamins and minerals, such as B vitamins, vitamin D, iron, and zinc, are critical for energy metabolism, mood regulation, and protecting the brain from oxidative stress. To ensure optimal brain function, guidelines recommend a balanced diet rich in fruits, vegetables, whole grains, and lean proteins, providing adequate intake of essential micronutrients. A well-rounded diet, combined with other healthy lifestyle factors, forms a comprehensive approach to maintaining brain health and cognitive function throughout life, reducing the risk of neurodegenerative disorders.

ABBREVIATIONS

AD	Alzheimer's Disease
ALS	Amyotrophic Lateral Sclerosis
BDNF	Brain-Derived Neurotrophic Factor
CNS	Central Nervous System
CoA	Coenzyme A
COX	Cytochrome c Oxidase
DHA	Docosahexaenoic Acid
FMN	Flavin Mononucleotide
FAD	Flavin Adenine Dinucleotide
GPx	Glutathione Peroxidase
HPA	Hypothalamic-Pituitary-Adrenal
LTP	Long-Term Potentiation
MCI	Mild Cognitive Impairment
MeDi	Mediterranean Diet
Mn	Manganese
MMT	Methylcyclopentadienyl Manganese Tricarbonyl
MUFA	Monounsaturated Fatty Acid
NMDA	N-Methyl-D-Aspartate
PD	Parkinson's Disease
PUFA	Polyunsaturated Fatty Acid
RNS	Reactive Nitrogen Species
ROS	Reactive Oxygen Species
SAM	S-Adenosylmethionine
SOD	Superoxide Dismutase
TrxR	Thioredoxin Reductase
VDR	Vitamin D Receptor

REFERENCES

Affrald RJ, Narayan S. A Review: Oligodendrocytes in Neuronal Axonal Conduction and Methods for Enhancing Their Performance. Int J Neurosci [Internet]. 2024 [cited 2025 Jan 27]; Available from: https://pubmed.ncbi.nlm.nih.gov/38850232/

Angelino D, Godos J, Ghelfi F, Tieri M, Titta L, Lafranconi A, et al. Fruit and Vegetable Consumption and Health Outcomes: An Umbrella Review of Observational Studies. Int J Food Sci Nutr [Internet]. 2019 Aug 18 [cited 2025 Jan 27];70(6):652–67. Available from: https://pubmed.ncbi.nlm.nih.gov/30764679/

Annweiler C, Dursun E, Féron F, Gezen-Ak D, Kalueff AV, Littlejohns T, et al. "Vitamin D and cognition in older adults": Updated International Recommendations. J Intern Med. 2015;277(1):45–57.

Ballaz SJ, Rebec GV. Neurobiology of Vitamin C: Expanding the Focus from Antioxidant to Endogenous Neuromodulator. Pharmacol Res [Internet]. 2019 Aug 1 [cited 2025 Jan 27];146. Available from: https://pubmed.ncbi.nlm.nih.gov/31229562/

Barua S, Kuizon S, Junaid MA. Folic Acid Supplementation in Pregnancy and Implications in Health and Disease. J Biomed Sci [Internet]. 2014 Aug 19 [cited 2025 Jan 27];21(1). Available from: https://pubmed.ncbi.nlm.nih.gov/25135350/

Bauer JM, Morley JE. The Relevance of Healthy Diets for the Prevention of Frailty and Cognitive Impairment. Curr Opin Clin Nutr Metab Care [Internet]. 2018 Jan 1 [cited 2025 Jan 28];21(1):1–3. Available from: https://pubmed.ncbi.nlm.nih.gov/29049044/

Boccardi V, Comanducci C, Baroni M, Mecocci P. Of Energy and Entropy: The Ineluctable Impact of Aging in Old Age Dementia. Int J Mol Sci [Internet]. 2017 Dec 9 [cited 2025 Jan 27];18(12). Available from: https://pubmed.ncbi.nlm.nih.gov/29232829/

Cardoso BR, Hare DJ, Bush AI. The Role of Selenium in Neurodegenerative Diseases. Biometals in Neurodegenerative Diseases: Mechanisms and Therapeutics [Internet]. 2017 Apr 28 [cited 2025 Jan 27];35–49. Available from: https://research.monash.edu/en/publications/the-role-of-selenium-in-neurodegenerative-diseases

Carr AC, Maggini S. Vitamin C and Immune Function. Nutrients [Internet]. 2017 Nov 1 [cited 2025 Jan 27];9(11). Available from: https://pubmed.ncbi.nlm.nih.gov/29099763/

Chan, R., Chan, D., & Woo, J. (2013). A cross sectional study to examine the association between dietary patterns and cognitive impairment in older Chinese people in Hong Kong. The journal of nutrition, health & aging, 17, 757-765.

Conkbayir Mine. The Neuroscience of the Developing Child : Self-Regulation for Wellbeing and a Sustainable Future [Internet]. Routledge; 2023 [cited 2025 Jan 27]. 276 p. Available from: https://www.routledge.com/The-Neuroscience-of-the-Developing-Child-Self-Regulation-for-Wellbeing-and-a-Sustainable-Future/Conkbayir/p/book/9781032355764

Corley, J., Starr, J. M., McNeill, G., & Deary, I. J. (2013). Do dietary patterns influence cognitive function in old age?. International psychogeriatrics, 25(9), 1393-1407.

Crichton, G. E., Bryan, J., Hodgson, J. M., & Murphy, K. J. (2013). Mediterranean diet adherence and self-reported psychological functioning in an Australian sample. Appetite, 70, 53-59.

Dakshinamurti K, Dakshinamurti S, Czubryt MP. Vitamin B6: Effects of Deficiency, and Metabolic and Therapeutic Functions. Handbook of Famine, Starvation, and Nutrient Deprivation: From Biology to Policy [Internet]. 2019 Mar 6 [cited 2025 Jan 27];1589–611. Available from: https://link.springer.com/referenceworkentry/10.1007/978-3-319-55387-0_81

Deibel SH, Zelinski EL, Keeley RJ, Kovalchuk O, McDonald RJ. Epigenetic Alterations in the Suprachiasmatic Nucleus and Hippocampus Contribute to Age-Related Cognitive Decline. Oncotarget [Internet]. 2015 [cited 2025 Jan 27];6(27):23181–203. Available from: https://pubmed.ncbi.nlm.nih.gov/26252151/

Dominguez LJ, Barbagallo M. The Relevance of Nutrition for the Concept of Cognitive Frailty. Curr Opin Clin Nutr Metab Care [Internet]. 2017 [cited 2025 Jan 27];20(1):61–8. Available from: https://pubmed.ncbi.nlm.nih.gov/27749714/

Eerman K, Brodaty H. Tocopherol (Vitamin E) in Alzheimer's Disease and Other Neurodegenerative Disorders. CNS Drugs [Internet]. 2004 [cited 2025 Jan 27];18(12):807–25. Available from: https://pubmed.ncbi.nlm.nih.gov/15377170/

Falkingham M, Abdelhamid A, Curtis P, Fairweather-Tait S, Dye L, Hooper L. The Effects of Oral Iron Supplementation on Cognition in Older Children and Adults: A Systematic Review and Meta-analysis. Nutr J [Internet]. 2010 [cited 2025 Jan 27];9(1). Available from: https://pubmed.ncbi.nlm.nih.gov/20100340/

Féart, C., Samieri, C., Rondeau, V., Amieva, H., Portet, F., Dartigues, J. F., ... & Barberger-Gateau, P. (2009). Adherence to a Mediterranean diet, cognitive decline, and risk of dementia. Jama, *302*(6), 638-648.

Fleming RE, Ponka P. Iron Overload in Human Disease. New England Journal of Medicine [Internet]. 2012 Jan 26 [cited 2025 Jan 27];366(4):348–59. Available from: https://pubmed.ncbi.nlm.nih.gov/22276824/

Fontes A, Pierson H, Bierła JB, Eberhagen C, Kinschel J, Akdogan B, et al. Copper Impairs the Intestinal Barrier Integrity in Wilson disease. Metabolism [Internet]. 2024 Sep 1 [cited 2025 Jan 28];158. Available from: https://pubmed.ncbi.nlm.nih.gov/38986805/

Guéant JL, Guéant-Rodriguez RM, Kosgei VJ, Coelho D. Causes and Consequences of Impaired Methionine Synthase Activity in Acquired and Inherited Disorders of Vitamin B12 Metabolism. Crit Rev Biochem Mol Biol [Internet]. 2022 [cited 2025 Jan 27];57(2):133–55. Available from: https://pubmed.ncbi.nlm.nih.gov/34608838/

Harrison FE, May JM. Vitamin C Function in the Brain: Vital Role of the Ascorbate Transporter SVCT2. Free Radic Biol Med [Internet]. 2009 Mar 15 [cited 2025 Jan 28];46(6):719–30. Available from: https://pubmed.ncbi.nlm.nih.gov/19162177/

Kaler SG. ATP7A-Related Copper Transport Diseases-Emerging Concepts and Future Trends. Nat Rev Neurol [Internet]. 2011 Jan [cited 2025 Jan 27];7(1):15–29. Available from: https://pubmed.ncbi.nlm.nih.gov/21221114/

Katsiardanis K, Diamantaras AA, Dessypris N, Michelakos T, Anastasiou A, Katsiardani KP, et al. Cognitive Impairment and Dietary Habits Among Elders: The Velestino Study. J Med Food [Internet]. 2013 Apr 1 [cited 2025 Jan 27];16(4):343–50. Available from: https://pubmed.ncbi.nlm.nih.gov/23514229/

Kerns JC, Arundel C, Chawla LS. Thiamin Deficiency in People With Obesity. Adv Nutr [Internet]. 2015 [cited 2025 Jan 27];6(2):147–53. Available from: https://pubmed.ncbi.nlm.nih.gov/25770253/

Lahoda Brodska H, Klempir J, Zavora J, Kohout P. The Role of Micronutrients in Neurological Disorders. Nutrients [Internet]. 2023 Oct 1 [cited 2025 Jan 28];15(19):4129. Available from: https://pmc.ncbi.nlm.nih.gov/articles/PMC10574090/

Liu JJ, Crous-Bou M, Giovannucci E, de Vivo I. Coffee Consumption Is Positively Associated with Longer Leukocyte Telomere Length in the Nurses' Health Study. J Nutr [Internet]. 2016 [cited 2025 Jan 27];146(7):1373–8. Available from: https://pubmed.ncbi.nlm.nih.gov/27281805/

Livingston G, Sommerlad A, Orgeta V, Costafreda SG, Huntley J, Ames D, et al. Dementia Prevention, Intervention, and Care. Lancet [Internet]. 2017 Dec 16 [cited 2025 Jan 27];390(10113):2673–734. Available from: https://pubmed.ncbi.nlm.nih.gov/28735855/

Loef M, Walach H. Fruit, Vegetables and Prevention of Cognitive Decline or Dementia: A Systematic Review of Cohort Studies. J Nutr Health Aging [Internet]. 2012 [cited 2025 Jan 27];16(7):626–30. Available from: https://pubmed.ncbi.nlm.nih.gov/22836704/

Lozoff B, Beard J, Connor J, Felt B, Georgieff M, Schallert T. Long-Lasting Neural and Behavioral Effects of Iron Deficiency in Infancy. Nutr Rev [Internet]. 2006 [cited 2025 Jan 27];64(5 Pt 2). Available from: https://pubmed.ncbi.nlm.nih.gov/16770951/

Martínez-Lapiscina EH, Clavero P, Toledo E, Estruch R, Salas-Salvadó J, San Julián B, et al. Mediterranean Diet Improves Cognition: The PREDIMED-NAVARRA Randomised Trial. J Neurol Neurosurg Psychiatry [Internet]. 2013 [cited 2025 Jan 27];84(12):1318–25. Available from: https://pubmed.ncbi.nlm.nih.gov/23670794/

Masuoka N, Yoshimine C, Hori M, Tanaka M, Asada T, Abe K, et al. Effects of Anserine/Carnosine Supplementation on Mild Cognitive Impairment with APOE4. Nutrients [Internet]. 2019 Jul 1 [cited 2025 Jan 27];11(7). Available from: https://pubmed.ncbi.nlm.nih.gov/31319510/

Mezzaroba L, Alfieri DF, Colado Simão AN, Vissoci Reiche EM. The Role of Zinc, Copper, Manganese and Iron in Neurodegenerative Diseases. Neurotoxicology [Internet]. 2019 Sep 1 [cited 2025 Jan 27];74:230–41. Available from: https://pubmed.ncbi.nlm.nih.gov/31377220/

Morley JE. Peptides and Aging: Their Role in Anorexia and Memory. Peptides (NY) [Internet]. 2015 Mar 5 [cited 2025 Jan 27];72:112–8. Available from: https://pubmed.ncbi.nlm.nih.gov/25895851/

Morley JE. An Overview of Cognitive Impairment. Clin Geriatr Med [Internet]. 2018 Nov 1 [cited 2025 Jan 27];34(4):505–13. Available from: https://pubmed.ncbi.nlm.nih.gov/30336985/

Muñoz P, Humeres A. Iron Deficiency on Neuronal Function. Biometals [Internet]. 2012 Aug [cited 2025 Jan 27];25(4):825–35. Available from: https://pubmed.ncbi.nlm.nih.gov/22639188/

Ngandu T, Lehtisalo J, Solomon A, Levälahti E, Ahtiluoto S, Antikainen R, et al. A 2 Year Multidomain Intervention of Diet, Exercise, Cognitive Training, and Vascular Risk Monitoring Versus Control to Prevent Cognitive Decline in At-Risk Elderly People (FINGER): A Randomised Controlled Trial. Lancet [Internet]. 2015 Jun 6 [cited 2025 Jan 27];385(9984):2255–63. Available from: https://pubmed.ncbi.nlm.nih.gov/25771249/

Pantothenic Acid | Linus Pauling Institute | Oregon State University [Internet]. [cited 2025 Jan 28]. Available from: https://lpi.oregonstate.edu/mic/vitamins/pantothenic-acid

Panza F, Solfrizzi V, Giannini M, Seripa D, Pilotto A, Logroscino G. Nutrition, Frailty, and Alzheimer's Disease. Front Aging Neurosci [Internet]. 2014 [cited 2025 Jan 27];6(AUG). Available from: https://pubmed.ncbi.nlm.nih.gov/25206332/

Parletta N, Milte CM, Meyer BJ. Nutritional Modulation of Cognitive Function and Mental Health. J Nutr Biochem [Internet]. 2013 May [cited 2025 Jan 27];24(5):725–43. Available from: https://pubmed.ncbi.nlm.nih.gov/23517914/

Penniston KL, Tanumihardjo SA. The Acute and Chronic Toxic Effects of Vitamin A. Am J Clin Nutr [Internet]. 2006 Feb 1 [cited 2025 Jan 27];83(2):191–201. Available from: https://pubmed.ncbi.nlm.nih.gov/16469975/

Plevin D, Plevin D, Plevin D, Galletly C. The Neuropsychiatric Effects of Vitamin C Deficiency: A Systematic Review. BMC Psychiatry [Internet]. 2020 Jun 18 [cited 2025 Jan 27];20(1). Available from: https://pubmed.ncbi.nlm.nih.gov/32552785/

Prakash A, Bharti K, Majeed ABA. Zinc: Indications in Brain Disorders. Fundam Clin Pharmacol [Internet]. 2015 Apr 1 [cited 2025 Jan 27];29(2):131–49. Available from: https://pubmed.ncbi.nlm.nih.gov/25659970/

Psaltopoulou T, Sergentanis TN, Panagiotakos DB, Sergentanis IN, Kosti R, Scarmeas N. Mediterranean Diet, Stroke, Cognitive Impairment, and Depression: A Meta-Analysis. Ann Neurol [Internet]. 2013 Oct [cited 2025 Jan 27];74(4):580–91. Available from: https://pubmed.ncbi.nlm.nih.gov/23720230/

Qin, B., Adair, L. S., Plassman, B. L., Batis, C., Edwards, L. J., Popkin, B. M., & Mendez, M. A. (2015). Dietary patterns and cognitive decline among Chinese older adults. Epidemiology, 26(5), 758-768.

Radlowski EC, Johnson RW. Perinatal Iron Deficiency and Neurocognitive Development. Front Hum Neurosci [Internet]. 2013 Sep 23 [cited 2025 Jan 27];7(SEP). Available from: https://pubmed.ncbi.nlm.nih.gov/24065908/

Recchia A, Debetto P, Negro A, Guidolin D, Skaper SD, Giusti P. α-Synuclein and Parkinson's Disease. The FASEB Journal [Internet]. 2004 Apr 1 [cited 2025 Jan 29];18(6):617–26. Available from: https://onlinelibrary.wiley.com/doi/full/10.1096/fj.03-0338rev

Sajovic J, Meglič A, Glavač D, Markelj, Hawlina M, Fakin A. The Role of Vitamin A in Retinal Diseases. Int J Mol Sci [Internet]. 2022 Feb 1 [cited 2025 Jan 27];23(3). Available from: https://pubmed.ncbi.nlm.nih.gov/35162940/

Samieri, C., Grodstein, F., Rosner, B. A., Kang, J. H., Cook, N. R., Manson, J. E., ... & Okereke, O. I. (2013). Mediterranean diet and cognitive function in older age. Epidemiology, 24(4), 490-499.

Samieri, C., Okereke, O. I., Devore, E. E., & Grodstein, F. (2013). Long-term adherence to the Mediterranean diet is associated with overall cognitive status, but not cognitive decline, in women. The Journal of nutrition, 143(4), 493-499.

Sharma, P., Gaur, V.K., Srivastava, J.K. (2019). Diet and Nutrition in Alzheimer's Disease and Healthy Aging. In: Ashraf, G., Alexiou, A. (eds) Biological, Diagnostic and Therapeutic Advances in Alzheimer's Disease. Springer, Singapore. https://doi.org/10.1007/978-981-13-9636-6_10

Shepard RD, Nugent FS. Early Life Stress- and Drug-Induced Histone Modifications Within the Ventral Tegmental Area. Front Cell Dev Biol [Internet]. 2020 Sep 30 [cited 2025 Jan 27];8. Available from: https://pubmed.ncbi.nlm.nih.gov/33102491/

Shirazi HA, Rasouli J, Ciric B, Rostami A, Zhang GX. 1,25-Dihydroxyvitamin D3 Enhances Neural Stem Cell Proliferation and Oligodendrocyte differentiation. Exp Mol Pathol [Internet]. 2015 Apr 1 [cited 2025 Jan 27];98(2):240–5. Available from: https://pubmed.ncbi.nlm.nih.gov/25681066/

Singh B, Parsaik AK, Mielke MM, Erwin PJ, Knopman DS, Petersen RC, et al. Association of Mediterranean Diet With Mild Cognitive Impairment and Alzheimer's Disease: A Systematic Review and Meta-Analysis. J Alzheimers Dis [Internet]. 2014 [cited 2025 Jan 27];39(2):271–82. Available from: https://pubmed.ncbi.nlm.nih.gov/24164735/

Sinigaglia-Coimbra R, Lopes AC, Coimbra CG. Riboflavin Deficiency, Brain Function, and Health. Handbook of Behavior, Food and Nutrition. 2011;2427–49.

Smith AD, Refsum H, Bottiglieri T, Fenech M, Hooshmand B, McCaddon A, et al. Homocysteine and Dementia: An International Consensus Statement. J Alzheimers Dis [Internet]. 2018 [cited 2025 Jan 27];62(2):561–70. Available from: https://pubmed.ncbi.nlm.nih.gov/29480200/

Solfrizzi V, Panza F, Frisardi V, Seripa D, Logroscino G, Imbimbo BP, et al. Diet and Alzheimer's Disease Risk Factors or Prevention: The Current Evidence. Expert Rev Neurother [Internet]. 2011 May [cited 2025 Jan 27];11(5):677–708. Available from: https://pubmed.ncbi.nlm.nih.gov/21539488/

Spira AP, Chen-Edinboro LP, Wu MN, Yaffe K. Impact of Sleep on the Risk of Cognitive Decline and Dementia. Curr Opin Psychiatry [Internet]. 2014 [cited 2025 Jan 27];27(6):478–83. Available from: https://pubmed.ncbi.nlm.nih.gov/25188896/

Talegawkar SA, Bandinelli S, Bandeen-Roche K, Chen P, Milaneschi Y, Tanaka T, et al. A higher Adherence to a Mediterranean-Style Diet is Inversely Associated With the Development of Frailty in Community-Dwelling Elderly Men and Women. J Nutr [Internet]. 2012 Dec [cited 2025 Jan 27];142(12):2161–6. Available from: https://pubmed.ncbi.nlm.nih.gov/23096005/

Tangney, C. C., Kwasny, M. J., Li, H., Wilson, R. S., Evans, D. A., & Morris, M. C. (2011). Adherence to a Mediterranean-type dietary pattern and cognitive decline in a community population. *The American journal of clinical nutrition*, *93*(3), 601-607.

Teleanu DM, Niculescu AG, Lungu II, Radu CI, Vladâcenco O, Roza E, et al. An Overview of Oxidative Stress, Neuroinflammation, and Neurodegenerative Diseases. Int J Mol Sci [Internet]. 2022 Jun 1 [cited 2025 Jan 27];23(11). Available from: https://pubmed.ncbi.nlm.nih.gov/35682615/

Tönnies E, Trushina E. Oxidative Stress, Synaptic Dysfunction, and Alzheimer's Disease. J Alzheimers Dis [Internet]. 2017 [cited 2025 Jan 28];57(4):1105–21. Available from: https://pubmed.ncbi.nlm.nih.gov/28059794/

Van Giau V, Wu SY, Jamerlan A, An SSA, Kim SY, Hulme J. Gut Microbiota and Their Neuroinflammatory Implications in Alzheimer's Disease. Nutrients [Internet]. 2018 Nov 14 [cited 2025 Jan 29];10(11). Available from: https://pubmed.ncbi.nlm.nih.gov/30441866/

Wallert M, Börmel L, Lorkowski S. Inflammatory Diseases and Vitamin E-What Do We Know and Where Do We Go? Mol Nutr Food Res [Internet]. 2021 Jan 1 [cited 2025 Jan 28];65(1). Available from: https://pubmed.ncbi.nlm.nih.gov/32692879/

Wołoszynowska-Fraser MU, Kouchmeshky A, McCaffery P. Vitamin A and Retinoic Acid in Cognition and Cognitive Disease. Annu Rev Nutr [Internet]. 2020 Aug 21 [cited 2025 Jan 28];40:247–72. Available from: https://pubmed.ncbi.nlm.nih.gov/32966186/

Yamanaka R, Shindo Y, Oka K. Magnesium Is a Key Player in Neuronal Maturation and Neuropathology. Int J Mol Sci [Internet]. 2019 Jul 2 [cited 2025 Jan 28];20(14). Available from: https://pubmed.ncbi.nlm.nih.gov/31336935/

Yuan Z, Cai S, Yan C, Rao S, Cheng S, Xu F, et al. Research Progress on the Physiological Mechanism by Which Selenium Alleviates Heavy Metal Stress in Plants: A Review. Agronomy 2024, Vol 14, Page 1787 [Internet]. 2024 Aug 14 [cited 2025 Jan 28];14(8):1787. Available from: https://www.mdpi.com/2073-4395/14/8/1787/htm

Yuda N, Tanaka M, Yamauchi K, Abe F, Kakiuchi I, Kiyosawa K, et al. Effect of the Casein-Derived Peptide Met-Lys-Pro on Cognitive Function in Community-Dwelling Adults Without Dementia: A Randomized, Double-Blind, Placebo-Controlled Trial. Clin Interv Aging [Internet]. 2020 [cited 2025 Jan 28];15:743–54. Available from: https://pubmed.ncbi.nlm.nih.gov/32546992/

Zapico SC, Ubelaker DH. mtDNA Mutations and Their Role in Aging, Diseases and Forensic Sciences. Aging Dis [Internet]. 2013 [cited 2025 Jan 28];4(6):364–80. Available from: https://pubmed.ncbi.nlm.nih.gov/24307969/

Zbeida, M., Goldsmith, R., Shimony, T., Vardi, H., Naggan, L., & Shahar, D. R. (2014). Mediterranean diet and functional indicators among older adults in non-Mediterranean and Mediterranean countries. The Journal of nutrition, health and aging, 18(4), 411-418.

Zbeida M, Goldsmith R, Shimony T, Yardi H, Naggan L, Shahar DR. Mediterranean Diet and Functional Indicators Among Older Adults in Non-Mediterranean and Mediterranean Countries. J Nutr Health Aging [Internet]. 2014 [cited 2025 Jan 28];18(4):411–8. Available from: https://pubmed.ncbi.nlm.nih.gov/24676323/

Zempleni J, Suttie JW, Gregory JF, Patrick I, Stover J. Handbook of Vitamins. Handbook of Vitamins [Internet]. 2013 Jul 29 [cited 2025 Jan 28]; Available from: https://www.taylorfrancis.com/books/mono/10.1201/b15413/handbook-vitamins-janos-zempleni-john-suttie-jesse-gregory-iii-patrick-stover

Zimmermann MB, Boelaert K. Iodine Deficiency and Thyroid Disorders. Lancet Diabetes Endocrinol [Internet]. 2015 Apr 1 [cited 2025 Jan 28];3(4):286–95. Available from: https://pubmed.ncbi.nlm.nih.gov/25591468/

Zoidis E, Seremelis I, Kontopoulos N, Danezis GP. Selenium-Dependent Antioxidant Enzymes: Actions and Properties of Selenoproteins. Antioxidants (Basel) [Internet]. 2018 May 1 [cited 2025 Jan 28];7(5). Available from: https://pubmed.ncbi.nlm.nih.gov/29758013/

14 Malnutrition and Cognitive Disorders

Clinical Perspectives

Rajesh Tamatta, Charu Jaiswal, Varsha Pai, Ishika Singh, and Abhishek Kumar Singh

14.1 INTRODUCTION

Cognitive disorders are brain disorders accompanied by a decline in cognitive capabilities such as learning, memory, perception, and problem-solving, and the normal functioning of the brain is severely affected. These disorders are a consequence of damage to the brain due to injury, illness, drug abuse, toxin exposure, or drug abuse. The decline in social, occupational, and intellectual work associated with rain is rapid and serious when cognitive decline results from injury to the brain or a stroke. On the other hand, cognitive disorders such as Alzheimer's disease (AD) and Parkinson's disease are slow and progressive but lead to more devastating effects on brain health (Narayana et al., 2023). With an increase in the elderly population, the incidence of cognitive decline is also increasing (Blazer & Wallace, 2016). Although age is considered to be the most common and major risk factor for cognitive disorders, various other factors, including malnutrition, have also been shown to play crucial roles.

Although there is no universal definition of the term malnutrition, it has been most appropriately defined as an imbalance, excess, or lack of a diverse range of nutrients, leading to severe damage to body function and composition as well as resulting in clinical pathologies (Saunders & Smith, 2010). Malnourishment can result from either caloric deficiency or a decline in protein, vitamin, or mineral intake. It can also be a consequence of the overconsumption of calories, vitamins, proteins, etc. Research conducted on the elderly population with cognitive disorders revealed that 8.4% of the individuals were malnourished, whereas 42.7% of the individuals were found to be at risk of malnourishment (Verlaan et al., 2017).

Malnutrition can be categorized broadly into undernutrition and overnutrition. Undernutrition is most strongly associated with malnutrition, which is a deficiency of macronutrients (fats, carbohydrates, and proteins), also known as protein-energy undernutrition, and micronutrients (vitamins and minerals). Therefore, a person can suffer from malnutrition if their nutritional intake in the diet is not appropriate or if the body has impaired metabolism. On the other hand, overnutrition, although substantial, is less commonly referred to in the context of malnutrition. This has recently been incorporated into the existing definition of malnutrition by the WHO to acknowledge the harmful effects (for instance, obesity and other non-transmissible diseases, as well as toxicity associated with the over intake of specific micronutrients) of overconsumption of essential nutrients.

Nutrition is intricately related to the process of brain development and function, and malnourishment can obstruct the proper development of the brain. It can affect the growth and maturation of brain cells, the formation of synapses, and the differentiation of cells into mature neurons or other brain cells. The impact of malnutrition can either be through a direct effect on brain structure and development or indirectly through psychological and mood changes along with an effect on motor abilities. Children with malnutrition present altered cell numbers and migration, development of the hippocampus, and neurotransmission in the brain. In addition, it can lead to decreased immune

DOI: 10.1201/9781003500025-14

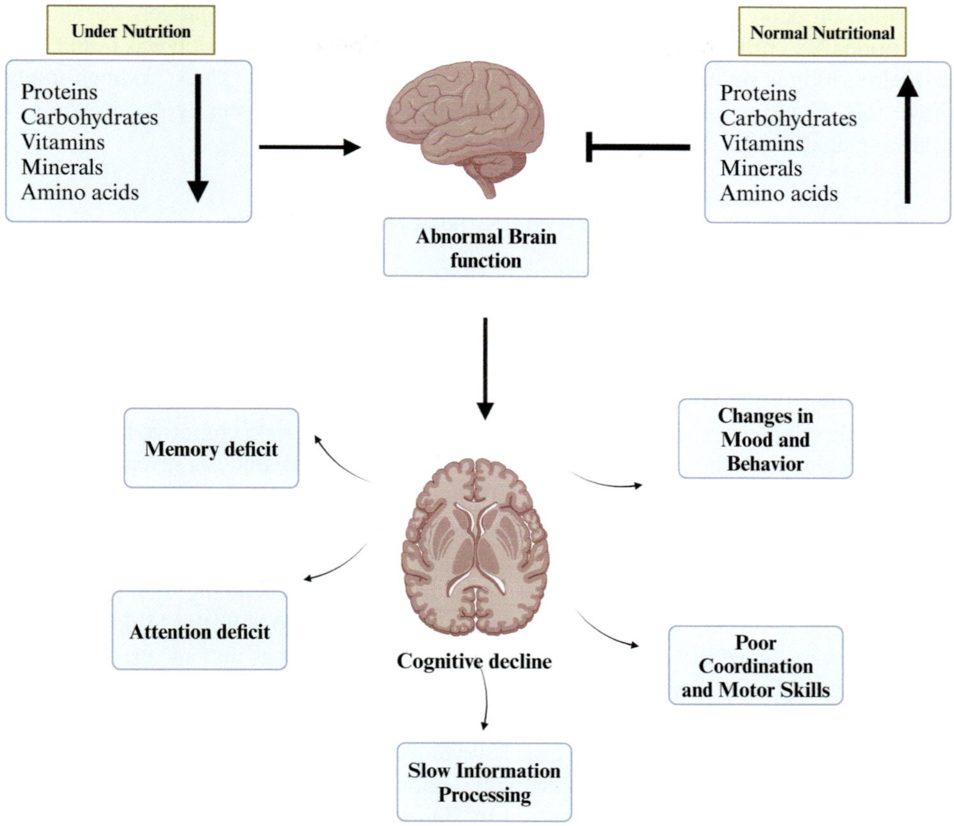

FIGURE 14.1 Effects of malnutrition on cognitive health. Undernutrition causes abnormal brain function, which can be mitigated by nutrient-rich diet supplementation. Abnormal brain function negatively impacts cognitive function. Signs of cognitive decline include memory deficits, attention deficits, poor coordination skills, changes in mood and behavior, and slow information processing.

tolerance and the ability to fight infections, thus affecting the overall health and well-being of an individual (Prado & Dewey, 2014) (Figure 14.1).

14.2 MALNUTRITION IN COGNITION

A significant percentage of the elderly population suffers from dementia, a widespread mental health illness that is characterized as a progressive neurodegenerative syndrome that impairs cognitive function. Dementia is thought to affect 5–7% of people over 60 years of age worldwide, with low- and middle-income nations bearing disproportionately high burdens (Prince et al., 2013). By 2050, the high global prevalence—55 million in 2020—is expected to double every 20 years to 139 million (GBD 2019 Dementia Forecasting Collaborators, 2022). Dementia is a top public health issue due to the aging population worldwide and the substantial effects that dementia and cognitive decline have on social and medical situations (Frankish & Horton, 2017).

Late-life depression is a complex condition that extends beyond being merely depression in older adults and is often considered a component of geriatric syndrome. This multifaceted syndrome involves interconnected comorbidities, making it essential to understand these relationships to develop effective prevention strategies. Malnutrition is common in individuals with geriatric syndrome and is recognized as a potential indicator of late-life depression. However, their individual and combined effects on depression remain poorly understood. A study involving the effect of malnutrition on depression revealed that depressive symptoms were negatively correlated with poorer nutrition. A significant association between malnutrition and depression was found, suggesting that malnutrition is significantly associated with late-life depression. These findings emphasize the critical role of addressing nutritional health in managing and preventing depression in older adults (Hwang et al., 2022).

14.2.1 Case Studies on the Effect of Malnutrition on Cognition

A study in Mediterranean countries examined the nutritional and mental health status of older adults in Lebanon aged 60 years and above who were at risk of dependency, poverty, or social exclusion. Using secondary data from home visits conducted by trained social workers between April 2022 and February 2023, the study assessed participants via validated tools such as the Mini Nutritional Assessment-Short Form (MNA-SF), the Ascertain Dementia Scale (AD8), and the Geriatric Depression Scale (GDS-15) and revealed high rates of malnutrition risk, cognitive impairment, and depression, especially among socially vulnerable groups. Statistical analysis revealed that depression and cognitive impairment were significant predictors of malnutrition, with decreased food intake identified as a critical link between poor mental health and nutritional decline. The results emphasize the urgent need for integrated public health strategies that include routine nutritional and mental health screenings, particularly those that target older adults facing social exclusion. The study advocates for culturally sensitive, community-level interventions in the Middle East, where older adults often reside within extended family homes. These findings underscore the importance of addressing mental health challenges as part of comprehensive nutritional care to improve overall health outcomes in aging populations (Karam et al., 2024).

Another study, which is a part of the China Hainan Centenarian Cohort Study (CHCCS), investigated cognitive decline and its associations with demographic, genetic, and nutritional factors in centenarians and oldest-old adults (≥80 years) in Hainan Province between June 2014 and June 2016. A total of 412 centenarians and 534 oldest-old adults were included after applying criteria that excluded individuals with neurodegenerative diseases or incomplete Mini-Mental State Examination (MMSE) data. Data collection involved comprehensive interviews, physical examinations, and blood tests, which focused on variables such as medical history, nutritional status (BMI, prognostic nutritional index [PNI], MNA-SF), and cognitive function, which was assessed via the MMSE. Statistical analyses revealed critical factors influencing MMSE scores and cognitive decline. The study revealed that age was the strongest predictor of cognitive decline, with additional contributions from demographic factors (e.g., sex and education level), poor nutritional status, and metabolic disorders. Malnutrition emerged as a significant risk factor, characterized by low PNI and MNA-SF scores, which were especially pronounced among centenarians and participants with cognitive impairment. Nutritional deficiencies, particularly inadequate protein and nutrient intake are linked to neuronal energy deficits, heightened susceptibility to neurodegeneration, and central nervous system dysfunction. Socioeconomic and environmental challenges further exacerbate poor nutritional outcomes, with markers such as hypoalbuminemia indicating increased mortality risk. Protective dietary patterns, such as the Mediterranean diet rich in omega-3 fatty acids, vitamins, and flavonoids, have the potential to mitigate inflammation, oxidative stress, and cognitive decline. These findings underscore the critical interplay between nutrition and cognitive health in aging populations, highlighting the urgent need for targeted nutritional interventions and community-level strategies to combat cognitive decline and promote healthy aging among Chinese centenarians and oldest-old adults (Feng et al., 2022). Childhood malnutrition negatively affects health, development, and productivity later in life, with underweight children often exhibiting cognitive impairments. Research has evaluated the effects of a nutrition-focused intervention on the cognitive development of preschool-aged, malnourished children in a few villages in the Karnataka district of Udupi. A cluster-randomized controlled trial assessed the impact of a nutrition-focused intervention on cognitive development in preschool children with moderate to severe underweight in rural

Udupi, Karnataka, India. A total of 253 participants (127 intervention, 126 control) were recruited, with eligibility requiring children to be moderately or severely underweight. Twelve villages were randomized into six intervention groups and six control groups. Baseline assessments recorded the weight, height, and nutritional status of the children. Cognitive development, evaluated via a validated tool covering six areas (e.g., memory, comprehension, and visual perception), revealed significant improvements in the intervention group compared with the control group. Malnutrition was found to adversely affect cognitive development, but educating caregivers on proper feeding practices and balanced diets mitigated these effects, highlighting the critical role of caregiver education and nutritional interventions in promoting cognitive growth among undernourished children.

14.3 BRAIN–GUT MICROBIOTA AXIS (BGA)

Diet is a major factor in establishing microbiota composition throughout a lifetime. The brain–gut microbiota relationship is a complex communication network associated with the central nervous system (CNS), the gut, and microbes. This axis comprises multiorgan bidirectional signaling between gut microbes and the brain, digestive system, and cognitive and mental health. The vagus nerve serves as a major channel for signaling between the brain and the gut. An overview of the impacts of malnutrition on the brain–gut microbiota and cognitive health is as follows: Gut microbiota reduction: Malnourished individuals show a decrease in gut microbe diversity and the depletion of beneficial microbes such as *Bifidobacterium* and *Lactobacillus,* affecting the immune response and metabolism. Increased pathogenic microbes exacerbate inflammation and have negative effects on health. Changes in the microbiota influence vagal signaling, affecting brain health (Frostad, 2022; Kamada et al., 2013). Immune mediators: The gut microbiota secretes metabolites such as short-chain fatty acids and modulates neuroinflammation and blood–brain barrier permeability. Malnutrition alters metabolite production and induces inflammatory processes by increasing blood–brain barrier permeability (Frostad, 2022). Owing to malnourishment, dysbiosis (imbalanced microbes) significantly impacts the brain–gut signaling axis and leads to mental disorders such as depression and anxiety (Croci et al., 2011) (Figure 14.2).

14.4 MECHANISM OF COGNITIVE DISORDERS LINKED TO MALNUTRITION

Nutrition-deficient diets have significant negative impacts on cognitive health through multiple mechanisms. Different nutrients and major mechanisms that play crucial roles in the development of cognitive disorders through malnutrition are discussed below.

14.4.1 Micronutrient

Moreover, micronutrient deficiencies significantly impact cognitive health, leading to the development of cognitive disorders. Deficiencies in the vitamin B complex hinder the synthesis of several neurotransmitters, such as dopamine and serotonin (Black, 2003; Singh et al., 2023). Deficiencies in antioxidative vitamins such as vitamin C and vitamin E increase oxidative stress (Mustafa Khalid et al., 2022). Low levels of zinc and vitamin D increase inflammatory reactions in the body (Mustafa Khalid et al., 2022). An abnormal myelin sheath around neurons results from lower levels of vitamin B12 (Black, 2003; Mustafa Khalid et al., 2022). Zinc deficiency results in disrupted synaptic plasticity. Neural tube defects and cognitive impairments in offspring are a result of lower levels of zinc in pregnant women (Swaminathan et al., 2013).

14.4.2 Carbohydrate Deficiencies

Glucose and carbohydrates are the main sources of energy for brain cells. Carbohydrate deficiencies can lead to impairment of brain cells, consequently leading to cognitive dysfunction,

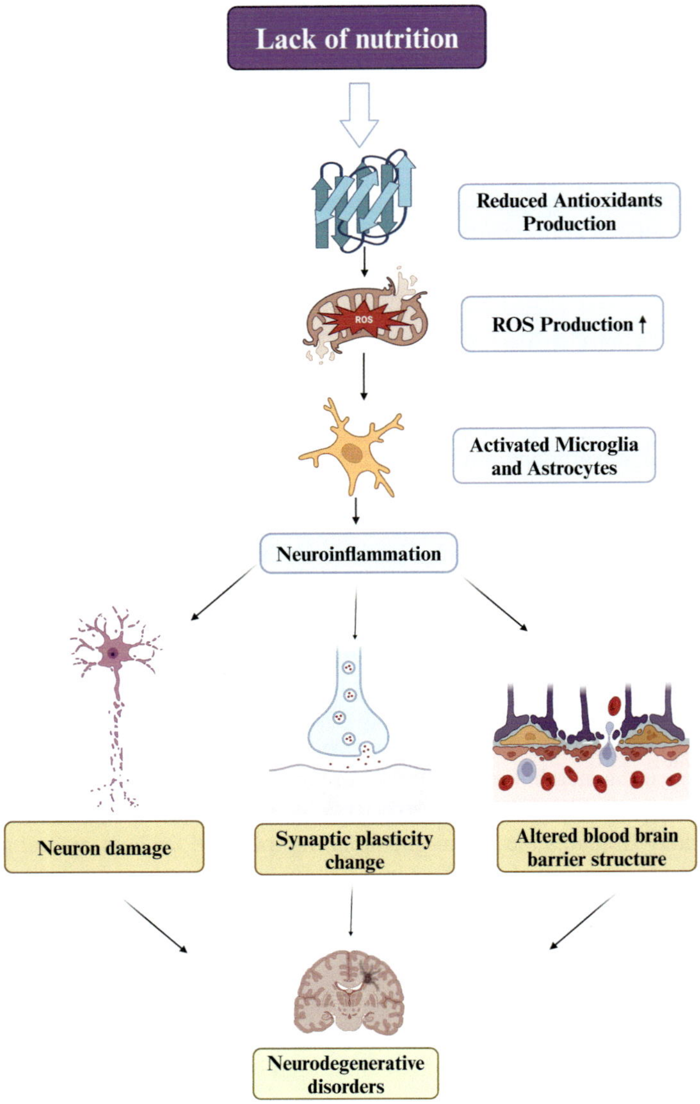

FIGURE 14.2 In the brain–gut axis, in malnourished individuals, impaired neurotransmitter production, abnormal brain function, elevated short-chain fatty acids, and an imbalance in the gut microbiome and the growth of pathogenic bacteria are observed. However, in undernourished conditions, normal neurotransmitter production enhances gut microbiome diversity and alleviates short-chain fatty acid levels.

synaptic impairment, and neurogenesis. The link between carbohydrate deficiency and cognitive disorders is as follows: (1) Energy deficiency: The brain primarily depends on glucose as a source of energy; deficiencies and limited availability of glucose alter the energy source, and the brain switches to an alternative energy source, such as ketone bodies. Ketone bodies are less efficient and are energy-deficient, leading to cognitive impairments, especially in memory and learning. Low glucose can reduce ATP production, leading to slower synaptic transmission and synaptic plasticity (Cunnane et al., 2016; Dienel, 2019; Ripoli et al., 2020). In a study conducted by Chugani (1998) in pediatrics, children experienced developmental delay and learning disabilities due to severe carbohydrate restriction (Chugani, 1998). (2) Affects neurotransmitter production: Low glucose availability hinders acetyl-CoA production, which is essential for the production of the neurotransmitter acetylcholine, which is associated with memory and learning (Szutowicz et al., 2013). (3) Neurogenesis and synaptogenesis impairment: Inadequate glucose availability to brain cells, impaired neurogenesis, and synaptogenesis caused by impaired insulin-like growth factor are linked to the progression of neurodegenerative disorders (Croci et al., 2011; Wrigley et al., 2017).

14.4.3 Protein Deficiency

Proteins maintain the structural integrity of the brain, play a role in neurotransmitter synthesis, and maintain cognitive health. Severe protein malnourishment causes Kwashiorkor disease. Moreover, Marasmus is caused by protein and calorie deficiency.

Protein-energy malnutrition is another disorder caused by protein and energy deficiency. Restricted protein levels significantly impact cognitive health. (1) Altered neurotransmitter synthesis: Proteins are necessary for neurotransmitter synthesis; according to research, a protein-deficient diet results in insufficient amino acid availability and impairs neurotransmitter synthesis, such as dopamine and serotonin, leading to cognitive disorders (Gao et al., 2022). (2) Impaired brain structures: Recent studies have shown that low protein intake and significantly altered brain structure can lead to brain atrophy in regions associated with cognition, such as the hippocampus (Gao et al., 2022). In addition to carbohydrates, protein deficiency results in a reduction in synaptic plasticity. (3) Synaptic plasticity: A protein-deficient diet reduces brain-derived neurotrophic factor (BDNF), which is crucial for neuronal growth, development, and maintenance. Low levels of BDNF reduce synaptic plasticity (Cattaneo et al., 2016). Oxidative stress: Little research has been conducted due to protein deficiency, and fewer antioxidants, such as glutathione peroxidase enzymes, are available to mitigate oxidative stress, increasing the vulnerability of brain cells to oxidative damage. Increased oxidative stress contributes to neurodegenerative disorders such as AD (Bramble et al., 2024). Oxidative stress exacerbates excitotoxicity by increasing the release of the excitatory neurotransmitter glutamate, activating NMDA receptors, and increasing intracellular Ca^{2+} levels. Increased Ca^{2+} levels cause ROS production, resulting in neurodegenerative disorders (Saha et al., 2017).

14.5 MALNUTRITION INDUCED OXIDATIVE STRESS, NEUROINFLAMMATION, AND COGNITIVE DEFICITS

Owing to inadequate antioxidative defenses (antioxidant vitamins and antioxidative enzymes), individuals present elevated levels of reactive oxygen species (ROS), reactive nitrogen species (RNS), nitrogen oxides (NOX), and free radicals such as alkoxyl radicals and hydroxyl radicals. Oxidative stress can cause the production of NADPH peroxidase, which exacerbates ROS production and enhances inflammation (Fabisiak & Patel, 2022). In 2014, studies conducted in children aged between six months and five years revealed that the levels of antioxidant enzymes, such as peroxidase and glutathione peroxidase, were significantly reduced, while the levels of malondialdehyde in the serum were increased (Khare et al., 2014). Elevated oxidative stress alters the intracellular signaling cascade, causing a dysregulated inflammatory response. These findings indicate that the activation and proliferation of $CD4^+$ T cells depend on the redox balance. An imbalance in stress conditions leads to the activation and differentiation of $CD4^+$ immune cells into Th1 and Th17 subpopulations, contributing to the development of neuroinflammation (Solleiro-Villavicencio & Rivas-Arancibia, 2018). NOX expression is associated with neuroinflammation (Fabisiak & Patel, 2022). High mobility group box 1 (HMGB1) is a damage-associated molecular pattern (DAMP) secreted by damaged neurons that is responsible for the activation of microglia via pattern recognition receptor (PRR) CR3 and TLR4 signaling (Fabisiak & Patel, 2022). Overproduction of ROS-activated microglia and astrocytes in the central nervous system (CNS) causes the release

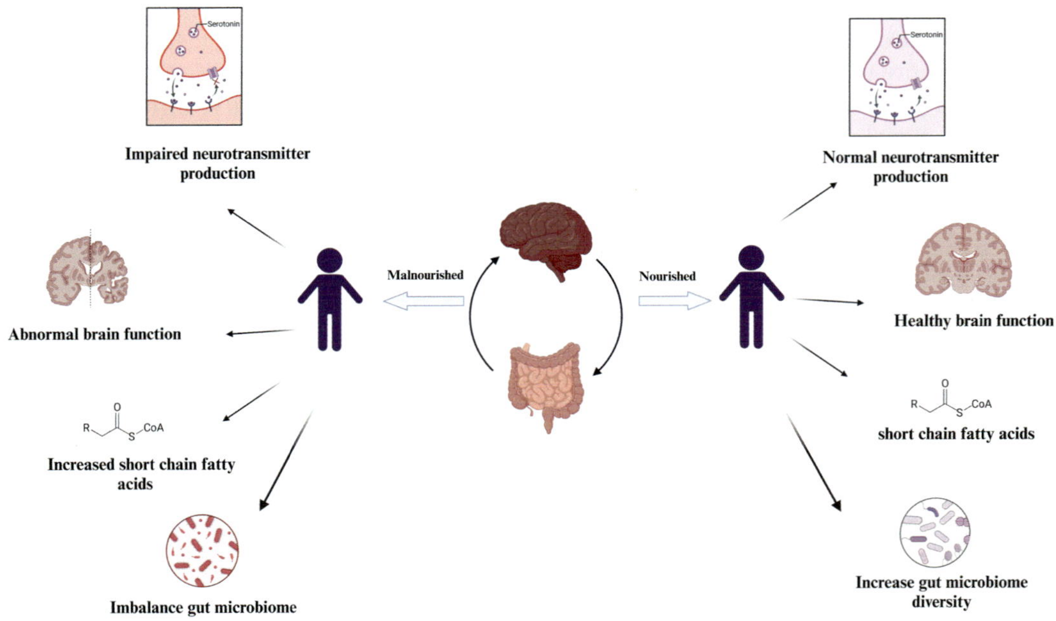

FIGURE 14.3 Relationships among malnutrition, neuroinflammation, and cognitive disorders. Malnutrition causes a significant decrease in the levels of antioxidants, increasing ROS production in cells. This leads to the activation of microglia and astrocytes. Neuroinflammation leads to neuron damage, a change in synaptic plasticity, and altered blood–brain barrier structure. The development of neurodegenerative disease negatively impacts cognitive function.

of proinflammatory cytokines such as interleukin-1 beta (IL-1β) and tumor necrosis factor-alpha (TNF-α) (Fabisiak & Patel, 2022; Solleiro-Villavicencio & Rivas-Arancibia, 2018). A dysregulated inflammatory pathway is responsible for an uncontrolled inflammatory response (Fabisiak & Patel, 2022; Li et al., 2024) (Figure 14.3).

14.6 PREVENTION AND THERAPEUTIC STRATEGIES

Malnutrition-linked cognitive disorders must be prevented and treated via multimodal strategies such as therapeutic intervention, dietary intervention, and supplementation. Frequent malnutrition tests for populations at risk, such as children and elderly individuals, allow for early detection and timely interventions. Awareness program to educate caregivers and patients about the importance of a nutrition-rich diet. The consumption of adequate amounts of vitamins may offer a protective function against AD (Polis & Samson, 2024). The introduction of fortified food to the population is prone to nutritional deficiencies. Recent studies have shown mixed results concerning the effects of nutritional supplements with omega-3 fatty acids and antioxidants on cognitive health (Panza et al., 2014). A diet with sufficient nutrients such as vitamin B12, omega-3 fatty acids, and antioxidants mitigates cognitive decline and supports brain health (Chen et al., 2022). A combination of medical treatment with nutritional support and cognitive therapy approaches is used to manage both cognitive decline and malnutrition (Li et al., 2024). Oral nutritional supplements such as high protein or energy-dense supplements are used for individuals at risk of malnutrition. According to the latest findings, adherence to the Mediterranean diet with a combination of healthy fruits, fats, and vegetables is associated with a reduced risk of developing cognitive impairments and slowing AD progression in older adults (Panza et al., 2014; Polis & Samson, 2024). Clinical studies have demonstrated that individuals with dietary insufficiency in folate who received folic acid supplementation experienced slower cognitive decline over three years than those who received a placebo. These findings suggest that folate deficiency can have a significant positive effect on cognitive health (Shea et al., 2015).

14.7 CONCLUSION

Individuals with early-onset cognitive disorders are more susceptible to micronutrient, calorie, and protein deficiencies because of aging, and cognitive impairments that influence food intake and metabolism. Malnutrition and cognitive impairment are strongly correlated and present serious health risks for people of all ages. It is crucial to implement comprehensive prevention measures to address the health problems associated with malnutrition. These include body weight monitoring, routine screening for malnutrition, and the provision of daily life support. A comprehensive assessment that addresses nutritional inadequacies and underlying causes should direct a customized management approach after malnutrition, or its risk has been established. It may be possible to reverse the cognitive deterioration caused by micronutrient deficiencies, such as those in vitamins B and D, if they are treated promptly. However, severe dementia typically results in irreversible cognitive abnormalities. The benefits of good dietary patterns, such as the Mediterranean diet, which has been associated with superior cognitive outcomes, are highlighted by the current research, which advises against vitamin B and D treatment to avoid cognitive decline in people without shortages. Such diets may lower the incidence of dementia, according to certain longitudinal studies, but more studies with longer follow-ups are needed.

To prevent and manage various cognitive disorders, future efforts should focus on incorporating strategies focused on nutrition into healthcare systems. Personalized approaches, including tailored dietary plans and supplementation, could address specific nutritional deficits in high-risk groups such as children and elderly individuals. Advances in biomarker research may provide better tools for the early identification of nutritional deficiencies and cognitive decline. Additionally, interdisciplinary collaboration among healthcare providers, caregivers, and policymakers is crucial to enhancing awareness of and accessibility to nutritional support. Long-term research into the interplay between diet, malnutrition, and cognitive health will be essential for developing evidence-based guidelines and public health initiatives to combat this global challenge.

ABBREVIATIONS

AD	Alzheimer's disease
AD8	Ascertain Dementia Scale
BDNF	Brain-derived neurotrophic factor
DAMP	Damage-associated molecular pattern
GDS-15	Geriatric Depression Scale
HMGB1	High mobility group box 1
*IL-1*β	Interleukin-1 beta
MMSE	Mini-Mental State Examination
MNA-SF	Mini Nutritional Assessment-Short Form
NOX	Nitrogen oxides
PRR	Pattern recognition receptor
PNI	Prognostic nutritional index
RNS	Reactive nitrogen species
ROS	Reactive oxygen species
TNF-α	Tumor necrosis factor-alpha

GLOSSARY

MMSE Score	Mini-Mental State Examination (MMSE) is a tool that can be used to systematically and thoroughly assess the mental health
Dysbiosis	Dysbiosis is the substitution of the normal gut flora to a dysfunctional array of organisms that can promote disease states
PRRs	PRRs are a class of receptors that can directly recognize the specific molecular structures on the surface of pathogens, apoptotic host cells, and damaged senescent cells.
Geriatric syndrome	Geriatric syndromes refer to a group of clinical conditions in older adults that do not fit neatly into specific disease categories but significantly impact quality of life and functional ability

ACKNOWLEDGMENTS

The authors acknowledge the support provided by Manipal Academy of Higher Education, Manipal, Karnataka, India.

CONFLICT OF INTEREST

The authors declare that they have no conflict of interest.

REFERENCES

Black, M. M. (2003). Micronutrient Deficiencies and Cognitive Functioning. *The Journal of Nutrition*, *133*(11), 3927S–3931S. https://doi.org/10.1093/jn/133.11.3927S

Blazer, D. G., & Wallace, R. B. (2016). Cognitive Aging: What Every Geriatric Psychiatrist Should Know. *The American Journal of Geriatric Psychiatry: Official Journal of the American Association for Geriatric Psychiatry*, *24*(9), 776–781. https://doi.org/10.1016/j.jagp.2016.06.013

Bramble, M. S., Fourcassié, V., Vashist, N., Roux-Dalvai, F., Zhou, Y., Bumoko, G., Kasendue, M. L., Spencer, D., Musasa Hanshi-Hatuhu, H., Kambale-Mastaki, V., Manalo, R. V. M., Mohammed, A., McIlwain, D. R., Cunningham, G., Summar, M., Boivin, M. J., Caldovic, L., Vilain, E., Mumba-Ngoyi, D., … & Droit, A. (2024). Glutathione Peroxidase 3 Is a Potential Biomarker for Konzo. *Nature Communications*, *15*(1), 7811. https://doi.org/10.1038/s41467-024-52136-5

Cattaneo, A., Cattane, N., Begni, V., Pariante, C. M., & Riva, M. A. (2016). The Human BDNF Gene: Peripheral Gene Expression and Protein Levels as Biomarkers for Psychiatric Disorders. *Translational Psychiatry*, *6*(11), e958–e958. https://doi.org/10.1038/tp.2016.214

Chen, L., Jiao, J., & Zhang, Y. (2022). Therapeutic Approaches for Improving Cognitive Function in the Aging Brain. *Frontiers in Neuroscience*, *16*. https://doi.org/10.3389/fnins.2022.1060556

Chugani, H. T. (1998). A Critical Period of Brain Development: Studies of Cerebral Glucose Utilization with PET. *Preventive Medicine*, *27*(2), 184–188. https://doi.org/10.1006/pmed.1998.0274

Croci, L., Barili, V., Chia, D., Massimino, L., van Vugt, R., Masserdotti, G., Longhi, R., Rotwein, P., & Consalez, G. G. (2011). Local Insulin-Like Growth Factor I Expression Is Essential for Purkinje Neuron Survival at Birth. *Cell Death & Differentiation*, *18*(1), 48–59. https://doi.org/10.1038/cdd.2010.78

Cunnane, S. C., Courchesne-Loyer, A., Vandenberghe, C., St-Pierre, V., Fortier, M., Hennebelle, M., Croteau, E., Bocti, C., Fulop, T., & Castellano, C.-A. (2016). Can Ketones Help Rescue Brain Fuel Supply in Later Life? Implications for Cognitive Health during Aging and the Treatment of Alzheimer's Disease. *Frontiers in Molecular Neuroscience*, *9*. https://doi.org/10.3389/fnmol.2016.00053

Dienel, G. A. (2019). Brain Glucose Metabolism: Integration of Energetics with Function. *Physiological Reviews*, *99*(1), 949–1045. https://doi.org/10.1152/physrev.00062.2017

Fabisiak, T., & Patel, M. (2022). Crosstalk between Neuroinflammation and Oxidative Stress in Epilepsy. *Frontiers in Cell and Developmental Biology*, *10*. https://doi.org/10.3389/fcell.2022.976953

Feng, L., Chu, Z., Quan, X., Zhang, Y., Yuan, W., Yao, Y., Zhao, Y., & Fu, S. (2022). Malnutrition Is Positively Associated with Cognitive Decline in Centenarians and Oldest-Old Adults: A Cross-Sectional Study. *EClinicalMedicine*, *47*, 101336. https://doi.org/10.1016/j.eclinm.2022.101336

Frankish, H., & Horton, R. (2017). Prevention and Management of Dementia: A Priority for Public Health. *The Lancet*, *390*(10113), 2614–2615. https://doi.org/10.1016/S0140-6736(17)31756-7

Frostad, S. (2022). Are the Effects of Malnutrition on the Gut Microbiota–Brain Axis the Core Pathologies of Anorexia Nervosa? *Microorganisms*, *10*(8), Article 8. https://doi.org/10.3390/microorganisms10081486

Gao, R., Yang, Z., Yan, W., Du, W., Zhou, Y., & Zhu, F. (2022). Protein Intake from Different Sources and Cognitive Decline over 9 Years in Community-Dwelling Older Adults. *Frontiers in Public Health*, *10*. https://doi.org/10.3389/fpubh.2022.1016016

GBD 2019 Dementia Forecasting Collaborators. (2022). Estimation of the Global Prevalence of Dementia in 2019 and Forecasted Prevalence in 2050: An Analysis for the Global Burden of Disease Study 2019. *The Lancet. Public Health*, *7*(2), e105–e125. https://doi.org/10.1016/S2468-2667(21)00249-8

Hwang, G., Cho, Y. H., Kim, E. J., Woang, J. W., Hong, C. H., Roh, H. W., & Son, S. J. (2022). Differential Effects of Sleep Disturbance and Malnutrition on Late-Life Depression Among Community-Dwelling Older Adults. *Frontiers in Psychiatry*, *13*. https://doi.org/10.3389/fpsyt.2022.820427

Kamada, N., Chen, G. Y., Inohara, N., & Núñez, G. (2013). Control of Pathogens and Pathobionts by the Gut Microbiota. *Nature Immunology*, *14*(7), 685–690. https://doi.org/10.1038/ni.2608

Karam, G., Abbas, N., El Korh, L., Abi Saad, A., Nasreddine, L., & Ouaijan, K. (2024). The Association of Cognitive Impairment and Depression with Malnutrition among Vulnerable, Community-Dwelling Older Adults: A Large Cross-Sectional Study. *Geriatrics*, *9*(5), Article 5. https://doi.org/10.3390/geriatrics9050122

Khare, M., Mohanty, C., Das, B. K., Jyoti, A., Mukhopadhyay, B., & Mishra, S. P. (2014). Free Radicals and Antioxidant Status in Protein Energy Malnutrition. *International Journal of Pediatrics*, *2014*, 1–254396. https://doi.org/10.1155/2014/254396

Li, J., Haj Ebrahimi, A., & Ali, A. B. (2024). Advances in Therapeutics to Alleviate Cognitive Decline and Neuropsychiatric Symptoms of Alzheimer's Disease. *International Journal of Molecular Sciences*, *25*(10), Article 10. https://doi.org/10.3390/ijms25105169

Li, J., Wang, Y., Xiong, K., & Gao, C. (2024). Editorial: Neuroinflammation and Cognitive Impairment. *Frontiers in Aging Neuroscience*, *16*. https://doi.org/10.3389/fnagi.2024.1453772

Mustafa Khalid, N., Haron, H., Shahar, S., & Fenech, M. (2022). Current Evidence on the Association of Micronutrient Malnutrition with Mild Cognitive Impairment, Frailty, and Cognitive Frailty among Older Adults: A Scoping Review. *International Journal of Environmental Research and Public Health*, *19*(23), Article 23. https://doi.org/10.3390/ijerph192315722

Narayana, D. B. A., Joshi, H., & Tiwari, V. H. S. (2023). Chapter Two—Overview of Approaches in Ayurveda for Neurological Health and Disorders. In M. Muralidhara & P. S. Rajini (Eds.), *Ayurvedic Herbal Preparations in Neurological Disorders* (pp. 41–88). Academic Press. https://doi.org/10.1016/B978-0-443-19084-1.00009-0

Panza, F., Solfrizzi, V., Tortelli, R., Resta, F., Sabbà, C., & Logroscino, G. (2014). Prevention of Late-Life Cognitive Disorders: Diet-Related Factors, Dietary Patterns, and Frailty Models. *Current Nutrition Reports*, *3*(2), 110–129. https://doi.org/10.1007/s13668-014-0080-8

Polis, B., & Samson, A. O. (2024). Enhancing Cognitive Function in Older Adults: Dietary Approaches and Implications. *Frontiers in Nutrition*, *11*. https://doi.org/10.3389/fnut.2024.1286725

Prado, E. L., & Dewey, K. G. (2014). Nutrition and Brain Development in Early Life. *Nutrition Reviews*, *72*(4), 267–284. https://doi.org/10.1111/nure.12102

Prince, M., Bryce, R., Albanese, E., Wimo, A., Ribeiro, W., & Ferri, C. P. (2013). The Global Prevalence of Dementia: A Systematic Review and Metaanalysis. *Alzheimer's & Dementia: The Journal of the Alzheimer's Association*, *9*(1), 63–75. https://doi.org/10.1016/j.jalz.2012.11.007

Ripoli, C., Spinelli, M., Natale, F., Fusco, S., & Grassi, C. (2020). Glucose Overload Inhibits Glutamatergic Synaptic Transmission: A Novel Role for CREB-Mediated Regulation of Synaptotagmins 2 and 4. *Frontiers in Cell and Developmental Biology*, *8*. https://doi.org/10.3389/fcell.2020.00810

Saha, S. K., Lee, S. B., Won, J., Choi, H. Y., Kim, K., Yang, G.-M., Dayem, A. A., & Cho, S. (2017). Correlation between Oxidative Stress, Nutrition, and Cancer Initiation. *International Journal of Molecular Sciences*, *18*(7), Article 7. https://doi.org/10.3390/ijms18071544

Saunders, J., & Smith, T. (2010). Malnutrition: Causes and Consequences. *Clinical Medicine*, *10*(6), 624–627. https://doi.org/10.7861/clinmedicine.10-6-624

Shea, T. B., Rogers, E., & Remington, R. (2015). Chapter 77 – Nutritional Approaches to Mitigating Cognitive Decline and Maintaining Function in Alzheimer's Disease: Transgenic Mouse Models, Combinatorial Approaches, and Dietary Supplementation with Folate and S-Adenosyl Methionine. In C. R. Martin & V. R. Preedy (Eds.), *Diet and Nutrition in Dementia and Cognitive Decline* (pp. 837–845). Academic Press. https://doi.org/10.1016/B978-0-12-407824-6.00077-X

Singh, S., Awasthi, S., Kumar, D., Sarraf, S. R., Pandey, A. K., Agarwal, G. G., Awasthi, A., Mathew, J. L., Kar, S., Nair, S., Rao, C. R., Pande, H., Mahanta, B. N., Bharti, B., Singh, C. M., Singh, K., Bhat, M. A., & Mahdi, A. A. (2023). Micronutrients and Cognitive Functions among Urban School-Going Children and Adolescents: A Cross-Sectional Multicentric Study from India. *PLoS ONE*, *18*(2), e0281247. https://doi.org/10.1371/journal.pone.0281247

Solleiro-Villavicencio, H., & Rivas-Arancibia, S. (2018). Effect of Chronic Oxidative Stress on Neuroinflammatory Response Mediated by CD4+T Cells in Neurodegenerative Diseases. *Frontiers in Cellular Neuroscience*, *12*. https://doi.org/10.3389/fncel.2018.00114

Swaminathan, S., Edward, B. S., & Kurpad, A. V. (2013). Micronutrient Deficiency and Cognitive and Physical Performance in Indian Children. *European Journal of Clinical Nutrition*, *67*(5), 467–474. https://doi.org/10.1038/ejcn.2013.14

Szutowicz, A., Bielarczyk, H., Jankowska-Kulawy, A., Pawełczyk, T., & Ronowska, A. (2013). Acetyl-CoA the Key Factor for Survival or Death of Cholinergic Neurons in Course of Neurodegenerative Diseases. *Neurochemical Research*, *38*(8), 1523–1542. https://doi.org/10.1007/s11064-013-1060-x

Verlaan, S., Ligthart-Melis, G. C., Wijers, S. L. J., Cederholm, T., Maier, A. B., & de van der Schueren, M. A. E. (2017). High Prevalence of Physical Frailty among Community-Dwelling Malnourished Older Adults – A Systematic Review and Meta-Analysis. *Journal of the American Medical Directors Association*, *18*(5), 374–382. https://doi.org/10.1016/j.jamda.2016.12.074

Wrigley, S., Arafa, D., & Tropea, D. (2017). Insulin-Like Growth Factor 1: At the Crossroads of Brain Development and Aging. *Frontiers in Cellular Neuroscience*, *11*. https://doi.org/10.3389/fncel.2017.00014

15 Genetic Factors, Molecular Mechanisms, and Therapeutic Advances in ALS

Surya Prakash Pandey, Kannan Badri Narayanan, Krishna Kumar Singh, Sung Soo Han, and Rakesh Bhaskar

15.1 INTRODUCTION

ALS is a progressive neurodegenerative condition in which the upper and lower motor neurons degenerate [1]. Motor neurons carry impulses from the spinal cord and brain to muscles throughout the body [2, 3]. Muscle weakness, atrophy, and eventually complete paralysis set in as ALS progresses in ALS patients. The etiology of ALS is multi-factorial with a complex interplay between genetic, molecular, and environmental factors (Figure 15.1). Most ALS occurs sporadically, but an extremely high percentage has a genetic etiology [3]. Clinical heterogeneity, rate of progression, and molecular pathology have all impeded understanding of the basic mechanisms. Recent advances in genomics, molecular biology, and neuroimaging have improved our ability to distinguish these mechanisms and identify potential targets for intervention. ALS is divided into two broad categories: Familial ALS (fALS), accounting for approximately 5%–10% of all cases, and sporadic ALS (sALS), accounting for the remaining 90%–95% [4, 5]. fALS usually manifests an autosomal dominant mode of inheritance, although autosomal recessive and X-linked modes have been encountered [6]. Conversely, sALS is encountered in patients without a family history. Few genes identified in fALS have also been found in sporadic ALS patients, suggesting that the same molecular mechanisms could be at work in both. Genetic of ALS SOD1 encodes for superoxide dismutase 1 (SOD1), a degradative enzyme for reactive oxygen species (ROS). Mutant SOD1 proteins are prone to misfolding and aggregation that can disrupt cellular homeostasis and lead to motor neuron cell death. Other genes of significance for ALS are C9orf72, TAR DNA-binding protein (TARDBP), and fused-in sarcoma (FUS), each producing different but often interplay pathogenic pathways [7].

In addition to these known genes, there are some other genes linked to ALS, including angiogenin (ANG), valosin-containing protein (VCP), ubiquilin-2 (UBQLN2), Optineurin (OPTN), and TANK-binding kinase 1 (TBK1). All these genes are involved in protein homeostasis, autophagy, vesicle trafficking, and inflammation [7, 8]. Modifiable lifestyle and environmental factors also contribute to ALS susceptibility. These include toxins (pesticides, heavy metals, etc.), traumatic brain injury, strenuous physical activity, and military service. Although they are supported by epidemiologic data, the etiologic role of these factors in either disease onset or progression is at present not very well understood. Environmental exposures are likely to interact with genetic susceptibility to drive the disease process. At the molecular level, ALS is characterized by the convergence of myriad neurodegenerative pathways. In particular, SOD1, TDP-43, and FUS misfolding and protein aggregation induce disruption of cellular compartments and impairment of the normal protein quality control system [9, 10]. Mitochondrial dysfunction is another key feature, resulting in energy depletion, increased ROS generation, and activation of the intrinsic apoptotic pathway [11]. Impaired axonal transport, a process necessary to maintain long-distance neuronal health, is often observed in ALS and can be due to defects in the cytoskeleton or motor proteins. Excitotoxicity due to hyperactive glutamatergic signaling is another major reason [12, 13]. Motor neurons are

DOI: 10.1201/9781003500025-15

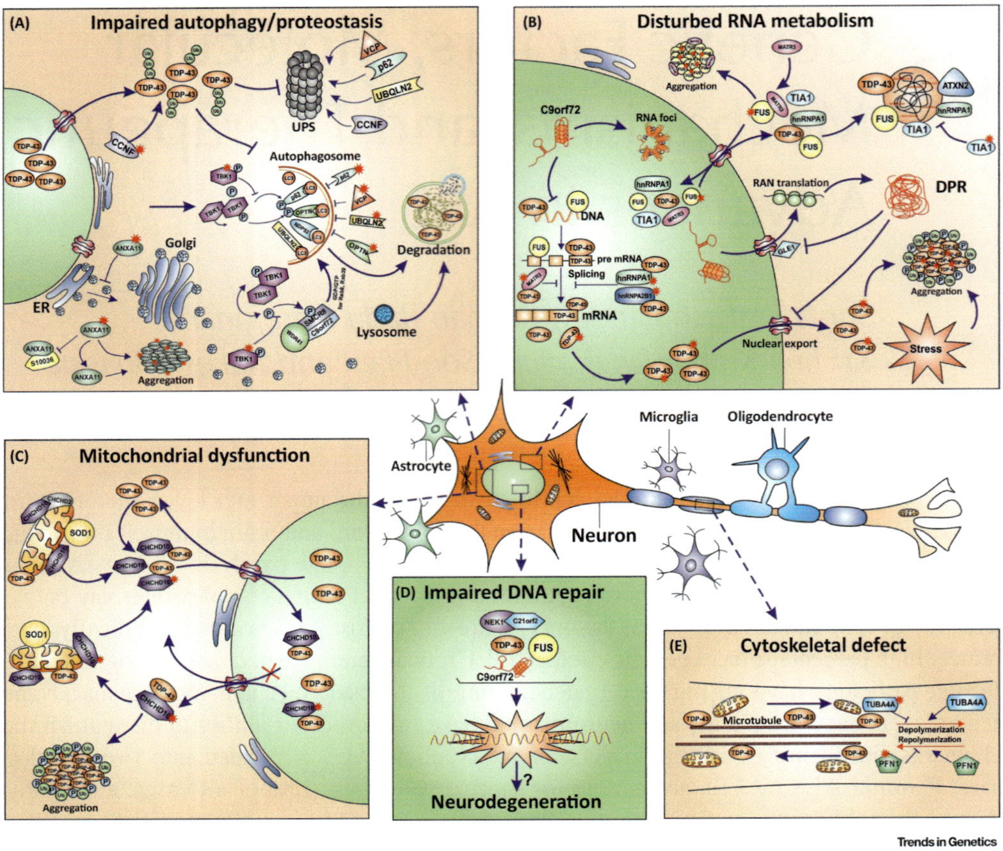

FIGURE 15.1 This schematic illustrates the diverse cellular and molecular mechanisms contributing to ALS pathogenesis: (A) Impaired Autophagy/Proteostasis: Defects in the ubiquitin-proteasome system (UPS) and autophagy disrupt protein degradation, leading to accumulation of misfolded proteins, such as TDP-43, FUS, and SOD1. Mutations in genes like CCNF, UBQLN2, OPTN, TBK1, VCP, and p62 impair proteostasis, resulting in cytoplasmic aggregation and endoplasmic reticulum (ER) stress. (B) Disturbed RNA Metabolism: Abnormalities in RNA-binding proteins (TDP-43, FUS, hnRNPA1, hnRNPA2/B1, MATR3) disrupt RNA splicing, transport, and translation. C9orf72 repeat expansions produce RNA foci and dipeptide repeat proteins (DPRs) through RAN translation, contributing to cellular stress and protein aggregation. (C) Mitochondrial Dysfunction: Mutant SOD1 and mislocalized TDP-43 interfere with mitochondrial dynamics, causing oxidative stress, impaired energy production, and apoptosis. Disrupted interactions with mitochondrial proteins (e.g., CHCHD10, VDAC1) compromise neuronal viability. (D) Impaired DNA Repair: DNA damage repair pathways are compromised in ALS, with involvement of FUS, TDP-43, C9orf72, NEK1, and C21orf2. These impairments contribute to genome instability and may exacerbate neurodegeneration. (E) Cytoskeletal Defects: Mutations in cytoskeletal regulators (TUBA4A, PFN1) impair microtubule dynamics and axonal transport. TDP-43 aggregation further disrupts cytoskeletal integrity, leading to defective neuronal structure and function. Together, these interconnected mechanisms culminate in progressive motor neuron degeneration, underscoring the complex etiology of ALS and the need for multifaceted therapeutic strategies. Reproduced with open access permission [54].

particularly vulnerable to the toxic effects of glutamate due to high metabolic demands and limited capacity for maintaining calcium homeostasis. This creates intracellular calcium overload, mitochondrial damage, and protease and nuclease activation leading to the destruction of vital cellular structures. Neuroinflammation, as indicated by reactive astrocytes and activated microglia, further contributes to neuronal damage by releasing ROS and inflammatory cytokines [14].

Unlike the complex and multifaceted pathophysiology of ALS, only therapeutic interventions are just beginning to be identified few FDA-approved drugs such as edaravone and riluzole addressed glutamatergic transmission and oxidative stress [15]. These therapies are minimally effective highlighting the need for more targeted and personalized treatments. Emerging advances in gene therapy have offered encouraging new opportunities. Antisense oligonucleotides (ASOs) targeted against mutant SOD1 and C9orf72 transcripts have yielded encouraging findings with preclinical models and early clinical trials [16, 17]. Cell-based therapies, including neural stem cell or mesenchymal stem cell transplantation, are also being explored. The cells may exert neuroprotective effects by the release of growth factors, modulation of the immune system, or as replacements for dead neurons [18]. Immunomodulatory treatments targeting microglial activation and neuroinflammation are in various stages of development and represent another potential strategy to modulate disease progression. ALS is a disabling neurodegenerative disorder with etiology multifactorial genetic and molecular. Efforts to delineate mechanisms of motor neuron degeneration and to apply these findings to new therapeutic strategies continue. With advancing knowledge, precision medicine approaches will ultimately transform ALS treatment and improve outcomes for patients.

15.2 GENETIC FACTORS IN ALS

ALS is a late-onset progressive neurodegenerative disorder that affects the motor neurons of the brain and spinal cord, leading to weakness of the muscles, and eventual respiratory failure. ALS is most commonly classified into two groups based on the mode of inheritance: fALS, which accounts for approximately 5%–10% of all cases, and sporadic ALS (sALS), which accounts for the remaining 90%–95% [4]. Whereas the majority of instances of ALS are sporadic, genetic studies have revealed significant overlap between the molecular mechanisms of the two, suggesting that there is bound to be substantial genetic participation within the overall spectrum of ALS. fALS is usually autosomal dominant but can also appear autosomal recessive or X-linked [6, 19]. The discovery of specific genetic mutations in fALS has not only improved the specificity of diagnosis but has also provided valuable information on the overall pathophysiology of ALS [20, 21]. In contrast, sporadic ALS does not have any specific genetic pattern, yet there is still a genetic predisposition. A number of the genes described to date in fALS also have been discovered in sporadic ALS, suggesting that such forms of ALS may not be completely separate diseases but are on a continuum controlled by both genetic and environmental determinants. One of the important ALS genes is SOD1, which encodes the enzyme superoxide dismutase 1 [22]. The enzyme is required for the detoxification of superoxide radicals, a type of reactive oxygen. Mutations in SOD1 were described for the first time in 1993 and are responsible for approximately 20% of inherited ALS and 2% of sporadic ALS. More than 150 SOD1 mutations have been reported, which have had a range of effects on disease course and severity. A second major ALS-predisposing gene is the C9orf72 gene [22, 23]. The most common mutation in this gene is an abnormal expansion of a hexanucleotide repeat (GGGGCC) in the non-coding portion. This expansion is seen in 40%–50% of fALS and approximately 7%–10% of sporadic ALS in European populations. Pathogenesis involves both loss of wild-type C9orf72 protein function as well as gain of toxic function by RNA foci and dipeptide repeat protein generated by repeat-associated non-ATG translocation [8, 24]. This mutation also links ALS with frontotemporal dementia and puts on display the shared molecular and clinical spectrum of the two conditions.

TARDBP is a second gene associated with ALS [25]. The gene encodes TDP-43, a DNA/RNA binding protein involved in a range of RNA processing functions, such as splicing, transport, and translation. TARDBP mutations are relatively rare but have been identified both in familial and sporadic ALS [25, 26]. Pathological cytoplasmic TDP-43 protein aggregation is a common event in ALS which is found in more than 95% of the cases regardless of genetic etiology. Mislocalization and TDP-43 aggregation result in the loss of its normal nuclear function and increased toxic function in the cytoplasm, resulting in motor neuron dysfunction [27, 28]. A second gene, FUS, encodes an RNA-binding protein which is involved in the same RNA metabolic process as TDP-43.

FUS mutations result in mislocalization to the cytoplasm from the nucleus, and the formation of aggregates [29]. The aggregates are neurotoxic and result in interference with normal cell function. FUS mutations result in approximately 5% of fALS and less than 1% of sporadic ALS. Clinically, FUS-associated ALS is earlier in age of onset and more rapidly progressive compared to other forms. The functional and pathological overlap between FUS and TDP-43 highlights the importance of RNA processing in ALS pathogenesis [30, 31]. In addition to these core genes, many other genes are identified as having associations with ALS, each of which implicates further the complex nature of the disease [32]. The ANG gene encodes for angiogenin, a ribonuclease with cellular stress response functions. ANG gene mutations impact nuclear translocation and angiogenic activity, making the motor neurons vulnerable to degeneration [33]. VCP gene encodes a valosin-containing protein that functions in protein degradation and autophagy. VCP gene mutations impact these functions, causing the accumulation of misfolded proteins and increased cellular stress [34]. Another essential gene is OPTN, which encodes OPTN, a protein that is involved in autophagy and inflammatory signaling. Mutations in OPTN have been shown to affect mitophagy, the specific degradation of dysfunctional mitochondria, leading to increased oxidative stress and neuronal damage [35]. UBQLN2 encodes for ubiquitin-2, which is implicated in the ubiquitin-proteasome pathway. Mutations within UBQLN2 can impair the breakdown of proteins, leading to toxic protein misfolding accumulation [36]. TBK1, also has a function in regulating autophagy and innate immunity. TBK1 loss-of-function mutations disrupt the autophagy pathway and increase neuroinflammation, both of which contribute to ALS pathogenesis [37, 38]. TBK1, for example, is a kinase that regulates innate immune signaling and autophagy. Loss-of-function mutations in TBK1 are associated with increased neuroinflammation and impaired clearance of misfolded proteins, characteristics of ALS pathology [38].

Despite the heterogeneity of ALS genes, several unifying pathogenic themes have been identified. Possibly the most significant is the perturbation of RNA metabolism. TARDBP, FUS, and C9orf72 are all genes essential to RNA processing, and their malfunction results in pervasive transcriptional and post-transcriptional aberration. Findings of these genetic determinants are of important clinical significance. Though fALS is rare, knowledge of gene mutations in familial patients has taken our general knowledge about the disease to a new level. Among them are SOD1, C9orf72, TARDBP and FUS, and they have put RNA metabolism, proteostasis, autophagy and neuroinflammation in the frontline in causing motor neuron degeneration. With further research, it is hoped that an increased comprehension of these hereditary mechanisms will make effective therapy possible that may reverse the course of this currently incurable disease.

15.3 MOLECULAR MECHANISMS OF ALS

15.3.1 Molecular Mechanisms of ALS: Protein Aggregation

One of the most important molecular mechanisms in ALS is the abnormal aggregation of misfolded proteins in motor neurons. Protein aggregation is one of the hallmarks of most neurodegenerative diseases, and in ALS, it is critical for triggering and propagating neuronal pathology [39]. In sporadic and familial cases of the disease, proteins such as superoxide dismutase 1 (SOD1), TAR DNA-binding protein 43 (TDP-43), and FUS undergo structural changes that result in them being abnormally misfolded and aggregating in the cytoplasm or other cellular structures [40]. SOD1 was the first ALS-linked protein to be identified as aggregating in neurons. In ideal situations, SOD1 is an antioxidant enzyme that detoxifies cells by catalyzing the dismutation of superoxide radicals and protecting them from oxidative damage. However, mutations in the gene for SOD1 blow up the native stability of the protein, leading it to misfold and aggregate. Not merely accumulated trash, these aggregates disrupt cellular homeostasis [41]. They are capable of interfering with mitochondrial function, interrupting axonal transport, and inhibiting proteasomal degradation pathways, inducing motor neuron degeneration in a toxic gain-of-function mechanism. Another important protein in ALS pathology is TDP-43. Although familial forms caused by mutations of its gene, TARDBP, represent only a small proportion of ALS, cytoplasmic inclusions of TDP-43 are seen in virtually all familial and sporadic ALS cases [40, 42]. TDP-43 is normally a nuclear protein that plays a function in RNA metabolism, including splicing, transport, and regulation of translation (Figure 15.1A). In ALS, however, TDP-43 becomes hyperphosphorylated, ubiquitinated, and mislocalized from the nucleus to the cytoplasm, where it accumulates as insoluble material. Mislocalization results in the loss of normal nuclear function of TDP-43 and a toxic gain of function in the cytoplasm. The aggregates can sequester other RNA-binding proteins and RNA, further interfering with gene expression and protein synthesis [43].

FUS is a second ALS-associated RNA-binding protein that is both structurally and functionally similar to TDP-43. Mutations in the FUS gene have been shown to relocate this protein from the nucleus to the cytoplasm, where it aggregates in motor neurons [44, 45]. Similar to TDP-43, FUS plays a critical role in RNA metabolism and DNA repair, and mislocalization results in comparable loss of function and gain of toxicity. FUS aggregates possess the capability to disrupt stress granule dynamics, affect transcriptional regulation, and abolish the ability of the cell to recover from stress and DNA damage [46, 47]. The protein aggregates are not only a consequence of disease but rather a causative force in a pathologic cascade. The aggregates disrupt the cytoskeleton, inhibit autophagy, undermine synaptic function, and ultimately induce apoptosis in motor neurons. The ability of the cell to degrade misfolded proteins through quality control mechanisms such as the ubiquitin-proteasome pathway or autophagy is typically burdened or directly impaired by the presence of such aggregates [47]. As a result, cellular stress accumulates with a self-reinforcing cycle of protein misfolding and aggregation. Protein aggregation of SOD1, TDP-43, and FUS is the pervasive molecular characteristic of ALS [46, 48]. These aggregates interfere with a number of cellular functions like RNA metabolism, protein degradation, and mitochondrial function, ultimately leading to the death of motor neurons. Understanding how these clumps form and spread is important to understand ALS pathogenesis and offers promising avenues to devise therapies to stabilize proteins, increase mechanisms of degradation or prevent aggregation in the first place.

15.3.2 Molecular Mechanisms of ALS: RNA Dysregulation

RNA dysregulation is the most prevalent pathogenic process in ALS, particularly if RNA-binding protein mutations such as TDP-43 and FUS are involved [44]. These proteins play a crucial role in post-transcriptional regulation of gene expression, RNA processing, splicing, transport, localization, and translation [45, 47]. Mutations or mislocalization disrupt the precise control of RNA metabolism to cause an array of cellular dysfunctions and loss of motor neurons. TDP-43, coded by the TARDBP gene, is a multifunctional RNA-binding protein with the majority of its localization being nuclear and where it plays roles in regulating alternative splicing, RNA stability, and transcription. In ALS, TDP-43 is found in its abnormally aggregated form in the cytoplasm and is associated with its loss from the nucleus. Nuclear loss accounts for splicing abnormalities and inappropriate expression of important neuronal genes. At the same time, cytoplasmic aggregation of TDP-43 also disrupts RNA metabolism by accumulating into poisonous inclusion bodies that sequester RNA and other RNA-binding proteins (Figure 15.1B). Similarly, gene mutations in the FUS gene cause the FUS protein to abnormally mislocalize from the nucleus into the cytoplasm [46]. FUS normally regulates RNA splicing, microRNA processing, and transcriptional regulation. Following mislocation, FUS aggregates interfere with the normal functioning of the RNA-protein complex and change the profile of neural RNA. The changes result in the expression of synaptic and cytoskeletal proteins that are vital to neuronal function and integrity. Other than TDP-43 and FUS, other RNA-binding proteins such as hnRNPA1, hnRNPA2/B1, and MATR3 have also been found to be implicated in ALS [49]. These proteins are involved in RNA transport and splicing, and mutations in all of them result in pathological aggregate formation and regulation of RNA dysregulation. This dysfunction further promotes the disease process by causing defects in axonal transport, neuronal mRNA localization, and translation [50]. C9orf72 repeat expansion mutations also result in RNA dysregulation but through a different mechanism. Longer GGGGCC repeats in RNA transcripts can fold into stable secondary structures that entrap RNA-binding proteins in nuclear RNA foci [51]. This segregation alters the accessibility and function of the proteins, thus altering the normal processing and transport of other RNAs [52]. Additionally, repeat-associated non-AUG translation of the repeats generates dipeptide repeat proteins that put additional pressure on the RNA regulatory machinery [39, 46]. Abnormal RNA metabolism has several downstream effects on motor neurons. Mis-splicing can result in the production of aberrant proteins or depletion of necessary isoforms. RNA dysregulation is a common molecular mechanism for ALS caused by mutations and mislocalization of core RNA-binding proteins. These alterations disrupt important RNA processing events such as splicing, transport, and translation and ultimately disrupt gene expression programs essential for motor neuron survival [52, 53]. Further research into such mechanisms may lead to the development of targeted therapies to restore RNA homeostasis and prevent neurodegeneration.

15.3.3 Molecular Mechanisms of ALS: Mitochondrial Dysfunction, Impaired Axonal Transport, Neuroinflammation, and Excitotoxicity

Mitochondrial dysfunction is a central feature of ALS pathogenesis (Figure 15.1D). Mitochondria play a crucial role in energy production, calcium buffering, and regulation of apoptosis [55]. Familial and sporadic ALS share mitochondrial dysfunction in motor neurons in the form of edema, cristae damage, and respiratory chain impairment [56]. For example, mutant SOD1 is localized in the mitochondrial intermembrane space and causes oxidative damage, disruption in calcium homeostasis, and impaired ATP synthesis [55, 57]. In addition, mitochondrial DNA damage and dysfunctional mitophagy increase cellular stress [58]. The combined effects of these dysfunctions are energy depletion that threatens neuronal maintenance and survival, heightened production of ROS that is damaging to proteins and lipids, and activation of intrinsic apoptotic pathways that lead to the death of neurons [56, 57]. Axonal transport dysfunctions are a major factor that leads to ALS pathogenesis. Motor neurons rely on efficient transport mechanisms to transport organelles, proteins, and RNA along their long axons. This movement is mediated by molecular motor proteins such as kinesin and dynein, which carry cargo on microtubules. It has been shown that mutations in SOD1, FUS, and TDP-43 genes interfere with this mechanism of transport in ALS, resulting in cellular material accumulation within the cell body and distal axonal and synaptic terminal degradation [59]. Impaired axonal transport impairs the delivery of essential components required for synaptic function, axonal integrity, and neuronal repair. Neuroinflammation plays a dual role in ALS, initially protective but ultimately playing a part in neuronal damage [59, 60]. Activated microglia, resident immune cells of the central nervous system, are commonly found in locations of ALS patients and can potentially secrete both neuroprotective and neurotoxic mediators. Activated microglia secrete inflammatory cytokines, ROS, and nitric oxide when they are chronically activated and may thereby augment oxidative stress and neuronal damage [61, 62]. Astrocytes, which are responsible for supporting neuronal metabolism and extracellular homeostasis, also become abnormally altered in ALS. These glial responses create a neurotoxic environment that facilitates faster motor neuron degeneration. In some ALS models, ALS patients' astrocytes have been shown to induce motor neuron death through non-cell-autonomous mechanisms, further indicating glial dysfunction plays a pivotal role in the progression of the disease [59, 60].

Excitotoxicity, particularly glutamate, is another confirmed mechanism in ALS [63]. Glutamate is the primary excitatory neurotransmitter of the central nervous system, and its concentration is tightly regulated to prevent overstimulation of neurons. In ALS, extracellular glutamate concentrations are increased by impaired glutamate reuptake in astrocytes, particularly by EAAT2 transporter downregulation [63, 64]. Overactivated AMPA and NMDA receptors by excessive glutamate prevent calcium influx in motor neurons. Excessive intracellular calcium triggers enzymes like proteases and lipases that destroy cell structure and induce apoptosis [65]. Motor neurons are particularly susceptible to excitotoxicity due to their high metabolism, long axons, and weak calcium buffering capacity. This vulnerability increases the pathological effect of glutamate dysregulation and worsens the disease course. ALS emerges from an intersection of interdependent molecular pathways that include mitochondrial injury, axonal transport disorders, chronic neuroinflammation, and excitotoxicity [66]. All are pathogenic to cellular homeostasis, energy balance, and neural connectivity and cause motor neuron degeneration. These mechanisms must be uncovered to produce directed therapies aimed at the preservation of motor neuron function and the interruption of disease progression.

15.4 THERAPEUTIC ADVANCES

Despite a progressive neurodegenerative disease, ALS has made remarkable progress in the design of therapeutics over the past few decades. These approaches are not just aimed at retarding disease progression but also at modulating the molecular pathways implicated in motor neuron degeneration [67]. The present therapeutics involve pharmacological intervention, gene therapy, cell-based therapies, immunomodulation, modulation of protein homeostasis, and gene editing approaches [68]. Pharmacological therapies have been the gold standard for ALS treatment, and the most used and approved medications in the clinic are riluzole and edaravone [69]. Riluzole is a glutamatergic antagonist that reduces excitotoxicity, one of the key mechanisms of neuronal damage in ALS [67]. Riluzole can prolong survival by several months and is well tolerated by reducing the presynaptic release of glutamate and regulating voltage-gated sodium channels [70, 71]. Edaravone is an intravenous antioxidant that acts by scavenging ROS and reducing oxidative stress [72]. They have been shown in clinical trials to slow the rate of functional decline, especially in some ALS patients at the onset of the disease. Gene therapy represents an important advance in precision medicine for ALS. ASOs are a promising targeted approach for silencing disease-causing genes. The synthetic nucleotide strands bind to target mRNA sequences, either degrading them or blocking their translation into protein [73, 74]. ASOs have been developed to target SOD1, one of the first ALS genes identified, and have in clinical trials demonstrated the capacity to reduce mutant SOD1 protein levels and slow disease. Similarly, ASOs developed to target the C9orf72 repeat expansion aim to reduce the production of toxic RNA lesions and dipeptide repeat proteins [75, 76]. These approaches are particularly encouraging because they target the disease mechanism in genetically defined subgroups of patients with ALS [75, 77]. Cell-based therapies are also being explored as a means of replacing damaged motor neurons or providing trophic support to surviving neurons [78]. A variety of types of stem cells, including neural progenitor cells and mesenchymal stem cells, have been examined in preclinical and early clinical studies [79, 80]. They can secrete neurotrophic factors, immunomodulate the milieu, and induce neuroprotection. Replacement of functional motor neurons remains a considerable challenge, yet ongoing research is aimed at addressing delivery, cell survival, and safety [78, 80]. Immunomodulatory therapy targets the inflammatory milieu that results in motor neuron damage. ALS is characterized by chronic microglial activation and dysfunctional astrocytes that both release inflammatory cytokines and toxic substances. Therapeutics targeting glial function or mitigating harmful immune responses are being explored. Experimental therapeutics with promise include monoclonal antibodies, small molecules, and gene therapies that target reducing neuroinflammation and restoring a more supportive glial environment [69, 81].

Given the central role of protein aggregation in ALS, molecular chaperones and protein homeostasis regulators are being developed to enhance the clearance of misfolded and aggregated proteins [82, 83]. Strategies that improve autophagy and the ubiquitin-proteasome system aim to reduce the toxic burden of misfolded proteins, such as SOD1, TDP-43, and FUS [83, 84]. Pharmacological chaperones may also help stabilize protein folding and prevent aggregation at an early stage. These approaches may be particularly effective when combined with gene therapy to reduce the production of mutant proteins [82, 83]. Emerging strategies offer exciting possibilities for future ALS treatments.

One of the most promising areas is the application of gene editing technologies, such as CRISPR-Cas9 [85]. This powerful tool allows for the precise correction of disease-causing mutations at the DNA level. In preclinical studies, CRISPR has been used to successfully edit the SOD1 gene in mouse models, delaying disease onset and increasing survival [85]. Although challenges such as delivery and off-target effects remain, the potential for long-term correction of genetic defects is significant [86, 87]. Other new approaches include neuroprotective agents that target downstream mechanisms of neuronal damage. These small molecules aim to stabilize mitochondria, reduce excitotoxicity, increase calcium buffering, and protect against oxidative damage. Some agents focus on improving axonal transport or enhancing synaptic function. These treatments cannot reverse the genetic cause of ALS, but when combined with other treatments, they can provide symptom relief and slow progression. The development of ALS treatments is advancing rapidly with various strategies targeting different aspects of the disease process [88]. From pharmacological agents that reduce excitotoxicity and oxidative stress, to gene silencing techniques and gene editing, to cell-based and immunomodulatory therapies [89–92], each approach is contributing to the development of treatments for ALS. While no single treatment can yet stop or reverse disease progression, the integration of these strategies offers new hope for more effective and personalized therapies for ALS patients.

15.5 CONCLUSION

ALS is a multifactorial neurodegenerative disorder affected by a complicated interaction of genetic mutations, molecular dysfunctions, and cellular pathologies. Identification of key genes such as SOD1, C9orf72, TARDBP, and FUS has provided clues to crucial mechanisms in ALS, including protein misfolding, RNA dysregulation, mitochondrial dysfunction, disrupted axonal transport, neuroinflammation, and excitotoxicity. Such pathogenic processes are not monolithic but intersect to compromise motor neuron function and integrity, leading to progressive paralysis and death. While pharmacological treatments are currently of limited benefit, a great deal of progress has been made in the development of targeted therapies. Gene-silencing approaches, stem cell therapy, immunomodulation, and proteostasis regulation are being explored with increasing specificity and sophistication. Emerging technologies like CRISPR-Cas9 hold the potential to correct disease-causing mutations at their source. The intersection of these therapeutic approaches, founded on a deeper molecular understanding of ALS, is a step toward personalized medicine and holds promise for more effective interventions. Continued multidisciplinary research, combined with translation of findings into the clinic, will be key to altering ALS prognosis and improving the quality of life for those affected.

FUNDING STATEMENT

This research was supported by the National Research Foundation of Korea (NRF) (Grant No. RS-2023-00278268).

CONFLICT OF INTEREST

Authors declare no conflict of interest.

CREDIT AUTHORSHIP CONTRIBUTION STATEMENT

Surya Prakash Pandey and Rakesh Bhaskar: Methodology, Investigation, Visualization, Validation, Writing-Original draft preparation, Reviewing, and Editing. Kannan Badri Narayanan: Investigation, Supervision, Writing-Original draft preparation. Sung Soo Han: Visualization, Investigation, Resources, Reviewing, and Editing.

REFERENCES

1. P. Rojas, A.I. Ramírez, J.A. Fernández-Albarral, I. López-Cuenca, E. Salobrar-García, M. Cadena, L. Elvira-Hurtado, J.J. Salazar, R. de Hoz, J.M. Ramírez, Amyotrophic lateral sclerosis: A neurodegenerative motor neuron disease with ocular involvement, Front. Neurosci. 14 (2020). https://doi.org/10.3389/fnins.2020.566858
2. E. Tiryaki, H.A. Horak, ALS and other motor neuron diseases, Contin. Lifelong Learn. Neurol. 20 (2014) 1185–1207.
3. L.C. Wijesekera, P. Nigel Leigh, Amyotrophic lateral sclerosis, Orphanet J. Rare Dis. 4 (2009) 1–22.
4. V. McGuire, L.M. Nelson, Epidemiology of ALS, Neurol. Dis. Ther. 78 (2006) 17.
5. Z.-Y. Zou, Z.-R. Zhou, C.-H. Che, C.-Y. Liu, R.-L. He, H.-P. Huang, Genetic epidemiology of amyotrophic lateral sclerosis: A systematic review and meta-analysis, J. Neurol. Neurosurg. Psychiatry. 88 (2017) 540–549.
6. K. Boylan, Familial ALS, Neurol. Clin. 33 (2015) 807.
7. R. Balendra, A.M. Isaacs, C9orf72-mediated ALS and FTD: Multiple pathways to disease, Nat. Rev. Neurol. 14 (2018) 544–558.
8. J. Beckers, A.K. Tharkeshwar, P. Van Damme, C9orf72 ALS-FTD: Recent evidence for dysregulation of the autophagy-lysosome pathway at multiple levels, Autophagy. 17 (2021) 3306–3322.
9. R.L. Jirtle, M.K. Skinner, Environmental epigenomics and disease susceptibility, Nat. Rev. Genet. 8 (2007) 253–262.
10. D.J. Hunter, Gene–environment interactions in human diseases, Nat. Rev. Genet. 6 (2005) 287–298.
11. H. Cui, Y. Kong, H. Zhang, Oxidative stress, mitochondrial dysfunction, and aging, J. Signal Transduct. 2012 (2012) 646354.
12. R. Crabé, F. Aimond, P. Gosset, F. Scamps, C. Raoul, How degeneration of cells surrounding motoneurons contributes to amyotrophic lateral sclerosis, Cells. 9 (2020) 2550.
13. S. Raffaele, M. Boccazzi, M. Fumagalli, Oligodendrocyte dysfunction in amyotrophic lateral sclerosis: Mechanisms and therapeutic perspectives, Cells. 10 (2021) 565.
14. S. Zarei, K. Carr, L. Reiley, K. Diaz, O. Guerra, P.F. Altamirano, W. Pagani, D. Lodin, G. Orozco, A. Chinea, A comprehensive review of amyotrophic lateral sclerosis, Surg. Neurol. Int. 6 (2015) 171.
15. A.G. Al-kaf, A.A. Al-yahawi, Current Drugs Strategies for Treatment of Rare Neurodegenerative Diseases, in: Rare Neurodegener. Disord. Insights, IntechOpen, 2024.
16. R. Romano, C. Bucci, Antisense therapy: A potential breakthrough in the treatment of neurodegenerative diseases, Neural Regen. Res. 19 (2024) 1027–1035.
17. M. Cappella, P.-F. Pradat, G. Querin, M.G. Biferi, Beyond the traditional clinical trials for amyotrophic lateral sclerosis and the future impact of gene therapy, J. Neuromuscul. Dis. 8 (2021) 25–38.
18. M.A. Zayed, S. Sultan, H.O. Alsaab, S.M. Yousof, G.I. Alrefaei, N.H. Alsubhi, S. Alkarim, K.S. Al Ghamdi, S.A. Bagabir, A. Jana, Stem-cell-based therapy: The celestial weapon against neurological disorders, Cells. 11 (2022) 3476.
19. P.M. Andersen, A. Al-Chalabi, Clinical genetics of amyotrophic lateral sclerosis: What do we really know? Nat. Rev. Neurol. 7 (2011) 603–615.
20. J. van den Bos, N. Geevasinga, M. Higashihara, P. Menon, S. Vucic, Pathophysiology and diagnosis of ALS: Insights from advances in neurophysiological techniques, Int. J. Mol. Sci. 20 (2019) 2818.
21. A. Al-Chalabi, L.H. Van Den Berg, J. Veldink, Gene discovery in amyotrophic lateral sclerosis: Implications for clinical management, Nat. Rev. Neurol. 13 (2017) 96–104.
22. E. Abati, N. Bresolin, G. Comi, S. Corti, Silence superoxide dismutase 1 (SOD1): A promising therapeutic target for amyotrophic lateral sclerosis (ALS), Expert Opin. Ther. Targets. 24 (2020) 295–310.
23. S.J. Kaur, S.R. McKeown, S. Rashid, Mutant SOD1 mediated pathogenesis of amyotrophic lateral sclerosis, Gene. 577 (2016) 109–118.
24. S. Rossi, M. Cozzolino, M.T. Carrì, Old versus new mechanisms in the pathogenesis of ALS, Brain Pathol. 26 (2016) 276–286.
25. E. Kabashi, P.N. Valdmanis, P. Dion, D. Spiegelman, B.J. McConkey, C. Vande Velde, J.-P. Bouchard, L. Lacomblez, K. Pochigaeva, F. Salachas, TARDBP mutations in individuals with sporadic and familial amyotrophic lateral sclerosis, Nat. Genet. 40 (2008) 572–574.
26. R. Del Bo, S. Ghezzi, S. Corti, M. Pandolfo, M. Ranieri, D. Santoro, I. Ghione, A. Prelle, V. Orsetti, M. Mancuso, TARDBP (TDP-43) sequence analysis in patients with familial and sporadic ALS: Identification of two novel mutations, Eur. J. Neurol. 16 (2009) 727–732.
27. P. Corcia, P. Valdmanis, S. Millecamps, C. Lionnet, H. Blasco, K. Mouzat, H. Daoud, V. Belzil, R. Morales, N. Pageot, Phenotype and genotype analysis in amyotrophic lateral sclerosis with TARDBP gene mutations, Neurology. 78 (2012) 1519–1526.

28. V.M. Van Deerlin, J.B. Leverenz, L.M. Bekris, T.D. Bird, W. Yuan, L.B. Elman, D. Clay, E.M. Wood, A.S. Chen-Plotkin, M. Martinez-Lage, TARDBP mutations in amyotrophic lateral sclerosis with TDP-43 neuropathology: A genetic and histopathological analysis, Lancet Neurol. 7 (2008) 409–416.
29. S. Lattante, G.A. Rouleau, E. Kabashi, TARDBP and FUS mutations associated with amyotrophic lateral sclerosis: Summary and update, Hum. Mutat. 34 (2013) 812–826.
30. X. Xiao, M. Li, Z. Ye, X. He, J. Wei, Y. Zha, FUS gene mutation in amyotrophic lateral sclerosis: A new case report and systematic review, Amyotroph. Lateral Scler. Front. Degener. 25 (2024) 1–15.
31. J. Scekic-Zahirovic, H.E. Oussini, S. Mersmann, K. Drenner, M. Wagner, Y. Sun, K. Allmeroth, S. Dieterlé, J. Sinniger, S. Dirrig-Grosch, Motor neuron intrinsic and extrinsic mechanisms contribute to the pathogenesis of FUS-associated amyotrophic lateral sclerosis, Acta Neuropathol. 133 (2017) 887–906.
32. T.G. Moens, S. Da Cruz, M. Neumann, T.A. Shelkovnikova, N.A. Shneider, L. Van Den Bosch, Amyotrophic lateral sclerosis caused by FUS mutations: Advances with broad implications, Lancet Neurol. 24 (2025) 166–178.
33. Y. Zhang, Y. Li, S. Bin, X. Cheng, Q. Niu, A neglected gene: The role of the ANG gene in the pathogenesis of amyotrophic lateral sclerosis, Aging Dis. 16 (2024) 13.
34. S. Chu, X. Xie, C. Payan, U. Stochaj, Valosin containing protein (VCP): Initiator, modifier, and potential drug target for neurodegenerative diseases, Mol. Neurodegener. 18 (2023) 52.
35. S. Zhao, R. Chen, Y. Gao, Y. Lu, X. Bai, J. Zhang, Fundamental roles of the Optineurin gene in the molecular pathology of amyotrophic lateral sclerosis, Front. Neurosci. 17 (2023) 1319706.
36. E. Parobkova, R. Matej, Amyotrophic lateral sclerosis and frontotemporal lobar degenerations: Similarities in genetic background, Diagnostics. 11 (2021) 509.
37. J.A. Oakes, M.C. Davies, M.O. Collins, TBK1: A new player in ALS linking autophagy and neuroinflammation, Mol. Brain. 10 (2017) 1–10.
38. R.P. Toth, Investigating the role of TANK-binding kinase 1 in autophagy and amyotrophic lateral sclerosis (2022). https://doi.org/10.25949/19443248.v1
39. S. Lee, H.-J. Kim, Prion-like mechanism in amyotrophic lateral sclerosis: Are protein aggregates the key?, Exp. Neurobiol. 24 (2015) 1–7. https://doi.org/10.5607/en.2015.24.1.1
40. A.M. Blokhuis, E.J.N. Groen, M. Koppers, L.H. Van Den Berg, R.J. Pasterkamp, Protein aggregation in amyotrophic lateral sclerosis, Acta Neuropathol. 125 (2013) 777–794. https://doi.org/10.1007/s00401-013-1125-6
41. E. Duranti, C. Villa, Molecular investigations of protein aggregation in the pathogenesis of amyotrophic lateral sclerosis, Int. J. Mol. Sci. 24 (2022) 704.
42. E.M. Verde, V. Secco, A. Ghezzi, J. Mandrioli, S. Carra, Molecular mechanisms of protein aggregation in ALS-FTD: Focus on TDP-43 and cellular protective responses, Cells. 14 (2025). https://doi.org/10.3390/cells14100680
43. C. Colombrita, E. Onesto, F. Megiorni, A. Pizzuti, F.E. Baralle, E. Buratti, V. Silani, A. Ratti, TDP-43 and FUS RNA-binding proteins bind distinct sets of cytoplasmic messenger RNAs and differently regulate their post-transcriptional fate in motoneuron-like cells, J. Biol. Chem. 287 (2012) 15635–15647. https://doi.org/10.1074/jbc.M111.333450
44. J. Rummens, S. Da Cruz, RNA-binding proteins in ALS and FTD: From pathogenic mechanisms to therapeutic insights, Mol. Neurodegener. 20 (2025) 1–35. https://doi.org/10.1186/s13024-025-00851-y
45. E. Buratti, Trends in Understanding the Pathological Roles of TDP-43 and FUS Proteins BT – Frontotemporal Dementias: Emerging Milestones of the 21st Century, in: B. Ghetti, E. Buratti, B. Boeve, R. Rademakers (Eds.), Springer International Publishing, Cham, 2021: pp. 243–267. https://doi.org/10.1007/978-3-030-51140-1_15
46. A.D. Gitler, J. Shorter, RNA-binding proteins with prion-like domains in ALS and FTLD-U, Prion. 5 (2011) 179–187. https://doi.org/10.4161/pri.5.3.17230
47. A. Ratti, E. Buratti, Physiological functions and pathobiology of TDP-43 and FUS/TLS proteins, J. Neurochem. 138 (2016) 95–111. https://doi.org/10.1111/jnc.13625
48. K. Kapeli, G.A. Pratt, A.Q. Vu, K.R. Hutt, F.J. Martinez, B. Sundararaman, R. Batra, P. Freese, N.J. Lambert, S.C. Huelga, S.J. Chun, T.Y. Liang, J. Chang, J.P. Donohue, L. Shiue, J. Zhang, H. Zhu, F. Cambi, E. Kasarskis, S. Hoon, M. Ares, C.B. Burge, J. Ravits, F. Rigo, G.W. Yeo, Distinct and shared functions of ALS-associated proteins TDP-43, FUS and TAF15 revealed by multisystem analyses, Nat. Commun. 7 (2016). https://doi.org/10.1038/ncomms12143

49. J.A. Fifita, K.Y. Zhang, J. Galper, K.L. Williams, E.P. McCann, A.L. Hogan, N. Saunders, D. Bauer, I.S. Tarr, R. Pamphlett, G.A. Nicholson, D. Rowe, S. Yang, I.P. Blair, Genetic and pathological assessment of hnRNPA1, hnRNPA2/B1, and hnRNPA3 in familial and sporadic amyotrophic lateral sclerosis, Neurodegener. Dis. 17 (2017) 304–312. https://doi.org/10.1159/000481258
50. Y.C. Xue, C.S. Ng, P. Xiang, H. Liu, K. Zhang, Y. Mohamud, H. Luo, Dysregulation of RNA-binding proteins in amyotrophic lateral sclerosis, Front. Mol. Neurosci. 13 (2020) 1–11. https://doi.org/10.3389/fnmol.2020.00078
51. K. Xu, Y. Li, E.G. Allen, P. Jin, Therapeutic development for CGG repeat expansion-associated neurodegeneration, Front. Cell. Neurosci. 15 (2021) 1–8. https://doi.org/10.3389/fncel.2021.655568
52. W. Kim, D.Y. Kim, K.H. Lee, RNA-binding proteins and the complex pathophysiology of ALS, Int. J. Mol. Sci. 22 (2021) 1–18. https://doi.org/10.3390/ijms22052598
53. K. Kapeli, F.J. Martinez, G.W. Yeo, Genetic mutations in RNA-binding proteins and their roles in ALS, Hum. Genet. 136 (2017) 1193–1214. https://doi.org/10.1007/s00439-017-1830-7
54. H.P. Nguyen, C. Van Broeckhoven, J. van der Zee, ALS genes in the genomic era and their implications for FTD, Trends Genet. 34 (2018) 404–423. https://doi.org/10.1016/j.tig.2018.03.001
55. M. Cozzolino, M.T. Carrì, Mitochondrial dysfunction in ALS, Prog. Neurobiol. 97 (2012) 54–66. https://doi.org/10.1016/j.pneurobio.2011.06.003
56. F. Tafuri, D. Ronchi, F. Magri, G.P. Comi, S. Corti, SOD1 misplacing and mitochondrial dysfunction in amyotrophic lateral sclerosis pathogenesis, Front. Cell. Neurosci. 9 (2015) 1–12. https://doi.org/10.3389/fncel.2015.00336
57. I. Hervias, M.F. Beal, G. Manfredi, Mitochondrial dysfunction and amyotrophic lateral sclerosis, Muscle and Nerve. 33 (2006) 598–608. https://doi.org/10.1002/mus.20489
58. J.K. Sinha, K. Jorwal, K.K. Singh, S.S. Han, R. Bhaskar, S. Ghosh, The potential of mitochondrial therapeutics in the treatment of oxidative stress and inflammation in aging, Mol. Neurobiol. (2024). https://doi.org/10.1007/s12035-024-04474-0
59. N. D'Ambrosi, M. Cozzolino, M.T. Carrì, Neuroinflammation in amyotrophic lateral sclerosis: Role of redox (dys)Regulation, Antioxid. Redox Signal. 29 (2017) 15–36. https://doi.org/10.1089/ars.2017.7271
60. T. Philips, W. Robberecht, Neuroinflammation in amyotrophic lateral sclerosis: Role of glial activation in motor neuron disease, Lancet Neurol. 10 (2011) 253–263. https://doi.org/10.1016/S1474-4422(11)70015-1
61. C. Consilvio, A.M. Vincent, E.L. Feldman, Neuroinflammation, COX-2, and ALS – A dual role?, Exp. Neurol. 187 (2004) 1–10. https://doi.org/10.1016/j.expneurol.2003.12.009
62. W. Zhang, D. Xiao, Q. Mao, H. Xia, Role of neuroinflammation in neurodegeneration development, Signal Transduct. Target. Ther. 8 (2023). https://doi.org/10.1038/s41392-023-01486-5
63. P. van Cutsem, M. Dewil, W. Robberecht, L. van den Bosch, Excitotoxicity and amyotrophic lateral sclerosis, Neurodegener. Dis. 2 (2006) 147–159. https://doi.org/10.1159/000089620
64. A.C. Ludolph, T. Meyer, M.W. Riepe, The role of excitotoxicity in ALS – What is the evidence?, J. Neurol. 247 (2000) I7–I16. https://doi.org/10.1007/s004150050552
65. P.R. Heath, P.J. Shaw, Update on the glutamatergic neurotransmitter system and the role of excitotoxicity in amyotrophic lateral sclerosis, Muscle and Nerve. 26 (2002) 438–458. https://doi.org/10.1002/mus.10186
66. L. Van Den Bosch, P. Van Damme, E. Bogaert, W. Robberecht, The role of excitotoxicity in the pathogenesis of amyotrophic lateral sclerosis, Biochim. Biophys. Acta – Mol. Basis Dis. 1762 (2006) 1068–1082. https://doi.org/10.1016/j.bbadis.2006.05.002
67. R.J. Mead, N. Shan, H.J. Reiser, F. Marshall, P.J. Shaw, Amyotrophic lateral sclerosis: A neurodegenerative disorder poised for successful therapeutic translation, Nat. Rev. Drug Discov. 22 (2023) 185–212. https://doi.org/10.1038/s41573-022-00612-2
68. L. Tzeplaeff, S. Wilfling, M.V. Requardt, M. Herdick, Current state and future directions in the therapy of ALS, Cells. 12 (2023). https://doi.org/10.3390/cells12111523
69. M. Bucchia, A. Ramirez, V. Parente, C. Simone, M. Nizzardo, F. Magri, S. Dametti, S. Corti, Therapeutic development in amyotrophic lateral sclerosis, Clin. Ther. 37 (2015) 668–680. https://doi.org/10.1016/j.clinthera.2014.12.020
70. K. Shibuya, R. Otani, Y.I. Suzuki, S. Kuwabara, M.C. Kiernan, Neuronal hyperexcitability and free radical toxicity in amyotrophic lateral sclerosis: Established and future targets, Pharmaceuticals. 15 (2022) 1–18. https://doi.org/10.3390/ph15040433
71. M. Bissaro, S. Moro, Rethinking to riluzole mechanism of action: The molecular link among protein kinase CK1δ activity, TDP-43 phosphorylation, and amyotrophic lateral sclerosis pharmacological treatment, Neural Regen. Res. 14 (2019) 2083–2085. https://doi.org/10.4103/1673-5374.262578

72. T. Masuda, M. Shimazawa, H. Hara, Retinal diseases associated with oxidative stress and the effects of a Free Radical Scavenger (Edaravone), Oxid. Med. Cell. Longev. 2017 (2017). https://doi.org/10.1155/2017/9208489
73. H. Kim, S. Kim, D. Lee, D. Lee, J. Yoon, H. Lee, Oligonucleotide therapeutics and their chemical modification strategies for clinical applications, J. Pharm. Investig. 54 (2024) 415–433. https://doi.org/10.1007/s40005-024-00669-8
74. E. Doxakis, Therapeutic antisense oligonucleotides for movement disorders, Med. Res. Rev. 41 (2021) 2656–2688. https://doi.org/10.1002/med.21706
75. J.L. Schwartz, K.L. Jones, G.W. Yeo, Repeat RNA expansion disorders of the nervous system: Post-transcriptional mechanisms and therapeutic strategies, Crit. Rev. Biochem. Mol. Biol. 56 (2021) 31–53. https://doi.org/10.1080/10409238.2020.1841726
76. M.S.C. Mis, S. Brajkovic, F. Tafuri, N. Bresolin, G.P. Comi, S. Corti, Development of therapeutics for C9ORF72 ALS/FTD-related disorders, Mol. Neurobiol. 54 (2017) 4466–4476. https://doi.org/10.1007/s12035-016-9993-0
77. A. Bhuiyan, Therapeutic approaches for repeat expansion diseases: A comprehensive review, Explor. Neuroprotective Ther. 4 (2024) 485–496. https://doi.org/10.37349/ent.2024.00095
78. E. Hedlund, M.P. Hefferan, M. Marsala, O. Isacson, Cell therapy and stem cells in animal models of motor neuron disorders, Eur. J. Neurosci. 26 (2007) 1721–1737. https://doi.org/10.1111/j.1460-9568.2007.05780.x
79. S.F. Abdul Wahid, Z.K. Law, N.A. Ismail, N.M. Lai, Cell-based therapies for amyotrophic lateral sclerosis/motor neuron disease, Cochrane Database Syst. Rev. (2019). https://doi.org/10.1002/14651858.CD011742.pub3
80. E. Abati, N. Bresolin, G.P. Comi, S. Corti, Preconditioning and cellular engineering to increase the survival of transplanted neural stem cells for motor neuron disease therapy, Mol. Neurobiol. 56 (2019) 3356–3367. https://doi.org/10.1007/s12035-018-1305-4
81. A. DeLoach, C. Michael, K. Arianna, M. Kiaei, A retrospective review of the progress in amyotrophic lateral sclerosis drug discovery over the last decade and a look at the latest strategies, Expert Opin. Drug Discov. 10 (2015) 1099–1118. https://doi.org/10.1517/17460441.2015.1067197
82. E. Duranti, C. Villa, Molecular investigations of protein aggregation in the pathogenesis of amyotrophic lateral sclerosis, Int. J. Mol. Sci. 24 (2023). https://doi.org/10.3390/ijms24010704
83. R. Malik, M. Wiedau, Therapeutic approaches targeting protein aggregation in amyotrophic lateral sclerosis, Front. Mol. Neurosci. 13 (2020) 1–6. https://doi.org/10.3389/fnmol.2020.00098
84. B. Kalmar, L. Greensmith, Cellular chaperones as therapeutic targets in ALS to restore protein homeostasis and improve cellular function, Front. Mol. Neurosci. 10 (2017) 1–10. https://doi.org/10.3389/fnmol.2017.00251
85. W. Duan, M. Guo, L. Yi, Y. Liu, Z. Li, Y. Ma, G. Zhang, Y. Liu, H. Bu, X. Song, C. Li, The deletion of mutant SOD1 via CRISPR/Cas9/sgRNA prolongs survival in an amyotrophic lateral sclerosis mouse model, Gene Ther. 27 (2020) 157–169. https://doi.org/10.1038/s41434-019-0116-1
86. H.X. Deng, H. Zhai, Y. Shi, G. Liu, J. Lowry, B. Liu, ÉB. Ryan, J. Yan, Y. Yang, N. Zhang, Z. Yang, E. Liu, Y.C. Ma, T. Siddique, Efficacy and long-term safety of CRISPR/Cas9 genome editing in the SOD1-linked mouse models of ALS, Commun. Biol. 4 (2021) 1–11. https://doi.org/10.1038/s42003-021-01942-4
87. T. Gaj, D.S. Ojala, F.K. Ekman, L.C. Byrne, P. Limsirichai, D.V. Schaffer, In vivo genome editing improves motor function and extends survival in a mouse model of ALS, Sci. Adv. 3 (2017) 1–10. https://doi.org/10.1126/sciadv.aar3952
88. F. De Marchi, L. Ivan, B. Alessandro, D. Luca, O. Marco, P. Elisa, T. Danilo, V. Edvige, F. Daniela, L. Mazzini, Recent therapeutic advances in the treatment and management of amyotrophic lateral sclerosis: The era of regenerative medicine, Expert Rev. Neurother. (2025) 1–17. https://doi.org/10.1080/14737175.2025.2508781
89. H. Sharma, K.B. Narayanan, S. Ghosh, K.K. Singh, P. Rehan, A.D. Amist, R. Bhaskar, J.K. Sinha, Nanotherapeutics for meningitis: Enhancing drug delivery across the blood-brain barrier, Biomimetics. 10 (2025) 1–22. https://doi.org/10.3390/biomimetics10010025
90. A. Nagayach, R. Bhaskar, S. Ghosh, K.K. Singh, S.S. Han, J.K. Sinha, Advancing the understanding of diabetic encephalopathy through unravelling pathogenesis and exploring future treatment perspectives, Ageing Res. Rev. 100 (2024) 102450. https://doi.org/10.1016/j.arr.2024.102450
91. S. Ghosh, S. Ghosh, H. Sharma, R. Bhaskar, S.S. Han, J.K. Sinha, Harnessing the power of biological macromolecules in hydrogels for controlled drug release in the central nervous system: A review, Int. J. Biol. Macromol. 254 (2024) 127708. https://doi.org/10.1016/j.ijbiomac.2023.127708
92. S. Ghosh, R. Bhaskar, R. Mishra, M. Arockia Babu, M. Mohammad Abomughaid, N. Kumar Jha, J.K. Sinha, Neurological insights into brain-targeted cancer therapy and bioinspired microrobots, Drug Discov. Today. 29 (2024) 104105. https://doi.org/10.1016/j.drudis.2024.104105

16 Molecular Basis of Glioblastoma and Emerging Therapies for Brain Tumor

Prachi Agarwal, Shampa Ghosh, Krishna Kumar Singh, Mukund P Singh, Kirthanashri S Vasanthan, and Jitendra Kumar Sinha

16.1 INTRODUCTION

Brain tumors are definitive abnormal growth of brain cells or surrounding structures, including the meninges, cranial nerves, and pituitary gland. These tumors unequivocally lead to increased intracranial pressure, cause significant edema, diminish blood flow, and result in the displacement as well as degeneration of healthy tissue that is vital for essential bodily functions[1]. Brain tumors are the second leading cause of cancer-related deaths in children and young adults. Although relatively rare, they pose significant challenges in identifying risk factors. Their location in the brain makes treatment difficult, with options like surgery, radiation, and chemotherapy often leading to side effects and incomplete tumor removal[2]. Brain tumors can be life-threatening due to limited skull space, so it's crucial to develop methods for early detection and prevention. Understanding the associated risk factors is essential for minimizing damage from the tumor and its treatment. Brain tumors can be classified as either *primary* (originating within the brain) or *secondary* (metastatic, originating from other parts of the body). Primary brain tumors are further divided into *benign* (non-cancerous, slow-growing), and *malignant* (cancerous, aggressive) types[3]. Primary brain tumors, aka Gliomas, are believed to originate from neurological stem or progenitor cells, classified based on their histological appearance into three prominent categories, i.e., astrocytic, ependymal, or oligodendroglial. These tumors are also graded according to the varying degrees of malignancy (I–IV) assigned by the WHO[4]. The slow-growing gliomas are non-cancerous, while the malignant ones invade the healthy brain tissue, spreading rapidly within the brain, making it difficult to treat using standard therapies. It can also be categorized based on the genetic and molecular characteristics, which affect prognosis and treatment[5].

Glioblastoma multiforme (GBM) is an aggressive type of glioma and a grade four astrocytoma depending on the size and location. It can present symptoms like headaches (in 50% of the cases), which might usually worsen in the mornings due to raised intracranial pressure. About 30–40% of patients experience seizures and cognitive changes such as memory impairment, confusion, or neurological deficits (motor weakness and vision) problems are also common. Nausea and vomiting can occur from raised pressure. Symptoms usually progress over weeks to months as the tumor grows. Diagnosis of GBM is primarily based on newer imaging techniques, particularly MRI with gadolinium enhancement, revealing a typical "ring-enhancing" lesion with central necrosis and surrounding edema. CT scans are used when MRI facilities are unavailable, revealing a hypodense mass with edema[6].

A better and clearer diagnosis is only possible with histopathological findings through a stereotactic biopsy, which usually reveals characteristics such as high cellular density, necrosis, and enhanced mitotic activity. Molecular analysis, including isocitrate dehydrogenase (IDH) mutation,

DOI: 10.1201/9781003500025-16

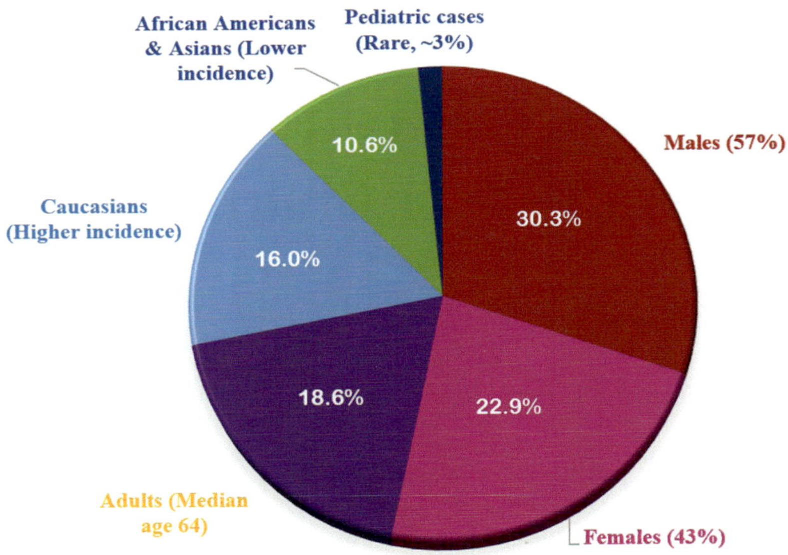

FIGURE 16.1 Demographic distribution of GBM incidence, illustrating the relative proportions based on sex, age, ethnicity, and pediatric prevalence.

distinguishes between aggressive IDH wild-type tumors and IDH-mutant GBM, which generally have a favorable prognosis[7].

O6-methylguanine-DNA-methyltransferase (MGMT) promoter methylation correlates with better response to temozolomide (TMZ) chemotherapy, whereas epidermal growth factor receptor (EGFR) amplification and phosphatase and tensin homolog (PTEN) mutations are other markers that give clues regarding tumor behavior and response to treatment. A thorough neurological examination assesses the cognitive, motor, and sensory function of the patient and aids in tracking the progression of the disease. It is important to distinguish GBM from other central nervous system (CNS) tumors, including primary CNS, lymphoma, metastatic tumors, and low-grade gliomas, and non-neoplastic lesions like brain abscesses. A successful strategy integrates with clinical assessment, imaging, histopathological examination, and molecular studies, allowing proper diagnosis and treatment plan for GBM. This provides a targeted approach to managing the multifaceted disease[7].

Researchers assessed the accuracy and reproducibility of a method that assessed the grading method of glioma for ordinary astrocytomas, which encompasses fibrillary, protoplasmic, gemistocytic, anaplastic astrocytomas (AAs), and glioblastomas. The grading was according to four criteria: nuclear atypia, mitosis, endothelial proliferation, and necrosis, resulting in grades 1–4[8]. Survival analysis showed a significant association between tumor grade and survival, with median survival times of 4 years for grade 2, 1.6 years for grade 3, and 0.7 years for grade 4. Patients with grade 1 tumors had longer survival times[9].

GBM exhibits a notable demographic and epidemiological pattern, as shown in Figure 16.1, which is vital to determine and understand the disease burden and guide the direction of future therapeutic research and clinical strategies.

16.2 MOLECULAR PATHOGENESIS OF GLIOBLASTOMA

Glial tumors are categorized into two major types based on their invasiveness: "diffuse gliomas," which aggressively infiltrate surrounding brain tissue, and those exhibiting more "circumscribed" growth patterns. Diffuse gliomas invade normal brain parenchyma unambiguously and consistently recur, even after gross total resection[10]. Research has focused on the mutations and genetic events driving tumors from their original cells to malignancy. Primary GBMs feature key mutations such as loss of heterozygosity at chromosome 10q (70%), EGFR amplification (36%), P16/INK4A deletion (31%), and PTEN mutations (25%). Secondary GBMs often arise from primary gliomas with P53 mutations, one of the most common mutations in cancer. Additionally, IDH mutations are seen as early events in glioma genesis, particularly in GBMs that develop from low-grade lesions. These mutations are crucial to the pathways regulating cellular survival and proliferation[11].

An overview of the molecular pathogenesis is summarized in Figure 16.2, highlighting the key genetic and epigenetic alterations that drive tumor initiation, progression, and resistance to therapy.

16.2.1 Epigenetic Modifications

16.2.1.1 DNA Methylation

DNA methylation is an epigenetic modification—a chemical change to DNA that affects gene expression without altering the DNA sequence itself. In this process, a methyl group (CH_3) is added to specific sites on the DNA molecule, usually to a cytosine base in a cytosine-guanine (CpG) dinucleotide. This modification typically occurs in regions known as CpG islands, which are stretches of DNA with a high frequency of CpG sites, often found near gene promoters (regions that initiate gene transcription). DNA methylation plays a pivotal role in the pathogenesis of GBM[12]. Hypermethylation of tumor suppressor genes and hypomethylation of oncogenes can disrupt critical cellular pathways, leading to uncontrolled cell proliferation, invasion, and resistance to therapy[13]. Recent studies have revealed the significance of epigenetic alterations in GBM. For instance, the silencing of the MGMT gene, a DNA repair enzyme, through promoter methylation can enhance the efficacy of alkylating agents[14]. Genome-wide analyses have identified numerous genes whose expression is regulated by DNA methylation in GBM, providing insights into the underlying molecular mechanisms. By integrating DNA methylation and gene expression data, researchers can identify novel biomarkers and therapeutic targets[15].

For example, the glioma CpG island methylator phenotype (G-CIMP) phenotype is found to be linked to different clinical features and responses to the treatment. The dynamic relationship between epigenetic factors and genetic mutation needs to be studied to make further steps toward effective therapeutic strategies for GBM9[16].

The study carried out by Ma et al. was to check how DNA methylation heterogeneity relates to the glioma tumor immune microenvironment. They found mean levels of intermediate methylation much greater in tumor tissue than in individual immune cell types, which indicates that the complex interactions occur between the tumor cells and the infiltrating immune cells, increase heterogeneity in DNA methylation[17]. It was also found that infiltration of CD4+ and CD8+ T cells have a positive correlation with DNA methylation heterogeneity.

Patients who had lower intermediate methylation levels proved to have a better prognosis and slow disease progression. A new prognostic risk score called cellular mesenchymal–hypoxic response (CMHR) was created based on the methylation levels of certain CpG sites[18]. Patients with low CMHR scores had specific molecular alterations, and hence, TP53 and alpha-thalassemia/mental retardation syndrome X-linked (ATRX) mutations led to poorer survival outcomes. In addition, the CMHR score was associated with T-lymphocyte infiltration and could predict IDH status glioma subtype as well as histological phenotype[19].

Methylation significantly influences tumor behavior in GBM by modulating gene expression. Regulation of G-CIMP phenotype is one of the major expressions of methylation in the modulation of gene expression. Methylation marks G-CIMP with hypermethylation of certain genes that may influence dynamics in tumor growth as well as treatment response. This defines G-CIMP tumors into two subtypes: low and high G-CIMP. G-CIMP high tumors usually have a better prognosis, while G-CIMP low tumors demonstrate shorter overall survival, indicative of their different biological and clinical characteristics[19].

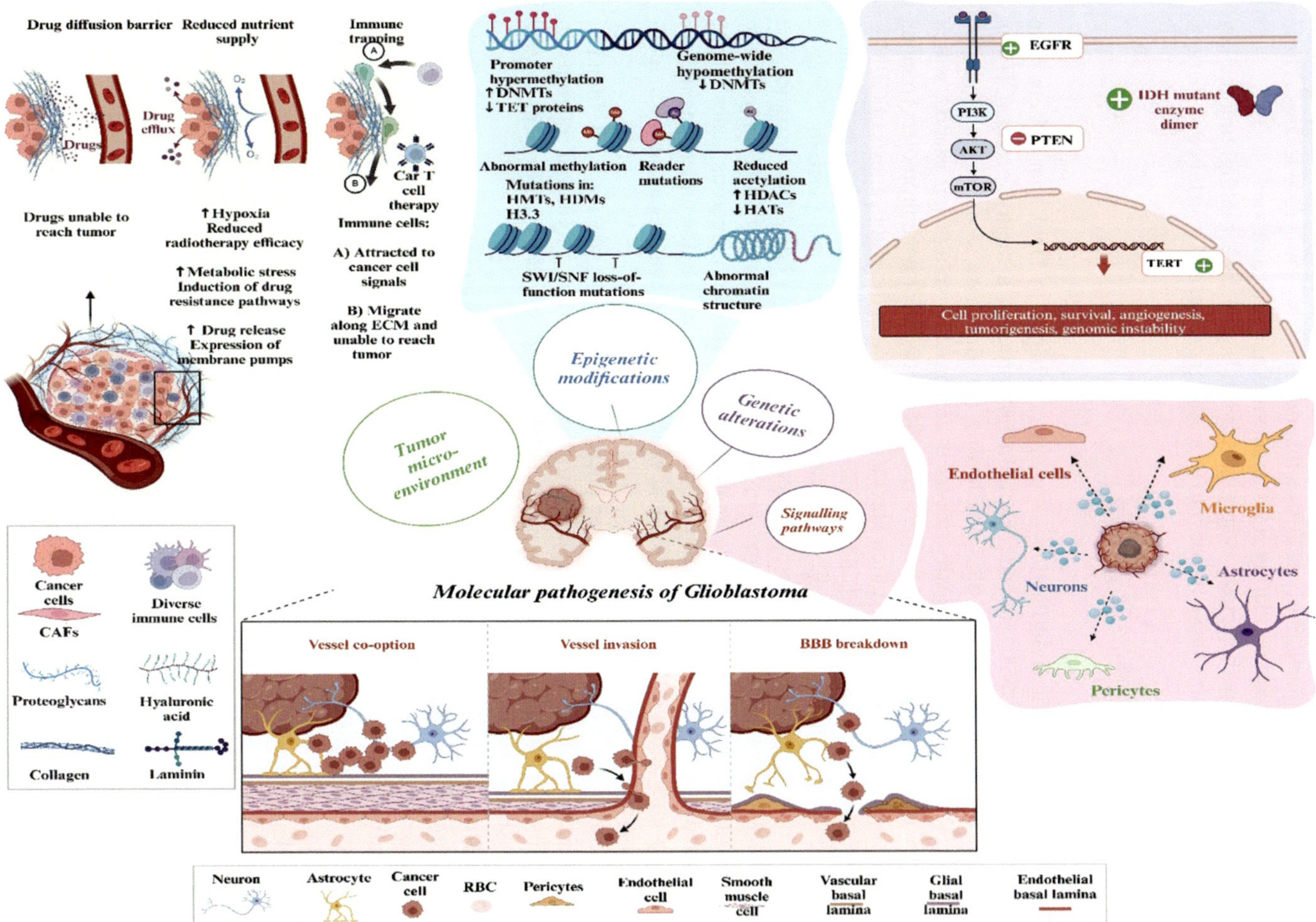

FIGURE 16.2 Molecular pathogenesis of GBM: the figure demonstrates a complex molecular mechanism that leads to glioblastoma progression. The key contributing factors include the tumor microenvironment, genetic alterations, signaling pathways, and epigenetic modifications. The image details how glioblastoma cells evade drug delivery through the BBB, exhibiting vessel co-option and invasion, and manipulating the immune cells and ECM. Dysregulation of signaling pathways (EGFR, PI3K/AKT/mTOR, and IDH mutations) and chromatin alterations are also presented, emphasizing the multifactorial nature of glioblastoma pathogenesis.

The methylation landscape in GBM is closely associated with mutations in the IDH gene. IDH-mutant gliomas frequently present with G-CIMP, which in turn informs tailored therapeutic strategies and influences patient outcomes. Additionally, the methylation status of the MGMT promoter plays a pivotal role in treatment efficacy, particularly in IDH mutation. Methylated MGMT promoters are linked to better responses to alkylating chemotherapy agents[20].

16.2.2 Genetic Alterations in Glioblastoma

16.2.2.1 Isocitrate Dehydrogenase (IDH) Mutation

IDH enzymes, namely IDH1, IDH2, and IDH3, are critical components of metabolic processes like the Krebs cycle. IDH1 is located in the cytoplasm, while IDH2 and IDH3 operate in the mitochondria. IDH1 and IDH2 depend on nicotinamide adenine dinucleotide phosphate (NADP)+, whereas IDH3 utilizes NAD+ to convert isocitrate into α-ketoglutarate (α-KG). Notably, mutations in IDH, especially at crucial arginine sites, are strongly associated with several cancers, including gliomas and acute myeloid leukemia. In gliomas, IDH mutations are predominant in lower-grade tumors and significantly correlate with improved survival rates[21]. For instance, patients with IDH-mutant glioblastoma have an average survival of about 31 months, compared to just 15 months for those with the wild-type form. These mutations shift the enzymatic activity toward the production of D-2-hydroxyglutarate (D-2-HG) instead of α-KG, altering cellular metabolism and driving tumor growth. While IDH mutations are typically associated with a more favorable prognosis, they also heighten the risk of recurrence and transformation into aggressive cancers. This makes IDH mutations crucial targets for diagnostic and therapeutic oncology strategies[22].

Study by Kurshunov et al. explored the relationship between CDKN2A and IDH mutations in gliomas, an integral aspect of tumor evolution. The loss of CDKN2A, common in many cancers, adds to the process of tumorgenesis; however, it is not until later that IDH mutations in gliomas typically signal more favorable initial outcomes that sped-up tumorigenesis due to loss of CDKN2A leading frequently to progression into higher grade tumors that secondary GBM form that this was realized[22]. The resultant mutation phenotype from these two alterations is one of hypermutations rendering malignancies that are quite aggressive and likely comparatively unresponsive to therapy. Mutated IDH1 produces D-2-HG which is an oncometabolite capable of interfering with cellular metabolism thereby potentially undermining the function of CDKN2A as a tumor suppressor. Together, the loss of CDKN2A and IDH-driven changes result in a more aggressive tumor phenotype and poorer clinical outcomes for patients[23].

Expression of mutant IDH1 in astrocytes induces a shift toward a stem cell-like state, characterized by lowered levels of glial fibrillary acidic protein (GFAP) (an astrocyte marker) and elevated expression of nestin, a neural stem cell (NSC) marker[24]. This mutation prevents normal differentiation even under conditions that typically stimulate astrocytic and neuronal differentiation. This undifferentiated state may be linked to modifications in methylation patterns, particularly through the inhibition of the H3K9 demethylase lysine demethylase 4C (KDM4C), which further supports a progenitor-like phenotype necessary for glioma initiation. However, it's not fully understood whether these changes in cellular behavior are solely due to methylation shifts, as IDH inhibition *in vitro* promotes differentiation without impacting the methylome[25].

In addition, mutant IDH1 impacts hypoxia pathways by suppressing hypoxia-inducible factor 1α (HIF-1α) through enhanced activity of prolyl hydroxylases, leading to its degradation. Reduced HIF-1α in IDH1-mutant gliomas may make cells more vulnerable to oxidative damage, potentially fostering early mutations. Despite this initial proliferative boost, lower HIF-1α levels might result in a less aggressive tumor profile, as HIF-1α is associated with enhanced tumor cell growth and blood vessel formation in glioblastomas. Thus, IDH1 mutations may initially support cancer cell growth while also shaping a potentially less aggressive tumor phenotype, which has implications for the clinical behavior of IDH-mutant gliomas[26].

16.2.2.2 Epidermal Growth Factor Receptor (EGFR) Amplification and Mutation

GBMs are classified as primary or secondary whereas the primary GBMs are characterized by the overexpression or amplification of the EGFR. Studies show that EGFR is amplified in approximately 40% of cases, overexpressed in about 60%, and exhibits deletions or mutations in 24–67% of cases[27]. The amplification and overexpression of EGFR are uniformly distributed throughout the tumor tissue and are rarely observed in normal brain tissue, making EGFR-targeted therapies particularly advantageous for GBM treatment. This receptor is activated by ligands such as TGF-α and EGF, leading to receptor dimerization and the activation of critical signaling pathways (RAS/RAF/MAPK, PI3K/AKT, and signal transducers and activators of transcription [STATs]) that regulate cell growth and survival. Abnormal EGFR activation in cancer is often caused by overexpression, gene amplification, or mutations, driving cancer cell proliferation and spread. In GBM, EGFR is the most commonly amplified oncogene, found in approximately 40% of cases, often linked to protein overexpression. Additionally, around half of GBMs with EGFR amplification have mutations like EGFRvIII, which results in constant activation of EGFR and excessive cell proliferation due to a deletion in the ligand-binding domain[28].

A study by van den Bent et al. found that EGFR amplification is stable in most tumors, with 84% showing consistent levels from diagnosis to recurrence. This suggests that EGFR amplification can reliably predict recurrent tumors, allowing EGFR-targeted therapies to be used across different disease stages without frequent reassessment. In contrast, EGFRvIII expression is more variable. About half of the tumors that expressed EGFRvIII at diagnosis lost this expression upon recurrence. While 79% of GBMs maintained EGFRvIII expression, the loss in some recurrent cases indicates caution is needed when using initial EGFRvIII data to guide treatment. These results emphasize the necessity of re-evaluating EGFRvIII status at recurrence to ensure the relevance of therapies targeting this mutant receptor[29].

Several other EGFR gene alterations have been identified in glioblastomas that include deletions in the carboxy-terminal domain (exons 25–28), fusions with SEPT14 or phosphoserine phosphatase (PSPH) that retain the tyrosine kinase domain, and cases with varying gene amplification. Notable findings include EGFR hypermethylation linked to tumor evolution and recurrence, 2 single-nucleotide polymorphisms (SNPs) in introns 4 and 13 associated with increased glioma risk, and alterations at 7p11.2. The EGFRc958 deletion in amino acids 521–603, amplification, and various point mutations have been observed. Additionally, EGF-related alterations include a polymorphism in the 5′-untranslated region associated with postoperative recurrence and seven SNPs correlating with reduced survival[30].

In a study involving adult patients with AA anaplastic oligodendroglioma (AO), and anaplastic oligoastrocytoma (AOA), Yang et al. found that patients who experienced preoperative seizures had a higher prevalence of EGFR amplification, frontal lobe involvement, and left cerebral hemisphere involvement. Additionally, these patients exhibited lower Ki-67 expression in univariate and multivariate analyses[31]. The EGFR variant, EGFRx, is crucial for glioblastoma stem cells (GSCs) as it supports key activities like self-renewal and tumorigenesis. Unlike the EGFRvIII variant, EGFRx lacks significant parts of the extracellular ligand-binding domain, specifically exons 2–14, which prevents it from binding to external ligands. However, it remains active due to its ability to spontaneously dimerize, activating the STAT5 pathway essential for maintaining the GSC phenotype and promoting glioblastoma progression.

Given its heightened activity in glioblastoma cells compared to normal brain tissue, EGFRx is a promising therapeutic target. Disrupting the EGFRx-STAT5 signaling axis could limit GSC survival and growth, offering new treatment options that specifically address this variant and its signaling pathway, potentially overcoming limitations of existing EGFR-targeted therapies[32].

Recent research has shown that EGFR can localize to the nucleus and mitochondria, particularly in cancers like glioblastoma. In the nucleus, EGFR acts as a transcriptional co-regulator, interacting with DNA repair proteins such as DNA-PK and proliferating cell nuclear antigen (PCNA). This interaction enhances DNA damage repair, helping cancer cells resist treatments like radiation and chemotherapy. Additionally, nuclear EGFR influences genes related to cell proliferation and survival, contributing to tumor aggressiveness. In the mitochondria, EGFR supports metabolic adaptations that allow tumor cells to thrive in low-oxygen and nutrient-deprived environments. This function helps glioblastoma cells survive and progress even when resources are limited. These additional locations of EGFR in GBM establish the biology of EGFR in distinct intracellular contexts for coordination in dictating tumor progression, which further complicates pathway-directed therapies[27].

16.2.2.3 Phosphatase and Tensin Homolog (PTEN) Mutations and Loss

PTEN is an important suppressor of tumor and functions as a protein and lipid phosphatase that plays a vital role in suppressing cell growth, survival, and cell cycle through inhibiting PI3K/AKT signaling pathway. As a lipid phosphatase, PTEN dephosphorylates lipids, including phosphatidylinositol-3,4,5-triphosphate (PIP3) to generate phosphatidylinositol-4,5-bisphosphate (PIP2). It is relevant because this will then prevent cell disorderly proliferation therefore need PTEN for the cellular homeostasis[33].

In GBM, if the PTEN gene is altered or mutated, it can no longer suppress tumors. This results in unregulated function of Phosphoinositide 3-kinase (PI3K), increasing PIP3 levels, and initiating the Protein kinase B (AKT) signaling pathway. Overactivation of AKT contributes to tumor growth by promoting cell growth, blocking normal cell demise, and prolonging cell lifespan. PTEN deficiency also activates mTOR, which allows the tumor to satisfy its high energy demands and divide rapidly. In addition, inactivity of PTEN also facilitates tumor growth and reduces the efficacy of treatment with agents such as EGFR inhibitors[34].

PTEN loss of function in GBM leads to tumor progression and development of resistance to therapy. For example, tumors with no PTEN do not respond to EGFR inhibitors since the PI3K/AKT pathway remains active. Second, PTEN mutations cause the tumor to become more adaptable and aggressive, and make therapy for glioblastoma more difficult[35].

The study by Dogan et al. studies the consequences of mutations of PTEN in immune presentation of the tumor through the cGAS-STING pathway, which triggers the antitumor immune response. PTEN, a tumor suppressor, normally regulates immune functions by dephosphorylating IRF3, crucial for type I interferon synthesis. Mutations in PTEN disrupt this function, reducing activation of key pathway components such as STING, IRF3, and NF-κB, leading to an immunosuppressive environment characterized by lower CD8+ T-cell infiltration and diminished immune activity. Additionally, PTEN mutations alter STING trafficking by affecting vesicular transport involving Rab7 and TBK1, further enhancing immune evasion in GBM. PTEN loss also induces galectin-9 secretion, promoting M2 macrophage polarization and supporting glioma progression. The findings suggest that targeting STING in PTEN-deficient GBMs may help convert the "cold" tumor microenvironment into a "hot" one, potentially improving immune responses to therapy. However, PTEN loss could reduce the effectiveness of STING agonists. Despite some limitations, the study sheds light on PTEN's immunosuppressive role in GBM and offers avenues for targeted therapy[36].

Kovalenko et al. explore the regulation and role of the long non-coding RNA lncPTENP1-S, a transcript of the PTEN pseudogene, in GBM. DNA methylation plays a crucial role in gene regulation, typically leading to transcriptional repression, but certain promoter methylations, such as for FOXA, hTERT, and BIRC5, increase gene expression. Similarly, methylation of the PTENP1 CpG island (CpGI) promotes lncPTENP1-S expression by potentially blocking a transcriptional repressor complex that includes the tumor suppressor PTEN. Interestingly, PTEN's role is pivotal in this regulatory mechanism, as it allows a transcriptional repressor to interact with the unmethylated PTENP1 CpGI island, highlighting PTEN's function in transcription modulation. lncPTENP1-S appears to act independently of PTEN in GBM, enhancing glioma stem cell (GSC) stemness, reducing cell adhesion, and contributing to GBM invasiveness—traits that imply an oncogenic role[33].

Baeza et al. explored the mechanisms of PTEN loss in glioblastoma (GBM) and its implications for tumor development. The loss of PTEN leads to increased cell proliferation, survival, and invasion, and can occur through mutations, loss of heterozygosity (LOH) on chromosome 10q23, and promoter methylation. Approximately one-third of glioblastoma biopsies and cell lines showed PTEN promoter methylation, but the relationship between methylation and PTEN expression was inconsistent. Some glioblastomas with PTEN mutations and LOH retained PTEN expression, while others had clusters of cells with PTEN loss associated with methylation. This suggests that PTEN methylation and expression loss may happen later in glioblastoma progression[37].

The PTEN pseudogene on chromosome 9p21, which closely resembles the PTEN gene, may explain PTEN expression and methylation inconsistencies. Its promoter is similar enough that methylation-specific PCR primers can amplify both genes. Although typically inactive, the pseudogene is actively transcribed in glioma cell lines, potentially compensating for PTEN mutations. A fusion protein from the pseudogene shows phosphatase activity, suggesting it could disrupt PTEN function. This study confirms that PTEN promoter methylation in GBM biopsies is specific to the PTEN gene, as different primer sets indicate that the signals do not come from the pseudogene. The findings reveal that PTEN regulation in glioblastoma is complex, and the PTEN pseudogene may influence GBM development, which could affect how PTEN mutations and expression data are interpreted in tumor studies[38].

16.2.2.4 Telomerase Reverse Transcriptase (TERT) Promoter Mutations

Telomerase reverse transcriptase (TERT) is a crucial enzyme in telomerase, a ribonucleoprotein complex that maintains telomeres, the protective caps at chromosome ends. As cells divide, their telomeres shorten, leading to aging or cell death. Telomerase adds telomeric DNA repeats, vital for continuously dividing cells like stem cells and certain immune cells. TERT expression in normal somatic cells is low or inactive, causing telomeres to shorten and limiting cell lifespan. In many cancers, TERT is upregulated due to mutations in the TERT promoter, allowing cells to maintain telomere length and divide indefinitely. This phenomenon is observed in various malignancies, including glioblastomas, melanomas, and certain bladder, thyroid, and CNS tumors, often linked to aggressive growth and poor prognosis[39].

Immortalized cancer cells overcome telomere-related crises mainly by activating telomerase through TERT promoter mutations or by using the alternative lengthening of telomeres (ALT) pathway, often linked to ATRX mutations. In gliomas, telomerase activation is caused by TERT mutations, while ALT leads to ATRX mutations. ALT tumors frequently demonstrate highly complex chromosomal rearrangements, suggestive of genomic instability. Loss of ATRX is a powerful prognostic marker and also associates with worse prognosis[40]. In GBM, tumors depend on either telomerase activation (through TERT mutation) or ALT (ATRX mutation), but not both. It influences prognosis and treatment possibilities. TERT mutations are typically associated with poorer survival outcomes, while ATRX mutations suggest an alternative aggressive pathway. The detection of TERT or ATRX mutations in GBM is useful in designing targeted therapies, for example, telomerase inhibitors in TERT-mutant tumors or ALT-targeted treatments in ATRX-mutant cases. Recent studies show that TERT promoter mutations are prevalent in over 50 cancer types, making them the most common mutation in many cancers[41].

TERT lets the cells overcome the Hayflick limit and suppress apoptosis. In GBM, elevation of telomerase activity, frequently as a result of TERT promoter mutations, leads to cellular immortalization and correlates with the progression and prognosis of the tumor. The frequency of TERT promoter mutation, and especially C228T and C250T, is very high in GBM, approximately 73%. These mutations lead to enhanced telomerase activity and allow cancer cells to grow without sense or limitation. This study examines the impact of specific TERT promoter (TERTp) mutations—C228T and C250T—on GBM prognosis, with a focus on their interaction with MGMT promoter methylation status. While TERTp mutations generally did not affect prognosis, the C250T mutation was associated with a longer median progression-free survival (mPFS) and overall survival (mOS) compared to the C228T mutation and TERTp wild-type (WT) tumors. This beneficial impact was independent of MGMT methylation, with patients carrying both the C250T mutation and MGMT unmethylation showing improved outcomes relative to those with the C228T mutation.

Gorria et al. highlight inconsistencies in prior research, likely due to differences in methodology and consideration of other prognostic factors. For instance, some studies reported improved outcomes in GBM with combined TERTp mutations and MGMT methylation, but failed to observe these effects in tumors without MGMT methylation. Further, TERTp mutations were found to affect telomerase activity and transcription through the creation of an ETS binding site, with the C228T mutation increasing TERT mRNA expression more significantly than C250T.

Tumors with TERTp mutations showed enrichment in pathways related to telomere and chromosomal mechanisms, with greater enrichment in C228T compared to C250T, which might explain the poorer prognosis associated with C228T. Unlike TERTp-WT tumors, which showed no specific pathway enrichment, these findings suggest that telomere lengthening plays a more active oncogenic role in TERTp-mutated GBM[42].

Mutations in TERT are typically mutually exclusive with IDH1 mutations, co-occurring in only a small fraction of GBM cases, and are often accompanied by alterations in other genes like TP53 and EGFR. Additionally, the rs2853669 SNP in the TERT promoter has been observed in GBM samples, but it does not significantly affect survival outcomes compared to the presence of TERT promoter mutations alone. TERT mutations are notably linked with poorer survival rates, especially in older patients[43].

16.2.3 Signaling Pathways Involved

16.2.3.1 TP53 Mutations

TP53, or p53, is an essential tumor suppressor gene that maintains cellular integrity by regulating the cell cycle, apoptosis, DNA repair, and senescence. Known as the "guardian of the genome," it prevents genetic mutations by halting the cell cycle in response to stress or DNA damage. On activation, p53 activates genes that promote DNA repair and ensure that damaged cells are repaired or removed[44].

TP53 mutations are frequent in a number of cancers, including GBM, and prevent the cells from undergoing essential regulatory processes, which results in uncontrolled tumor expansion. In GBM, the most aggressive brain cancer, TP53 mutations are frequently observed and have a strong effect on the behavior and prognosis of the disease. Unlike the loss-of-function mutations that are seen in tumor suppressor genes, TP53 mutations in GBM are frequently activated by gain-of-function (GOF). These mutations not only inactivate p53's tumor suppressor functions but also provide it with oncogenic activities that drive tumor growth, invasion, and therapy resistance. Mutant p53 may also favor survival signaling, promote invasiveness, and alter the tumor microenvironment by promoting inflammation and immune cell recruitment[45]. These factors lead to therapy-resistant tumors and emphasize the importance of TP53 as a target for glioblastoma research and therapy[46].

Recent findings have suggested that mutant p53 gains oncogenic capabilities through GOF mutations, which are absent in the wild-type protein. These mutations promote cell proliferation, invasiveness, and angiogenesis, which further increases the malignancy of tumors. Suppressing GOF mutant p53 can impede cancer progression, affirming its role as an independent oncoprotein[47].

A study by Ham et al. shows that mutations in TP53 GOF promote inflammation and are linked to poorer patient outcomes, as inflammation significantly impacts cancer progression by increasing tumor risk and promoting growth-supporting cytokines. In later stages, NFκB activation enhances cell proliferation and resistance to chemotherapy through pro-inflammatory cytokines. These mutations upregulate inflammatory genes like CCL2 and TNFA, attracting microglia and myeloid immune cells. This aligns with a tumor-associated myeloid signature in various cancers, suggesting that TP53 GOF mutations drive cancer progression by creating an inflammatory microenvironment[48].

16.2.3.2 RAS/RAF/MEK/ERK Pathway

The RAS-RAF-MEK-ERK signaling pathway, also known as the mitogen-activated protein kinase (MAPK) pathway, is a critical cascade that regulates key cellular processes such as proliferation, differentiation, survival, and apoptosis. Activated by extracellular signals like growth factors, the pathway begins with activating rat sarcoma viral oncogene homolog (RAS), a GTPase, which recruits and activates rapidly accelerated fibrosarcoma (RAF) kinases. RAF then phosphorylates and activates MEK, which activates ERK through phosphorylation. Once activated, ERK translocates to the nucleus, regulating transcription factors to modulate gene expression[49]. This is an essential pathway for cellular process. Yet its dysregulation is a common feature of many cancers, typically as a result of mutations in RAS or BRAF, which cause cells to grow and survive unchecked. Inhibiting this pathway with inhibitors, such as BRAF or MEK inhibitors, has proven to be successful for treating cancers such as melanoma, although resistance to therapeutics is an issue. Apart from cancer, mutations in this pathway have also been linked to developmental diseases, attesting to its significance in health and disease[50].

Gliomas, including GBMs, overexpress a range of receptor tyrosine kinases (RTKs) and their ligands, which promote malignancy by stimulating cell proliferation, motility, apoptosis suppression, and angiogenesis. Hyperactive RTK signaling, in particular that of EGFR, PDGFR, and VEGFR, is a mark of 88% of glioblastoma. Mutations from aberrations, gain of genes, overexpression, and structural changes of RTKs cause autocrine or paracrine growth factor signaling loops. The Cancer Genome Atlas has confirmed the significant target pathways involved are RTK/RAS/PI3K, p53, and retinoblastoma. Activation of mitogenic pathways, especially RAS-RAF-MEK-ERK, plays a central role in glioma progression and is driven by genomic alterations in pathway components. Additionally, tropomyosin receptor kinases (Trks), which mediate neurotrophin signaling, have emerged as significant targets in glioma therapy due to their involvement in RAS/MAPK and PI3K/AKT signaling[51]. AA and GBM are resistant to conventional therapies like TMZ due to hyperactive survival pathways, including RAS/RAF/MEK/ERK and PLCγ1/PKC, regulated by TrkB receptor signaling. TMZ induces apoptosis primarily via caspase-3 activation, through intrinsic pathways in AA and extrinsic in GBM. Combining TMZ with pathway inhibitors significantly enhances apoptosis compared to single-agent treatments[52].

The study by Zając et al. through their research found that sorafenib inhibits Raf kinase and ERK activity, promoting apoptosis, while U-73122, a PLCγ1 inhibitor, induces extrinsic apoptosis via caspase-3/8 activation. LOXO-101 targets TrkB signaling, boosting apoptosis in GBM. Combination therapies, such as PLCγ1 and Raf inhibitors with TMZ for AA and LOXO-101, sorafenib, and TMZ for GBM, showed superior efficacy[53]. These treatments modulate the Bcl-2:beclin-1 complex, shifting from autophagy to apoptosis. The findings highlight the promise of multi-targeted therapies to overcome resistance, improving apoptosis induction and treatment outcomes in aggressive gliomas[54].

As the study demonstrated by Jeuken et al. RAS/RAF/MAPK pathway plays a critical role in glioma oncogenesis, with activation largely driven by copy number gains rather than mutations. Elevated RAS expression and RAS-GTP levels are observed in GBM, and mouse models confirm the oncogenic potential of HRAS. Mutations in NRAS, KRAS, HRAS, or BRAF are rare in gliomas (2/93 cases), whereas copy number gains involving these genes occur in 57% of cases, particularly affecting BRAF (38%). Gains in upstream growth factors and receptors are common and often coincide with pathway activation, supported by increased MAPK-P staining. Higher-grade gliomas (WHO grade III/IV) and specific genetic subtypes, such as +7/−10, frequently show these gains. As CGH analysis captures only large-scale alterations, the role of smaller or alternative regulatory mechanisms may be underestimated. These findings highlight the importance of targeting copy number-driven RAS/RAF/MAPK activation in glioma therapies[55].

16.2.3.3 NOTCH and WNT Pathway

The Notch signaling pathway is an evolutionarily conserved mechanism crucial for various cellular and developmental processes, including proliferation, apoptosis, stem cell maintenance, and differentiation. In mammals, the pathway involves four Notch receptors (Notch1–4) and two ligand families, Delta-like (Dll1, –3, –4) and Jagged (Jagged1, –2). These transmembrane proteins interact either in *trans* (on neighboring cells) to activate signaling or in *cis* (on the same cell) to inhibit it. Notch signaling is vital in the CNS, regulating NSCs in neurogenic niches like the subventricular zone (SVZ) and dentate gyrus (DG)[56].

Notch signaling levels dictate NSC behavior: low levels promote proliferation and neuronal differentiation, while high levels induce quiescence. Notch1 primarily supports active NSC proliferation, which declines with age, while Notch2 and Notch3, mediated by RBPJK activity, regulate quiescent NSC fate. Loss of RBPJK leads to progenitor depletion and reduced neurogenesis. Notch also influences neuronal migration, with reelin-DAB1 signaling stabilizing NICD, and altering neuron morphology and migration.

Notch activation by Delta ligands resists neurogenic signals during neurogenesis, guiding NSCs toward glial differentiation, excluding oligodendrocytes. This role extends to gliogenesis, promoting differentiation via bHLH transcription factors. Emerging models suggest Notch prevents all NSC differentiation, leaving fate decisions to prevailing external cues.[57]

The Notch signaling pathway is crucial in GBM progression, influencing tumor growth, stem cell maintenance, and therapy resistance. Elevated Notch1/Notch4 and their ligands (Dll1, Dll4, Jagged1) drive proliferation, invasion, and angiogenesis, supported by crosstalk with AKT, NF-κB, and hypoxia-inducible factors. Notch sustains GSCs, particularly in hypoxic and vascular niches, promoting aggressiveness. Molecular subtypes show distinct Notch patterns, with enrichment in mesenchymal and classical GBMs, while proneural GBMs exhibit high Notch1 and IDH mutation-associated Dll3 expression. Epigenetic alterations, like low HEY1 promoter methylation, and dysregulated miRNAs (e.g., miR-34a, miR-148a) modulate Notch activity. Therapeutic strategies include γ-secretase inhibitors (GSIs), such as RO4929097, which reduce GSC proliferation and enhance radiation and chemotherapy effects, though resistance mechanisms, including Hedgehog pathway activation and angiogenesis-independent adaptations, limit efficacy. New approaches, like CB-103 and novel agents (e.g., arsenic trioxide, fibulin-3 inhibitors), aim to overcome these challenges. Despite hurdles, targeting Notch offers potential in GBM therapy[58].

The canonical Wnt signaling pathway, regulated by β-catenin, governs processes like morphogenesis, differentiation, and proliferation, with its dysregulation driving cancers, including gliomas. In its inactive state, β-catenin undergoes degradation via a destruction complex (Axin, APC, GSK3β). Wnt ligand interaction with Frizzled/LRP disrupts this process, stabilizing β-catenin and enabling its nuclear accumulation to activate Tcf/Lef transcription factors. In gliomas, β-catenin is overactivated through mechanisms like FoxM1-dependent nuclear localization, mutations in degradation complex components, and hypermethylation of Wnt antagonists (e.g., Dkk1, sFRPs). Aberrant Wnt signaling sustains GSCs, contributing to tumor aggressiveness and therapy resistance. Therapeutic strategies target β-catenin, its interactions, or associated pathways, including EGFR crosstalk, but challenges remain due to pathway complexity and tumor heterogeneity[59].

Aberrant activation of the WNT/β-catenin signaling pathway is crucial in gliomas. In a study involving 74 gliomas of varying grades and 24 glioblastoma cell lines, WNT3a, β-catenin, and TCF4 were analyzed. WNT3a was overexpressed at both protein and mRNA levels in high-grade astrocytic tumors and cell lines. Cytoplasmic β-catenin expression was observed in high-grade gliomas, with evidence of nuclear translocation, though no activating mutations in the β-catenin gene (CTNNB1) were detected. TCF4 expression correlated with proliferation markers Ki-67/MIB-1 and cyclin D1. The expression levels of WNT3a, cytoplasmic β-catenin, and TCF4 were strongly associated with higher tumor grade and poorer prognosis in glioma patients[60].

16.2.4 Role of Tumor Microenvironment (TME)

Gliomas invade brain tissue uniquely, moving through the extracellular matrix (ECM) rather than metastasizing via lymphatic or vascular routes. This process resembles NSC migration during brain development and involves mesenchymal movement, requiring ECM degradation, cytoskeletal remodeling, and integrin-mediated adhesion. GBM hijacks mechanisms from NSC niches, such as the SVZ and subgranular zone (SGZ), where stemness is maintained through interactions with the ECM, glial cells, and blood vessels[61].

16.2.4.1 Immune Evasion and Tumor Progression

GSCs are central to this invasion, with their self-renewal properties contributing to tumor growth and recurrence. TME plays a critical role, comprising glioma cells, immune cells, endothelial cells, and stromal components, which interact dynamically to promote invasion, proliferation, and angiogenesis. Glioma cells remodel the ECM using proteases such as MMP2 and MMP9, aided by proteins like hyaluronan and tenascins, which facilitate migration. GSCs reside in specialized hypoxic niches and maintain tumor growth by promoting self-renewal and preventing differentiation. These cells are marked by CD133 expression and are highly resistant to therapy[62].

Hypoxic and acidic conditions within the tumor stimulate GSC self-renewal and invasive behavior through pathways like Notch and HIF signaling. Glioma cells migrate along blood vessels, disrupting the blood-brain barrier and gaining access to nutrients while altering normal brain vasculature. Immune cells, including microglia and tumor-associated macrophages (TAMs), release cytokines, growth factors, and ECM-modifying proteins, further enhancing glioma invasiveness. Additionally, interstitial fluid flow, driven by tumor pressure, and chemokine gradients such as CXCL12/CXCR4 guide glioma cell movement toward the tumor edge. These mechanisms highlight the complex interplay between glioma cells and their microenvironment in driving invasion and progression[63].

16.2.4.2 Immune Cell Infiltration

A complex network of cellular and non-cellular components that support tumor progression and immune evasion. Key cellular components include TAMs, often polarized to an M2-like immunosuppressive phenotype, dendritic cells, myeloid-derived suppressor cells (MDSCs), neutrophils, B and T lymphocytes, and natural killer (NK) cells. While B and T lymphocytes and NK cells are part of the immune response, their activity is frequently suppressed in the glioblastoma microenvironment. Glioblastoma cells interact extensively with these immune and stromal cells, secreting soluble factors such as cytokines and chemokines that drive inflammation, angiogenesis, and immunosuppression. The ECM provides structural support and biochemical signals, aiding in tumor invasion. Abnormal vasculature, supported by pericytes, creates hypoxic zones that further promote tumor aggressiveness. This intricate interplay between tumor cells and their microenvironment highlights the need for therapies targeting both glioblastoma cells and their supportive niche[64].

Fan et al. identified the extracellular pH sensor GPR65 as a key regulator in the glioblastoma tumor microenvironment, promoting the M2 polarization of TAMs and contributing to tumor progression. GPR65-driven acid production by glioma cells acidifies the microenvironment, fostering immunosuppression. Radiomics analyses reveal that gliomas with higher acidity, particularly IDH wild-type tumors, exhibit worse survival outcomes, emphasizing the link between tumor acidity and aggressiveness. Enrichment analyses associate GPR65 with immune suppression pathways, notably IL-10 signaling, highlighting its role in shaping an immunosuppressive milieu[65].

16.2.4.3 Angiogenesis and Abnormal Vasculature

GBM cells secrete vascular endothelial growth factor (VEGF), a potent driver of angiogenesis, leading to the formation of new blood vessels within the tumor microenvironment. However, unlike normal vasculature, these newly formed blood vessels are disorganized, leaky, and poorly functional, which impairs efficient oxygen delivery and contributes to the development of hypoxic regions within the tumor. This hypoxia exacerbates tumor aggressiveness and therapeutic resistance. Furthermore, tumor endothelial cells directly interact with GSCs through Notch signaling mediated by the Delta-like ligand 4 (Dll4). This interaction promotes the self-renewal and proliferation of GSCs, reinforcing their role in tumor progression and the maintenance of the GBM microenvironment[66].

16.2.4.4 Hypoxia and Metabolic Reprogramming

GBM is characterized by a hypoxic microenvironment, with oxygen levels ranging from 2.5% to as low as 0.1%. Hypoxia stabilizes hypoxia-inducible factors (HIF-1α and HIF-2α), which activate a suite of genes involved in angiogenesis, glycolysis, and cell survival, thereby facilitating tumor adaptation to low oxygen conditions. This hypoxic state drives a metabolic shift from oxidative phosphorylation to glycolysis, increasing glucose uptake and lactate production. This metabolic reprogramming supports the rapid growth and survival of GBM cells in oxygen-deprived regions, further contributing to tumor progression and resistance to therapy[67].

16.3 CURRENT STANDARD THERAPIES FOR GLIOBLASTOMA

Standard therapies for glioblastoma include surgical resection, radiotherapy, and chemotherapy, often combined for maximum effect. Surgical resection aims to remove as much of the tumor as possible, but complete removal is nearly impossible due to the invasive nature of glioblastoma, leading to rapid recurrence. However, there have been some advances in the surgical procedures that are opted for the treatment. Clarke et al. conducted a multicenter phase 3 study in Germany that assessed using 5-aminolevulinic acid (5-ALA), a fluorescent label, to enhance surgical resection of GBM. The trial included patients with ring-enhancing tumors believed to be fully resectable. Participants were randomized to either standard white light or fluorescence guidance using 5-ALA during surgery. The results showed that 5-ALA significantly improved the rate of complete resection, with 65% of patients in the fluorescence-guided group achieving gross total resection compared to 35% in the conventional group[68].

Laser interstitial thermal therapy (LITT) is a relatively newer method compared to standard therapies for GBM. While it is not yet considered a first-line treatment, it is increasingly being explored as an alternative or adjunct to traditional therapies such as surgery, radiation, and chemotherapy. LITT uses a laser to heat and destroy tumor tissue, and MRI typically guides it. It is mainly used in cases where surgery may be challenging, such as for tumors in difficult-to-reach locations or for patients who are not candidates for traditional surgery. Though promising, it is still considered investigational and is typically used in clinical trials or for patients undergoing other treatments. The method has been gaining attention for its potential to provide a minimally invasive approach[69].

Radiotherapy is employed to target the remaining tumor cells post-surgery. Advances in radiotherapy, such as intensity-modulated radiotherapy (IMRT), volumetric-modulated arc therapy (VMAT), and image-guided radiotherapy (IGRT), have significantly improved the precision and effectiveness of glioblastoma treatment. IMRT enables precise dose delivery to irregularly shaped tumors, sparing critical structures like the brainstem and hippocampus, while IGRT enhances accuracy by correcting planning and setup errors through daily imaging techniques, such as cone-beam CT. VMAT offers comparable dosimetric benefits to IMRT but with shorter treatment times, making it increasingly favored despite limited outcome studies. Hypofractionated regimens, which involve delivering higher doses over fewer sessions, reduce treatment duration and costs but have not demonstrated significant survival benefits over standard fractionation. Decreased clinical target volume margins, reduced from the traditional 2 cm to 5 mm, have shown comparable outcomes with less damage to surrounding tissues. For elderly patients with limited life expectancy, short-course radiotherapy regimens, like 40 Gy in 15 fractions, are effective and better tolerated, and combining radiotherapy with TMZ has improved survival without compromising quality of life. Novel imaging techniques, such as amino acid PET tracers like [18F]-fluoroethyl-L-tyrosine (FET), enhance tumor delineation and help distinguish pseudoprogression from true progression, improving radiotherapy planning and monitoring. Re-irradiation has also emerged as a viable option for recurrent glioblastoma, utilizing stereotactic techniques or IMRT/VMAT, often combined with bevacizumab to reduce toxicity, achieving a median survival of 9–11 months in selected patients. These innovations emphasize the progression toward personalized, precision radiotherapy, optimizing outcomes while minimizing side effects, though further research is essential to refine treatment protocols and enhance survival rates. Despite all the benefits, the effectiveness of radiotherapy is reduced in hypoxic tumor regions, and it can cause significant damage to healthy brain tissue[70].

Chemotherapy, primarily with TMZ, inhibits tumor growth. TMZ is an alkylating agent used in combination with radiotherapy and as adjuvant therapy. TMZ's efficacy is influenced by the methylation status of the MGMT promoter; methylated tumors show better response due to reduced DNA repair capability. Alternative chemotherapy regimens include nitrosoureas-like lomustine, particularly for recurrent glioblastoma. Despite its widespread use, TMZ faces challenges, including resistance mediated by CSCs and detoxifying proteins like MGMT. In elderly or frail patients, TMZ may be used alone instead of radiotherapy[71]. Overall, while chemotherapy can extend survival, outcomes remain poor, necessitating new strategies to overcome resistance and improve efficacy. But it is frequently challenged by over-expression of DNA repair enzyme, such as MGMT, and the poor ability of drugs to penetrate across the Blood–brain barrier (BBB). GSCs possess multifactorial mechanisms of resistance to chemotherapy, including TMZ, and are regulated by both intrinsic and extrinsic factors, including local drug concentrations and the tumor microenvironment. One of the resistance factors is represented by detoxifying proteins, including MGMT, whose high expression is associated with a worse treatment response. Furthermore, chemotherapy is not capable of effectively disposing off CSCs, because often chemotherapy fails due to poor drug delivery at the tumor site, allowing the presence of CSCs that can induce regrowth of the tumor. The sensitization of glioblastoma CSCs to chemotherapy is a matter of debate, and controversial data in the literature hinder the design of treatment schedules. Consequently, the prognosis for glioblastoma patients is quite bleak, with a median survival time of 14.6 months, which suggests the necessity of novel therapeutic strategies[72].

GBM immunotherapy aims to leverage the immune system to attack tumors but is challenged with several barriers, such as GBM's immunosuppressive microenvironment and low mutational load. Recent strategies in immunotherapy involve ICIs against PD-1, PD-L1, and CTLA-4 with great success in some cancers, but low in GBM. New targets such as LAG-3, TIM-3, and TIGIT are under investigation, as well as combination therapies to enhance responses. Vaccination therapies aim to stimulate immune responses against tumor-specific or tumor-associated antigens, though challenges like antigen heterogeneity and instability hinder their effectiveness. Multi-antigen vaccines are under development to enhance anti-tumor immunity and reduce immune evasion. Adoptive T cell therapies, including tumor-infiltrating lymphocytes (TILs) and chimeric antigen receptor T cells (CAR-T), show promise but are limited by issues such as antigen escape, variable antigen expression, and the immunosuppressive tumor environment. Efforts to develop next-generation CAR-T cells targeting multiple antigens are ongoing. Virus-based therapies, including oncolytic viruses (OVs) and viral vectors, selectively infect tumor cells or deliver therapeutic genes, inducing immune responses and disrupting the tumor microenvironment. Despite these advancements, GBM remains resistant to immunotherapy, underscoring the need for innovative strategies to overcome immune evasion and enhance treatment efficacy[73].

16.4 EMERGING TARGETED GENE AND CELL THERAPIES

Despite the progress in surgery, radiation therapy, and chemotherapy with TMZ, the outcome of GBM continues to be poor, with a median survival of less than 15 months. The dismal results are due to the highly heterogeneous nature of the tumor, its ability to evade immune surveillance, and the BBB, which limits delivery of drugs to the brain. These obstacles underscore the necessity for novel treatment strategies to which these GBM features are vulnerable[74].

Traditional treatments frequently are not as effective, as GBM cells rapidly adjust and ultimately grow resistant to treatments such as chemotherapy and radiation. In addition, the various types of cells in GBM, a phenomenon called tumor heterogeneity, pose a challenge for effective treatment of all of the tumor subtypes. Additionally, GBM establishes an immunosuppressive environment, which also prevents the immune system from fighting the tumor. The BBB is also a physical and biochemical barrier that effectively restricts most of these therapy agents from entering in the brain compartment[75].

Emerging therapeutic strategies offer hope in overcoming these challenges. Unlike traditional approaches, these trends are more precise, leveraging advances in molecular biology and technology to target GBM more effectively. Targeted therapies and immunotherapies, for instance, address the unique genetic and molecular features of GBM, potentially bypassing resistance mechanisms. Nanotechnology and other advanced drug delivery systems improve the penetration of therapeutic agents across the BBB, ensuring that treatments reach the tumor more efficiently[76]. Furthermore, combination therapies, which employ multiple mechanisms simultaneously, have shown promise in producing synergistic effects that enhance overall efficacy. Importantly, many of these approaches aim to reduce systemic toxicity, improving patients' quality of life during treatment.

The following section will explore these emerging trends in GBM treatment, highlighting their potential to revolutionize the management of this devastating disease. From immunotherapies and nanotechnology to targeted and gene-based approaches, these innovations aim to address the limitations of conventional treatments and offer new hope to patients[77].

16.4.1 Oncolytic Viral Therapies

Oncolytic virotherapy is an innovative cancer treatment that uses viruses to infect and kill cancer cells while sparing healthy tissues selectively. These OVs are either genetically modified or naturally occurring, designed to exploit the vulnerabilities of cancer cells, such as their impaired antiviral defenses and rapid growth. Once inside the tumor, the virus replicates, causing the infected cells to rupture, a process called oncolysis. This destruction releases tumor-specific antigens, which activate the immune system to attack remaining cancer cells, thereby combining direct tumor killing with immunotherapy[78].

This dual mechanism offers several advantages, including precision targeting of cancer cells, minimal damage to normal tissues, and the potential to stimulate systemic anti-tumor immunity. Examples of OVs include modified herpes simplex virus (e.g., T-VEC), adenoviruses (e.g., DNX-2401), and poliovirus derivatives (e.g., PVS-RIPO), many of which are in clinical trials for glioblastoma and other cancers. Despite its promise, challenges remain, including the need to overcome the blood-brain barrier in brain tumors, immune clearance of the virus, and tumor heterogeneity. Nevertheless, oncolytic virotherapy is a promising treatment approach, particularly for aggressive cancers like glioblastoma, offering hope for improved outcomes when combined with other therapies[76].

A recent 1/2 clinical trial conducted by Nassiri et al. investigated the safety and efficacy of combining the OV DNX-2401 with the immune checkpoint inhibitor pembrolizumab in patients with recurrent glioblastoma. The study included 49 patients treated with a single intratumoral dose of DNX-2401 followed by pembrolizumab administered intravenously. The primary endpoints were safety and objective response rates, while secondary endpoints included overall survival and clinical benefit[78].

Results showed that the treatment was well-tolerated, with no dose-limiting toxicities. The 12-month overall survival rate reached 52.7%, significantly surpassing the 20% historical benchmark. The objective response rate was 10.4%, including complete and partial tumor responses, though it did not meet the primary efficacy threshold. A clinical benefit, defined as stable disease or better, was observed in 56.2% of patients. Notably, two patients achieved durable complete responses lasting over 45 months.

Exploratory analyses suggested that the tumor microenvironment, characterized by moderate immune cell infiltration and checkpoint protein expression, influenced treatment outcomes. The study concluded that DNX-2401 combined with pembrolizumab is a promising therapeutic strategy for recurrent glioblastoma, warranting further investigation to optimize efficacy[78].

Another clinical study by Ling et al. explored the efficacy and safety of CAN-3110, an oncolytic herpes simplex virus type 1 (oHSV), for treating recurrent glioblastoma (rGBM) and high-grade gliomas (rHGG). The study enrolled 41 patients, utilizing intratumoral injections of the engineered virus, which selectively replicates in tumor cells via a nestin promoter, sparing healthy brain tissue. The primary endpoints included safety, immunological responses, and survival outcomes.

Key findings highlighted the safety of CAN-3110, with no dose-limiting toxicities reported. Patients with positive pre-existing herpes simplex virus 1 (HSV1) serology showed significantly improved median overall survival (14.2 months) compared to seronegative patients (7.8 months). The treatment elicited strong immune activation, increasing CD4+ and CD8+ tumor-infiltrating lymphocytes, especially in HSV1-seropositive patients. Immunological analyses revealed that the virus altered the tumor microenvironment, converting it from immunosuppressive to pro-inflammatory, which correlated with extended survival.

Notably, the virus persisted in tumors for several months, a rare feature compared to previous OV therapies. The study concluded that CAN-3110 effectively activates immune responses and holds promise as a therapeutic modality for rGBM, particularly for HSV1-seropositive patients, warranting further investigation in larger trials[79].

16.4.2 CRISPR-Cas9 and Gene Editing Approaches

Clustered regularly interspaced short palindromic repeats-associated protein 9 (CRISPR-Cas9), a natural immune defense in bacteria and archaea, recognizes and eliminates viral DNA through CRISPR arrays and the Cas9 endonuclease. This mechanism inspired a revolutionary gene-editing technology widely used to study GBM development and progression. CRISPR-Cas9 comprises guide RNA and Cas9 enzyme, where the guide RNA directs Cas9 to target DNA for precise editing. Advances like Cas9 nickase enable single-strand DNA breaks, reducing off-target effects[80].

In GBM research, CRISPR-Cas9 components are derived from lentiviral or plasmid vectors and delivered via viral or lipid-mediated methods. Cas9 creates double-strand DNA breaks, repaired through non-homologous end joining (NHEJ) or homology-directed repair (HDR). NHEJ is imprecise, while HDR uses a designed template for precise insertion. Transfected cells are selected using puromycin or FACS, with qPCR or Western blot validating gene editing. This technology facilitates precise GBM gene function studies, advancing therapeutic strategies[81].

CRISPR-Cas9 genome editing has become an indispensable tool in GBM research, enabling precise identification and manipulation of genes linked to critical hallmarks of tumor development, progression, and recurrence. This technology has revealed the central role of genes like Dazl and the regulatory region SRR2 in GBM's aggressive behavior. Knockdown of Dazl, a germline gene upregulated in late-stage GBM, significantly reduces cell proliferation, migration, invasion, and stem cell marker expression (CD133, Oct4, Nanog, Sox2) while increasing apoptosis and chemosensitivity to chemotherapeutics like TMZ and doxorubicin. Similarly, deletion of the SRR2 regulatory region disrupts SOX2 expression, leading to impaired self-renewal, tumor growth, and increased expression of cell cycle inhibitors p21CIP1 and p27KIP1, further reducing the tumorigenic potential of GBM cells[82].

In the autophagy pathway, genes like ATG5 and ATG7 have been shown to play dual roles, either promoting or inhibiting cell death depending on the cellular context, underscoring the complex nature of autophagy regulation in GBM therapy resistance. Investigating apoptosis regulators such as TRIM45 and FAT1 revealed opposing effects on programmed cell death, with TRIM45 knockout suppressing apoptosis and promoting tumor growth, while FAT1 knockout enhances apoptotic activity, offering potential therapeutic leverage. In angiogenesis, knockdowns of NOTCH1, DDX39B, and PDPN impair vascular development, with PDPN identified as a novel biomarker for therapeutic response[66].

Furthermore, CRISPR-Cas9-mediated knockouts of immune regulatory genes have illuminated new strategies for enhancing immune responses against GBM, although detailed mechanisms remain fully elucidated. These findings collectively demonstrate the potential of CRISPR-Cas9 not only to dissect the molecular underpinnings of GBM but also to inform the development of targeted treatments. By identifying and modulating genes with unique and overlapping roles in key GBM pathways, this technology offers a promising avenue for innovative therapies that address tumor heterogeneity, enhance chemosensitivity, and reduce recurrence[23].

Application of CRISPR-Cas9 genome editing to identifying genes that correlate with different hallmarks of GBM. CRISPR-Cas9-mediated gene knockdown, knockout, and overexpression have been used to identify several genes with unique and overlapping roles in GBM development, progression, and recurrence. The listed genes with their references are the ones already edited using CRISPR-Cas9 genome editing technology in GBM research.

16.4.3 Stem Cell-Based Delivery Systems

Stem cell-based therapy represents a pioneering approach to treat glioblastomas by effectively targeting cancer stem cells (CSCs) that drive tumor growth and resistance against traditional therapies. MSCs and NSCs are particularly notable because of their ability to home in on tumors. Glioblastomas attract MSCs through angiogenic cytokines such as VEGF, IL-8, and TGF-β, making these stem cells ideal carriers for therapeutic agents. Engineered MSCs that deliver IL-2 have demonstrated significant efficacy in glioma models. NSCs can also be modified to express prodrug-activating systems such as cytosine deaminase paired with 5-fluoroscytosine and thymidine kinase (TK) paired with ganciclovir. This targeted delivery can effectively reduce tumor volumes. In addition to this delivery, stem cells can secrete BMP4, which reduces the tumorogenic potential of CSCs by activating suppressor of mothers against decapentaplegic (SMAD) signaling pathways that inhibit their proliferation. This strategic targeting helps prevent tumor recurrence[83]. Overall, stem cell-based therapies of advantages over conventional treatments, including targeted specificity, reducing damage to healthy brain tissue, and enhanced ability to overcome chemoresistance while leveraging the tumor-homing capabilities of stem cells along with an advanced delivery system. These therapies provide an innovative solution that can significantly improve outcomes for patients. In glioblastoma therapy, stem cells are particularly effective in addressing major challenges such as the BBB and the blood-tumor barrier, which restricts the efficacy of conventional treatments. Engineered MSCs and NSCs can be utilized to deliver therapeutic agents, including OVs, prodrug-activating enzymes, and anti-angiogenic agents, directly to the tumor site. These therapies not only inhibit tumor progression but also specifically target GSCs, which are responsible for tumor resistance, recurrence, and progression; moreover, the integration of nanotechnology with stem cell therapy. The effectiveness of this approach, nanoparticles loaded with chemotherapeutic drugs or imaging agents, can be delivered via stem cells for controlled release and precise targeting, further improving therapeutic outcomes. Therefore, the presence of stem cell-based therapy is a highly promising strategy to overcome the limitation of current glioblastoma treatments while offering sustained, localized and effective tumor eradication.

A study by Huang et al. explored the use of adipose-derived stem cells (ADSCs) to deliver nanotherapeutics specifically to brain tumors for thermo-chemotherapy. The nanoparticles were created using a combination of poly(γ-glutamic acid-co-distearyl γ-glutamate), poly(lactic-co-glycolic acid), paclitaxel (PTX), and oleic acid-coated superparamagnetic iron oxide nanoparticles. These had a size of 110 nm and a drug loading content of 8.4 wt%. The study found that when ADSCs engulfed these nanotherapeutics, they were relatively harmless to host cells at a PTX concentration of 30 μM over 48 h without specific stimulation. However, combining ADSCs with a high-frequency magnetic field led to a four-fold increase in therapeutic effectiveness for brain astrocytoma (ALTS1C1)-bearing mice compared to the standard chemotherapy drug, TMZ. Immunohistochemical analysis confirmed the therapy's effective delivery and strong cytotoxic effects against tumor cells *in vivo*. This research underscores the potential of ADSC-mediated chemo-thermal therapy for brain tumors[82].

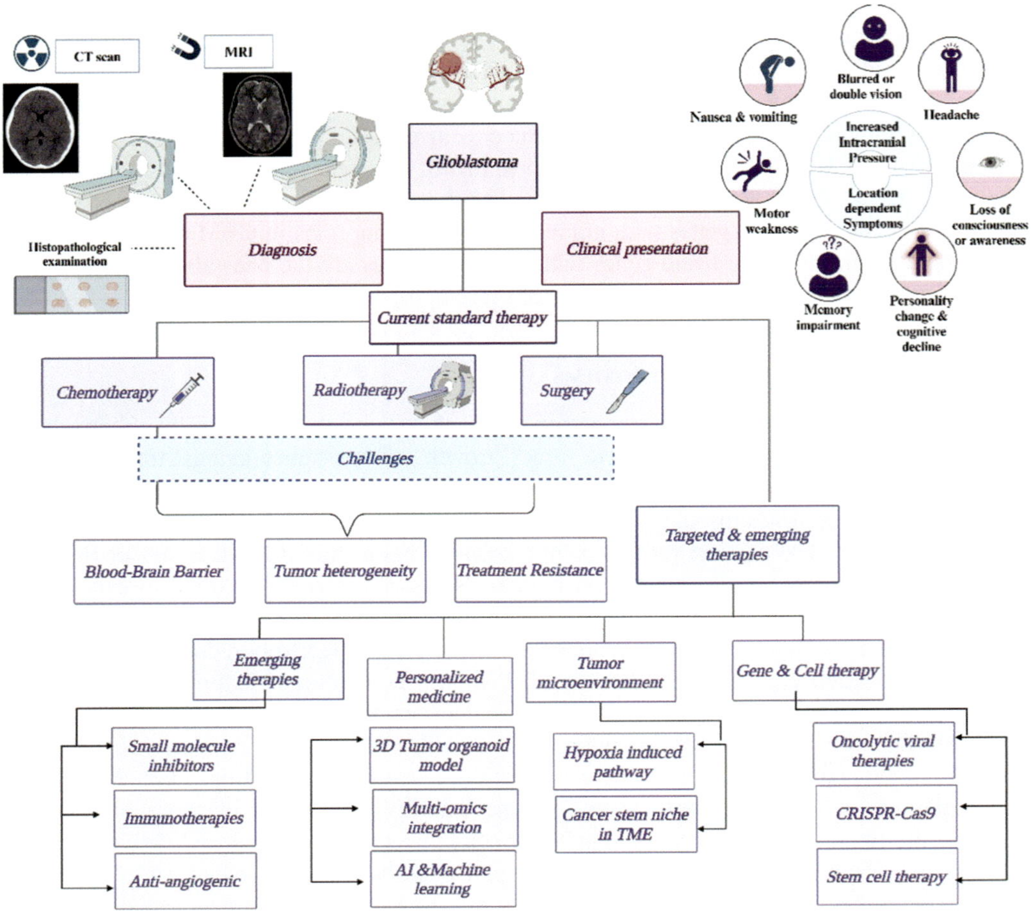

FIGURE 16.3 Diagnosis, symptoms, treatment, and advancing therapies for glioblastoma.

Yudintceva et al. emphasize the potential of MSCs in glioblastoma treatment due to their ability to migrate toward tumors. In vitro tests showed that MSCs effectively migrate to C6 glioblastoma cells, making them suitable for targeted therapies. Intravenous delivery of superparamagnetic iron oxide nanoparticle (SPION)-labeled MSCs resulted in significant tumor accumulation and improved visibility during high-field magnetic resonance imaging. These findings highlight MSCs as a promising drug-delivery platform for enhancing glioblastoma treatment outcomes[84].

Understanding the therapeutic profile of GBM, standard and emerging therapies are increasingly being integrated to form a more comprehensive and personalized treatment strategy. This combined strategy not only aims to debulk the tumor and control immediate progression but also to address the infiltrative and heterogenous nature of GBM at the molecular level. Figure 16.3, outlines how traditional and innovative modalities are being aligned to improve the clinical outcomes. Such synergy reflects a translational shift toward the refined version of ongoing research and clinical trials that have been happening.

16.5 CLINICAL TRIALS

GBM remains one of the most lethal and treatment-resistant brain tumors, as illustrated in Figure 16.4. With only 4 FDA-approved drugs since 1920 and minimal improvement in the outcomes, clinical trials play a pivotal role in advancing the understanding and treatment of this

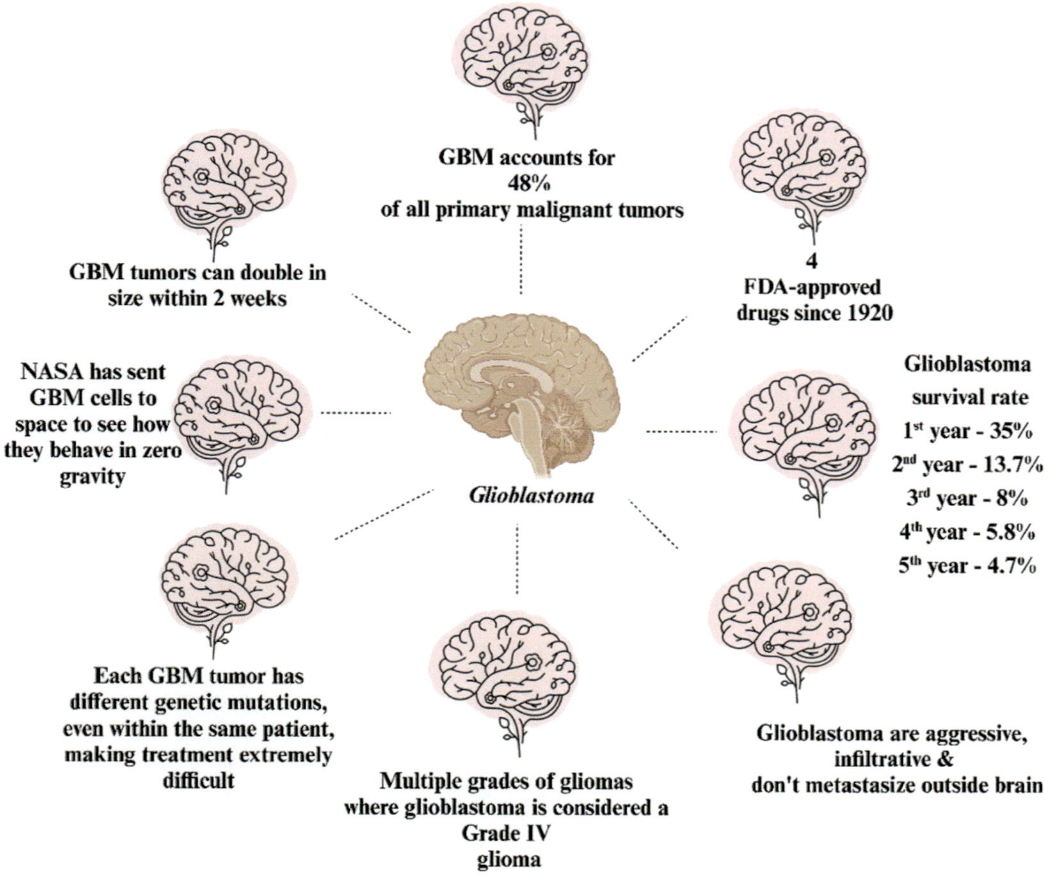

FIGURE 16.4 Epidemiological and factual overview of glioblastoma.

aggressive malignancy. Traditional therapy, such as surgical resection, radiotherapy, and TMZ chemotherapy, has not significantly improved long-term survival due to GBM's aggressive nature, rapid proliferation, and resistance[85]. The complexities of GBM necessitate innovative strategies with a significant focus on exploring the tumor microenvironment to enhance treatment efficacy[86]. Novel approaches such as gene therapy, immunotherapy, epigenetic therapy, oncologic virus therapy, anti-angiogenic therapy, and boron neutron capture therapy are being investigated, though their clinical efficacy remains uncertain[87]. Immunotherapy, including immune checkpoint blockade, CAR-T therapy, oncolytic viral therapy, and vaccine therapy, offers new hope for improving GBM outcomes. Ongoing studies are using combinatorial therapies with the aim of minimizing adverse side effects and augmenting antitumor immune responses[88].

Nanotechnology presence opportunities for targeted drug delivery to overcome the BBB and enhance therapeutic precision. The application of nanobiotechnology in targeted drug delivery is a prominent area of research, aiming to selectively eradicate tumor cells while sparring the surrounding brain tissue[89]. Despite significant progress in understanding GBM biology and developing new therapies, the translation of basic science achievements into clinical practice has been slow, emphasizing the need for more effective strategies[90].

The use of viral therapies represents a more promising avenue, leveraging viral mechanisms to selectively lyse neoplastic cells or deliver therapeutic transgenes, potentially revolutionizing treatment approaches[91]. The OVs represent a novel immunotherapy treatment modality that is gaining interest in the clinical community because of their promising level of antitumor activity and the fact that they are not subjected to the same resistance mechanisms that limit the use of chemotherapy.

Despite decades of research, the prognosis of GBM remains grim, with a median overall survival of approximately 14–16 months and a 5-year survival rate of less than 7%, even with the aggressive standard therapy[92]. This stark reality underscores the critical need for innovative, biology-driven interventions. The emergence of personalized medicine, molecular diagnostics, and next-generation delivery systems is redefining the therapeutic landscape.

Table 16.1 gives an insight into the clinical trials and translational research initiatives aimed at addressing the therapeutic challenges of GBM.

TABLE 16.1
Clinical Trials Evaluating Emerging Therapeutic Strategies in GBM

Trial Phase	Objective	Intervention	Target Population	Mechanism of Action	Endpoints	Study Type
Phase I/II	Test safety, side effects, and optimal dose of selinexor with radiation therapy in newly diagnosed DIPG or H3 K27 M-mutant HGG	Selinexor + radiation therapy	Children and young adults with DIPG or H3 K27 M-mutant HGG	Selinexor blocks CRM1 (nuclear export inhibitor), combined with radiation, to shrink tumors	MTD, safety, side effects, tumor response, and efficacy	Interventional (open-label)
Phase II	Compare safety, side effects, and effectiveness of anti-LAG-3 (relatlimab) and anti-PD-1 (nivolumab) vs. lomustine in recurrent glioblastoma	Relatlimab + nivolumab vs. lomustine	Patients with recurrent glioblastoma	Relatlimab and nivolumab are immune checkpoint inhibitors; lomustine damages DNA to kill cancer cells	Safety, tolerability, progression-free survival (PFS), overall survival (OS)	Interventional (randomized)
Phase I	Determine maximum tolerated dose (MTD) of a novel drug in glioblastoma	Drug A + standard therapy	Patients with recurrent glioblastoma	Inhibition of specific molecular targets (e.g., EGFR or VEGF pathways)	MTD, safety profile, preliminary efficacy	Interventional (open-label)
Phase II (NCT04574577, NCT 03696355)	Evaluate efficacy of a targeted therapy in glioblastoma	Targeted Therapy (e.g., EGFR inhibitor)	Patients with EGFR-amplified glioblastoma	Blocks EGFR signaling, reducing tumor growth	Progression-free survival (PFS), overall response rate	Interventional (non-randomized)
Phase III (NCT04330241)	Compare standard treatment with experimental therapy	Experimental drug + standard care	Newly diagnosed glioblastoma patients	Enhances standard therapy by targeting tumor-specific pathways	Overall survival (OS), quality of life	Interventional (randomized)

(Continued)

TABLE 16.1 (*Continued*)
Clinical Trials Evaluating Emerging Therapeutic Strategies in GBM

Trial Phase	Objective	Intervention	Target Population	Mechanism of Action	Endpoints	Study Type
Phase II (NCT02852655)	Assess the combination of immunotherapy with chemotherapy	Checkpoint inhibitor + chemotherapy	Recurrent glioblastoma patients	Enhances immune response against tumor cells	Tumor response rate, immune markers	Interventional (Randomized)
Phase I/II (NCT02026271, NCT03750071)	Explore gene therapy approaches for glioblastoma	Gene Therapy	Glioblastoma patients	Delivers therapeutic genes to tumor cells	Safety, feasibility, and preliminary efficacy	Interventional (Open-Label)
Phase I	Establish MTD and assess safety of allogenic adipose-derived mesenchymal stem cells (AMSCs) in recurrent glioblastoma patients undergoing craniotomy	AMSCs (intratumoral) + Craniotomy	Patients with recurrent glioblastoma or astrocytoma	AMSCs may reduce tumor cell growth, residual tumor cell survival, and chemotherapy resistance	MTD, safety, progression-free survival (PFS), overall survival (OS), systemic immune response	Interventional (open-label)

16.6 CONCLUSION

GBM remains a highly aggressive and recalcitrant malignancy, marked by profound molecular heterogeneity, diffuse infiltration, and intrinsic resistance to conventional therapies. This chapter has dissected the molecular framework of GBM pathogenesis, emphasizing critical genetic drivers such as TP53, EGFR, IDH1/2, and PTEN, as well as epigenetic alterations like MGMT promoter methylation and global histone modification patterns. These abnormalities cover dysregulated signaling networks, including PI3K/AKT/mTOR, RTK/RAS, NOTCH, and Wnt, that sustain tumor cell proliferation, stemness, and resistance to apoptosis.

Beyond the intrinsic changes in cells, GBM is shaped by a dynamic tumor microenvironment that includes hypoxia, metabolic stress, immune evasion mechanisms, and stromal interactions. The immunosuppressive niche dominated by TAMs, regulatory T cells, and altered cytokine signaling poses a major barrier to immunotherapeutic efficacy. Hypoxic gradients mediated by HIF-1α signaling further potentiate invasion, angiogenesis, and therapy resistance. Current treatment paradigms, maximal surgical resection followed by radiotherapy and TMZ, have limited impact on long-term survival with inevitable tumor recurrence. However, emerging strategies such as immune checkpoint blockade, personalized neoantigen vaccines, CAR-T therapy, oncolytic virotherapy, and epigenetic modulators are under active investigation, with several progressing through early-phase clinical trials. Innovations in nanoparticle-based drug delivery, blood-brain barrier modulation, and single-cell multi-omics are also enhancing our ability to dissect and therapeutically target GBM complexity at an unprecedented resolution.

Looking ahead, the future of GBM treatment lies in harnessing precision oncology, integrating comprehensive molecular profiling, spatial transcriptomics, and patient-derived models to stratify patients and tailor multimodal interventions. Interdisciplinary efforts bridging tumor biology, immunology, bioengineering, and computational modeling will be essential. As we continue to unravel the intricate molecular circuitry and adaptive mechanisms of GBM, novel avenues for durable targeted and personalized therapies will emerge, offering renewed hope for transforming patient outcomes in this formidable disease.

REFERENCES

1. Zülch KJ. *Brain Tumors: Their Biology and Pathology*. Springer Science & Business Media; 2012.
2. Girardi F, Allemani C, Coleman MP. Worldwide Trends in Survival from Common Childhood Brain Tumors: A Systematic Review. *J Glob Oncol*. 2019;(5):1–25. doi:10.1200/JGO.19.00140
3. Agrawal A. *Brain Tumors*. BoD – Books on Demand; 2022.
4. Weller M, Wick W, Aldape K, et al. Glioma. *Nat Rev Dis Primer*. 2015;1(1):1–18. doi:10.1038/nrdp.2015.17
5. Finch A, Solomou G, Wykes V, Pohl U, Bardella C, Watts C. Advances in Research of Adult Gliomas. *Int J Mol Sci*. 2021;22(2):924. doi:10.3390/ijms22020924
6. Alifieris C, Trafalis DT. Glioblastoma Multiforme: Pathogenesis and Treatment. *Pharmacol Ther*. 2015;152:63–82. doi:10.1016/j.pharmthera.2015.05.005
7. Mansour MA, Khalil DF, Ayad AA. Glioblastoma Masquerading as a Cystic Brain Lesion: A Case Report and Evidence-Based Review. *Int J Surg Case Rep*. 2023;106:108277. doi:10.1016/j.ijscr.2023.108277
8. Yano H, Miwa K, Nakayama N, et al. Differentiation of Astrocytoma Between Grades II and III Using a Combination of Methionine Positron Emission Tomography and Magnetic Resonance Spectroscopy. *World Neurosurg X*. 2023;19:100193. doi:10.1016/j.wnsx.2023.100193
9. Chiappa A, Bertani E, Pravettoni G, et al. Aggressive Surgical Approach for Treatment of Primary and Recurrent Retroperitoneal Soft Tissue Sarcoma. *Indian J Surg*. 2018;80(2):154–162. doi:10.1007/s12262-018-1722-7
10. Aldape K, Zadeh G, Mansouri S, Reifenberger G, von Deimling A. Glioblastoma: Pathology, Molecular Mechanisms and Markers. *Acta Neuropathol (Berl)*. 2015;129(6):829–848. doi:10.1007/s00401-015-1432-1
11. Ramos AD, Magge RS, Ramakrishna R. Molecular Pathogenesis and Emerging Treatment for Glioblastoma. *World Neurosurg*. 2018;116:495–504. doi:10.1016/j.wneu.2018.04.021
12. Singh SR, Bhaskar R, Ghosh S, et al. Exploring the Genetic Orchestra of Cancer: The Interplay between Oncogenes and Tumor-Suppressor Genes. *Cancers*. 2025;17(7):1082. doi:10.3390/cancers17071082
13. Stojchevski R, Sutanto EA, Sutanto R, et al. Translational Advances in Oncogene and Tumor-Suppressor Gene Research. *Cancers*. 2025;17(6):1008. doi:10.3390/cancers17061008
14. Gangoso E, Southgate B, Bradley L, et al. Glioblastomas Acquire Myeloid-Affiliated Transcriptional Programs via Epigenetic Immunoediting to Elicit Immune Evasion. *Cell*. 2021;184(9):2454–2470.e26. doi:10.1016/j.cell.2021.03.023
15. Daia, L, Liud, Z, Zhua, Y, Mab L. Genome-Wide Methylation Analysis of Circulating Tumor DNA: A New Biomarker for Recurrent Glioblastoma. *Heliyon*. 2023;9(3):e14339. Accessed June 5, 2025. https://www.cell.com/heliyon/fulltext/S2405-8440(23)01546-3
16. Drexler R, Khatri R, Sauvigny T, et al. A Prognostic Neural Epigenetic Signature in High-Grade Glioma. *Nat Med*. 2024;30(6):1622–1635. doi:10.1038/s41591-024-02969-w
17. Ma S, Pan X, Gan J, et al. DNA Methylation Heterogeneity Attributable to a Complex Tumor Immune Microenvironment Prompts Prognostic Risk in Glioma. *Epigenetics*. 2024;19(1):2318506. doi:10.1080/15592294.2024.2318506
18. Chai RC, Chang YZ, Wang QW, et al. A Novel DNA Methylation-Based Signature Can Predict the Responses of MGMT Promoter Unmethylated Glioblastomas to Temozolomide. *Front Genet*. 2019;10. doi:10.3389/fgene.2019.00910
19. Li J, Taich ZJ, Goyal A, et al. Epigenetic Suppression of EGFR Signaling in G-CIMP+ Glioblastomas. *Oncotarget*. 2014;5(17):7342–7356. doi:10.18632/oncotarget.2350
20. Ashkan K, Baig Mirza A, Soumpasis C, et al. MGMT Promoter Methylation: Prognostication beyond Treatment Response. *J Pers Med*. 2023;13(6):999. doi:10.3390/jpm13060999
21. Salvati M, Tariciotti L, Brunetto GMF, et al. Glioblastoma: Molecular Profile and Immunophenotypic Analysis as Prognostic Tools for Tailored Therapy and Decision Making in a Recent Surgical Series. *Interdiscip Neurosurg*. 2020;20:100697. doi:10.1016/j.inat.2020.100697
22. Han S, Liu Y, Cai SJ, et al. IDH Mutation in Glioma: Molecular Mechanisms and Potential Therapeutic Targets. *Br J Cancer*. 2020;122(11):1580–1589. doi:10.1038/s41416-020-0814-x

23. Huang LE. Friend or foe—IDH1 Mutations in Glioma 10 Years on. *Carcinogenesis.* 2019;40(11):1299–1307. doi:10.1093/carcin/bgz134
24. Pavlou MAS, Grandbarbe L, Buckley NJ, Niclou SP, Michelucci A. Transcriptional and Epigenetic Mechanisms Underlying Astrocyte Identity. *Prog Neurobiol.* 2019;174:36–52. doi:10.1016/j.pneurobio.2018.12.007
25. Wu RS. *Influence of Hypoxia on Epigenetic Regulation and Differentiation in Glioblastoma.* University of Toronto. Published online June 2022. Accessed June 16, 2025. http://hdl.handle.net/1807/139068
26. Cano Galiano A. *Identifying and Targeting Metabolic Vulnerabilities of IDH Mutant Gliomas.* ORBilu-University of Luxembourg. Published online April 29, 2021. Accessed November 6, 2024. https://orbilu.uni.lu/handle/10993/47405
27. Pulivarthi J. EGFR Mutations and Signaling Pathways in Glioblastoma: Implications for Pathogenesis and Therapeutic Targeting. *UC Merced Undergrad Res J.* 2024;16(2). doi:10.5070/M416263507
28. Viana-Pereira M, Lopes JM, Little S, et al. Analysis of EGFR Overexpression, EGFR Gene Amplification and the EGFRvIII Mutation in Portuguese High-Grade Gliomas. *Anticancer Res.* 2008;28(2A):913–920.
29. van den Bent MJ, Gao Y, Kerkhof M, et al. Changes in the EGFR Amplification and EGFRvIII Expression between Paired Primary and Recurrent Glioblastomas. *Neuro-Oncol.* 2015;17(7):935–941. doi:10.1093/neuonc/nov013
30. Maire CL, Ligon KL. Molecular Pathologic Diagnosis of Epidermal Growth Factor Receptor. *Neuro-Oncol.* 2014;16(Suppl_8):viii1–viii6. doi:10.1093/neuonc/nou294
31. Yang P, You G, Zhang W, et al. Correlation of Preoperative Seizures with Clinicopathological Factors and Prognosis in Anaplastic Gliomas: A Report of 198 Patients from China. *Seizure Eur J Epilepsy.* 2014;23(10):844–851. doi:10.1016/j.seizure.2014.07.003
32. Huang W, Li J, Zhu H, et al. A Novel EGFR Variant EGFRx Maintains Glioblastoma Stem Cells through STAT5. *Neuro-Oncol.* 2024;26(1):85–99. doi:10.1093/neuonc/noad153
33. Chen CY, Chen J, He L, Stiles BL. PTEN: Tumor Suppressor and Metabolic Regulator. *Front Endocrinol.* 2018;9. doi:10.3389/fendo.2018.00338
34. Alimohammadi M, Rahimzadeh P, Khorrami R, et al. A Comprehensive Review of the PTEN/PI3K/Akt Axis in Multiple Myeloma: From Molecular Interactions to Potential Therapeutic Targets. *Pathol Res Pract.* 2024;260:155401. doi:10.1016/j.prp.2024.155401
35. Barzegar Behrooz A, Talaie Z, Jusheghani F, Łos MJ, Klonisch T, Ghavami S. Wnt and PI3K/Akt/mTOR Survival Pathways as Therapeutic Targets in Glioblastoma. *Int J Mol Sci.* 2022;23(3):1353. doi:10.3390/ijms23031353
36. Dogan E, Yildirim Z, Akalin T, et al. Investigating the Effects of PTEN Mutations on cGAS-STING Pathway in Glioblastoma Tumours. *J Neurooncol.* 2024;166(2):283–292. doi:10.1007/s11060-023-04556-4
37. Baeza N, Weller M, Yonekawa Y, Kleihues P, Ohgaki H. PTEN Methylation and Expression in Glioblastomas. *Acta Neuropathol (Berl).* 2003;106(5):479–485. doi:10.1007/s00401-003-0748-4
38. Liu M, Zhang Y, Jian Y et al. The Regulations of Telomerase Reverse Transcriptase (TERT) in Cancer. *Cell Death Dis.* 2024;15(1):90. doi:10.1038/s41419-024-06454-7
39. The regulations of telomerase reverse transcriptase (TERT) in cancer | Cell Death & Disease. Accessed May 13, 2025. https://www.nature.com/articles/s41419-024-06454-7
40. Kim S, Chowdhury T, Yu HJ, et al. The Telomere Maintenance Mechanism spectrum and Its Dynamics in Gliomas. *Genome Med.* 2022;14(1):88. doi:10.1186/s13073-022-01095-x
41. Bell RJA, Rube HT, Xavier-Magalhães A, et al. Understanding TERT Promoter Mutations: A Common Path to Immortality. *Mol Cancer Res.* 2016;14(4):315–323. doi:10.1158/1541-7786.MCR-16-0003
42. Gorria T, Crous C, Pineda E, et al. The C250T Mutation of TERTp Might Grant a Better Prognosis to Glioblastoma by Exerting Less Biological Effect on Telomeres and Chromosomes Than the C228T Mutation. *Cancers.* 2024;16(4):735. doi:10.3390/cancers16040735
43. Tomoszková S, Škarda J, Lipina R. Potential Diagnostic and Clinical Significance of Selected Genetic Alterations in Glioblastoma. *Int J Mol Sci.* 2024;25(8):4438. doi:10.3390/ijms25084438
44. Toufektchan E, Toledo F. The Guardian of the Genome Revisited: p53 Downregulates Genes Required for Telomere Maintenance, DNA Repair, and Centromere Structure. *Cancers.* 2018;10(5):135. doi:10.3390/cancers10050135
45. Song B, Yang P, Zhang S. Cell Fate Regulation Governed by p53: Friends or Reversible Foes in Cancer Therapy. *Cancer Commun Lond Engl.* 2024;44(3):297–360. doi:10.1002/cac2.12520
46. Liu S, Liu T, Jiang J, Guo H, Yang R. p53 Mutation and Deletion Contribute to Tumor Immune Evasion. *Front Genet.* 2023;14. doi:10.3389/fgene.2023.1088455

47. Alvarado-Ortiz E, De La Cruz-López KG, Becerril-Rico J, Sarabia-Sánchez MA, Ortiz-Sánchez E, García-Carrancá A. Mutant p53 Gain-of-Function: Role in Cancer Development, Progression, and Therapeutic Approaches. *Front Cell Dev Biol*. 2021;8:607670. doi:10.3389/fcell.2020.607670
48. Ham SW, Jeon HY, Jin X, et al. TP53 Gain-of-Function Mutation Promotes Inflammation in Glioblastoma. *Cell Death Differ*. 2019;26(3):409–425. doi:10.1038/s41418-018-0126-3
49. Toyooka Y, Aoki K, Usami FM, Oka S, Kato A, Fujimori T. Generation of Pulsatile ERK Activity in Mouse Embryonic Stem Cells Is Regulated by Raf Activity. *Sci Rep*. 2023;13(1):9465. doi:10.1038/s41598-023-36424-6
50. Dwijaksara NLB, Andromeda S, Gusti AWR, Siswanto PA, Devi NLPML, Arnita NPS. Innovations in Bioink Materials and 3D Bioprinting for Precision Tissue Engineering. *Metta J Ilmu Multidisiplin*. 2024;4(3):101–118. doi:10.37329/metta.v4i3.3520
51. Tilak M, Holborn J, New LA, Lalonde J, Jones N. Receptor Tyrosine Kinase Signaling and Targeting in Glioblastoma Multiforme. *Int J Mol Sci*. 2021;22(4):1831. doi:10.3390/ijms22041831
52. Zhu Y, Chen Z, Kim SN, et al. Characterization of Temozolomide Resistance Using a Novel Acquired Resistance Model in Glioblastoma Cell Lines. *Cancers*. 2022;14(9):2211. doi:10.3390/cancers14092211
53. Zając A, Sumorek-Wiadro J, Maciejczyk A, et al. LY294002 and Sorafenib as Inhibitors of Intracellular Survival Pathways in the Elimination of Human Glioma Cells by Programmed Cell Death. *Cell Tissue Res*. 2021;386(1):17–28. doi:10.1007/s00441-021-03481-0
54. Zając A, Sumorek-Wiadro J, Maciejczyk A, et al. The Engagement of Ras/Raf/MEK/ERK and PLCγ1/PKC Pathways Regulated by TrkB Receptor in Resistance of Glioma Cells to Elimination upon Apoptosis Induction. *Neuropharmacology*. 2025;262:110204. doi:10.1016/j.neuropharm.2024.110204
55. Jeuken J, van den Broecke C, Gijsen S, Boots-Sprenger S, Wesseling P. RAS/RAF Pathway Activation in Gliomas: The Result of Copy Number Gains Rather than Activating Mutations. *Acta Neuropathol (Berl)*. 2007;114(2):121–133. doi:10.1007/s00401-007-0239-0
56. Lampada A, Taylor V. Notch Signaling as a Master Regulator of Adult Neurogenesis. *Front Neurosci*. 2023;17:1179011. doi:10.3389/fnins.2023.1179011
57. Ma H, Yu B, Kong L, Zhang Y, Shi Y. Neural Stem Cells Over-Expressing Brain-Derived Neurotrophic Factor (BDNF) Stimulate Synaptic Protein Expression and Promote Functional Recovery Following Transplantation in Rat Model of Traumatic Brain Injury. *Neurochem Res*. 2012;37(1):69–83. doi:10.1007/s11064-011-0584-1
58. Bazzoni R, Bentivegna A. Role of Notch Signaling Pathway in Glioblastoma Pathogenesis. *Cancers*. 2019;11(3):292. doi:10.3390/cancers11030292
59. Paul I, Bhattacharya S, Chatterjee A, Ghosh MK. Current Understanding on EGFR and Wnt/-Catenin Signaling in Glioma and Their Possible Crosstalk. *Genes Cancer*. 2013;4(11–12):427–446. doi:10.1177/1947601913503341
60. Denysenko T, Annovazzi L, Cassoni P, Melcarne A, Mellai M, Schiffer D. WNT/β-Catenin Signaling Pathway and Downstream Modulators in Low- and High-Grade Glioma. *Cancer Genom Proteom*. 2016;13(1):31–45.
61. Li X, Kim HJ, Yoo J, et al. Distant Origin of Glioblastoma Recurrence: Neural Stem Cells in the Subventricular Zone Serve as a Source of Tumor Reconstruction after Primary Resection. *Mol Cancer*. 2025;24(1):64. doi:10.1186/s12943-025-02273-2
62. Li YR, Fang Y, Lyu Z, Zhu Y, Yang L. Exploring the Dynamic Interplay between Cancer Stem Cells and the Tumor Microenvironment: Implications for Novel Therapeutic Strategies. *J Transl Med*. 2023;21(1):686. doi:10.1186/s12967-023-04575-9
63. Manini I, Caponnetto F, Bartolini A, et al. Role of Microenvironment in Glioma Invasion: What We Learned from In Vitro Models. *Int J Mol Sci*. 2018;19(1):147. doi:10.3390/ijms19010147
64. Understanding Neovascularization in Glioblastoma: Insights from the Current Literature. Accessed June 12, 2025. https://www.mdpi.com/1422-0067/26/6/2763
65. Fan J, Liu J, Zhang B, et al. GPR65 Contributes to Constructing Immunosuppressive Microenvironment in Glioma. *Neurosurg Rev*. 2024;47(1):417. doi:10.1007/s10143-024-02633-4
66. Shabani M, Javanshir HT, Bereimipour A, Sadrabadi AE, Jalili A, Nayernia K. Contradictory Effect of Notch1 and Notch2 on Phosphatase and Tensin Homolog and Its Influence on Glioblastoma Angiogenesis. *Galen Med J*. 2021;10:e2091.
67. Persano L, Rampazzo E, Basso G, Viola G. Glioblastoma Cancer Stem Cells: Role of the Microenvironment and Therapeutic Targeting. *Biochem Pharmacol*. 2013;85(5):612–622. doi:10.1016/j.bcp.2012.10.001
68. Clarke J, Butowski N, Chang S. Recent Advances in Therapy for Glioblastoma. *Arch Neurol*. 2010;67(3):279–283. doi:10.1001/archneurol.2010.5

69. van Solinge TS, Nieland L, Chiocca EA, Broekman MLD. Advances in Local Therapy for Glioblastoma—Taking the Fight to the Tumour. *Nat Rev Neurol.* 2022;18(4):221–236. doi:10.1038/s41582-022-00621-0
70. Gzell C, Back M, Wheeler H, Bailey D, Foote M. Radiotherapy in Glioblastoma: The Past, the Present and the Future. *Clin Oncol.* 2017;29(1):15–25. doi:10.1016/j.clon.2016.09.015
71. Jezierzański M, Nafalska N, Stopyra M, et al. Temozolomide (TMZ) in the Treatment of Glioblastoma Multiforme—A Literature Review and Clinical Outcomes. *Curr Oncol.* 2024;31(7):3994–4002. doi:10.3390/curroncol31070296
72. Eramo A, Ricci-Vitiani L, Zeuner A, et al. Chemotherapy Resistance of Glioblastoma Stem Cells. *Cell Death Differ.* 2006;13(7):1238–1241. doi:10.1038/sj.cdd.4401872
73. Salvato I, Marchini A. Immunotherapeutic Strategies for the Treatment of Glioblastoma: Current Challenges and Future Perspectives. *Cancers.* 2024;16(7):1276. doi:10.3390/cancers16071276
74. Ghosh S, Vashisth K, Ghosh S, Han SS, Bhaskar R, Sinha JK. From Sleep to Cancer to Neurodegenerative Disease: The Crucial Role of Hsp70 in Maintaining Cellular Homeostasis and Potential Therapeutic Implications. *J Biomol Struct Dyn.* 2024;42(18):9812–9823. doi:10.1080/07391102.2023.2252509
75. Ghosh S, Bhaskar R, Mishra R, et al. Neurological Insights into Brain-Targeted Cancer Therapy and Bioinspired Microrobots. *Drug Discov Today.* 2024;29(9):104105. doi:10.1016/j.drudis.2024.104105
76. Sachdeva B, Sachdeva P, Negi A, et al. Chitosan Nanoparticles-Based Cancer Drug Delivery: Application and Challenges. *Drugs.* 2023;21(4):211. doi:10.3390/md21040211
77. Nguyen TTT, Greene LA, Mnatsakanyan H, Badr CE. Revolutionizing Brain Tumor Care: Emerging Technologies and Strategies. *Biomedicines.* 2024;12(6):1376. doi:10.3390/biomedicines12061376
78. Nassiri F, Patil V, Yefet LS, et al. Oncolytic DNX-2401 Virotherapy Plus Pembrolizumab in Recurrent Glioblastoma: A Phase 1/2 Trial. *Nat Med.* 2023;29(6):1370–1378. doi:10.1038/s41591-023-02347-y
79. Ling AL, Solomon IH, Landivar AM, et al. Clinical Trial Links Oncolytic Immunoactivation to Survival in Glioblastoma. *Nature.* 2023;623(7985):157–166. doi:10.1038/s41586-023-06623-2
80. Sharma G, Sharma AR, Bhattacharya M, Lee SS, Chakraborty C. CRISPR-Cas9: A Preclinical and Clinical Perspective for the Treatment of Human Diseases. *Mol Ther.* 2021;29(2):571–586. doi:10.1016/j.ymthe.2020.09.028
81. Mormino A, Garofalo S. Dialogue among Lymphocytes and Microglia in Glioblastoma Microenvironment. *Cancers.* 2022;14(11):2632. doi:10.3390/cancers14112632
82. Huang W.-C., Lu IL, Chiang W-H, et al. Tumortropic Adipose-Derived Stem Cells Carrying Smart Nanotherapeutics for Targeted Delivery and Dual-Modality Therapy of Orthotopic Glioblastoma. *J Controlled Release.* 2017;254:119–130. doi:10.1016/j.jconrel.2017.03.035
83. Ghosh AK, Ghosh A, Das PK. Nanotechnology Meets Stem Cell Therapy for Treating Glioblastomas: A Review. *ACS Appl Nano Mater.* 2024;7(3):2430–2460. doi:10.1021/acsanm.3c04714
84. Yudintceva N, Lomert E, Mikhailova N, et al. Targeting Brain Tumors with Mesenchymal Stem Cells in the Experimental Model of the Orthotopic Glioblastoma in Rats. *Biomedicines.* 2021;9(11):1592. doi:10.3390/biomedicines9111592
85. Cai X, Sughrue ME. Glioblastoma: New Therapeutic Strategies to Address Cellular and Genomic Complexity. *Oncotarget.* 2017;9(10):9540–9554. doi:10.18632/oncotarget.23476
86. Rawojć K, Ahmed MM, Mukhtiar A, Łukowiak M, Kisielewicz K. Nanomedicine-Enhanced Radiotherapy for Glioblastoma: Advances in Targeted Therapy and Adaptive Treatment Strategies. *Pharmaceutics.* 2025;17(4):508. doi:10.3390/pharmaceutics17040508
87. Zhang H, Wang R, Yu Y, Liu J, Luo T, Fan F. Glioblastoma Treatment Modalities Besides Surgery. *J Cancer.* 2019;10(20):4793–4806. doi:10.7150/jca.32475
88. Rong L, Li N, Zhang Z. Emerging Therapies for Glioblastoma: Current State and Future Directions. *J Exp Clin Cancer Res CR.* 2022;41(1):142. doi:10.1186/s13046-022-02349-7
89. Yeini E, Ofek P, Albeck N, et al. Targeting Glioblastoma: Advances in Drug Delivery and Novel Therapeutic Approaches. *Adv Ther.* 2021;4(1):2000124. doi:10.1002/adtp.202000124
90. Veliz I, Loo Y, Castillo O, Karachaliou N, Nigro O, Rosell R. Advances and Challenges in the Molecular Biology and Treatment of glioblastoma—Is There Any Hope for the Future? *Ann Transl Med.* 2015;3(1):7. doi:10.3978/j.issn.2305-5839.2014.10.06
91. Webb MJ, Sener U, Vile RG. Current Status and Challenges of Oncolytic Virotherapy for the Treatment of Glioblastoma. *Pharmaceuticals.* 2023;16(6):793. doi:10.3390/ph16060793
92. Giammona A, Commisso M, Bonanomi M, et al. A Novel Strategy for Glioblastoma Treatment by Natural Bioactive Molecules Showed a Highly Effective Anti-Cancer Potential. *Nutrients.* 2024;16(15):2389. doi:10.3390/nu16152389

17 Cognitive Decline and Ageing

Bhaskar Sharma, Renu Negi, Hannah Priscilla, and Prabhakar Singh

EXECUTIVE SUMMARY

- Cognitive decline is a hallmark of ageing, influenced by genetic, epigenetic, lifestyle, and environmental factors.
- Key mechanisms include oxidative stress, neuroinflammation, synaptic dysfunction, and impaired neurogenesis.
- Cognitive function can be assessed through neuropsychological tools, biomarkers, and neuroimaging modalities.
- A combination of pharmacological, non-pharmacological, and emerging therapies shows promise in mitigating cognitive decline.
- Prevention strategies are most effective when started early, emphasising lifestyle modifications and personalised care.

17.1 INTRODUCTION

Cognitive decline includes age-related deterioration in cognitive processes such as memory, attention, processing speed, executive function, and language fluency. While a modest slowdown in cognition is typical with ageing, significant deficits that disrupt daily functioning may indicate pathological processes such as mild cognitive impairment (MCI) or dementia (Gonzales et al., 2022; Murman, 2015).

In normal ageing, fluid abilities such as processing speed, working memory, and reasoning tend to decline, while crystallised abilities like vocabulary and general knowledge often remain stable or even improve until very late in life (Murman, 2015). Conversely, pathological decline, such as that seen in Alzheimer's disease (AD), is marked by disruptions in neuroplasticity, the accumulation of amyloid-β plaques and tau tangles, and progressive neuronal and synaptic loss (Gonzales et al., 2022). Emerging research indicates that even before the onset of clinical symptoms, individuals may exhibit neuropathological changes decades earlier, including hippocampal atrophy and cerebrovascular dysfunction (Brito et al., 2023; Gonzales et al., 2022). Moreover, conditions like chronic traumatic encephalopathy (CTE), often linked to repeated head impacts, may accelerate the trajectory of decline (Broglio, Eckner, Paulson, & Kutcher, 2012).

17.1.1 Public Health Relevance in an Ageing Global Population

The global demographic transition towards an ageing population has profound implications. By 2050, the number of people aged 65 and older is projected to double, reaching nearly 1.5 billion worldwide (Bhatti, Reddy, Reddy, & Bhatti, 2020; Castiglione et al., 2016). In the United States, nearly one in three individuals in their 80s shows signs of dementia (Gonzales et al., 2022). Beyond clinical burdens, the societal cost is substantial. The Alzheimer's Association projected that the total cost of Alzheimer's and other dementias in the U.S. will rise from $290 billion in 2019 to over $1.1 trillion by 2050 (Bhatti et al., 2020). This surge poses challenges for healthcare infrastructure, caregivers, and economic systems alike. Socioeconomic and racial disparities also affect cognitive ageing. For instance, Black Americans exhibit higher dementia incidence rates than White

DOI: 10.1201/9781003500025-17

Americans, even after adjusting for genetic risk, a disparity largely attributed to differences in educational attainment and socioeconomic conditions (Weuve et al., 2018). This chapter explores the nature of cognitive decline in ageing from both biological and behavioural perspectives. It distinguishes between normal and pathological decline, examines underlying mechanisms such as neuroinflammation and mitochondrial dysfunction, and reviews evidence from human and animal studies. Subsequent sections address modifiable risk factors, including lifestyle, nutrition, and sensory deficits, as well as public health strategies and translational research aimed at delaying or preventing cognitive decline (Bhatti et al., 2020; Castiglione et al., 2016).

17.2 NEUROBIOLOGICAL BASIS OF COGNITIVE AGEING

17.2.1 Age-Related Changes in Brain Structure and Function

As individuals age, one of the most prominent neural changes is cortical atrophy, especially in the prefrontal cortex, hippocampus, and temporal lobes. While early assumptions attributed these reductions to widespread neuron death, recent stereological analyses reveal that neuron count remains relatively stable during normal ageing. The volume loss instead results from reduced dendritic arborisation, spine density, and synaptic numbers, particularly in highly plastic pyramidal neurons. This shrinkage has significant consequences for working memory, processing speed, and executive function, all of which show an age-associated decline (Murman, 2015).

In addition to grey matter loss, ageing is accompanied by microstructural changes in white matter. White matter hyperintensities (WMHs), observed via magnetic resonance imaging (MRI), reflect vascular-related pathology such as chronic hypoperfusion, gliosis, and demyelination. These lesions increase in prevalence with age and are particularly common in periventricular and frontal deep white matter regions. WMHs disrupt connectivity in critical executive control networks and have been correlated with executive dysfunction, processing speed deficits, and a higher risk of dementia, even in cognitively normal adults (Brito et al., 2023).

One of the most sensitive markers of ageing-related cognitive decline is synaptic loss. Synaptophysin immunohistochemistry reveals reductions in presynaptic terminals, especially in the hippocampus and association cortices, areas crucial for memory encoding and integration. This loss occurs even before overt neuronal death, implicating it as an early pathological feature of cognitive ageing. Simultaneously, the ageing brain experiences heightened neuroinflammation. Activated microglia and astrocytes secrete inflammatory cytokines such as IL-6 and TNF-α, which disrupt synaptic plasticity, impair long-term potentiation (LTP), and promote oxidative stress. This pro-inflammatory state accelerates synaptic degradation and may facilitate the progression from normal ageing to neurodegenerative disease (Gonzales et al., 2022).

17.2.2 Role of Neuroplasticity and Compensation

Despite these structural and molecular insults, the ageing brain retains substantial neuroplastic capacity. Functional neuroimaging studies have documented a posterior-anterior shift in ageing (PASA), where older adults recruit prefrontal regions more extensively to compensate for deficits in posterior brain areas. This compensatory mechanism reflects a functional reorganisation that supports task performance in the face of structural decline. Similarly, the Hemispheric Asymmetry Reduction in Older Adults (HAROLD) model describes how ageing is accompanied by increased bilateral cortical activation, particularly in the frontal lobes. This reflects the brain's adaptive response to reduced efficiency or local decline in specific neural systems. While hippocampal neurogenesis diminishes with age, it continues into late life under the influence of favourable environmental and behavioural factors. Physical exercise, enriched environments, cognitive training, and adequate sleep enhance synaptic density and neurotrophic factors like brain-derived neurotrophic factor (BDNF), reinforcing neuroplasticity and mitigating decline (Brito et al., 2023; Murman, 2015).

17.2.3 Involvement of Neurotransmitter Systems

Age-related cognitive decline is further modulated by alterations in key neurotransmitter systems, particularly:

Dopamine: Dopaminergic tone and receptor density, especially D1 and D2 receptors, decline significantly with age, affecting executive function, reward processing, and working memory. Decreased dopamine contributes to slower response times and diminished cognitive flexibility (Gonzales et al., 2022).

Acetylcholine: The cholinergic system, essential for learning, attention, and memory consolidation, suffers due to degeneration of basal forebrain cholinergic neurons. This decline is a hallmark of both normal ageing and AD. Pharmacological efforts in AD, including acetylcholinesterase inhibitors, aim to alleviate symptoms by enhancing cholinergic transmission (Bhatti et al., 2020).

Glutamate: Age impairs glutamatergic signalling, especially N-methyl-D-aspartate (NMDA) receptor function, critical for synaptic plasticity and LTP. Excessive or dysregulated glutamate release may cause excitotoxicity, contributing to synapse loss and neural dysfunction. Ageing brains may also develop neuronal hyperexcitability, a precursor to epileptiform activity and network instability, observed even in early-stage AD (Gonzales et al., 2022).

17.3 DOMAINS OF COGNITIVE DECLINE IN AGEING

Cognitive decline associated with ageing is not monolithic; it manifests across multiple cognitive domains with variable onset and progression. Each domain is supported by distinct but overlapping neural circuits, and understanding these domains is crucial for early diagnosis and tailored interventions.

17.3.1 Memory

Memory is one of the most prominently affected domains in ageing and encompasses several subtypes: *working memory*, *episodic memory*, and *semantic memory*.

Working memory is responsible for the temporary holding and manipulation of information and is anchored in the frontoparietal network. Age-related decline in working memory is consistently observed and is linked to reductions in processing speed and diminished capacity to inhibit irrelevant information (Brown, Brockmole, Gow, & Deary, 2012). Studies using tasks like the n-back or digit span backwards consistently show reduced performance in older adults compared to younger cohorts. *Episodic memory*, or memory for personally experienced events, is among the earliest and most vulnerable functions affected by age. This form of memory depends heavily on the hippocampus and medial temporal lobes, areas known to undergo atrophy with age. Older adults demonstrate deficits in both the encoding and retrieval of episodic content, which contributes to difficulties in recalling specific dates, events, or the context in which something was learned (Kirova, Bays, & Lagalwar, 2015). In contrast, *semantic memory*, which stores general knowledge and facts about the world, is relatively preserved during early ageing and declines more gradually. This dissociation is supported by neuroimaging findings that reveal relative sparing of lateral temporal cortices, which support semantic retrieval, even in the face of hippocampal atrophy Nevertheless, semantic fluency can decline in more advanced stages of neurodegeneration, especially AD.

Executive functions refer to high-level cognitive processes that enable goal-directed behaviour, including planning, problem-solving, cognitive flexibility, inhibition, and multitasking. These abilities are mediated primarily by the *prefrontal cortex* and its subcortical connections, including the *thalamus* and *basal ganglia*. Age-related degradation of these regions, along with declines in dopaminergic signalling, contributes to diminished executive capacity. Older adults often exhibit deficits in planning tasks (e.g., Tower of London), inhibitory control (e.g., Stroop test), and task-switching (e.g., Wisconsin Card Sorting Test). These impairments may not only interfere with complex daily tasks but also contribute to reduced adaptability and increased susceptibility to cognitive overload (Roheger, Kalbe, & Liepelt-Scarfone, 2018). Executive dysfunction is also a critical early feature in MCI and Parkinson's disease, often preceding memory loss (Rattanavichit, Chaikeeree, Boonsinsukh, & Kitiyanant, 2022).

Attention includes the capacity to maintain focus (sustained attention), switch between tasks (alternating attention), and manage competing stimuli (divided attention). Older adults show reduced performance in divided and selective attention tasks, especially under conditions of distraction or multitasking. A key limiting factor is *processing speed*, defined as the rate at which cognitive operations can be performed. Processing speed declines consistently with age and underpins performance in nearly all other cognitive domains. Neurobiologically, reduced processing speed is attributed to age-related degradation of *white matter tracts*, particularly in frontoparietal networks. This slowing is a strong predictor of future decline and differentiates healthy ageing from pathological states, as it affects everything from learning new material to decision-making.

17.3.2 Language and Visuospatial Abilities

Language functions—including naming, fluency, syntax, and comprehension—are generally preserved in early ageing, but subtle deficits may appear with advancing age or disease. Older adults may demonstrate word-finding difficulties or slower retrieval, although vocabulary size remains stable or even increases. In pathological ageing, particularly AD, naming and semantic fluency are affected early and can be assessed using tools like the *Boston Naming Test Visuospatial abilities* involve spatial perception, orientation, and mental rotation. These functions are supported by the *parietal lobes* and *occipital–temporal visual pathways*. Declines in visuospatial functioning can lead to difficulties in navigation, recognising faces or objects, and misplacing items. Older adults often report getting lost in familiar places or having trouble parking or driving. Visuospatial deficits are among the first signs in conditions like *posterior cortical atrophy*, and they also feature prominently in *Parkinson's disease dementia* (Lawrence, Myerson, & Hale, 1998).

17.3.3 Comparative Progression Across Domains

The progression of cognitive decline is not uniform across domains. *Processing speed* and *episodic memory* often decline earliest, followed by executive functions and attention, while *semantic memory*, *language comprehension*, and *crystallised intelligence* tend to remain stable until late stages (Gallais, Gagnon, Mathieu, & Richer, 2017). Visuospatial skills can show early impairment depending on the underlying neuropathology. Longitudinal meta-analyses confirm that multi-domain impairment, particularly involving executive function and memory, best predicts conversion from MCI to AD (Prado, Watt, Treeby, & Crowe, 2019). Understanding these domain-specific trajectories aids in refining diagnostic tools and personalising intervention strategies.

17.4 DISTINGUISHING NORMAL AGEING FROM NEURODEGENERATIVE CONDITIONS

Cognitive ageing spans a continuum from subtle, non-pathological changes to progressive neurodegenerative disorders. In normal ageing, cognitive changes such as slower processing speed and reduced working memory capacity are common but typically do not interfere with daily activities (Murman, 2015). In contrast, MCI represents an intermediate stage between normal ageing and dementia, characterised by measurable cognitive deficits that exceed age-related expectations but do not significantly impair daily functioning. Amnestic MCI, in particular, is associated with memory deficits and carries a 10–15% annual conversion rate to AD (Murman, 2015; Petersen, 2011).

AD is the most common neurodegenerative disorder among older adults. It is marked by progressive memory loss, executive dysfunction, language impairment, and eventual loss of independence. Pathologically, AD is characterised by extracellular amyloid-beta plaques and intracellular neurofibrillary tangles composed of hyperphosphorylated tau protein. These pathologies often begin years or decades before the appearance of symptoms. Advances in biomarkers, including amyloid PET, cerebrospinal fluid (CSF) measures of tau and amyloid, and structural MRI showing medial temporal atrophy, have facilitated the early detection of AD pathology. The National Institute on Ageing–Alzheimer's Association (NIA-AA) research framework proposes a four-stage continuum of disease progression, incorporating biomarkers and clinical symptomatology (Gonzales et al., 2022; Murman, 2015).

Vascular Cognitive Impairment (VCI) results from cerebrovascular injuries such as infarcts, small vessel disease, and chronic cerebral hypoperfusion. Unlike the gradual progression of AD, VCI may present acutely (e.g., post-stroke) or subacutely with executive dysfunction and psychomotor slowing. The concept of the neurovascular unit (NVU)—which includes neurons, endothelial cells, glia, and pericytes—has been crucial in understanding how vascular dysfunction impairs neurocognitive processes. Disruption of the NVU compromises the blood-brain barrier and contributes to neuroinflammation and synaptic damage, increasing dementia risk (Gonzales et al., 2022).

Beyond AD and VCI, other forms of dementia also play key roles in differential diagnosis. Dementia with Lewy Bodies (DLB) is distinguished by fluctuating cognition, well-formed visual hallucinations, REM sleep behaviour disorder, and Parkinsonian motor features. The underlying pathology includes alpha-synuclein deposits in cortical and subcortical regions. Frontotemporal Dementia (FTD) typically affects younger adults (ages 45–65) and presents with early changes in behaviour, personality, or language, depending on the subtype. FTD is associated with tau or TDP-43 proteinopathies, in contrast to the amyloid pathology of AD. Both DLB and FTD often overlap with other pathologies, complicating diagnosis (Brito et al., 2023).

A major challenge in distinguishing normal ageing from neurodegenerative conditions lies in the diagnostic overlap and variability in biomarker expression. Neuropathological studies reveal that many older adults with no clinical dementia still harbour significant levels of amyloid and tau pathology. Indeed, over 80% of cognitively normal elderly individuals show some AD-type neuropathology on autopsy (Murman, 2015). This has led to the theory of "cognitive reserve," where individual differences in brain resilience, shaped by factors such as education, occupation, and lifestyle, modulate the clinical expression of pathology. As such, two individuals with similar levels of pathology may show vastly different cognitive outcomes.

17.5 GENETIC BIOMARKER FOR COGNITIVE DECLINE WITH AGEING

Age-related cognitive decline refers to the gradual reduction in cognitive capacity that occurs with increasing age. While ageing itself is one of the risk factors for dementia. It is important to distinguish between normal ageing and pathological ageing. Cognitive ageing is linked with various age-related disorders such as PD, AD, amyotrophic lateral sclerosis (ALS), etc. Both the genetic factor and age contribute to cognitive decline. Numerous genetic factors have been identified to contribute to this process, such as apolipoprotein E (APOE), BDNF, COMT and epigenetic factors were significantly correlated with age-related cognitive decline (Figure 17.1). Among these, the APOE gene has a significant correlation with age-related cognitive decline. A meta-analysis of 77 studies further supported this, illustrating that the APOE4 isoform is involved in cognitive functions such as overall cognitive ability and episodic memory (Wisdom, Callahan, & Hawkins, 2011). Additionally, the APOE rs405509 and APOE rs440446 single-nucleotide polymorphism (SNPs) were found more likely to be associated with non-pathological

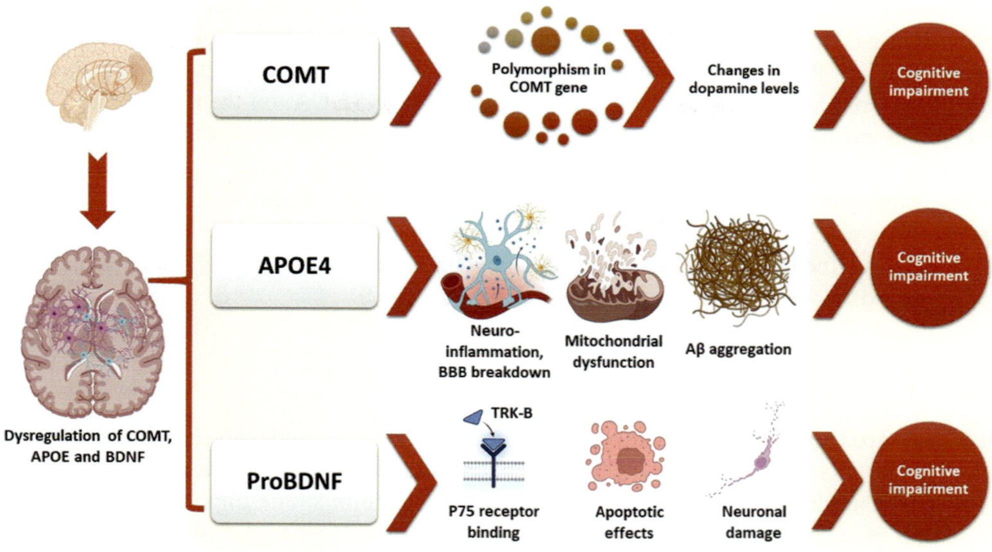

FIGURE 17.1 Representing the dysregulation of genetic factors like COMT, APOE, BDNF, leading to age-related cognitive impairment.

cognitive ageing, independent of APOE major isoforms, in a Finnish population (Rantalainen et al., 2016). However, a replication study conducted in a Taiwanese population found no significant association between the APOE gene and age-related cognitive decline. Furthermore, the BDNF emerges as another candidate gene involved in age-related cognitive decline among the elderly (Harris & Deary, 2011). It encodes a protein from nerve growth factor that has been shown to influence cognitive deficits associated with normal ageing in healthy individual as evaluated by Z-score across three cognitive domain such as motor function, memory, perceptual speed (Laing et al., 2012). Supporting this, Ward et al. reported the interaction between APOE and BDNF, which together predicted the cognitive effect in healthy older adults (Ward et al., 2014). Another key gene, COMT, Furthermore, COMT located on chromosome 22q11.21, which is involved in the metabolism of dopamine. The Val158Met polymorphism in COMT has been associated with cognitive ageing in Taiwanese and German populations (Liu et al., 2008; Papenberg et al., 2014). Furthermore, Dixon et al. demonstrated that the APOE and COMT genes function as complementary biomarkers for normal cognitive ageing and the early stage of MCI in older adults (Dixon et al., 2014). The CR1 gene, which encodes a complement reporter protein involved in immune response, has also been implicated in cognitive ageing (Chibnik et al., 2011). Specifically, the rs6656401 SNP in the tCR1 gene, which is one of the loci vulnerable to AD, is also associated with cognitive impairment in normal ageing. Beyond various genetic variants, epigenetic mechanisms such as DNA methylation play an important role in cognitive ageing (Mather, Kwok, Armstrong, & Sachdev, 2014; Xu, 2015). DNA methylation is mediated by the DNA methyltransferase enzyme that suppresses transcription. Supporting its involvement in cognitive ageing, a study using repeated measures and composite scores for annual cognitive testing reported that the rs11887120 SNP in the DNMT3A gene was associated with cognitive decline (Chouliaras et al., 2015).

17.6 ASSESSMENT OF COGNITIVE DECLINE

17.6.1 Standardised Neuropsychological Tools

Traditional neuropsychological assessments remain the clinical foundation for diagnosing cognitive decline. Among these, the Mini-Mental State Examination (MMSE) and Montreal Cognitive

Assessment (MoCA) are the most widely utilised tools in clinical and research settings. The MMSE offers a rapid evaluation of cognitive domains such as orientation, registration, attention, recall, and language, but it is less sensitive to MCI, particularly in highly educated individuals (Chun, Seward, Patterson, Melton, & MacDonald-Wicks, 2021). The MoCA, on the other hand, includes more complex executive functions and visuospatial tasks, enhancing its sensitivity to early cognitive changes (Chehrehnegar et al., 2020). The Wechsler Adult Intelligence Scale (WAIS), though more time-consuming, is a gold standard for comprehensive cognitive profiling. It measures verbal comprehension, perceptual reasoning, working memory, and processing speed. Studies have shown a 1.2% annual decline in WAIS scores in ageing populations, compared to a 0.6% decline in MMSE, reflecting its broader cognitive scope (Suchy-Dicey et al., 2022).

17.6.2 Digital Cognitive Assessment Technologies

Digital tools are transforming cognitive screening by enabling standardised, scalable, and frequently remote assessments. These include computer- and tablet-based applications, virtual reality (VR) environments, and wearables. Digital tests can capture subtle impairments in domains like reaction time, attention, and working memory—metrics often missed by paper-based tools. They are increasingly validated as proxies for clinical ratings and functional outcomes (Curcic et al., 2022; Lawson et al., 2023).For instance, MoCA and MMSE have now been digitised and integrated into cloud-based cognitive screening platforms, some of which use gamified tasks to improve engagement and longitudinal tracking. Importantly, digital assessments reduce inter-rater variability and offer higher repeatability, which is crucial in longitudinal studies (Curcic et al., 2022).

17.6.3 Longitudinal vs. Cross-Sectional Evaluation

Longitudinal studies provide a direct view of cognitive trajectories over time, distinguishing between stable ageing and pathological decline. For example, one large-scale study reported that annual declines in cognitive test scores ranged from 0.6% (MMSE) to 1.2% (WAIS), highlighting inter-tool sensitivity to change over time (Suchy-Dicey et al., 2022). Longitudinal designs also account for intra-individual variability and better identify early predictors of dementia. By contrast, cross-sectional assessments offer a snapshot of cognitive function at one point in time and are more vulnerable to cohort effects and confounding variables such as education and socioeconomics (Chun et al., 2021). Nonetheless, cross-sectional studies remain valuable for identifying group-level cognitive trends and comparing diagnostic tools across populations (Chehrehnegar et al., 2020).

17.6.4 Functional Assessments and ADL Scales

Cognitive evaluations are incomplete without assessing Activities of Daily Living (ADLs) and Instrumental ADLs (iADLs), which measure how cognitive decline translates into functional impairment. ADL scales like the Katz Index and Lawton–Brody Scale evaluate the ability to perform daily tasks, from dressing and bathing to managing finances and medications. Emerging evidence supports the use of digital ADL assessments, which passively track behaviours like mobility, screen interaction, and speech via smart devices, offering objective and real-time functional status updates (Lawson et al., 2023). ADL scores also correlate with MoCA and MMSE in dementia and post-stroke cohorts, emphasising their diagnostic and prognostic value For instance, a cross-sectional study linked lower MoCA and MMSE scores with impaired iADL performance in elderly stroke survivors, suggesting that combined cognitive-functional assessments can better inform care decisions (Hamdan et al., 2025).

17.7 INTERVENTION STRATEGIES

Cognitive decline associated with ageing is a complex and multifactorial process that necessitates a holistic and integrative approach to management. A diverse array of intervention strategies—ranging from pharmacological treatments and lifestyle modifications to dietary supplements, cognitive training, and emerging neuromodulation techniques—have been explored to preserve cognitive function, delay deterioration, and improve quality of life in older adults (Table 17.1). This section

TABLE 17.1
Evidence-Based Intervention Strategies for Cognitive Decline in Ageing

Category	Intervention	Mechanism/Effect	Key References
1. Pharmacological Interventions	Cholinesterase inhibitors (Donepezil, Rivastigmine, Galantamine)	Enhance cholinergic transmission; modest improvement in memory & ADLs	Murman (2015)
	Memantine (NMDA antagonist)	Reduces excitotoxicity in moderate-to-severe Alzheimer's disease	Bhatti et al. (2020)
	Anti-amyloid & anti-tau agents	Target disease pathology (experimental)	Bhatti et al. (2020)
2. Non-Pharmacological Interventions	Cognitive training (e.g., the ACTIVE study)	Improves reasoning, memory, delays functional decline	Willis et al. (2006)
	Physical exercise (aerobic)	Enhances hippocampal volume, cerebral perfusion, mitochondrial health	Bhatti et al. (2020) and Brito et al. (2023)
	Lifestyle modifications (diet, sleep, social engagement)	Reduces inflammation and vascular risk; maintains cognitive reserve	Bhatti et al. (2020) and Ngandu et al. (2015)
3. Nutraceutical and Dietary Approaches	Omega-3 fatty acids, B-complex vitamins	Support membrane integrity, neurotransmission; reduce oxidative stress	Bhatti et al. (2020)
	Curcumin	Antioxidant, anti-inflammatory, reduces amyloid aggregation	Ghosh et al., (2015) and Williams et al. (2011)
	EGCG, Resveratrol	Improves neuroprotection, reduces oxidative stress and Aβ aggregation	Wen et al. (2017)
4. Psychosocial and Behavioural Interventions	Mindfulness-based stress reduction (MBSR), art therapy	Improves mood, reduces loneliness and depression, indirectly supports cognition	Murman (2015)
5. Sleep and Circadian Rhythm Regulation	Improved sleep hygiene, light exposure	Enhances glymphatic clearance of Aβ and tau; regulates neuroendocrine rhythms	Gonzales et al. (2022) and Randhi et al. (2023)
6. Neurostimulation and Neuroplasticity Enhancers	TMS (Transcranial Magnetic Stimulation)	Enhances cortical excitability and frontoparietal connectivity	Cotelli et al. (2008)
	tDCS (Transcranial Direct Current Stimulation)	Boosts learning and working memory; synergistic with cognitive training	Coffman, Trumbo, and Clark (2012)
7. Neurogenesis Enhancers	BDNF mimetics, serotonergic agents, Wnt pathway modulators	Stimulate hippocampal neurogenesis; improve memory	Britt and Bonci (2013) and Gómez-Pinilla (2008)
	Exercise, flavonoids (e.g., omega-3, polyphenols)	Induce neurotrophic factor expression (BDNF, IGF-1)	Bhatti et al. (2020)
	Young plasma infusion, intranasal insulin	Experimental systemic approaches to enhance plasticity	Yuen et al. (2017)

provides a comprehensive overview of evidence-based interventions that aim not only to alleviate symptoms but also to target underlying neurobiological mechanisms contributing to cognitive ageing.

17.7.1 Pharmacological Treatments

Among pharmacological treatments, the most widely used medications include cholinesterase inhibitors such as donepezil, rivastigmine, and galantamine, which enhance cholinergic transmission and provide modest benefits in memory and daily functioning for patients with mild to moderate AD (Murman, 2015). Additionally, memantine, an NMDA receptor antagonist, has been approved for moderate to severe stages of Alzheimer's, helping to stabilise symptoms by reducing glutamate-induced excitotoxicity. However, while these drugs can offer symptomatic relief, they do not modify disease progression. Trials aimed at disease-modifying therapies, such as anti-amyloid and anti-tau agents, have so far yielded limited clinical success (Bhatti et al., 2020).

17.7.2 Non-Pharmacological approaches

Non-pharmacological interventions have gained significant attention. It plays an important role in alleviating age-related cognitive decline, often complementing pharmacological treatment in terms of safety and long-term efficacy. There are various non-pharmacological approaches discussed below.

17.7.2.1 Cognitive Training and Stimulation

Such as memory exercises, problem-solving tasks, and computerised brain games, have been found to improve memory, reasoning, and processing speed. The Advanced Cognitive Training for Independent and Vital Elderly (ACTIVE) study demonstrated long-term improvements in trained cognitive domains and delayed functional decline up to five years post-intervention (Willis et al., 2006). Similarly, physical exercise, particularly aerobic training, has shown neuroprotective effects through increased hippocampal volume, enhanced cerebral blood flow, and improved mitochondrial function.

17.7.2.2 Physical Exercise and Lifestyle Modification

Regular exercise also modulates inflammation and oxidative stress, which are key contributors to cognitive decline (Bhatti et al., 2020).Lifestyle modification, including diet, sleep hygiene, social engagement, and chronic disease management, plays a foundational role in preserving cognition. The Finnish Geriatric Intervention Study to Prevent Cognitive Impairment and Disability (FINGER) trial revealed that a multidomain intervention—combining diet, physical activity, cognitive training, and vascular risk management—significantly improved cognitive outcomes in older adults at risk for dementia (Bhatti et al., 2020; Ngandu et al., 2015).

17.7.3 Nutraceuticals and Dietary Interventions

Nutraceuticals and dietary supplements such as omega-3 fatty acids, curcumin, and B vitamins have been studied for their potential to reduce inflammation, promote synaptic plasticity, and enhance neurogenesis. Although some trials show positive cognitive effects, particularly with omega-3 fatty acids and B-complex vitamins in those with nutrient deficiencies, overall findings remain mixed and inconclusive (Bhatti et al., 2020). In addition, phytochemicals such as curcumin are known to be as shown antioxidant effects that reduce the oxidative stress contributing to ageing and cognitive decline (Nelson & Tabet, 2015). The curcumin helps to reduce the amyloid aggregation or oxidative neurotoxicity by forming a complex with metal ions (Williams, Sorribas, & Howes, 2011). Its antioxidant properties reduce oxidative damage by enhancing GSH and decreasing lipid peroxide levels in brain tissue (Williams et al., 2011). Anti-inflammatory: reduces NF-κB, TNF-α, and COX-2 signalling pathways and inhibits IL-8, IL-6, IL-1β and IL-1α production via inhibition of the MAPK and NF-κB pathways (Ghosh, Banerjee, & Sil, 2015).

In addition, other nutraceuticals such as Epigallocatechin gallate (EGCG). It is an antioxidant that reduces oxidative damage by increasing mRNA expression of γ-glutamylcysteine ligase to increase cellular GSH, stimulation of PKC, and modulation of cell cycle genes results in improved cognitive function (Williams et al., 2011). (Williams et al., 2011). Resveratrol has been improving verbal learning in older adults with cognitive age decline Because it promotes the decomposition of Aβ aggregates, prevents fibril formation, destabilises preformed fAβ42 in vitro, inhibits Aβ42 aggregation and reduces Aβ plaque formation (Williams et al., 2011) and aggregation and toxicity of amyloidogenic proteins (Wen et al., 2017). In 1,462 women (>34-year study duration), wine consumption was associated with reduced dementia risk yet a separate study concluded that a high red wine intake reduces AD risk in men but increases risk in women.

17.7.3.1 Lifestyle Factors in Cognitive Ageing

Emerging lifestyle factors have been recognised as one of the major risk factors for cognitive decline with ageing. Sleep quality and circadian rhythm are crucial, as poor sleep tends to reduce the slow wave and REM sleep that are related to impaired glymphatic clearance of neurotoxic proteins like amyloid-β and tau, contributing to neurodegeneration and cognitive impairment. In addition, Gonzales, M.M., et al., reported that the slow wave sleep decreases with ageing and a disturbed sleep cycle. A-beta lipoprotein and tau accumulation were associated with lower slow-wave activity in non-rapid eye movement sleep. Its accumulation was reduced by getting enough sleep, which will reduce the oxidative stress in cells that, in turn, reduces A-beta lipoprotein accumulation (Randhi et al., 2023). Regular wake and sleep cycles and exposure to light help to maintain the circadian cycle and maintain cognitive function.

17.7.4 Psychosocial Interventions

Psychosocial interventions, including mindfulness-based stress reduction, art therapy, and social engagement programs, have been linked to improvements in mood and quality of life. These interventions can indirectly support cognition by reducing depression and loneliness, both of which are independent risk factors for cognitive decline (Murman, 2015). Among emerging therapies, non-invasive brain stimulation techniques like transcranial magnetic stimulation (TMS) and transcranial direct current stimulation (tDCS) are being explored for their potential to enhance cortical excitability and promote neuroplasticity. Preclinical studies also suggest that targeting adult hippocampal neurogenesis may reverse or slow cognitive ageing, though human applications remain in early experimental stages (Bhatti et al., 2020).

17.7.5 Emerging Therapies: Neurostimulation and Neurogenesis Enhancers

Emerging therapeutic approaches in cognitive ageing aim to not only mitigate symptoms but also engage the brain's latent plasticity to reverse or delay cognitive decline. Two of the most promising innovations in this area are *neurostimulation techniques* and *neurogenesis-enhancing interventions.*

Neurostimulation includes non-invasive brain stimulation methods such as *TMS* and *transcranial direct current stimulation (tDCS).* These techniques modulate cortical excitability and synaptic plasticity in targeted brain regions associated with cognitive function. Repetitive TMS (rTMS) has shown potential in improving attention and executive functions in older adults and patients with MCI by enhancing connectivity within the default mode and frontoparietal networks. Clinical trials have demonstrated that repeated stimulation over the dorsolateral prefrontal cortex can lead to short-term improvements in working memory and processing speed, although long-term efficacy remains under investigation (Cotelli, Manenti, Cappa, Zanetti, & Miniussi, 2008). tDCS, another form of neuromodulation, applies low-intensity electrical currents to modulate neuronal activity. It has shown promise in boosting learning capacity and memory consolidation when paired with cognitive training. The synergistic use of tDCS with behavioural interventions may enhance neuroplasticity in ageing brains, although variability in outcomes suggests a need for individualised protocols (Paul et al., 2018).

Alongside neuromodulation, efforts to *enhance adult hippocampal neurogenesis* have garnered attention. In the adult human brain, neurogenesis primarily occurs in the subgranular zone of the hippocampus. However, ageing significantly reduces the proliferation of neural progenitor cells. Recent research in animal models has shown that pharmacological agents such as BDNF mimetics, serotonergic agents, and Wnt signalling modulators can stimulate neurogenesis and improve cognitive performance (Britt & Bonci, 2013). Exercise and dietary factors—such as flavonoids, omega-3 fatty acids, and caloric restriction—also upregulate neurotrophic factors like BDNF and IGF-1, supporting neural regeneration and synaptic plasticity (Gómez-Pinilla, 2008). Moreover, intranasal insulin, ghrelin analogues, and young plasma infusion have been investigated as systemic modulators of neurogenesis and cognition. These interventions appear to act via both central and peripheral pathways, enhancing neurotrophic signalling and reducing neuroinflammation. While most of these approaches are in preclinical or early-phase clinical testing, they offer exciting possibilities for disease modification in cognitive ageing (Yuen et al., 2017).

17.8 CULTURAL AND SOCIETAL PERSPECTIVES

The experience and interpretation of cognitive ageing vary considerably across cultures and societies. Global variability in cognitive ageing reflects differences in nutrition, education, healthcare infrastructure, environmental exposures, and cultural beliefs about ageing. In high-income countries, older adults are more likely to receive early diagnoses and access to interventions. Conversely, in many low- and middle-income countries, cognitive impairment often goes unrecognised or is normalised as a part of ageing, leading to delayed or absent medical attention (Bhatti et al., 2020). Socioeconomic status (SES) significantly influences cognitive ageing outcomes. Individuals with lower SES face cumulative disadvantages such as reduced access to healthcare, lower educational attainment, chronic stress, and poor nutrition—all of which affect brain health. For example, disparities in education have been shown to mediate racial differences in cognitive function and dementia incidence, as demonstrated in the Chicago Health and Ageing Project (Weuve et al., 2018). The concept of cognitive reserve posits that higher education and cognitively stimulating environments can buffer against age-related decline and clinical manifestations of neuropathology.

Cultural perceptions of ageing and cognition also shape how societies respond to dementia. In some cultures, memory loss may be attributed to spiritual causes or accepted as a normal part of growing old, potentially delaying diagnosis and treatment. In other words, stigma around dementia can lead to social isolation and decreased quality of life for both patients and their caregivers. Understanding these cultural dynamics is crucial for designing effective public health interventions and improving access to care (Bhatti et al., 2020). Caregivers and healthcare systems play an essential role in managing cognitive ageing. Family caregivers often bear the primary responsibility for individuals with dementia, particularly in regions with limited formal care services. This role can be both rewarding and burdensome, with caregivers at increased risk for depression, anxiety, and economic hardship. Healthcare systems need to support caregivers through respite care, financial assistance, education, and integrated care models. Additionally, policies promoting routine cognitive screening, early intervention, and equitable resource allocation are essential in addressing the global burden of cognitive impairment (Bhatti et al., 2020).

17.9 RESEARCH FRONTIERS AND FUTURE DIRECTIONS

The study of cognitive ageing is advancing rapidly through the integration of neuroimaging technologies, artificial intelligence (AI), and personalised medical approaches. At the core of these innovations are neuroimaging and connectomics, which have transformed our understanding of age-related brain changes. Advanced imaging modalities such as structural MRI, diffusion tensor imaging (DTI), and functional MRI provide high-resolution insights into brain architecture and neural efficiency. These tools reveal age-related deterioration in white matter tracts and a reduction in global and local brain network connectivity, especially within the default mode network and executive control networks (Bacon et al., 2024). Connectomics allows for the comprehensive mapping of brain connectivity, showing that declines in network modularity and efficiency are key predictors of cognitive slowing (Chatterjee & Moradikor, 2024).

Simultaneously, AI-driven predictive models are becoming indispensable in the early detection of cognitive decline. Machine learning methods are capable of processing vast multi-modal datasets—including neuroimaging, behavioural scores, and genetic data—to predict trajectories of cognitive decline with increasing accuracy. For instance, AI-enhanced neuroimage analysis can differentiate between normal ageing and early neurodegenerative processes, often years before clinical symptoms emerge (Bacon et al., 2024). Such approaches have already demonstrated superiority over traditional methods in diagnostic sensitivity and specificity (Sultana, Rafi, Chowdhury, & Tariq, 2023). Moreover, explainable AI models are being developed to make transparent predictions and guide clinicians in risk stratification and patient management (Kumar et al., 2024).

The rise of precision medicine in neurology represents another critical development. Precision approaches utilise an individual's biological, environmental, and lifestyle data to tailor prevention and treatment. This methodology acknowledges the heterogeneity of cognitive ageing by focusing on individualised intervention strategies based on genotype (e.g., APOE status), metabolic profile, inflammation markers, and cognitive reserve capacity (Lorenzon, 2024) In clinical trials, precision frameworks have led to more effective targeting of therapeutic agents and stratified analysis that reveals subpopulation-level treatment responses (Gutman, Shmilovitch, Aran, & Shelly, 2024).

However, these advances raise critical ethical questions. The increasing ability to detect preclinical cognitive changes poses dilemmas regarding disclosure, stigma, and data privacy. Should asymptomatic individuals be informed of their risk if no effective treatment is available? How can AI systems protect patient autonomy while avoiding algorithmic bias or over-reliance? As cognitive enhancement technologies—ranging from nootropics to neurostimulation—become more mainstream, there is concern about potential inequities, pressure to enhance in healthy individuals, and the redefinition of normal ageing (Ienca & Ignatiadis, 2020; Sultana et al., 2023). These concerns highlight the importance of embedding ethical foresight into technological innovation.

17.10 CONCLUSION

The growing prevalence of cognitive decline among ageing populations around the world underscores the urgent need for a thorough understanding of both normal and pathological cognitive trajectories. Normal cognitive ageing involves subtle changes in memory, attention, and processing speed, while pathological trajectories, such as those seen in AD or other dementias, are marked by progressive and debilitating impairments that interfere with daily functioning. Differentiating between these trajectories is crucial for timely diagnosis and the development of targeted interventions.

This chapter has explored cognitive ageing from multiple interdisciplinary perspectives—biological, psychological, and sociocultural—highlighting how ageing impacts cognitive function in a multifactorial and dynamic manner. Biologically, key mechanisms underlying cognitive decline include age-related structural brain changes such as cortical thinning, hippocampal atrophy, white matter degradation, and synaptic loss. These changes are often compounded by alterations in key neurotransmitter systems, such as those involving dopamine and acetylcholine, which impair communication between brain regions. Additionally, the role of chronic neuroinflammation and increased oxidative stress has been emphasised, both of which contribute to neural damage and reduced neuroplasticity. Genetic susceptibility, particularly the presence of risk alleles like APOE ε4, also plays a critical role in predisposing individuals to earlier and more severe cognitive decline.

However, a growing body of research points to the influence of modifiable lifestyle factors in mediating these biological changes. Evidence suggests that regular physical activity, a balanced and nutrient-rich diet, cognitive engagement, and social interaction can significantly reduce the risk of cognitive deterioration and may even promote neural resilience. These modifiable behaviours represent accessible and cost-effective strategies to preserve cognitive function and delay the onset of clinical symptoms, especially when adopted in midlife or earlier. In response to the complexity of cognitive ageing, contemporary intervention strategies are increasingly multidomain. Pharmacological treatments such as cholinesterase inhibitors and NMDA receptor antagonists offer symptomatic relief in certain cases, but their impact is often limited. Therefore, a more promising approach lies in the integration of pharmacological agents with lifestyle modifications, psychosocial interventions, cognitive training, and emerging technological solutions like digital health platforms and non-invasive brain stimulation. Such comprehensive, person-centred interventions acknowledge the interdependence of biological and behavioural factors in shaping cognitive outcomes.

Looking ahead, future research must prioritise interdisciplinary collaboration, combining insights from neuroscience, geriatrics, behavioural psychology, data science, and ethics. Precision medicine—which involves tailoring interventions based on individual genetic, physiological, and behavioural profiles—holds particular promise in this domain. Likewise, the ethical implementation of AI in predicting cognitive trajectories and guiding intervention strategies presents both exciting opportunities and pressing concerns. It is essential to ensure that technological advancements do not widen existing disparities in healthcare but rather contribute to more inclusive and equitable outcomes. Finally, preserving cognitive health at a population level will require more than medical innovation—it will necessitate systemic changes in healthcare delivery, policy, and community engagement. Ensuring equitable access to cognitive care, providing support to caregivers, and respecting cultural differences in how ageing and cognition are perceived will be critical in building effective, humane, and sustainable models for ageing societies. Only through a holistic and inclusive approach can we hope to address the challenges of cognitive decline and enable older adults to maintain independence, dignity, and quality of life throughout the ageing process.

17.11 FUTURE CHALLENGES AND POLICY DIRECTIONS

As populations age globally, cognitive decline is poised to become a major burden for individuals, families, and healthcare systems. Addressing this challenge will require both biomedical innovation and socio-political reform. A key priority is ensuring equitable access to care. While high-income countries benefit from advanced diagnostics and interventions, many low- and middle-income regions lack basic infrastructure, screening tools, and trained personnel. Public health efforts must invest in scalable, low-cost cognitive assessments and digital health platforms to close these gaps. Integrating cognitive screening into primary care is essential, as many older adults remain undiagnosed until functional impairment occurs. Standardised assessments, early referrals, and family support systems can facilitate timely intervention and reduce long-term care costs. At the same time, greater attention must be paid to informal caregivers, who provide the majority of dementia care globally. National strategies should offer caregiver education, respite services, and financial support to alleviate emotional and economic strain.

Policy frameworks should also promote age-friendly environments and dementia-inclusive communities. Urban design must support safe mobility and cognitive engagement, while employment policies should accommodate cognitive health needs in ageing workforces. Emerging ethical concerns—such as those related to AI-driven cognitive monitoring and genetic risk disclosure—require clear guidelines to protect privacy, prevent discrimination, and ensure informed consent.

Finally, future research must foster interdisciplinary collaboration across neuroscience, public health, data science, and ethics. Priority should be given to diverse longitudinal studies and translational research that bridge lab findings with real-world applications. Meeting the complex challenges of cognitive ageing will depend on evidence-based, inclusive, and ethically grounded policies, supported by global cooperation and sustained investment in brain health.

REFERENCES

Bacon, E. J., He, D., Achi, N. A. D. A., Wang, L., Li, H., Yao-Digba, P. D. Z., … QI, S. (2024). Neuroimage analysis using artificial intelligence approaches: a systematic review. *Medical & Biological Engineering & Computing*, *62*(9), 2599–2627.

Bhatti, G. K., Reddy, A. P., Reddy, P. H., & Bhatti, J. S. (2020). Lifestyle modifications and nutritional interventions in aging-associated cognitive decline and Alzheimer's disease. *Frontiers in Aging Neuroscience*, 11, 369.

Brito, D. V. C., Esteves, F., Rajado, A. T., Silva, N., Araújo, I., Bragança, J., … Nóbrega, C. (2023). Assessing cognitive decline in the aging brain: lessons from rodent and human studies. *NPJ Aging*, *9*(1), 23.

Britt, J. P., & Bonci, A. (2013). Optogenetic interrogations of the neural circuits underlying addiction. *Current Opinion in Neurobiology*, *23*(4), 539–545.

Broglio, S. P., Eckner, J. T., Paulson, H. L., & Kutcher, J. S. (2012). Cognitive decline and aging: the role of concussive and subconcussive impacts. *Exercise and Sport Sciences Reviews*, *40*(3), 138–144.

Brown, L. A., Brockmole, J. R., Gow, A. J., & Deary, I. J. E. (2012). Processing speed and visuospatial executive function predict visual working memory ability in older adults. Experimental Aging Research, *38*(1), 1–19.

Castiglione, A., Benatti, A., Velardita, C., Favaro, D., Padoan, E., & Severi, D. (2016). Aging, cognitive decline and hearing loss: effects of auditory rehabilitation and training with hearing aids and cochlear implants on cognitive function and depression among older adults. *Journal Audiology and Neurotology*, *21*(Suppl. 1), 21–28.

Chatterjee, I., & Moradikor, N. (2024). *Integrating Neuroimaging, Computational Neuroscience, and Artificial Intelligence*. CRC Press.

Chehrehnegar, N., Nejati, V., Shati, M., Rashedi, V., Lotfi, M., Adelirad, F., & Foroughan, M. J. P. (2020). Early detection of cognitive disturbances in mild cognitive impairment: a systematic review of observational studies. *Psychogeriatrics*, *20*(2), 212–228.

Chibnik, L. B., Shulman, J. M., Leurgans, S. E., Schneider, J. A., Wilson, R. S., Tran, D., … Myers, A. J. J. A. (2011). CR1 is associated with amyloid plaque burden and age-related cognitive decline. *Annals of Neurology*, *69*(3), 560–569.

Chouliaras, L., Kenis, G., Visser, P. J., Scheltens, P., Tsolaki, M., Jones, R. W., … Wallin, Å. K. (2015). DNMT3A moderates cognitive decline in subjects with mild cognitive impairment: replicated evidence from two mild cognitive impairment cohorts. *Epigenomics*, *7*(4), 533–537.

Chun, C. T., Seward, K., Patterson, A., Melton, A., & MacDonald-Wicks, L. (2021). Evaluation of available cognitive tools used to measure mild cognitive decline: a scoping review. *Nutrients*, *13*(11), 3974.

Coffman, B. A., Trumbo, M. C., & Clark, V. P. (2012). Enhancement of object detection with transcranial direct current stimulation is associated with increased attention. *Randomized Controlled Trial*, *13*, 1–8.

Cotelli, M., Manenti, R., Cappa, S., Zanetti, O., & Miniussi, C. J. E. J. (2008). Transcranial magnetic stimulation improves naming in Alzheimer disease patients at different stages of cognitive decline. *European Journal of Neurology*, *15*(12), 1286–1292.

Curcic, J., Vallejo, V., Sorinas, J., Sverdlov, O., Praestgaard, J., Piksa, M., … Hannesdottir, K. (2022). Description of the method for evaluating digital endpoints in Alzheimer disease study: protocol for an exploratory, cross-sectional study. *JMIR Research Protocols*, *11*(8), e35442.

Dixon, R. A., DeCarlo, C. A., MacDonald, S. W., Vergote, D., Jhamandas, J., & Westaway, D. (2014). APOE and COMT polymorphisms are complementary biomarkers of status, stability, and transitions in normal aging and early mild cognitive impairment. *Frontiers in Aging Neuroscience*, *6*, 236.

Gallais, B., Gagnon, C., Mathieu, J., & Richer, L. (2017). Cognitive decline over time in adults with myotonic dystrophy type 1: a 9-year longitudinal study. *Neuromuscular Disorders*, *27*(1), 61–72.

Ghosh, S., Banerjee, S., & Sil, P. C. (2015). The beneficial role of curcumin on inflammation, diabetes and neurodegenerative disease: a recent update. *Food Chemical Toxicology*, *83*, 111–124.

Gómez-Pinilla, F. (2008). Brain foods: the effects of nutrients on brain function. *Nature Reviews Neuroscience*, *9*(7), 568–578.

Gonzales, M. M., Garbarino, V. R., Pollet, E., Palavicini, J. P., Kellogg, D. L., Kraig, E., & Orr, M. E. (2022). Biological aging processes underlying cognitive decline and neurodegenerative disease. *The Journal of Clinical Investigation*, *132*(10), e158453.

Gutman, B., Shmilovitch, A.-H., Aran, D., & Shelly, S. (2024). Twenty-five years of AI in neurology: the journey of predictive medicine and biological breakthroughs. *JMIR Neurotechnology*, *3*(1), e59556.

Hamdan, M., Zidan, S., Abualhaija, R., Vanoh, D., Mouamer, A., Taqatqa, H., … Badrasawi, M. (2025). Factors associated with mild cognitive impairment among Palestinian community-dwelling older adults: a cross-sectional study. *Cogent Psychology*, *12*(1), 2478715.

Harris, S. E., & Deary, I. J. (2011). The genetics of cognitive ability and cognitive ageing in healthy older people. *Trends in Cognitive Sciences*, *15*(9), 388–394.

Ienca, M., & Ignatiadis, K. (2020). Artificial intelligence in clinical neuroscience: methodological and ethical challenges. *AJOB Neuroscience*, *11*(2), 77–87.

Kirova, A.-M., Bays, R. B., & Lagalwar, S. (2015). Working memory and executive function decline across normal aging, mild cognitive impairment, and Alzheimer's disease. *BioMed Research International*, *2015*(1), 748212.

Kumar, R., Waisberg, E., Ong, J., Paladugu, P., Amiri, D., Saintyl, J., … Tavakkoli, A. (2024). Artificial intelligence-based methodologies for early diagnostic precision and personalized therapeutic strategies in neuro-ophthalmic and neurodegenerative pathologies. *Brain Sciences*, *14*(12), 1266.

Laing, K. R., Mitchell, D., Wersching, H., Czira, M. E., Berger, K., & Baune, B. T. (2012). Brain-derived neurotrophic factor (BDNF) gene: a gender-specific role in cognitive function during normal cognitive aging of the MEMO-Study? *Age*, *34*, 1011–1022.

Lawrence, B., Myerson, J., Hale, S. (1998). Differential decline of verbal and visuospatial processing speed across the adult life span. *Neuropsychology, and Cognition*, *5*(2), 129–146.

Lawson, L., Mc Ardle, R., Wilson, S., Beswick, E., Karimi, R., & Slight, S. P. (2023). Digital endpoints for assessing instrumental activities of daily living in mild cognitive impairment: systematic review. *Journal of Medical Internet Research*, *25*, e45658.

Liu, M.-E., Hong, C.-J., Liou, Y.-J., Tsai, Y.-L., Hsieh, C.-H., & Tsai, S.-J. (2008). Association study of a functional catechol-O-methyltransferase polymorphism and executive function in elderly males without dementia. Neuroscience Letters, *436*(2), 193–195.

Lorenzon, G. (2024). *Brain heterogeneity within aging and cognitive impairment: implications for precision medicine and prevention*. Karolinska Institutet.

Mather, K. A., Kwok, J. B., Armstrong, N., & Sachdev, P. S. (2014). The role of epigenetics in cognitive ageing. International Journal of Geriatric Psychiatry, *29*(11), 1162–1171.

Murman, D. L. (2015). The impact of age on cognition. Seminars in Hearing, *36*(3), 111–121.

Nelson, L., & Tabet, N. (2015). Slowing the progression of Alzheimer's disease; what works? *Ageing Research Reviews*, *23*, 193–209.

Ngandu, T., Lehtisalo, J., Solomon, A., Levälahti, E., Ahtiluoto, S., Antikainen, R., … Laatikainen, T. (2015). A 2 year multidomain intervention of diet, exercise, cognitive training, and vascular risk monitoring versus control to prevent cognitive decline in at-risk elderly people (FINGER): a randomised controlled trial. The Lancet, *385*(9984), 2255–2263.

Papenberg, G., Bäckman, L., Nagel, I. E., Nietfeld, W., Schröder, J., Bertram, L., & Lindenberger, U. (2014). COMT polymorphism and memory dedifferentiation in old age. Psychology and Aging, *29*(2), 374.

Paul, R., Dutta, A., Phukan, B. C., Mazumder, M. K., Justin-Thenmozhi, A., Manivasagam, T., … Borah, A. (2018). Accumulation of cholesterol and homocysteine in the nigrostriatal pathway of brain contributes to the dopaminergic neurodegeneration in mice. Neuroscience, *388*, 347–356.

Petersen, R. C. (2011). Mild cognitive impairment. The New England Journal of Medicine, *364*(23), 2227–2234.

Prado, C. E., Watt, S., Treeby, M. S., & Crowe, S. F. (2019). Performance on neuropsychological assessment and progression to dementia: a meta-analysis. *Psychological Assessment*, *34*(7), 954.

Randhi, B., Gutlapalli, S. D., Pu, J., Zaidi, M. F., Patel, M., Atluri, L. M., … Hamid, P. (2023). Sleep disorders in mild cognitive impairment. Cureus, *15*(3), e35863.

Rantalainen, V., Lahti, J., Henriksson, M., Kajantie, E., Tienari, P., Eriksson, J. G., & Raikkonen, K. (2016). APOE and aging-related cognitive change in a longitudinal cohort of men. Neurobiology of Aging, *44*, 151–158.

Rattanavichit, Y., Chaikeeree, N., Boonsinsukh, R., & Kitiyanant, K. (2022). The age differences and effect of mild cognitive impairment on perceptual-motor and executive functions. Frontiers in Psychology, *13*, 906898.

Roheger, M., Kalbe, E., & Liepelt-Scarfone, I. (2018). Progression of cognitive decline in Parkinson's disease. Journal of Parkinson's Disease, *8*(2), 183–193.

Suchy-Dicey, A. M., Oziel, K., Sawyer, C., Olufadi, Y., Ali, T., Fretts, A. M., ... Rhoads, K. L. (2022). Educational and clinical associations with longitudinal cognitive function and brain imaging in American Indians: the Strong Heart Study. *Neurology*, *99*(24), e2637–e2647.

Sultana, A., Rafi, A. H., Chowdhury, A. A. A., & Tariq, M. (2023). AI in neurology: predictive models for early detection of cognitive decline. Research and Emerging Trends in Disease Control, *17*(2), 335–349.

Ward, D. D., Summers, M. J., Saunders, N. L., Janssen, P., Stuart, K. E., & Vickers, J. C. (2014). APOE and BDNF Val66Met polymorphisms combine to influence episodic memory function in older adults. Behavioural Brain Research, *271*, 309–315.

Wen, P., Wen, Y., Zong, M.-H., Linhardt, R. J., & Wu, H. (2017). Encapsulation of bioactive compound in electrospun fibers and its potential application. Journal of Agricultural and Food Chemistry, *65*(42), 9161–9179.

Weuve, J., Barnes, L. L., De Leon, C. F. M., Rajan, K. B., Beck, T., Aggarwal, N. T., ... Evans, D. A. (2018). Cognitive aging in black and white Americans: cognition, cognitive decline, and incidence of Alzheimer disease dementia. Epidemiology, *29*(1), 151–159.

Williams, P., Sorribas, A., & Howes, M.-J. R. (2011). Natural products as a source of Alzheimer's drug leads. Natural Product Reports, *28*(1), 48–77.

Willis, S. L., Tennstedt, S. L., Marsiske, M., Ball, K., Elias, J., Koepke, K. M., ... Stoddard, A. M. (2006). Long-term effects of cognitive training on everyday functional outcomes in older adults. *JAMA*, *296*(23), 2805–2814.

Wisdom, N. M., Callahan, J. L., & Hawkins, K. A. (2011). The effects of apolipoprotein E on non-impaired cognitive functioning: a meta-analysis. *Aging*, *32*(1), 63–74.

Yuen, R. K. C., Merico, D., Bookman, M., Howe, J. L., Thiruvahindrapuram, B., Patel, R. V., ... Wang, Z. (2017). Whole genome sequencing resource identifies 18 new candidate genes for autism spectrum disorder. *Nature Neuroscience*, *20*(4), 602–611.

Xu, X. (2015). DNA methylation and cognitive aging. *Oncotarget*, *6*(16), 13922.

18 Geriatric Neurology
Challenges and Opportunities

Ashish Kumar, Sanjay Das, and Arpita Konar

18.1 INTRODUCTION

Aging is a natural biological phenomenon that accompanies structural disabilities and a gradual decline in the physiological competence of an organism[1]. Aging may be quantified as either chronological aging, which is the duration of life one has lived from birth, or biological aging, which is the functional capacity condition of an individual relative to his or her chronological age[2]. Even when chronological age is the same, biological age differs and is determined by an enormous number of intrinsic and extrinsic variables such as genetics, lifestyle, nutrition, activity, psychosocial stress, and environmental exposure[3]. Two individuals of the same 70 years old, for example, would be extremely different in terms of cognitive ability, mobility, and burden as they would have experienced different biological markers of aging.

Biological aging can be quantified by a variety of molecular and cellular biomarkers, including genomic instability, telomere attrition, epigenetic alterations, mitochondrial damage, loss of proteostasis, reduction in stem cells, defective intercellular communication, cellular senescence, and inflammation[1,4]. Telomere attrition, for instance, has a direct linear relationship with aging and age-related diseases like Alzheimer's and cardiovascular disease[1,5]. Meanwhile, epigenetic clocks based on DNA methylation profiles are increasingly being described as reliable measures to estimate biological age and forecast health-span[2,6]. Systemic alterations like attenuated neurogenesis, defective nutrient sensing, and expansion of senescent cells to promote frailty and heightened disease susceptibility are also involved in the aging process[1]. Lifestyle components like Mediterranean diet, aerobic fitness, sufficient sleep, and stress management activities like mindfulness have been found to have a positive effect on the markers of biological aging. On the other hand, long-term exposure to outdoor air pollution, smoking, UV radiation, and unhealthy diet can cause accelerated biological aging[3].

Of all the biological aging factors, brain aging is the most incapacitating. Brain aging entails structural, functional, and molecular deteriorations impairing cognitive and motor functions and posing risk for neurodegenerative diseases[7]. Neuronal and synaptic loss, cortical thinning, hippocampal shrinkage, and compromised white matter integrity in frontal and temporal lobes constitute the structural deformities[8]. Reduced hippocampal neurogenesis and neurotransmission indicate functional declines. Consequently, executive function and memory are disrupted, mood is modulated, leading to age-associated neurocognitive disorders[9]. Moreover, aggregation of misfolded proteins, including amyloid-beta, tau, alpha-synuclein, and mutant huntingtin contributed to the pathogenesis of Alzheimer's disease (AD), Parkinson's disease (PD), and Huntington's disease (HD), respectively[7,10].

Geriatric neurology is a key subspecialty of neurology focusing on the diagnosis, treatment, and long-term management of neurological disorders in the elderly. With an increasing aging population, geriatric neurology gains utmost priority in clinical management, health policies, caregiver education, and healthcare infrastructure tailored to the aging brain. This chapter is concerned with brain aging-associated neurological disorders, and challenges as well as opportunities in geriatric neurology to combat this global health concern (Figure 18.1).

DOI: 10.1201/9781003500025-18

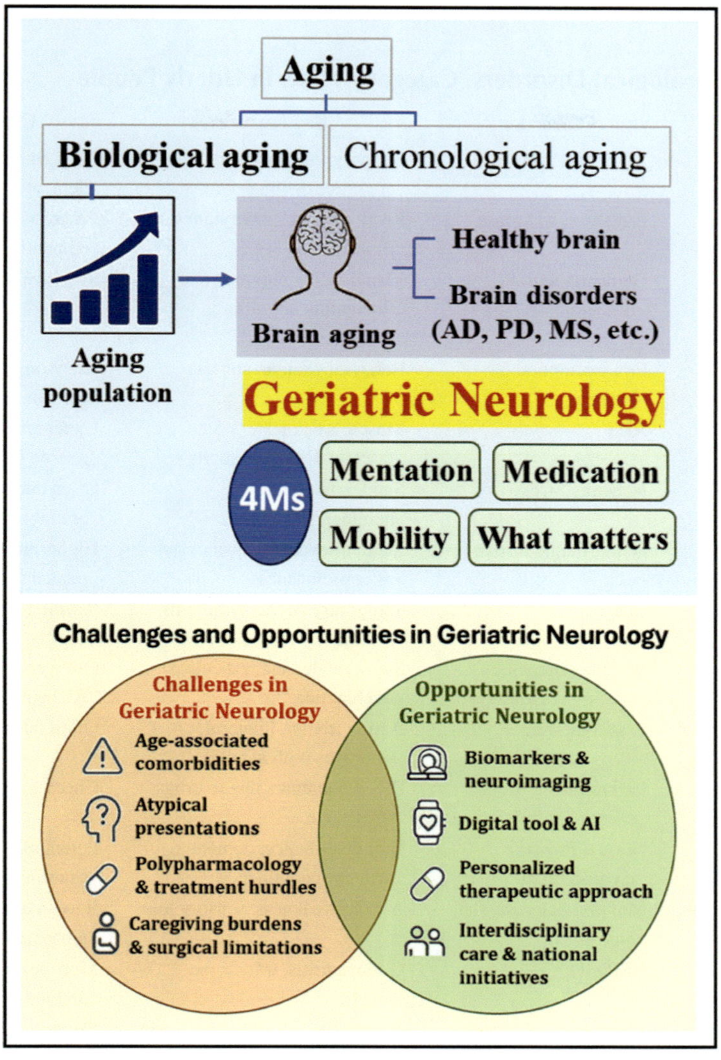

FIGURE 18.1 Geriatric neurology, its challenges and opportunities.

18.2 NEUROLOGICAL DISORDERS IN ELDERLY PEOPLE

Brain aging considerably elevates the risk of multiple neurological conditions such as delirium, mild cognitive impairment, PD, AD, normal pressure hydrocephalus, epilepsy, stroke, and headache syndromes[11,12]. Moreover, urinary incontinence, age-related hyponatremia, gait disturbances, and cervical and lumbar spondylosis are among the frequent causes of falls and fractures, which remain a major cause of morbidity in the aged people[13]. Cellular neurodegeneration due to aging is determined by protein aggregation, neuroinflammation, defective autophagy, mitochondrial damage, oxidative stress, epigenetic changes, and loss of ionic and calcium homeostasis[14]. Also, mutations in mitochondrial DNA and elevated reactive oxygen species decrease energy efficacy in neurons, leading to synaptic and neuronal loss[15]. Further, gene expression alterations in neurons as well as synapses are suggested to be governed by age-related changes in DNA methylation[16]. On the basis of the nature of the disease or the primary system involved, age-associated neurological diseases may be classified as follows and also summarized in Table 18.1.

TABLE 18.1
Summary of Neurological Disorders, Category Wise, in Elderly People

Category	Disorder	Key Features	Clinical Notes
Neurodegenerative disorders	Alzheimer's Disease	Memory loss, disorientation, cognitive decline	Most common cause of dementia
	Parkinson's Disease	Resting tremor, bradykinesia, rigidity	Progressive, may later include dementia
	Dementia with Lewy Bodies	Visual hallucinations, fluctuating cognition, parkinsonism	Sensitive to antipsychotics
	Frontotemporal Dementia	Behavior change, aphasia	Early-onset more common, but can affect elderly
	ALS	Muscle weakness, fasciculations, spasticity	Upper and lower motor neuron involvement
Cerebrovascular disorders	Ischemic Stroke	Sudden focal deficit, hemiparesis, aphasia	Often due to atherosclerosis or embolism
	Hemorrhagic Stroke	Sudden severe headache, loss of consciousness	Higher mortality; often related to hypertension
	TIA	Temporary stroke symptoms (<24 h)	Warning sign for future stroke
	Vascular Dementia	Stepwise decline, executive dysfunction	Often coexists with Alzheimer's
Movement disorders	Essential Tremor	Action tremor, bilateral, improves with alcohol	Often familial
	Parkinsonism (other causes)	PD-like features due to drugs, stroke, etc.	Check for secondary causes
	Normal Pressure Hydrocephalus	Gait disturbance, dementia, urinary incontinence	Potentially reversible with shunting
Demyelinating disorders	Multiple Sclerosis (late onset)	Ataxia, weakness, sensory loss	Rare in elderly, atypical presentation
	NMOSD	Optic neuritis, transverse myelitis	More severe than MS, AQP4 antibody associated
Peripheral nervous system disorders	Peripheral Neuropathy	Numbness, burning, tingling in feet/hands	Common in diabetes, B12 deficiency
	Guillain–Barré Syndrome	Ascending paralysis, areflexia	Emergency—can cause respiratory failure
	Carpal Tunnel Syndrome	Hand numbness (median nerve distribution)	Worse at night
Neuromuscular junction and muscle disorders	Myasthenia Gravis	Fluctuating weakness, ptosis, dysphagia	Late-onset MG increasing in elderly
	Polymyositis/ Dermatomyositis	Proximal weakness, rash (in DM)	Check for underlying malignancy
Inflammatory and infectious disorders	Herpes Zoster	Dermatomal rash, burning pain	Risk of postherpetic neuralgia
	Postherpetic Neuralgia	Chronic neuropathic pain	Hard to treat in elderly
	Encephalitis	Fever, confusion, seizures	HSV common cause
	Meningitis	Headache, fever, altered mental status	Often subtle in elderly
Psychiatric and functional disorders	Depression (Pseudodementia)	Apathy, forgetfulness, poor concentration	Improves with antidepressants
	Delirium	Acute confusion, fluctuating attention	Common post-op or with infections
	Late-Onset Psychosis	Hallucinations, delusions	Rule out organic causes

(Continued)

TABLE 18.1 (*Continued*)
Summary of Neurological Disorders, Category Wise, in Elderly People

Category	Disorder	Key Features	Clinical Notes
Neoplastic	Primary Brain Tumors	Seizures, headache, focal deficits	Meningiomas common, glioblastoma aggressive
	Metastatic Brain Tumors	Symptoms vary by location	From lung, breast, melanoma
	Paraneoplastic Syndromes	Subacute cerebellar ataxia, limbic encephalitis	Autoimmune; related to underlying cancer
Structural/Traumatic	Chronic Subdural Hematoma	Gradual confusion, weakness	From minor head trauma; reversible
	Cervical Spondylotic Myelopathy	Gait disturbance, spasticity, hand clumsiness	Spinal cord compression due to degenerative changes
	Traumatic Brain Injury	Confusion, headache, memory loss	Risk increases with falls

1. *Neurodegenerative diseases*: Neurodegenerative diseases are a category of progressive, irreversible conditions primarily marked by neuronal damage and its loss, which include AD, PD, dementia with Lewy bodies (DLB), HD, multiple sclerosis (MS), amyotrophic lateral sclerosis (ALS), etc. These diseases cause cognitive impairment, movement disorders, or altered behavior, each in different brain regions. Neurodegenerative diseases display misfolding aggregation in proteins, neuroinflammation, oxidative stress, and mitochondrial dysfunction[14,15,17]. These diseases are complex, i.e., gene mutations, polymorphisms, and epigenetic changes are found to be responsible for the cellular and molecular alterations[14,15]. However, treatment options are limited, focusing on symptomatic relief to enhance quality of life.
2. *Cerebrovascular diseases*: Primary cause of this category of diseases are alterations in the blood vessels and blood supply to the brain resulting in brain damage, e.g., stroke, vascular malformations, cerebral aneurysms, and transient ischemic attacks. In stroke, the disease may be caused by hemorrhagia (bleeding due to ruptured vessels) or ischemia (obstruction or blood clot). Cerebrovascular diseases display neurological damage leading to cognitive impairments, vision difficulties, speech problems, and paralysis, according to the specific impaired brain region. Major risk factors are diabetes, high cholesterol, hypertension, and heart disease[18].
3. *Movement disorders*: This category of diseases is primarily marked by motor dysfunctions such as irregular, involuntary, or voluntary movement, or no movement, related to impairments in specific brain region, like basal ganglia. Ataxia, Tourette syndrome, dystonia, chorea, essential tremor, and PD are the most common movement disorders. These diseases show slowed movements (bradykinesia), rigidity, tremors, and impairments in coordination and jerking. Possible causes of movement disorders include mutations, environmental factors, infections, neurodegeneration, brain injuries, or toxicity in brain. Medicines like L-DOPA, injection of botulinum toxin, or deep-brain stimulation may provide symptomatic relief[19].
4. *Demyelination disorders*: Major cause of these disorders is myelin sheath damage resulting in erroneous transmission of nerve impulses, e.g., acute disseminated encephalomyelitis, neuromyelitis optica spectrum disorder, chronic inflammatory demyelinating polyneuropathy, and MS. Besides demyelination, other conditions may also be displayed, like cognitive difficulties, numbness, muscle weakness, coordination problems, and visual disturbances. Medication such as immunomodulators

focuses on symptomatic relief, slowing down disease advancement, and reduction in neuroinflammation[20].

5. *Disorders of Peripheral Nervous System (PNS)*: PNS disorders, including Charcot-Marie-Tooth disease, Guillain-Barré syndrome, diabetic neuropathy, and peripheral neuropathy, are due to the diminished transmission of nerve impulses in the body[21,22]. Main symptoms are tingling, numbness, pain, muscle weakness, and decreased reflexes[23]. Genetic mutations, autoimmunity, infections, or metabolic disturbances are main attributes of the disorders related to this category[24,25]. Based on the causes, lifestyle changes, physical therapy, medicines, or immunotherapy may be provided for the management and slowing down the disease symptoms[26,27].
6. *Disorders of Muscles and Neuromuscular Junction*: These disorders represent a reduction in signal exchange between the nerves and the muscles or impairment of muscle function, leading to fatigue, weakness, cramps, or loss of muscles and ultimately, impaired movement. Muscle disorders include Duchenne muscular dystrophy and other myopathies, and Lambert-Eaton myasthenic syndrome and myasthenia gravis are examples of neuromuscular junction Disorders[28]. Based on the specificity of the disease, physiotherapy, corticosteroids, immunosuppressive therapy, or gene therapy may be prescribed[29,30].
7. *Inflammatory and Infectious Disorders*: This category of neurological disorders affects CNS, as well as PNS, and are caused by autoimmune conditions or pathogens. Autoimmune encephalitis, neuromyelitis optica spectrum disorder, and multiple sclerosis are included in inflammatory disorders. Brain abscesses, encephalitis, meningitis, and HIV or herpes simplex virus-related neurotropic infections are common examples of infectious diseases[31]. Symptoms encompass headache, fever, motor and sensory impairments, seizures, and cognitive deficits. Depending on the cause treatment options such as corticosteroids, immunosuppressive therapies, antimicrobial drugs, are provided to manage symptoms[32,33].
8. *Psychiatric and Functional Disorders with Neurological Overlap*: This category of disorders manifests neurological or sensory disturbances, lacking identifiable alterations in physiology or structure in the neurons. Common neurological symptoms of these disorders are seizures, tremors, or weakness. Schizophrenia, anxiety, depression, and conversion disorder with the above-mentioned are a few examples of psychiatric disorders having neurological overlaps[34,35]. Functional disorder presents authentic and incapacitating symptoms, which may be explained by the consequence of aberrant brain function rather than due to nerve injury[36]. Psychiatric conditions such as depression, anxiety, schizophrenia, and conversion disorder can also present with neurological features like cognitive impairment, psychomotor slowing, or non-epileptic seizures, creating diagnostic challenges[37,38]. Psychiatric and functional disorders are heterogeneous and have a complex interplay between pathological brain connectivity, neurobiological processes, and psychological stress[36,39]. Multidisciplinary approaches, such as physical rehabilitation, cognitive behavioral therapy, and psychiatric care, are provided to manage symptoms and improve quality of life[40].
9. *Neoplastic Disorders*: Neoplastic disorders of the nervous system involve the abnormal growth of cells forming tumors that may be benign or malignant, affecting the brain, spinal cord, or peripheral nerves. Primary tumors, such as gliomas, meningiomas, and schwannomas, originate within nervous system tissues, while secondary or metastatic tumors spread from cancers elsewhere in the body[41]. Symptoms vary depending on tumor location and size but may include headaches, seizures, motor or sensory deficits, cognitive changes, and increased intracranial pressure[41,42]. Treatment options include surgical resection, radiation therapy, chemotherapy, and targeted or immunotherapies[42], often used in combination to manage tumor growth and relieve symptoms. Prognosis depends on the tumor type, grade, location, and patient response to treatment[43].

10. *Traumatic and Structural Disorders*: Traumatic and structural disorders of the nervous system result from physical injury or abnormalities in the brain, spinal cord, or surrounding structures, often leading to acute or chronic neurological impairment. Traumatic brain injury (TBI) and spinal cord injury (SCI) are common examples, caused by external forces such as accidents, falls, or violence, and can result in cognitive deficits, paralysis, sensory loss, or behavioral changes[44]. Structural disorders include conditions like hydrocephalus, Chiari malformations, arteriovenous malformations (AVMs), and herniated disks, which disrupt normal anatomy and function through compression or altered cerebrospinal fluid dynamics[44]. Symptoms vary widely depending on the affected region but may include headaches, coordination issues, altered consciousness, or neuropathic pain. Management may require emergency care, surgical intervention, rehabilitation, and long-term support to optimize recovery and function[45].

18.3 IMPORTANCE OF GERIATRIC NEUROLOGY

Geriatric neurology mainly focuses on the neurological health of aging individuals, addressing disorders that are common but often underdiagnosed or mismanaged in this age group[46]. With increasing longevity, the population aged 65 years and older is expected to double to 1.5 billion by 2050 globally. Increase in age poses a risk of age-related neurological disorders, making it a major public health priority, and demands urgent focus on healthcare systems, caregivers, and policy frameworks[47,48]. Geriatric neurology emphasizes functional status, quality of life, and goal-directed outcomes, especially when managing multiple chronic conditions. For example, in a patient with coexisting dementia, heart failure, and osteoarthritis, geriatric neurology aims to optimize cognition and mobility rather than aggressively treat each disease independently. In geriatric neurology, the main focus is on what matters to the patient, including independence in daily living, social participation, and preservation of identity[49]. Moreover, geriatric neurology contributes significantly to policy development, research funding prioritization, and training of interdisciplinary teams. Countries like Japan, Germany, and Canada have integrated geriatric neurology into national aging strategies by supporting dementia clinics, caregiver support programs, and preventive screening for cognitive impairment[50,51]. An example of geriatric neurology in action is the “Age-Friendly Health Systems” initiative, which implements the 4Ms framework, mentation, mobility, medication, and what matters, to improve care for older adults. This framework has proven effective in reducing delirium, polypharmacy, falls, and hospital readmissions when applied in neurologically vulnerable elderly populations[52,53].

18.4 CHALLENGES IN GERIATRICS NEUROLOGY

The unparalleled advances in medical science have ushered in an opportunity for longer lifespans and better health profiles. Yet, this windfall has also ushered in with it another carga of neurological disease in later life. The rapid increase in elderly population across the globe predominantly in developed and developing nations has conferred tremendous challenges as well as innovative opportunities for geriatric neurology[54]. Age-associated neurological ailments are often comorbid with hypertension, diabetes, heart failure, and osteoarthritis, which complicate the diagnosis and disease management. For instance, a diabetic patient with cerebral stroke may have atypical symptoms and increased risk of complications[55]. Moreover, neurological disorders are sometimes manifested differently in the elderly referred to as atypical presentations. For example, urinary tract infection can lead to delirium rather than fever, and stroke can present with confusion or falls rather than hemiparesis[56].

Overlapping symptoms of age-related cognitive decline and early neurodegeneration leads to underdiagnosis and misdiagnosis and are often considered as features of advancing age delaying the intervention. Of note, depression may occasionally be misdiagnosed as dementia due to overlapping symptoms like loss of memory and interest[57]. Besides diagnostic challenges, consumption of multiple medications by the aged individuals causes treatment hurdles due to higher risk for drug interaction side effects and cognitive side effects. Anticholinergic drugs used for incontinence, for example, worsen dementia-associated cognitive symptoms[58].

Long-term neurologic illnesses like dementia require constant care. Caregivers within the family can suffer from emotional burnout, depression, and economic strain, especially in low-resource environments. Older adults also have difficulty with things such as compromised mobility, transportation issues, and disjointed care systems. A PD patient, for instance, might need coordinated care from a neurologist, physical therapist, and mental health professional—a system that is not necessarily established. Limitations of neurosurgical treatments are another obstacle confronted by the geriatric neurologists. Frailty and comorbid disease typically limit surgery options. Subdural hematoma patients who are elderly are not always candidates for surgery because of anesthesia risk and minimal potential for recovery postoperatively[59].

18.5 OPPORTUNITIES IN GERIATRIC NEUROLOGY

Neurodegenerative disorders often have very early onset despite clinical manifestations at advanced stages rendering the treatment arduous and difficult[60]. Therefore, strategies for early detection and prevention based on biomarker development and neuroimaging-based technological innovations are gaining extreme prominence. In this regard, positron emission tomography (PET) scans have enabled detections of amyloid deposition well ahead of symptom onset facilitating pharmacological and lifestyle interventions[61,62]. Artificial intelligence-based digital tools and devices have revolutionized geriatric health care. Wearables and telehealth tools, gait sensors that predict risk of falling, global positioning system (GPS) tracking, and cognitive training apps have enabled monitoring of dementia patients and also improved their safety[63,64].

Individual drug response and treatment resistance are one of the major causes of clinical trial failures in Alzheimer's and PD. Currently, personalized therapeutic approaches owing to advancements in genomics and pharmacogenetics have allowed individualized drug regimens. For example, selective serotonin reuptake inhibitors (SSRIs) are not recommended to elderly patients with serotonin syndrome and bleeding risks[65]. Exclusion of elderly in clinical trials also led to lacunae in drug trials. However, increasingly, older adults are included in research trials, such as the IDEAS (Imaging Dementia—Evidence for Amyloid Scanning) study, which demonstrated how amyloid PET scans impact the management of cognitive disorders in the elderly[66,67].

Neurological disorders are multifactorial and thus recent developments in interdisciplinary care models integrating opinions and expertise of neurologists, geriatricians, physiotherapists, and social workers have significantly improved interventional outcomes. Stroke treatment combining neurorehabilitation with cognitive therapy showed a higher success rate in recovery and quality of life[68]. In addition, national healthcare initiatives like National Programme for Health Care of the Elderly (NPHCE) in India and Dementia Strategies in the UK and Canada have provided accessible, age-friendly care through community clinics, training, and caregiver support[69].

Institute for Healthcare Improvement (IHI) of geriatric neurology domain has recently adopted a widely accepted, high quality and impactful model for neurological care based on 4Ms framework that entails mentation, medication, mobility and what matters. Mentation refers to proactive and comprehensive screening of individuals for dementia, delirium, and depression. Medication as the name suggests recommends use of age-appropriate drugs avoiding polypharmacy and ascertaining cognitive safety. Mobility ensures safe movement to reduce fall risk and maintaining independence. What Matters is aligning treatment with care upholding values, goals, and personal priorities[70]. In addition, innovative strategies for CNS drug delivery, such as intranasal administration, blood-brain barrier (BBB) modulation, nanocarriers, and polymeric nanoparticles, are being explored to enhance therapeutic efficacy with reduced systemic toxicity in older adults[71,72].

18.6 CONCLUSION

The increasing population of elderly people globally poses an unprecedented challenge and opportunity for contemporary neurology. Geriatric neurology as a subspecialty is increasingly making a contribution toward meeting the complex, multifactorial medical needs of the elderly. With age-related neurodegenerative conditions like Alzheimer's, Parkinson's, and vascular dementia on the rise, progress in imaging, diagnosis, biomarker studies, and treatment therapies opens new vistas for early diagnosis, preventive therapy, and personalized therapy.

An integrated care model based on the 4Ms framework (mentation, medication, mobility, and what matters) promises improved outcomes and improved quality of life. Coupled with the promise of nanomedicine, drug delivery systems, and non-invasive diagnostic technologies, these innovations all suggest the necessity of interdisciplinarity in clinical, technological, and policy spheres.

Addressing health disparities of access, ensuring age-friendly policy, and caregiving support will be critical to achieving success with this new frontier. Education for geriatric neurology specialists, grants for research funding, and trials on older populations are crucial in attempting to push back against the health issues of an aging world. Practice must remain up-to-date, using the instruments of modern medicine, but aware of the lived experience of the elderly.

REFERENCES

1. López-Otín C, Blasco MA, Partridge L, Serrano M, Kroemer G. The hallmarks of aging. *Cell.* 2013;153(6):1194–1217. doi:10.1016/j.cell.2013.05.039
2. Horvath S. DNA methylation age of human tissues and cell types. *Genome Biol.* 2013;14(10):R115. doi:10.1186/gb-2013-14-10-r115
3. López-Gil L, Pascual-Ahuir A, Proft M. Genomic instability and epigenetic changes during aging. *Int J Mol Sci.* 2023;24(18):14279. doi:10.3390/ijms241814279
4. López-Otín C, Blasco MA, Partridge L, Serrano M, Kroemer G. Hallmarks of aging: an expanding universe. *Cell.* 2023;186(2):243–278. doi:10.1016/j.cell.2022.11.001
5. Blackburn EH, Epel ES, Lin J. Human telomere biology: a contributory and interactive factor in aging, disease risks, and protection. *Science.* 2015;350(6265):1193–1198. doi:10.1126/science.aab3389
6. Horvath S, Raj K. DNA methylation-based biomarkers and the epigenetic clock theory of aging. *Nat Rev Genet.* 2018;19(6):371–384. doi:10.1038/s41576-018-0004-3
7. Yankner BA, Lu T, Loerch P. The aging brain. *Annu Rev Pathol.* 2008;3:41–66.
8. Resnick SM, Pham DL, Kraut MA, Zonderman AB, Davatzikos C. Longitudinal magnetic resonance imaging studies of older adults: a shrinking brain. *J Neurosci.* 2003;23(8):3295–3301. doi:10.1523/JNEUROSCI.23-08-03295.2003
9. Behrouzi A, Kelley MR, Fehrenbacher JC. Oxidative DNA damage: a role in altering neuronal function. *J Cell Signal.* 2022;3(3):160–166. doi:10.33696/signaling.3.079
10. O'Brien RJ, Wong PC. Amyloid precursor protein processing and Alzheimer's disease. *Annu Rev Neurosci.* 2011;34:185–204. doi:10.1146/annurev-neuro-061010-113613
11. Bellelli G, Brathwaite JS, Mazzola P. Delirium: a marker of vulnerability in older people. *Front Aging Neurosci.* 2021;13:626127. doi:10.3389/fnagi.2021.626127
12. Nasreddine ZS, Phillips NA, Bédirian V, et al. The Montreal Cognitive Assessment, MoCA: a brief screening tool for mild cognitive impairment. *J Am Geriatr Soc.* 2005;53(4):695–699. doi:10.1111/j.1532-5415.2005.53221.x
13. Guirguis-Blake JM, Perdue LA, Coppola EL, Bean SI. Interventions to prevent falls in older adults: updated evidence report and systematic review for the US Preventive Services Task Force. *JAMA.* 2024;332(1):58–69. doi:10.1001/jama.2024.4166
14. Heneka MT, Golenbock DT, Latz E. Innate immunity in Alzheimer's disease. *Nat Immunol.* 2015;16(3):229–236. doi:10.1038/ni.3102
15. Perez Ortiz JM, Swerdlow RH. Mitochondrial dysfunction in Alzheimer's disease: role in pathogenesis and novel therapeutic opportunities. *Br J Pharmacol.* 2019;176(18):3489–3507. doi:10.1111/bph.14585
16. Cacabelos R, Torrellas C. Epigenetics of aging and Alzheimer's disease: implications for pharmacogenomics and drug response. *Int J Mol Sci.* 2015;16(12):30483–30543. doi:10.3390/ijms161226236

17. Ross CA, Poirier MA. Protein aggregation and neurodegenerative disease. *Nat Med.* 2004;10(Suppl):S10–S17. doi:10.1038/nm1066
18. Gorelick PB, Scuteri A, Black SE, et al. Vascular contributions to cognitive impairment and dementia: a statement for healthcare professionals from the American Heart Association/American Stroke Association. *Stroke.* 2011;42(9):2672–2713. doi:10.1161/STR.0b013e3182299496
19. Kalia LV, Lang AE. Parkinson's disease. *Lancet.* 2015;386(9996):896–912. doi:10.1016/S0140-6736(14)61393-3
20. Thompson AJ, Baranzini SE, Geurts J, Hemmer B, Ciccarelli O. Multiple sclerosis. *Lancet.* 2018;391(10130):1622–1636. doi:10.1016/S0140-6736(18)30481-1
21. Pareyson D, Marchesi C. Diagnosis, natural history, and management of Charcot-Marie-Tooth disease. *Lancet Neurol.* 2009;8(7):654–667.
22. Vitrikas K, Hurst N. Disorders of the peripheral nervous system. In: Paulman PM, Taylor RB, Paulman AA, Nasir LS, eds. *Family Medicine.* Springer; 2022. doi:10.1007/978-3-030-54441-6_74
23. Beloribi-Djefaflia S, Attarian S. Treatment of Charcot-Marie-Tooth neuropathies. *Rev Neurol (Paris).* 2023;179(1):35–48. doi:10.1016/j.neurol.2022.11.006
24. Beekman R, van den Berg LH. Guillain-Barré syndrome. *Nat Rev Neurol.* 2016;12(12):723–733. doi:10.1038/nrneurol.2016.172
25. Feldman EL, Callaghan BC, Pop-Busui R, et al. Diabetic neuropathy. *Nat Rev Dis Primers.* 2019;5(1):41. doi:10.1038/s41572-019-0092-1
26. Léger JM, Guimarães-Costa R, Muntean C. Immunotherapy in peripheral neuropathies. *Neurotherapeutics.* 2016;13(1):96–107. doi:10.1007/s13311-015-0401-7
27. National Institute of Neurological Disorders and Stroke. *Peripheral neuropathy.* National Institutes of Health. Updated April 13, 2023. Accessed June 16, 2025. https://www.ninds.nih.gov/health-information/disorders/peripheral-neuropathy
28. Bushby K, Finkel R, Birnkrant DJ, et al. Diagnosis and management of Duchenne muscular dystrophy, part 1: diagnosis, and pharmacological and psychosocial management. *Lancet Neurol.* 2010;9(1):77–93. doi:10.1016/S1474-4422(09)70271-6
29. Jaretzki A 3rd, Barohn RJ, Ernstoff RM, et al. Myasthenia Gravis: recommendations for clinical research standards. Task Force of the Medical Scientific Advisory Board of the Myasthenia Gravis Foundation of America. *Neurology.* 2000;55(1):16–23. doi:10.1212/wnl.55.1.16
30. Gilhus NE, Tzartos S, Evoli A, Palace J, Burns TM, Verschuuren JJGM. Myasthenia gravis. *Nat Rev Dis Primers.* 2019;5(1):30. doi:10.1038/s41572-019-0079-y
31. Lennon VA, Kryzer TJ, Griesmann GE, et al. Anti-NMDA receptor encephalitis: case series and analysis of the effects of antibodies. *Lancet Neurol.* 2008;7(12):1091–1098. doi:10.1016/S1474-4422(08)70224-2
32. Wingerchuk DM, Banwell B, Bennett JL et al. International consensus diagnostic criteria for neuromyelitis optica spectrum disorders. *Neurology.* 2015;85(2):177–189. doi:10.1212/WNL.0000000000001729
33. Tunkel AR, Hartman BJ, Kaplan SL, et al. Practice guidelines for the management of bacterial meningitis. *Clin Infect Dis.* 2004;39(9):1267–1284. doi:10.1086/425368
34. LaFrance WC Jr, Reuber M, Goldstein LH et al. Nonepileptic seizures misdiagnosed as epilepsy in psychiatric and neurology practice. *Epilepsia.* 2013;54(Suppl 1):53–67. doi:10.1111/epi.12106
35. Reuber M, Elger CE. Psychogenic nonepileptic seizures: review and update. *Epilepsy Behav.* 2003;4(3):205–220.
36. Espay AJ, Aybek S, Carson A, et al. Current concepts in diagnosis and treatment of functional neurological disorders. *JAMA Neurol.* 2018;75(9):1132–1141. doi:10.1001/jamaneurol.2018.1264
37. Nakken EI, Grinde F, Vaaler A, Drange OK, Brodtkorb E, Sæther SG. Epilepsy and other seizure disorders in acute psychiatric inpatients. *BMC Psychiatry.* 2021;21(1):626. doi:10.1186/s12888-021-03619-y
38. Buchsbaum MS, Wu J, DeLisi LE, et al. Frontal cortex and basal ganglia metabolic rates assessed by positron emission tomography with [18F]fluorodeoxyglucose in affective illness. *J Affect Disord.* 1997;39(1):49–56. doi:10.1016/S0165-0327(99)00068-3
39. Hallett M, Stone J, Carson A. Functional neurological disorders: a primer for physiotherapists and psychologists. In: Hallett M, Stone J, Carson A, eds. *Handbook of Clinical Neurology.* Vol 168. Elsevier; 2020:275–284.
40. Stone J, Carson A, Duncan R, et al. UK guidelines for management of functional neurological disorders. *J Neurol Neurosurg Psychiatry.* 2020;91(9):923–936.
41. Ostrom QT, Cioffi G, Waite K, Kruchko C, Barnholtz-Sloan JS. CBTRUS statistical report: primary brain and other CNS tumors diagnosed in the US in 2020. *Neuro Oncol.* 2020;22(Suppl 1):iv1–iv96. doi:10.1093/neuonc/noaa200

42. Wen PY, Weller M, Lee EQ, et al. Glioblastoma in adults: EANO guideline. *Nat Rev Clin Oncol.* 2020;18(7):395–408.
43. Louis DN, Perry A, Wesseling P, et al. The 2021 WHO classification of tumors of the central nervous system: a summary. *Neuro Oncol.* 2021;23(8):1231–1251. doi:10.1093/neuonc/noab106
44. Maas AIR, Menon DK, Adelson PD, et al. Traumatic brain injury: integrated approaches to improve prevention, clinical care, and research. *Lancet Neurol.* 2017;16(12):987–1048. doi:10.1016/S1474-4422(17)30371-X
45. Fehlings MG, Vaccaro A, Wilson JR, et al. Early versus delayed decompression for traumatic cervical spinal cord injury: results of the Surgical Timing in Acute Spinal Cord Injury Study (STASCIS). *PLoS One.* 2012;7(2):e32037. doi:10.1371/journal.pone.0032037
46. Jacobs AH, Emmert K, Baron R, et al. Neurogeriatrics—a vision for improved care and research for geriatric patients with predominating neurological disabilities. *Z Gerontol Geriatr.* 2020;53(4):340–346. doi:10.1007/s00391-020-01734-1
47. United Nations Department of Economic and Social Affairs, Population Division. *World Population Ageing 2019: Highlights.* United Nations; 2019. doi:10.18356/9df3caed-en
48. GBD 2016 Neurology Collaborators. Global, regional, and national Burden of neurological disorders, 1990–2016: a systematic analysis for the Global Burden of Disease Study 2016. *Lancet Neurol.* 2019;18(5):459–480. doi:10.1016/S1474-4422(18)30499-X
49. Stancu P, Hentsch L, Seeck M, et al. Neurology of aging: adapting neurology provision for an aging population. *Neurodegener Dis.* 2025;25(1):14–20. doi:10.1159/000540972
50. Weller J, Budd M, Murray A, et al. Geriatric neurology initiatives in Germany: enhancing caregiver support and dementia care infrastructure. *J Alzheimers Dis.* 2023;87(2):455–466.
51. Stall NM, Morra D, Crowther MA, et al. Dementia care optimization in Canada: policy interventions and outcomes. *Can J Neurol Sci.* 2021;48(5):553–562.
52. Masel MC, Moss M, Tamayo A, et al. Implementing the 4Ms in neurology: effects on hospital readmissions and delirium incidence. *J Am Geriatr Soc.* 2021;69(5):1309–1315.
53. Mitchell G, Lobach KA, Stow R. Age-friendly 4Ms domains reduce polypharmacy and falls in hospitalized older adults. *JAMA Intern Med.* 2023;183(10):1120–1128.
54. Feigin VL, Nichols E, Alam T, et al. Global, regional, and national burden of neurological disorders, 1990–2016: a systematic analysis. *Lancet Neurol.* 2019;18(5):459–480.
55. Mani A, McAvinue L, Macken W, et al. Atypical stroke presentations in older adults: an observational study. *Stroke.* 2021;52(3):e70–e78.
56. Dutta A, et al. Urinary tract infection–induced delirium in elderly patients: a systematic review. *Geriatr Gerontol Int.* 2022;22(5):468–475.
57. Boller B, Becker JT. Cognitive decline in elderly depressed patients: distinguishing depression from dementia. *J Psychosom Res.* 2020;130:109914.
58. Carnahan RM, Johnson MM. Anticholinergic medication exposure and adverse cognitive outcomes in dementia. *J Am Geriatr Soc.* 2016;65(2):238–240.
59. McDuff SGR, et al. Surgical outcomes in frail elderly patients with subdural hematoma: a systematic review. *Neurosurgery.* 2018;83(3):490–498.
60. Jack CR Jr, Bennett DA, Blennow K, et al. NIA-AA research framework: toward a biological definition of Alzheimer's disease. *Alzheimers Dement.* 2018;14(4):535–562.
61. Betthauser TJ, Edison K, Ances BM. PET imaging for early detection of Alzheimer's disease biomarkers. *Curr Opin Neurol.* 2020;33(3):325–333.
62. Palmqvist S, Insel PS, Stomrud E, et al. Residual amyloid-beta clearance and PET-negativity in preclinical Alzheimer disease. *Brain.* 2024;147(2):497–508.
63. Mirelman A, Maidan I, Bernad-Elazari H, et al. Gait impairments in Parkinson's disease link to fall risk in elderly. *Front Aging Neurosci.* 2022;14:921506.
64. Clay I, Xu H, Ware E et al. AI and IoT–based wearable fall prevention in older adults: systematic review. *J Med Internet Res.* 2024;26(1):e54934.
65. Udeh-Momoh CT, Jones DK, Teo L et al. Personalized pharmacogenomics in late-life depression and dementia. *Clin Pharmacol Ther.* 2023;113(5):1170–1181.
66. Rabinovici GD, Gatsonis C, Apgar C, IDEAS Study Investigators, et al. Association of amyloid PET with care outcomes: IDEAS study. *JAMA.* 2019;322(11):1153–1160.
67. Rabinovici GD, et al. Amyloid PET impact on clinical management in memory clinics. *JAMA Neurol.* 2023;80(2):165–174.
68. Langhorne P, Bernhardt J, Kwakkel G. Stroke rehabilitation. *Lancet.* 2011;377(9778):1693–1702.

69. Rao GN, Andrade C. Dementia care efforts in India and Canada: programs and outcomes. *Int Psychogeriatr.* 2022;34(7):821–832.
70. Masel MC, Moss M, Tamayo A, et al. Implementing the 4Ms in neurology: effects on outcomes. *J Am Geriatr Soc.* 2021;69(5):1309–1315.
71. Lochhead JJ, Thorne RG. Intranasal delivery of biologics to CNS. *Pharm Res.* 2012;29(7):1640–1650.
72. Pardridge WM. Nanoparticles for nose-to-brain drug delivery: current state. *Front Cell Dev Biol.* 2023;11:1214450.

Index

Note: Page numbers in **bold** and *italics* represent tables and figures, respectively.

For Product Safety Concerns and Information please contact our
EU representative GPSR@taylorandfrancis.com Taylor & Francis
Verlag GmbH, Kaufingerstraße 24, 80331 München, Germany